Lecture Notes in Computer Science 11383

Commenced Publication in 1973
Founding and Former Series Editors:
Gerhard Goos, Juris Hartmanis, and Jan van Leeuwen

More information about this series at http://www.springer.com/series/7412

Alessandro Crimi · Spyridon Bakas
Hugo Kuijf · Farahani Keyvan
Mauricio Reyes · Theo van Walsum (Eds.)

Brainlesion:
Glioma, Multiple Sclerosis, Stroke and Traumatic Brain Injuries

4th International Workshop, BrainLes 2018
Held in Conjunction with MICCAI 2018
Granada, Spain, September 16, 2018
Revised Selected Papers
Part I

 Springer

Editors
Alessandro Crimi
University Hospital of Zurich
Zürich, Switzerland

Spyridon Bakas 🄳
University of Pennsylvania
Philadelphia, PA, USA

Hugo Kuijf
University Medical Center Utrecht
Utrecht, The Netherlands

Farahani Keyvan
National Cancer Institute
Bethesda, MD, USA

Mauricio Reyes 🄳
University of Bern
Bern, Switzerland

Theo van Walsum 🄳
Erasmus University Medical Center
Rotterdam, The Netherlands

ISSN 0302-9743   ISSN 1611-3349 (electronic)
Lecture Notes in Computer Science
ISBN 978-3-030-11722-1   ISBN 978-3-030-11723-8 (eBook)
https://doi.org/10.1007/978-3-030-11723-8

Library of Congress Control Number: 2018967942

LNCS Sublibrary: SL6 – Image Processing, Computer Vision, Pattern Recognition, and Graphics

This Springer imprint is published by the registered company Springer Nature Switzerland AG
The registered company address is: Gewerbestrasse 11, 6330 Cham, Switzerland

Preface

This volume contains articles from the Brain-Lesion Workshop (BrainLes), as well as the (a) International Multimodal Brain Tumor Segmentation (BraTS) challenge, (b) Ischemic Stroke Lesion Segmentation (ISLES) challenge, (c) grand challenge on MR Brain Image Segmentation (MRBrainS18), (d) Computational Precision Medicine (CPM) challenges, and (e) Stroke Workshop on Imaging and Treatment Challenges (SWITCH). All these events were held in conjunction with the Medical Image Computing for Computer Assisted Intervention (MICCAI) conference during September 16–20, 2018, in Granada, Spain.

The papers presented describe research of computational scientists and clinical researchers working on glioma, multiple sclerosis, cerebral stroke, traumatic brain injuries, and white matter hyper-intensities of presumed vascular origin. This compilation does not claim to provide a comprehensive understanding from all points of view; however, the authors present their latest advances in segmentation, disease prognosis, and other applications to the clinical context.

The volume is divided into seven parts: The first part comprises three invited papers summarizing the presentations of the keynote speakers; the second includes the paper submissions to the BrainLes workshop; the third through the seventh parts contain a selection of papers presenting methods that participated at the 2018 challenges of ISLES, MRBrainS, CPM, SWITCH, and BraTS, respectively.

The first chapter in these proceedings describes invited papers from the four keynote speakers of the MICCAI BrainLes 2018 workshop (www.brainlesion-workshop.org). The overarching aim of these papers is to give an updated review of the work done in (a) the domain of machine learning applied in neuro-oncology diagnostics, (b) connectomics of traumatic brain injury and brain tumors, (c) computational/memory considerations for deep learning in medical image analysis, and (d) computed tomography perfusion. The sequence of these papers reflects the order that they were presented during the workshop.

The aim of the second chapter, focusing on the BrainLes workshop submissions, is to provide an overview of new advances in medical image analysis in all of the aforementioned brain pathologies. Bringing together researchers from the medical image analysis domain, neurologists, and radiologists working on at least one of these diseases. The aim is to consider neuroimaging biomarkers used for one disease applied to the other diseases. This session did not have a specific dataset to be used.

The third chapter contains descriptions of a selection of algorithms that participated in the ISLES 2018 challenge. The purpose of this challenge was to directly compare methods for the automatic prediction of stroke lesion outcome from CT-perfusion imaging. A dataset consisting of CT-perfusion image volumes acquired at acute and 3-month follow-up was released for training. A dedicated test set of cases was used for evaluation. Test data were not released, but participants had to submit their segmentation results to: www.isles-challenge.org.

The fourth chapter includes a number of papers from MRBrainS 2018. The purpose of this challenge is to directly compare methods for segmentation of gray matter, white matter, cerebrospinal fluid, and other structures on 3T MRI scans of the brain, and to assess the effect of (large) pathologies on segmentation and volumetry. Over 30 teams participated and the challenge remains open for future submissions. An up-to-date ranking is hosted on: http://mrbrains18.isi.uu.nl/.

The fifth chapter presents a selection of papers from the leading participants in the two CPM 2018 challenges in brain tumors (http://miccai.cloudapp.net/competitions/). The "Combined MRI and Pathology Brain Tumor Classification" challenge used corresponding imaging and pathology data with the task of classifying a cohort of "low-grade" glioma tumors ($n = 52$) into two sub-types of oligodendroglioma and astrocytoma. This challenge presented a new paradigm in algorithmic challenges, where data and analytical tasks related to the management of brain tumors were combined to arrive at a more accurate tumor classification. In the challenge of "Segmentation of Nuclei in Digital Pathology," participants were asked to detect and segment all nuclei in a set of image tiles ($n = 33$) of glioblastoma and lower-grade glioma extracted from whole slide tissue images. Data from both challenges were obtained from The Cancer Genome Atlas/The Cancer Imaging Archive (TCGA/TCGA) repository.

Finally, the sixth chapter of these proceedings contains scientific contributions of the SWITCH workshop, which aims to bring together clinicians and medical imaging experts to discuss challenges and opportunities for medical imaging in stroke care and treatment. In 2018, three clinical keynote speakers addressed various aspects of stroke and ischemic stroke treatment: Prof. Aad van der Lugt discussed imaging biomarkers related to stroke, Prof. Matt Gounis shared his research on the development for stroke devices, and Prof. Roland Wiest presented stroke mimics and chameleons. The scientific contributions of the medical imaging field, addressing topics such as perfusion parameter estimation and the relation between diffusion MRI and microstructural changes in gray matter, were presented at the workshop in oral and poster presentations. All accepted full paper contributions are part of these proceedings.

The seventh chapter focuses on a selection of papers from the BraTS challenge participants. BraTS 2018 made publicly available a large ($n = 542$) manually annotated dataset of pre-operative brain tumor scans from 19 institutions, in order to gauge the current state-of-the-art in automated glioma segmentation using multi-parametric structural MRI modalities and to compare fairly between different methods. To pinpoint and evaluate the clinical relevance of tumor segmentation, BraTS 2018 also included the prediction of patient overall survival, via integrative analyses of radiomic features and machine learning algorithms (www.cbica.upenn.edu/BraTS2018.html).

We heartily hope that this volume will promote further exciting research on brain lesions.

December 2018

Alessandro Crimi
Spyridon Bakas

Organization

Main Organizing Committee
(Lead Organizers from Each Individual Event)

Spyridon Bakas — Center for Biomedical Image Computing and Analytics, University of Pennsylvania, USA

Alessandro Crimi — African Institute for Mathematical Sciences, Ghana

Keyvan Farahani — National Cancer Institute, National Institutes of Health, USA

Hugo Kuijf — University Medical Center Utrecht, The Netherlands

Mauricio Reyes — Biomedical Neuroimage Analysis Group, University of Bern, Switzerland

Theo van Walsum — Biomedical Imaging Group Rotterdam, Erasmus MC, The Netherlands

Challenges Organizing Committees

Brain Tumor Segmentation (BraTS) Challenge

Spyridon Bakas — University of Pennsylvania, USA

Christos Davatzikos — University of Pennsylvania, USA

Keyvan Farahani — National Cancer Institute (NCI), National Institutes of Health, USA

Jayashree Kalpathy-Cramer — Harvard Medical School, USA

Bjoern Menze — Technical University of Munich, Germany

Computational Precision Medicine (CPM) Challenges

Spyridon Bakas — University of Pennsylvania, USA

Hesham Elhalawani — University of Texas MD Anderson Cancer Center, USA

Keyvan Farahani — National Cancer Institute (NCI), National Institutes of Health, USA

John Freymann — Frederick National Laboratory for Cancer Research, USA

David Fuller — University of Texas MD Anderson Cancer Center, USA

Jayashree Kalpathy-Cramer — Harvard Medical School, USA

Justin Kirby — Frederick National Laboratory for Cancer Research, USA

Tahsin Kurc — Stony Brook Cancer Center, USA

Joel Saltz — Stony Brook Cancer Center, USA

Amber Simpson — Memorial Sloan Kettering Cancer Center, USA

Ischemic Stroke Lesion Segmentation (ISLES) Challenge

Søren Christensen	Stanford University, USA
Arsany Hakim	Inselspital Bern, Switzerland
Maarten G. Lansberg	Stanford University, USA
Mauricio Reyes	University of Bern, Switzerland
David Robben	KU Leuven, Belgium
Roland Wiest	Inselspital Bern, Switzerland
Stefan Winzeck	University of Cambridge, UK
Greg Zaharchuk	Stanford University, USA

Grand Challenge on MR Brain Segmentation 2018 (MRBrainS18)

Edwin Bennink	University Medical Center Utrecht, The Netherlands
Hugo Kuijf	University Medical Center Utrecht, The Netherlands

Stroke Workshop on Imaging and Treatment Challenges (SWITCH)

Adrian Dalca	CSAIL, MIT and MGH, Harvard Medical School, USA
Mauricio Reyes	University of Bern, Biomedical Neuroimage Analysis Group, Switzerland
David Robben	KU Leuven, Belgium
Theo van Walsum	Biomedical Imaging Group Rotterdam, Erasmus MC, The Netherlands
Roland Wiest	Support Center of Advanced Neuroimaging, Inselspital Bern, Switzerland

Program Committee

Chincisan Andra	University Hospital of Zurich, Switzerland
Meritxell Bach Cuadra	University of Lausanne, Switzerland
Jacopo Cavazza	Instituto Italiano di Tecnologia, Italy
Adrian Dalca	Massachusetts Institute of Technology, USA
Guray Erus	University of Pennsylvania, USA
Alvaro Gomariz	ETH Zürich, Switzerland
Moises Hernandez	University of Pennsylvania, USA
Ender Konukoglu	ETH-Zurich, Switzerland
Jana Lipkova	Technical University of Munich, Germany
Yusuf Osmanlioglu	University of Pennsylvania, USA
Saima Rathore	University of Pennsylvania, USA
Zahra Riahi Samani	University of Pennsylvania, USA
Aristeidis Sotiras	University of Pennsylvania, USA

Koen Van Leemput   Harvard Medical School, USA
Benedikt Wiestler   Technical University of Munich, Germany
Stefan Winzeck   University of Cambridge, UK

Sponsoring Institutions

Center for Biomedical Image Computing and Analytics, University of Pennsylvania, USA

Contents – Part I

Grand Challenge on MR Brain Segmentation

Computational Precision Medicine

Stroke Workshop on Imaging and Treatment Challenges

Contents – Part II

Invited Talk

Multimodal Patho-Connectomics of Brain Injury

Ragini Verma[✉], Yusuf Osmanlioglu, and Abdol Aziz Ould Ismail

Penn Patho-Connectomics Lab, Department of Radiology,
University of Pennsylvania, Philadelphia, PA 19104, USA
ragini@pennmedicine.upenn.edu,
yusuf.osmanlioglu@uphs.upenn.edu

Abstract. The paper introduces the concept of *patho-connectomics*, an injury-specific connectome creation and analysis paradigm, that treats injuries as a diffuse disease pervading the whole brain network. The foundation of the "patho-connectomic" ideology of analysis is that no part of the brain can function in isolation, and abnormality in the brain network is a combination of structural and functional anomalies. Brain injuries introduce anomalies in this brain network that could affect the quality of brain tissue, break a pathway, and lead to disrupted connectivity in neural circuits. This in turn affects functionality. Thus, *patho-connectomes* go beyond the traditional connectome and include information of tissue quality and structural and functional connectivity, forming a comprehensive map of the brain network. Information from diffusion and functional MRI are combined to create these patho-connectomes. The creation and analysis of patho-connectomes are discussed in the case of brain tumors, that suffers from the challenges of mass effect and infiltration of the peritumoral region, which in turn affect the surgical and radiation plan, and in traumatic brain injury, where the exact injury may be difficult to determine, but the effect is diffuse manifesting in heterogenous symptoms. A network-based approach to analysis of both these forms of injury will help determine the effect of pathology on the whole brain, while incorporating recovery and plasticity. Thus, patho-connectomics with a broad network perspective on brain injuries, has the potential to cause a major paradigm shift in their research of brain injuries, facilitating subject specific analysis and paving the way for precision medicine.

Keywords: Diffusion MRI · fMRI · Connectomes · Free water ·
Tractography · Brain tumors · Neoplasms · Traumatic brain injury

1 Introduction

Brain injury can be organic as in brain tumors, incipient and developmental as in autism and psychopathology, or acquired as in traumatic brain injury. Traditionally, diffusion and functional magnetic resonance imaging (dMRI and fMRI) have been used for group-based statistical analysis and biomarker creation, in non-focal diseases, based on imaging features derived from the data. In injuries with focal anomalies like brain

© Springer Nature Switzerland AG 2019
A. Crimi et al. (Eds.): BrainLes 2018, LNCS 11383, pp. 3–14, 2019.
https://doi.org/10.1007/978-3-030-11723-8_1

tumors, imaging has been used to localize pathology for focused treatment like surgery or radiation. However, the brain is a network, with no part that can truly be isolated.

dMRI offers a structural insight into the complex network architecture of the brain, based on the differential water diffusivity in tissue, with the white matter (WM) being highly organized and the cerebro-spinal fluid being pure water. It is therefore able to provide measures of tissue quality quantifying the anisotropy (directionality) and ease of diffusivity of water in tissue. Based on acquisition, dMRI can be fitted with a tensor model (as in Diffusion Tensor Imaging (DTI) [1, 2], with a short and clinically feasible acquisition) or a higher order model (as in high angular resolution diffusion imaging (HARDI) [3], with a longer and more research-oriented acquisition). Scalar measures of anisotropy and diffusivity derived from DTI [4], and invariants derived from HARDI [5] provide a characterization of tissue quality. Tractography, provides an insight into the fiber pathways in the brain [6], by tracking the directionality of water flow in tissue. Finally, by parcellating the brain into regions [7, 8], and tracking between them, we can create the structural connectome of the brain [9]. Resting state fMRI (rs-fMRI) provides information about the functional interaction between brain regions and can be used to create a functional connectome. Advances in dMRI and rs-fMRI analytics has enabled the systematic interrogation of the structural and functional networks of the brain, as well as the tissue and tractography maps, albeit, separately.

Brain injuries introduce anomalies in the brain's network. This could manifest as WM pathways being affected in the form of deteriorated quality, as well as a breakage in the pathway, leading to disrupted connectivity in neural circuits. This can in turn cause decrease or delay in functional activation. As brain regions are inter-connected, the effect of pathology on these measures is also inter-related, underlining further the need for a holistic approach to image analysis. This paves the way for a new philosophy of analysis, *patho-connectomics*, that goes beyond treating injuries as focal anomalies, but as a diffuse disease that is pervading the whole brain. A network-based approach to analysis will help determine the effect of pathology on the whole brain, while incorporating recovery and plasticity. This requires combining the information from structure and function, leading to a functionally guided structural network analysis of the brain. A *patho-connectome* therefore goes far beyond a traditional connectome, which is a simple map of connectivity, and its study involves a comprehensive analysis of the tissue quality, its connectivity, and the dynamics of the neural circuits in the face of plasticity. This has the potential to cause a major paradigm shift in the research of brain injuries, allowing a broader network perspective on them, as well as facilitating subject specific analysis and paving the way for precision medicine.

2   Patho-Connectomics of Tumors

Primary and secondary neoplasms may result in structural and functional impairment in the brain. Structurally, WM fiber pathways are displaced by tumor mass effect and disrupted by cancer infiltration. The functional framework in the brain is damaged by several mechanisms including the pathophysiological exchange in ion channels [10] and the presence of angiogenesis [11]. Traditionally, a combination of surgical resection, chemotherapy, and radiation therapy are used, in order to resect and restrict the tumor with the aim of increasing survival. However, the prognosis remains poor, largely because the etiology of brain neoplasms continues to be poorly understood, especially of primary tumors. Also, despite increased survival, the quality of life may deteriorate as a result of the resected WM. Hence, there is an urgent need for a "patho-connectomic" global view of treating neoplasms, that can predict global effects of local changes, as well as the behavioral deficits associated with a pathologic, surgical, or therapeutic change in connectivity. Such a connectomic approach will revolutionize the oncological protocol: (1) in surgical planning where a better characterization of the peritumoral region, and non-invasive markers that identify infiltration will aid in maximal resection with minimal deficit; and (2) in treatment monitoring by determining the effect of WM removal mediated and compensated by plasticity, enabling a better quality of life.

2.1   Characterizing Peritumoral Tissue

A characterization of the peritumoral tissue will aid in determining the extent of infiltration, as well as potential for recurrence. As tumor growth may be associated with several abnormal processes like leaking blood-brain barrier, change in the tissue due to infiltration, inflammation, increased permeability of blood vessels, etc., various MR imaging contrasts provide information about the edema (dMRI and FLAIR), vasculature and blood flow (perfusion, dynamic contrast enhanced imaging), metabolic tissue properties (spectroscopic imaging), and molecular water mobility within the tissue (diffusion and magnetization transfer imaging). It is expected that a multimodal radiomic marker characterizing the peritumoral tissue will aid in determining tumor type, malignancy grade, and the potential for recurrence, thereby replacing an invasive biopsy and shaping the treatment procedures. Peritumoral tissue has traditionally been characterized by the dMRI measures of fractional anisotropy (FA) and mean diffusivity (MD). However, as dMRI is representative of water diffusivity in tissue, and edema and infiltration pertain to variable water content and ease of diffusion, the characterization of these regions will gain from multicompartment tissue modeling. Maps of various compartments, especially free water volume fraction, have the potential to provide additional information about the tissue microstructure associated with edema and tumor infiltration, and can characterize changes associated with different tumor types as well as their genetic underpinnings.

Infiltration of tumor in the peritumoral region suggests a tissue type different from healthy tissue and edema and would gain from a specific modeling of this compartment. However, research in multi-compartment modeling is limited to healthy tissue without incorporating pathology as a compartment. Additionally, most of the microstructural modeling is on long acquisitions infeasible in the clinic. Models based on single shell clinical acquisitions [12] use initializations tuned for the healthy tissue. We have developed a free water elimination method based on the bi-compartment modeling of clinically feasible dMRI acquisitions [13]. As can be seen in Fig. 1, it estimates both the unhealthy peritumoral and the healthy regions well. Such a characterization of tissue microstructure should be able to distinguish tumor types, which till now needed an inclusion of information from various modalities. The variation in water content in the peri-tumoral regions between primary (glioblastoma multiforme (GBM)) and secondary (metastatic)) tumors is captured by the free water volume fraction maps, as the former has infiltration and the latter is mostly pure water. A statistical analysis comparing 87 GBMs and 54 metastatic tumors shows a significant difference between their volume fraction maps, as well as the extent of change in fractional anisotropy induced by the free water correction (Fig. 2).

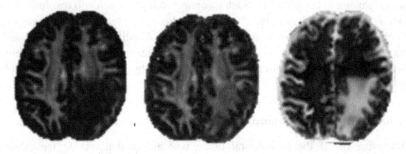

Fig. 1. Effect of Free Water (FW) Correction on tissue quality measures: FA map pre-FW correction shows decrease in edema (left), which is restored after correction (middle), with the FW map capturing the water and infiltration (right).

Thus, dMRI offers imaging contrasts that could permit more uniform sampling than heterogeneous biopsies and can be applied repeatedly to monitor therapy and facilitate a comprehensive diagnosis of brain tumors. Currently, clinical protocols comprise multiple modalities, and it may be possible to lower these by using different contrasts from DTI. Modeling the infiltrated tissue as a separate compartment, is the next big step in this area, that will revolutionize cancer research, by providing an alternate to invasive biopsy. However, obtaining ground truth in the form of biopsy of peritumoral tissue is difficult, as there is high heterogeneity in that region as the tissue could be a mix of edema, cancer and healthy tissue. With brain shift and swelling, pin pointing the exact location is difficult. Thus, the validation of such a tissue compartment remains the biggest challenge to this research.

2.2 Better Mapping of Structural Pathways

Optimal surgical resection or radiation planning of brain tumors hinges on the knowledge of the spatial extent of the tumor, the displacement and infiltration of fiber pathways, as well as a characterization of the peritumoral edema to obtain better delineation of fiber pathways. DTI is used in pre-surgical planning for the purposes of creating maps of eloquent tracts using tractography techniques, that will enable the surgeon to plan maximal resection with minimal damage. fMRI is used to provide additional evidence of presence of tracts. Tractography is based on the diffusion parameter of FA which is adversely affected by edema, as well as infiltration, which alter the tissue quality and hence these measures of diffusion, leading to potentially erroneous tracking. With the free water modeling of the peri-tumoral tissue, the

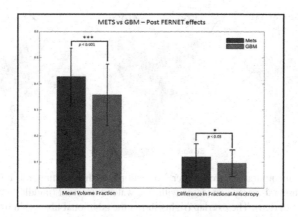

Fig. 2. Characterization of the peri-tumoral region using free water correction is able to distinguish tumor types: there is a significant difference between free water volume fraction maps (left) and difference in corrected FA (right) of GBMs and metastatic tumors.

corrected tissue shows improved tracking in peri-tumoral region (Fig. 3), irrespective of the type of tractography and length of tract. Manual identification of tracts for surgical planning is very challenging due to displaced and broken tracts, leading to the development of several automated tract segmentation algorithms. Most of these are based on geometric and shape features [14] that fail for displaced and broken tracts. This paved the way for a fiber clustering algorithm [15, 16] that is based on encoding the tracts based on their connectivity signatures, and subsequent clustering them using an atlas. The method is able to capture tumor and subject-wise variability, underlining its importance for precision medicine. A combination of the edema invariant tractography with this connectivity-based tract clustering will make the fiber extraction robust to disruption due to infiltration and displacement as a result of mass effect caused by the tumor. The biggest challenge of research projects in fiber tracking and clustering, is the validation. Animal models with similar pathology are hard to find, tracer-based results are an unfair comparison to tracking done on clinical resolution, and cortical stimulation is challenging to perform on deep white matter, and on tracts like language tracts.

3   Patho-Connectomes of Traumatic Brain Injury

dMRI has been used in traumatic brain injury (TBI) [17], mostly to demonstrate differences at the group level using measures of tissue quality like FA and MD, which are representative of anomalies in the axonal and myelin structure. Additionally, a few connectivity studies have revealed abnormalities [18–21] in the structural connectome. However, studies showing resting state fMRI abnormalities in brain injury [22], suggest that both structure and function are affected, and that TBI studies will gain from their combination.

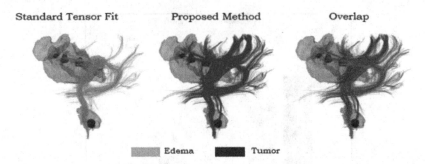

Fig. 3. Tracking incorporating free water correction of the peritumoral region: Tracking coverage in peri-tumoral regions increases with free water correction (middle), in comparison to the standard tensor fit (left). The overlap (right) shows the additional fibers (blue) that are tracked after free water correction. (Color figure online)

3.1 Measures on Injury in TBI

Due to the diffuse effect of TBI on WM, a large number of studies have used measures of tissue quality like FA and MD, as measures of injury in TBI. In most chronic TBI cases, there is an increase in FA and a decrease in MD. However, the results, and their interpretations have varied across studies due to the small sample sizes and high heterogeneity of the effect of trauma, even within the groups of mild, moderate and severe TBI. Higher order dMRI, have the ability to provide other measures called invariants [5], but long acquisition times and sequences that are clinically not viable, have hindered the popularity of higher order schemes. Due to the extent of knowledge provided by these invariants, it is expected that with the advent of deep learning, and with advanced acquisition in recent studies, these invariants will provide more sensitive measures of injury. Diffuse axonal injury in traumatic brain injury may be a combination of swelling (inflammation) and axonal disruption in the chronic stages, that presents itself in the form of changes in the free water. Figure 4 shows the extent of free water differences as a result of trauma to the brain, in a group of TBI patients, as compared to controls. The extent of free water may be a measure of damage and indicative of future cognitive decline.

In addition to the measures of tissue quality, connectomes, provide a wealth of information that can be used to derive measures to quantify injury post trauma [19]. However, measures devised in the literature are generic and their interpretability is difficult. Ideally, a measure should incorporate the vulnerability of the different parts of the brain, mediated by the amount of injury that causes a disruption in the connectivity of the brain. In that vein, we have developed a disruption index [18], that captures the extent of injury, in correlation with clinical and cognitive scores. Although the measure has been used to evaluate the extent of injury in TBI, the same can be used to determine the effect of surgery or radiation planning on the brain, as the extent of change can be determined, along with the vulnerability of the nodes with regard to a population of healthy controls. Future work involves extending this to functional connectomes.

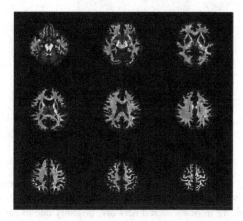

Fig. 4. Free-water increased in TBI brains as compared to healthy controls. These are the results of voxel-based analysis, when TBI patients are compared to controls. The colored voxel indicates significant differences relative to healthy controls.

3.2 Creation of Injury Connectome

The creation of the injury score from a traditional connectome underlines the potential of a whole brain map in quantifying the injury burden. Despite their glory, traditional connectomes are limited in that, they encode only pairwise connectivity information of the regions which is taken from a single imaging modality, overlooking much of the information that can be obtained from the network structure of the brain. These limitations can be overcome by introducing an enhanced injury connectome that includes regional features in addition to connectivity information, that are obtained from various modalities. This can further be supported by tools to discover similarity among such structures at various levels, from individual regions up to the whole network. Considering connectomes as annotated graphs and establishing similarity measures between them, that account for network wide connectivity as well as local features of regions, provide ways to obtain such an injury connectome.

As the first step along these lines, we evaluated structural and functional connectomes as weighted graphs and proposed using a novel similarity measure based on graph matching, enabling comparison across subjects in [20]. The main idea with this form of connectome matching is that, regions with similar connectivity signatures should get mapped to each other, with mismatches indicating an anomaly of connectivity (Fig. 5). In analyzing a TBI dataset consisting of structural and functional MRI of patients at three time points post injury, we observed significant group differences between patients and controls longitudinally. We also observed that, our graph matching based similarity measure correlates well with the clinical scores of patients, indicating the ability of the approach to capture underlying changes in pathology. This study lays the foundation for the analysis of structural and functional plasticity as it provides the means to measure connectomic similarity at various levels.

Taking a further step towards obtaining an injury connectome in [21], we further enriched the traditional connectome with graph theoretical measures such as node

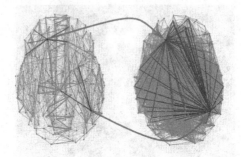

Fig. 5. Graph matching establishes a one-to-one mapping between nodes of two connectomes based on nodal connectivity patterns. Mismatches indicate dissimilarity between corresponding nodes across subjects arising from an altered network, possibly due to pathology. Hence, graph matching is a viable method to distinguish patients from controls as well as identifying brain regions that differ across groups or in individual subjects.

degree, strength, and centrality, in addition to the connectivity signature of nodes with respect to the rest of the network. We also extended our graph matching based similarity metric to enriched connectomes. When applied to a dataset with TBI patients on a classification task, we found that the enriched connectome along with the similarity measure outperforms traditional connectome in distinguishing patients from controls (Table 1). Our results further indicate that, graph theoretical features and the connectivity signature of nodes complement each other in classification, highlighting the importance of enriching the connectome with extra nodal information.

Table 1. Comparison of classification accuracy of TBI patients and controls between traditional and enriched connectomes (EC)

Scenario	Accuracy (%)
Traditional connectome	66.7
EC with graph theory measures alone	42.03
EC with edge weights alone	62.32
EC with graph theory measures and edge weights	72.46

A future direction will be to investigate different communication models in the brain, that could explain the functional changes related to the structure and can aid in incorporating the concept of plasticity better. Traditionally, all the analysis is carried out with the assumption that the brain adopts the "shortest path" mode of communication. But that does allow for the concept of "delayed" processing in the brain, that can occur as a result of injury. It is also difficult to explain the rewiring of the brain that must occur due to plasticity. This presents an exciting direction of future research.

3.3 Applying Injury Connectomes to Tumors

Creating an injury connectome for tumors will revolutionize neuro-oncology. An enriched connectome with information from tissue quality and fMRI, goes beyond the traditional use of fMRI which is used in surgical planning to compliment the extracted tracts from dMRI. It will enable the surgeon to take a connectomic approach to surgery in which the effect of the resection of tumor and healthy tissue leading to tumor region, can be known beforehand, enabling maximal resection with minimal current and future deficit, structurally, functionally and behaviorally. Such a connectome will help understand the mechanism of plasticity as the tumor grows, as well as after surgery, suggesting therapeutic prospects. This is expected to improve the quality of life considerably. However, creating such a connectome in the presence of tumors is challenging. The creation of a connectome requires the parcellation of the brain into regions, which has traditionally been achieved by registering an atlas to the brain. However, registration algorithms fail in the presence of a tumor, or at least in the vicinity of it as the layout of regions and the pathways connecting them are altered. The methods that avoid registration to an atlas by obtaining subject specific parcellation [7], still fail around the tumor. Hence, doing tracking based on edema invariant tractography, using it to create a connectivity map of the brain, and finally obtaining a parcellation is an open area of research. Success of graph matching in identifying anomalies in TBI [20, 21], suggests it as a promising tool to overcome the registration problem through mapping regions from a healthy brain into a brain with tumor. Figure 6 demonstrates how graph matching could be used for registration.

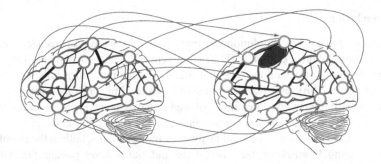

Fig. 6. Matching networks between brains, post- (left) and pre- (right) tumor resection: Using graph matching to overcome the registration problem in the presence of tumor, by finding a mapping between the nodes of a resected brain with nodes of brain with tumor.

Once such a connectome is created, measures of injury like the one described above, can be used to quantify the effect of WM removal, and that of radiation on the rest of the brain. In a longitudinal study on the effect of radiation on brain connectivity, we saw a decreased efficiency of the brain with increased dosage (Fig. 7). Analysis at the subnetwork level of eloquent (motor, visual and language) and behavioral (executive functioning, working memory and other cognitive networks) subnetworks will pave the way of quantifying cognitive changes temporally. Surgical and radiation planning will improve with such a connectome, as it will be possible to know how the surgical path will affect the brain post-surgery, as well as determine a radiation plan that is least invasive. This "patho-connectomic" view of surgery will therefore enrich therapeutic care, and subsequently the quality of life.

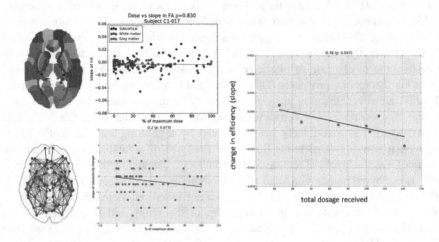

Fig. 7. Effect of radiation on local measures of tissue quality (left: top) and connectivity (left: bottom) does not show any changes with dosage. The global measure of efficiency of brain network shows a decrease with increasing dosage (right).

3.4 Conclusions

Connectomics [9, 23] has revolutionized brain research in the past decade, enabling research in sex differences [24, 25], development [26] and psychopathology [27]. However, despite its tremendous potential, traditional connectomics falls short in the analysis of brain injury and has not translated to the clinic. In this paper, we enumerated some of the major limitations of traditional connectomics in brain injury, highlighting the need for *patho-connectomics,* an injury specific connectomic paradigm, that would enable a multi-modal analysis of injured brains, both at the population level for creating biomarkers, but also at the individual level paving the way for precision medicine.

References

1. Basser, P.J., Jones, D.K.: Diffusion-tensor MRI: theory, experimental design and data analysis - a technical review. NMR Biomed. **15**(7–8), 456–467 (2002)
2. Pierpaoli, C., et al.: Diffusion tensor MR imaging of the human brain. Radiology **201**(3), 637–648 (1996)
3. Tuch, D.S., et al.: Diffusion MRI of complex neural architecture. Neuron **40**(5), 885–895 (2003)

4. Pierpaoli, C., Basser, P.J.: Toward a quantitative assessment of diffusion anisotropy. Magn. Reson. Med. **36**(6), 893–906 (1996)

5. Caruyer, E., Verma, R.: On facilitating the useof HARDI in population studies by creating rotation-invariant markers. Med. Image Anal. **20**(1), 87–96 (2015)

6. Caruyer, E., et al.: A comparative study of 16 tractography algorithms for the corticospinal tract: reproducibility and subject-specificity. In: ISMRM (2014)

7. Honnorat, N., Parker, D., Tunç, B., Davatzikos, C., Verma, R.: Subject-specific structural parcellations based on randomized AB-divergences. In: Descoteaux, M., Maier-Hein, L., Franz, A., Jannin, P., Collins, D.L., Duchesne, S. (eds.) MICCAI 2017. LNCS, vol. 10433, pp. 407–415. Springer, Cham (2017). https://doi.org/10.1007/978-3-319-66182-7_47

8. Descoteaux, M., et al.: Deterministic and probabilistic tractography based on complex fibre orientation distributions. IEEE Trans. Med. Imaging **28**(2), 269–286 (2009)

9. Hagmann, P., et al.: MR connectomics: principles and challenges. J. Neurosci. Methods **194** (1), 34–45 (2010)

10. Molenaar, R.J.: Ion channels in glioblastoma. In: ISRN Neurology (2011)

11. Wang, N., Jain, R.K., Batchelor, T.T.: New directions in anti-angiogenic therapy for glioblastoma. Neurotherapeutics **14**(2), 321–332 (2017)

12. Pasternak, O., et al.: Free water elimination and mapping from diffusion MRI. Magn. Reson. Med.: Official J. Soc. Magn. Reson. Med./Soc. Magn. Reson. Med. **62**, 717–730 (2009)

13. Ismail, A.A.O., et al.: Characterizing peritumoral tissue using DTI-Based free water elimination. In: Crimi, A., et al. (eds.) BrainLes 2018. LNCS, vol. 11383, pp. 123–131. Springer, Cham (2018)

14. Garyfallidis, E., et al.: Recognition of white matter bundles using local and global streamline-based registration and clustering. Neuroimage **170**, 283–293 (2017)

15. Tunc, B., et al.: Automated tract extraction via atlas based adaptive clustering. Neuroimage **102**(P2), 596–607 (2014)

16. Tunc, B., et al.: Individualized map of white matter pathways: connectivity-based paradigm for neurosurgical planning. Neurosurgery **79**(4), 568–577 (2016)

17. Hulkower, M.B., et al.: A decade of DTI in traumatic brain injury: 10 years and 100 articles later. AJNR Am. J. Neuroradiol. **34**, 2064–2074 (2013)

18. Solmaz, B., et al.: Assessing connectivity related injury burden in diffuse traumatic brain injury. Hum. Brain Mapp. **38**(6), 2913–2922 (2017)

19. Kim, J., et al.: Disrupted structural connectome is associated with both psychometric and real-world neuropsychological impairment in diffuse traumatic brain injury. J. Int. Neuropsychol. Soc. **20**(9), 887–896 (2014)

20. Osmanlıoğlu, Y., Alappatt, J.A., Parker, D., Kim, J., Verma, R.: A graph based similarity measure for assessing altered connectivity in traumatic brain injury. In: Crimi, A., et al. (eds.) BrainLes 2018. LNCS, vol. 11383, pp. 189–198. Springer, Cham (2018)

21. Osmanlıoğlu, Y., et al.: A graph representation and similarity measure for brain networks with nodal features. In: Stoyanov, D., et al. (eds.) GRAIL/Beyond MIC -2018. LNCS, vol. 11044, pp. 14–23. Springer, Cham (2018). https://doi.org/10.1007/978-3-030-00689-1_2

22. Hillary, F.G., et al.: The rich get richer: brain injury elicits hyperconnectivity in core subnetworks. PLoS ONE **9**(8), e104021 (2014)

23. Sporns, O.: The human connectome: origins and challenges. Neuroimage **80**, 53–61 (2013)

24. Ingalhalikar, M., et al.: Sex differences in the structural connectome of the human brain. Proc. Natl. Acad. Sci. U.S.A. **111**(2), 823–828 (2014)

25. Tunc, B., et al.: Establishing a link between sex-related differences in the structural connectome and behaviour. Phil. Trans. R. Soc. Lond. B Biol. Sci. **371**(1688), 20150111 (2016)
26. Fair, D.A., et al.: Functional brain networks develop from a "local to distributed" organization. PLoS Comput. Biol. **5**(5), e1000381 (2009)
27. van den Heuvel, M.P., et al.: Abnormal rich club organization and functional brain dynamics in schizophrenia. JAMA Psychiatry **70**(8), 783–792 (2013)

CT Brain Perfusion: A Clinical Perspective

Arsany Hakim[(⊠)] [iD] and Roland Wiest

Support Center for Advanced Neuroimaging,
University Institute of Diagnostic and Interventional Neuroradiology,
Bern University Hospital, Inselspital, University of Bern, Bern, Switzerland
arsany_hakim@yahoo.com

Abstract. Computed tomography perfusion (CTP) is an important exam performed in neuroradiology that adds functional information regarding hemodynamics to that obtained from morphological imaging and thereby supports clinical decision-making in several vascular and non-vascular conditions.

This paper outlines the clinical applications of CTP, its advantages over MRI and disadvantages. Factors affecting the results of CTP will also be discussed. Finally, a clinically oriented overview of the calculated perfusion parameters and their value will be provided.

Keywords: CT · Perfusion · Stroke

1 Clinical Applications of CTP

CTP is applied in different conditions, and its uses are divided into vascular and non-vascular applications.

1.1 Vascular Applications

CTP is necessary to identify infarction core and penumbra and their mismatch in cases with acute cerebral ischemia, which is not possible using conventional scans; it is therefore very important in triaging patients and deciding whether recanalization therapy (i.e., thrombectomy or thrombolysis) is needed. Many institutes consequentially implement CTP (or MR perfusion) as part of their stroke-imaging algorithm, based on evidence and/or experience [1, 2]. It is particularly valuable in patients with extended time window, such as those exhibiting symptoms for longer than six hours or with an unknown symptom onset (such as wake-up stroke), to approximate the risk benefit ratio for aggressive stroke therapy and to identify eligible patients in whom CTP can be used to estimate the tissue at risk [3].

Patients with classic thromboembolic stroke may benefit from CTP, as can those with other conditions that may lead to ischemic lesions, such as sickle cell anemia [4], which can cause vascular occlusion, or myomoya disease [5] and other types of occlusive diseases. The detection of perfusion disturbance appears to be helpful in predicting the risk of developing stroke in patients with extracranial internal carotid stenosis or occlusion [6].

© Springer Nature Switzerland AG 2019
A. Crimi et al. (Eds.): BrainLes 2018, LNCS 11383, pp. 15–24, 2019.
https://doi.org/10.1007/978-3-030-11723-8_2

Vasculitis results in narrowing of intracranial vessels, leading to a reduction in the blood supply to the brain, and in these cases perfusion can help detect affected territories [5].

Perfusion imaging is also helpful in evaluating the stages of hypoxic ischemic injury after resuscitation [7, 8], and assessing the ischemic effects of hypotensive cerebral syndrome [9].

Sudden onset of neurological symptoms can potentially be due to stroke but can also be present in other disorders that cause stroke-like symptoms; some of these symptoms are not easy differentiated from stroke clinically and therefore called stroke mimics. The most important members in this group of diseases are seizures and migraine in which perfusion imaging plays an important role not only in excluding stroke but also in demonstrating the pattern characteristic of these diseases [10, 11].

Because CTP can be used to evaluate macro- and microvascular circulation, it is important to be added to CT angiography in patients with delayed cerebral ischemia and vasospasm after subarachnoid hemorrhage [12] because it can predict which patients that may develop ischemia [13], identify the extent and degree of ischemia [14], helps in selecting patients for treatment [15], including either triple-H therapy (hypertension, hypervolemia, and hemodilution) or catheter-based management, such as intra-arterial vasodilator and balloon angioplasty.

CTP can also help to confirm brain death [16].

1.2   Non-vascular Applications

Neoplastic
CTP parameters, especially CBV and permeability, can be used to differentiate low-from high-grade gliomas, predict progression from low- to high-grade glioma, and differentiate between high-grade glioma and lymphoma [17, 18]. Perfusion imaging also plays a role in monitoring patients under therapy, as it can differentiate radionecrosis from recurrent tumor [17]. It is also helpful in differentiating neoplastic lesions, such as high-grade tumors, from inflammatory lesions, such as tumefactive demyelinating lesions, a process that is difficult when using only conventional imaging.

Traumatic
Another non-vascular indication is traumatic squeal, in which CTP provides some benefits. It has a higher sensitivity than non-enhanced CT for detecting cerebral contusions; It can show changes related to mass effects caused by edema, swelling, and extra-axial hematomas related to trauma; and the number of territories involved with perfusion changes may help to predict outcomes in severe trauma [19].

Degenerative
CTP can play a role in degenerative diseases of the brain by detecting decreased perfusion in characteristic locations [20].

2   Advantages and Disadvantages of CTP

CTP has many advantages, CT scanners are widely available and more affordable than MRI. CT is also readily available in emergency and acute settings and suitable for intensive care patients without the need for the special equipment required to monitor the patient during MRI. The high temporal resolution of CTP is advantageous in uncooperative patient and CT can be performed in patients with different types of implants that are contraindicated in MRI, such as a pacemaker. A change in CT density observed after the application of contrast agent is linearly related to the concentration of iodine, and this allows a more robust quantification with absolute measures of perfusion parameters [21]. CTP benefits from the high spatial resolution of the CT scanner.

However, small lesions can be missed by a CT exam. Because of the usual beam hardening artifacts, visualization of the posterior fossa is not optimal in CT exams. The limited soft tissue contrast in CT is also a drawback that renders MRI more superior in evaluating brain tissues. The application of contrast agents could be contraindicated in certain patients, such as those with kidney failure or allergy. However the most important drawback of CTP is the high radiation dose, which differs according to the technique applied and the coverage required on the Z-axis; this limits the use of such exams, especially in younger patients. High doses of radiation can be delivered to the radiosensitive eye lens during CTP.

Several strategies can be used to reduce the radiation dose in CT; these include reducing the tube voltage and/or current. Ultra-low-dose CTP decreases radiation by approximately 20%, but should be combined with denoising techniques [22]. Modern scanners that use iterative reconstruction allow the radiation dose to be reduced without an associated reduction in image quality. Reducing the frequency of image acquisition and the number of CTPs performed per patient are important factors. CTP should be performed only when indicated, and whenever possible, other radiation-free methods should be used, such as MR perfusion [23].

3   Acquisition and Post-processing

A variety of factors may affect the results of CTP.

3.1   Factors Related to Acquisition

Injection Rate
Typically, 4–7 ml/s flow of contrast agent given via an antecubital vein is advised, but this is not feasible in every patient, such as those with a smaller venous caliber. A higher injection rate leads to a maximum but brief peak, which is more suitable for rapid data collection during acquisition. A lower injection rate leads to a shorter but prolonged peak, which is more suitable for slower data collection [24]. Therefore, both the injection rate and the frequency of data collection should be jointly considered.

Duration of Image Acquisition

Generally, an acquisition duration of 45 s is enough to include the tissue attenuation curve (TAC) in patients without a cardiac condition, but there is a risk of acquiring an incomplete TAC in patients with low cardiac output, atrial fibrillation or severe vascular stenosis. To avoid these instances, a prolonged duration time (60–90 s) is advised.

Frequency of Image Acquisition

As was previously stated, the injection rate and frequency of data collection should both be optimized to record the maximum enhancement. Using a low frequency may miss the peak point of the curve, while using a high frequency rate can lead to increased radiation exposure. Accordingly, a multiphase protocol with variation in image acquisition over three different phases is sometimes advised [25].

Coverage

The volume of brain included in the perfusion exam depends on the detector width, which can reach 16 cm in 256 and 320-slice scanners. Patients scanned with a lower detector number can be scanned twice at two different levels if clinically indicated. Some scanners include a toggle table with shuttle mode (instead of a static table during scanning), and this allows for larger coverage.

3.2   Factors Related to Mathematical Modeling

Deconvolution and Non-deconvolution

There are two major methods for mathematical calculations; deconvolution and non-deconvolution.

The non-deconvolution method is based on Fick's principle of conservation of mass, and the Mullani Gould formula, which neglects venous return, and therefore requires a high injection rate. This method is easier to calculate, but the non-venous return assumption is considered an oversimplification. Absolute quantification is not possible with this method [21].

There are two different methods of performing deconvolution techniques; parametric and non- parametric. Non-parametric methods include the Fourier transformation, which is very sensitive to noise, and singular value decomposition (SVD), which is the most common method used in perfusion calculations. This type of deconvolution contains other subtypes, such as standard SVD, oscillation SVD, circular SVD and Bayesian methods.

Delay and Dispersion

In cases of vessel occlusion, there will be a delay in contrast approaching the tissue voxel, and there may be dispersion of the contrast bolus before it reaches the occlusion site [26]. The opinions of authors differ in this matter, with some advising delay correction, and others suggesting that delay is a part of the pathological process and should not be corrected. Therefore, some software corrects for delay (delay-invariant), while others do not (delay-sensitive). Methods that can be used to correct for delay include Fourier transformation, block circulant decomposition matrix, the use of an arterial input function (AIF) obtained from smaller vessel near the region of interest, and curve fitting [21].

3.3   Factors Related to Post-processing

Motion

Uncooperative patients tend to move during the exam, and this can lead to motion artifacts, the loss of accurate results, or in some instances an inability to process the data. Thus motion correction, which is usually performed by software packages developed for perfusion analysis, is an important step.

Vessel Definition

To avoid partial volume, the artery used for the AIF and the vein used for venous reference should be perpendicular to the slice orientation; hence the anterior cerebral artery is usually used for AIF, and the superior sagittal sinus is used for a venous reference. Most modern software performs this step automatically.

Accordingly, the results of CTP can vary widely according to the previously mentioned factors. There are various models to perform this processing, each of which produces different results. Therefore, care should be taken, especially in multicenter studies, as the results obtained in different centers may be not comparable.

4   Interpretation of Perfusion Parameters

Several parameters can be generated from CT perfusion, and these vary among software packages. In this section, an overview is provided of the value of each parameter and how it can be interpreted from a clinical point of view. Understanding of each parameter will improve interpretation of imaging findings. Each parameter should not be interpreted separately, but should instead be interpreted with other parameters and morphological images.

4.1   Cerebral Blood Flow (CBF)

CBF is defined as the volume of blood moving through a given unit volume of brain per unit time [24]. On the TAC, it is the upward slope of the curve. A more vertical line indicates a faster flow. This parameter is measured in ml of blood/100 g brain tissue/s. The flow, according to Ohm's law (applied to fluids), is directly proportional to the pressure (or pressure gradient) and indirectly proportional to the resistance [27]. Consequently, the flow will decrease when the difference between the arterial and venous sides is reduced, as is observed in cases of arterial occlusion, or when resistance increases, which can be caused by the vascular wall in cases of vasospasm or by the surrounding structures in cases of brain edema or hydrocephalus, and by other causes that can increased intracranial pressure. Another important factor that affects CBF is the vascular diameter; according to the Hagen-Poiseuille equation the volumetric flow is directly proportional to the fourth power of the internal radius of a tube [28]. Thus, a reduction in the vascular diameter by half leads to a 16-fold decrease in the flow, which results in a reduction in CBF when autoregulation is exhausted. A reduction in the vascular diameter occurs in cases of stenosis, such as that observed in vasospasm and other vaso-occlusive diseases.

In summary, CBF can be affected by pressure gradients, resistance and vessel diameter, which are altered in many conditions and is therefore one of the most important parameters.

4.2   Cerebral Blood Volume (CBV)

CBV is the total volume of blood in a given unit volume of brain [24]. In the TAC CBV is the area under the curve represents the total amount of contrast (blood) in the tissue of interest (ROI) regardless of time. Generally, it indicates whether blood reaches a tissue, even when it does so in a reduced rate or delayed manner [29]. It is measured in ml of blood/100 g brain tissue.

In certain conditions the flow of blood may be significantly reduced, even though the volume may be normal or even increased. This is because of autoregulation phenomena, which are physiological process that aims to adjust hemodynamics when there are changes in cerebral blood pressure. Additionally, the hypoxia and hypoglycemia caused by hypoperfusion can cause vasodilatation, leading to increased blood volume.

4.3   Time to Peak (TTP)

TTP is defined as the time taken for the contrast to achieve maximum enhancement. In the TAC, this is the time from when the contrast injection is initiated until the peak of the curve is reached. TTP is very sensitive to flow changes. However, many factors (technical, or patient-related), such as a lower injection rate, or low cardiac output, can prolong TTP without the presence of cranial pathology. Hence, some vendors define TTP in an alternative method in which the time is calculated from the earliest enhancement of the cerebral arteries until the peak is reached on the tissue curve. This method (tracer delay-insensitive or delay-invariant) reduces extracranial factors that may cause delay and thereby increases specificity.

4.4   Time to Maximum (Tmax)

Tmax is defined as the time to maximum of the residue function. Hence, Tmax is also a "time to peak" but after processing the residue. This is one of the more complex parameters, because it is affected by many factors. It reflects bolus delay but is also affected by temporal dispersion and, to a lesser extent, mean transit time. These factors increase the complexity of analyzing this parameter as a prolonged Tmax can represent any one of these factors, and this should be taken into consideration when interpreting results. However, a significant delay in Tmax is most likely not caused by a prolonged MTT.

Tmax is considered a measure of macrovascular parameters [30] and one of the important parameters in stroke imaging. It is understood that elevated Tmax in acute ischemia coexists with hypoperfusion and delayed poor collateral supply. In addition, regions with highly elevated Tmax, even when well-perfused, are the most vulnerable to further perfusion pressure reduction.

Quantifying Tmax is helpful in stroke management, in which a threshold between 4–6 s appears to be optimal for the early identification of critical hypoperfused brain tissues [31].

4.5   Time to Start (TTS)

TTS indicates the interval between contrast material injection and the beginning of contrast enhancement. This is the time between the start of contrast administration until the bolus arrival time (BAT) in seconds and is affected by multiple factors, as in TTP. Thus, the delay-invariant method can increase specificity by calculating the time from the beginning of the earliest enhancement of the cerebral artery until BAT at the tissue side. When using this definition (time spent from beginning enhancement of the arteries until the contrast reaches the tissue), this is the time spent on the arterial side (mostly inside the large- and medium-sized vessels), so TTS can be considered as a marker of macrovascular structures that reflects the bolus arrival delay [32].

4.6   Mean Transit Time (MTT)

MTT is, with CBF and CBV, one of the three basic parameters used in CTP. It may be calculated by the central volume principle, CBF = CBV/MTT (i.e., MTT = CBV/CBF). It is the average transit time of blood through a given brain region [9]. It also represents the time spent between inflow and outflow of contrast on a tissue curve. In other words, it indicates the time spent in capillary vessels, and can therefore be considered a marker of microvascular circulation. It can be used to detect changes due to microvascular alterations and lesions due to vasospasms of the microvasculature as it is highly sensitive to hemodynamic disturbances.

MTT is inversely proportional to perfusion pressure. Compensatory vasodilatation (which is preserved in the penumbra) occurs when cerebral blood pressure drops, leading to prolongation of the MTT. Therefore, MTT is a useful parameter for identifying penumbras.

4.7   Time to Drain (TTD)

TTD describes the time to washout. It is the time from when the enhancement is started until the outflow of contrast in the tissue curve; in other words, TTD represents the sum of TTS and MTT, making this parameter very sensitive to both macro- and microvascular disturbances. This parameter appears to condense both types of pathologies into one image. Therefore, it is sensitive to different types of changes in hemodynamics, and a normal TTD predicts normal perfusion with a high probability [33].

4.8   Flow Extraction Products (Extraction Flow Products)

This parameter is also called the volume transfer constant (K^{trans}) [34] and reflects the passage of contrast between the blood and the extravascular extracellular space [35]. K^{trans} is considered a marker of blood brain barrier disturbances and depends on the flow of a tracer (F) in addition to permeability (P) and the vascular surface area (S).

Surface area and permeability cannot be separated in practice, so they are commonly called permeability-surface products (PS). Thus K^{trans} can vary according to these two factors (F and PS) [36]. This parameter is important in neoplastic conditions.

5 Conclusion

New methods for automated image analysis are currently developed in many domains. For appropriate clinical application of artificial intelligence technologies in this domain, it remains important to understand processing steps in CTP and factors that affect final lesion load prediction so that solutions can be reached that will help clinicians in decision-making even in challenging situations.

References

1. Jung, S., et al.: Stroke guidelines of the Bern stroke network physicians on duty. http://www.neurologie.insel.ch/fileadmin/neurologie/neurologie_users/Unser_Angebot/Dokumente/Stroke_Guidelines_2018.pdf. Accessed 11 Oct 2018
2. Gilberto González, R., et al.: The Massachusetts General Hospital acute stroke imaging algorithm: an experience and evidence based approach. J. Neurointerv. Surg. **5**, i7–i12 (2013). https://doi.org/10.1136/neurintsurg-2013-010715
3. Powers, W.J., et al.: 2018 guidelines for the early management of patients with acute ischemic stroke: a guideline for healthcare professionals from the American Heart Association/American Stroke Association. Stroke **49**, e46–e99 (2018). https://doi.org/10.1161/STR.0000000000000158
4. Thust, S.C., Burke, C., Siddiqui, A.: Neuroimaging findings in sickle cell disease. Br. J. Radiol. **87**(1040) (2014). https://doi.org/10.1259/bjr.20130699
5. Orrison, W.W., et al.: Whole-brain dynamic CT angiography and perfusion imaging. Clin. Radiol. **66**(6), 566–574 (2011). https://doi.org/10.1016/j.crad.2010.12.014
6. Gaudiello, F., et al.: Sixty-four-section CT cerebral perfusion evaluation in patients with carotid artery stenosis before and after stenting with a cerebral protection device. Am. J. Neuroradiol. **29**(5), 919–923 (2008). https://doi.org/10.3174/ajnr.A0945
7. Huang, B.Y., Castillo, M.: Hypoxic-ischemic brain injury: imaging findings from birth to adulthood. RadioGraphics **28**(2), 417–439 (2008). https://doi.org/10.1148/rg.282075066
8. Salzman, K.L., Shah, L.M.: Adult hypoxic ischemic injury. https://my.statdx.com/document/adult-hypoxic-ischemic-injury/e7fbe6b9-eef2-4494-a64c-be639a1d2639?searchTerm=Hypoxic-IschemicEncephalopathy. Accessed 19 Oct 2018
9. Derdeyn, C.P., et al.: Severe hemodynamic impairment and border zone-region infarction. Radiology **220**(1), 195–201 (2001)
10. Gelfand, J.M., Wintermark, M., Josephson, S.A.: Cerebral perfusion-CT patterns following seizure. Eur. J. Neurol. **17**(4), 594–601 (2010). https://doi.org/10.1111/j.1468-1331.2009.02869.x
11. Floery, D., et al.: Acute-onset migrainous aura mimicking acute stroke: MR perfusion imaging features. Am. J. Neuroradiol. **33**(8), 1547–1552 (2012). https://doi.org/10.3174/ajnr.A3020

12. Vulcu, S., et al.: Repetitive CT perfusion for detection of cerebral vasospasm-related hypoperfusion in aneurysmal subarachnoid hemorrhage. World Neurosurg. (18), 32263–32270 (2018). https://doi.org/10.1016/j.wneu.2018.09.208
13. Zhang, H., Zhang, B., Li, S., Liang, C., Xu, K., Li, S.: Whole brain CT perfusion combined with CT angiography in patients with subarachnoid hemorrhage and cerebral vasospasm. Clin. Neurol. Neurosurg. 115(12), 2496–2501 (2013). https://doi.org/10.1016/j.clineuro.2013.10.004
14. Binaghi, S., et al.: CT angiography and perfusion CT in cerebral vasospasm after subarachnoid hemorrhage. AJNR: Am. J. Neuroradiol. 28(4), 750–758 (2007). PMID: 17416833
15. Sanelli, P.C., et al.: Can CT perfusion guide patient selection for treatment of delayed cerebral ischemia? Adv. Comput. Tomogr. 2, 4–12 (2013). https://doi.org/10.4236/act.2013.21002
16. Shankar, J.J.S., Vandorpe, R.: CT perfusion for confirmation of brain death. Am. J. Neuroradiol. 34(6), 1175–1179 (2013). https://doi.org/10.3174/ajnr.A3376
17. Jain, R.: Perfusion CT imaging of brain tumors: an overview. Am. J. Neuroradiol. 32(9), 1570–1577 (2011). https://doi.org/10.3174/ajnr.a2263
18. Haldorsen, I.S., Espeland, A., Larsson, E.-M.: Central nervous system lymphoma: characteristic findings on traditional and advanced imaging. Am. J. Neuroradiol. 32(6), 984–992 (2011). https://doi.org/10.3174/ajnr.A2171
19. Wintermark, M., Sanelli, P.C., Anzai, Y., Tsiouris, A.J., Whitlow, C.T.: Imaging evidence and recommendations for traumatic brain injury: advanced neuro-and neurovascular imaging techniques. J. Am. Coll. Radiol. 12(2), e1–e14 (2015). https://doi.org/10.1016/j.jacr.2014.10.014
20. Tang, Z., et al.: Low-dose cerebral CT perfusion imaging (CTPI) of senile dementia: diagnostic performance. Arch. Gerontol. Geriatr. 56(1), 61–67 (2013). https://doi.org/10.1016/J.ARCHGER.2012.05.009
21. Konstas, A.A., Goldmakher, G.V., Lee, T.-Y., Lev, M.H.: Theoretic basis and technical implementations of CT perfusion in acute ischemic stroke, part 1: theoretic basis. AJNR: Am. J. Neuroradiol. 30(4), 662–668 (2009). https://doi.org/10.3174/ajnr.A1487
22. Othman, A.E., et al.: Impact of image denoising on image quality, quantitative parameters and sensitivity of ultra-low-dose volume perfusion CT imaging. Eur. Radiol. 26(1), 167–174 (2016). https://doi.org/10.1007/s00330-015-3853-6
23. Hakim, A., Vulcu, S., Dobrocky, T., Z'Graggen, W.J., Wagner, F.: Using an orbit shield during volume perfusion CT: is it useful protection or an obstacle? Clin. Radiol. 73(9), 834.e1–834.e8 (2018). https://doi.org/10.1016/j.crad.2018.05.003
24. Konstas, A.A., Wintermark, M., Lev, M.H.: CT Perfusion imaging in acute stroke. Neuroimaging Clin. N. Am. 21(2), 215–238 (2011). https://doi.org/10.1016/J.NIC.2011.01.008
25. Konstas, A.A., Goldmakher, G.V., Lee, T.-Y., Lev, M.H.: Theoretic basis and technical implementations of CT perfusion in acute ischemic stroke, part 2: technical implementations. AJNR: Am. J. Neuroradiol. 30(5), 885–892 (2009). https://doi.org/10.3174/ajnr.A1492
26. Lin, L., et al.: Correction for delay and dispersion results in more accurate cerebral blood flow ischemic core measurement in acute stroke. Stroke 49(4), 924–930 (2018). https://doi.org/10.1161/STROKEAHA.117.019562
27. Klabunde, R.: Hemodynamics (pressure, flow, and resistance). https://www.cvphysiology.com/Hemodynamics/H001.htm. Accessed 5 Oct 2018
28. Wikipedia: Hagen – Poiseuille equation. https://en.wikipedia.org/wiki/Hagen–Poiseuille_equation. Accessed 5 Oct 2018

29. Shapiro, M.: Perfusion primer. http://neuroangio.org/neuroangio-topics/perfusion-primer/. Accessed 4 Oct 2018
30. Calamante, F., Christensen, S., Desmond, P.M., Østergaard, L., Davis, S.M., Connelly, A.: The physiological significance of the time-to-maximum (Tmax) parameter in perfusion MRI. Stroke **41**(6), 1169–1174 (2010). https://doi.org/10.1161/STROKEAHA.110.580670
31. Olivot, J.-M., et al.: Optimal Tmax threshold for predicting penumbral tissue in acute stroke. Stroke **40**(2), 469–475 (2009). https://doi.org/10.1161/STROKEAHA.108.526954
32. Dolatowski, K., et al.: Volume perfusion CT (VPCT) for the differential diagnosis of patients with suspected cerebral vasospasm: qualitative and quantitative analysis of 3D parameter maps. Eur. J. Radiol. **83**(10), 1881–1889 (2014). https://doi.org/10.1016/j.ejrad.2014.06.020
33. Thierfelder, K.M., et al.: Whole-brain CT perfusion: reliability and reproducibility of volumetric perfusion deficit assessment in patients with acute ischemic stroke. Neuroradiology **55**(7), 827–835 (2013). https://doi.org/10.1007/s00234-013-1179-0
34. Tofts, P.S., et al.: Estimating kinetic parameters from dynamic contrast-enhanced t1-weighted MRI of a diffusable tracer: standardized quantities and symbols. J. Magn. Reson. Imaging **10**(3), 223–232 (1999). https://doi.org/10.1002/(SICI)1522-2586(199909)10:3%3c223:AID-JMRI2%3e3.0.CO;2-S
35. Patankar, T.F., et al.: Is volume transfer coefficient (Ktrans) related to histologic grade in human gliomas? Am. J. Neuroradiol. **26**(10), 2455–2465 (2005). PMID: 16286385
36. Elster, A.: Ktrans & permeability. http://mriquestions.com/k-trans–permeability.html. Accessed 5 Oct 2018

Adverse Effects of Image Tiling
on Convolutional Neural Networks

G. Anthony Reina[(✉)] and Ravi Panchumarthy

Artificial Intelligence Products Group, Intel Corporation, Hillsboro, OR, USA
g.anthony.reina@intel.com
https://ai.intel.com/

Abstract. Convolutional neural network models perform state of the art accuracy on image classification, localization, and segmentation tasks. A fully convolutional topology, such as U-Net, may be trained on images of one size and perform inference on images of another size. This feature allows researchers to work with images too large to fit into memory by simply dividing the image into small tiles, making predictions on these tiles, and stitching these tiles back together as the prediction of the whole image.

We compare how a tiled prediction of a U-Net model compares to a prediction that is based on the whole image. Our results show that using tiling to perform inference results in a significant increase in both false positive and false negative predictions when compared to using the whole image for inference. We are able to modestly improve the predictions by increasing both tile size and amount of tile overlap, but this comes at a greater computational cost and still produces inferior results to using the whole image.

Although tiling has been used to produce acceptable segmentation results in the past, we recommend performing inference on the whole image to achieve the best results and increase the state of the art accuracy for CNNs.

1 Introduction

Since their resurgence in 2012 convolutional neural networks (CNN) have rapidly proved to be the state-of-the-art method for computer-aided diagnosis in medical imaging and have led to improved accuracy in classification, localization, and segmentation tasks [1,2]. However, memory constraints have often limited training on large 2D and 3D images due to the size of the activation maps held for the backward pass during gradient descent [3]. Two methods are commonly used to manage these memory limitations: (1) images are often downsampled to a lower resolution and/or (2) images are broken into smaller tiles [4].

Fully convolutional networks are a natural fit for tiling methods because they can be trained on one image size and perform inference on another. These networks can perform inference on arbitrarily large images by breaking the large image into smaller, overlapping tiles [5]. However, we question whether this overlapping tiles approach is indeed as accurate as simply performing inference on

© Springer Nature Switzerland AG 2019
A. Crimi et al. (Eds.): BrainLes 2018, LNCS 11383, pp. 25–36, 2019.
https://doi.org/10.1007/978-3-030-11723-8_3

the whole image. In this report, we design an experiment where the whole image can fit within the memory limitations and compare whole image inference to the overlapping tiles approach.

2   Methods

2.1   Brain Tumor Segmentation Dataset (BraTS)

The brain tumor segmentation (BraTS) challenge created a publicly-available multi-institutional dataset for benchmarking and quantitatively evaluating the performance of computer-aided segmentation algorithms to detect gliomal brain tumors from MRI [6–9]. In this study we use the 2018 BraTS dataset which is comprised of pre-operative MRI scans from 285 patients at 19 institutions (https://www.med.upenn.edu/sbia/brats2018/data.html). The scans were performed on 1T, 1.5T, or 3 T multimodal MRI machines and all the ground truth labels were manually annotated by expert, board-certified neuroradiologists.

2.2   U-Net

We implemented a U-Net topology which predicts tumor segmentation masks from the raw MRI slices [5] (Fig. 1). U-Net is a fully convolutional network based on an encoder-decoder architecture. Because of its design U-Net is agnostic to image size. Training and inference can be performed on different sized images.

Our model takes as input a single T2 Fluid Attenuated Inversion Recovery (FLAIR) slice from the BraTS dataset and outputs an equivalently-sized

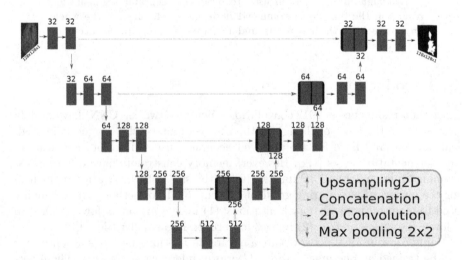

Fig. 1. The U-Net topology used in this study. We reduced the number of kernels in each layer by half from the original paper [5] and added dropout just before the 3rd and 4th max pooling layers.

mask predicting the whole tumor. The contracting path captures context (via max pooling) and the expanding path enables localization (via upsampling). Unlike the standard encoder-decoder, each feature map in the expanding path is concatenated with a corresponding feature map from the contracting path, augmenting downstream feature maps with spatial information acquired using smaller receptive fields. Intuitively, this allows the network to consider features at various spatial scales. Since its introduction in 2015, U-Net has quickly become one of the standard deep learning topologies for image segmentation. We modified the published topology by reducing the number of feature maps by half and adding dropout (0.2) just before the 3rd and 4th max pooling layers.

2.3 Training

We divided the BraTS 2018 dataset into a training/test split of approximately 85/15. Although we are considering the 2D slices from the MRI to be independent images for the model, we ensured that the 2D slices from a single study were contained in only one of the two datasets in order to prevent data leakage. There were 35,960 image/mask pairs in the training set and 8,215 in the test set. The FLAIR channels from each 2D slice were normalized by subtracting the mean pixel value of the slice and dividing by the standard deviation of the pixel values from the slice. The original slices were 240×240 pixels (*i.e.* whole image). A random crop of 128×128 pixels was taken from the normalized FLAIR slices and their corresponding ground truth masks. We performed randomized flipping (up/down and left/right) and 90 degree rotation of the training set images.

The Dice coefficient was used to measure the quality of the tumor predictions. Dice is defined as:

$$\frac{2 \times |Truth \cap Prediction| + smooth}{|Truth| + |Prediction| + smooth}$$

where *Truth* is the ground truth tumor mask and *Prediction* is the predicted tumor mask. A smoothing factor of 1.0 is added to both numerator and denominator for numerical stability in the case of non-existent ground truth masks.

The model was implemented in Keras 2.2 and TensorFlow 1.10. The complete source code can be found at https://github.com/NervanaSystems/topologies/tree/master/tiling_experiments. Stochastic gradient descent with the Adam optimizer (learning rate = 0.0001) was used to minimize $-\log Dice$. A batch size of 128 was used during training. We created a batch generator which randomly selected cropped images/masks from the training set for each batch. We trained 30 epochs (280 steps per epoch) and saved the model that produced the best Dice loss on the test dataset.

2.4 Tiling

Tiling is typically applied when using large images due to the memory limitations of the hardware (Fig. 2). For the current experiment we specifically chose the topology and dataset so that it would fit within 16 GB of RAM at inference time for a batch size of at least 1. Our goal is compare inference based on the whole image to inference based on an overlapping tiled version of the whole image.

To perform the overlapping tiling at inference time, our algorithm cropped 128×128 patches from the whole image at uniformly spaced offsets in both the horizontal and vertical dimensions. Inference was performed individually on the patch and the prediction mask was added to a running mean prediction of the whole tumor mask. As highlighted in Fig. 2, (*Middle*), two tiles may produce different predictions for the pixels they share in their overlap. This results in a lower confidence (green pixels) prediction for those pixels rather than the highly positive (yellow) or highly negative (blue) confidence present in the predictions of the individual tiles (and when predicting the whole image) (Fig. 2, *Right*).

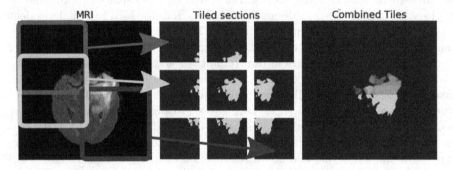

Fig. 2. An illustration of tiling. At training time random crops of the images/masks are used to build the model. At inference time, a series of overlapping crops from the whole image (depicted by the red, yellow, blue tiles) are input separately into the model and the multiple outputs are then reassembled and averaged to generate a prediction of the whole image. For the combined tile prediction (*Right*): yellow = high probability of tumor; blue = low probability of tumor; green = moderate probability of tumor (Color figure online)

3   Results

3.1   Single, Center Crop Tile

Figure 3 shows the result from using just a single, center 128×128 crop of the whole image to do prediction. For this case, a center crop removes the border from the MRI and retains most of the brain. The entire ground truth mask is contained within the cropped region. Nevertheless, the prediction using the entire image (including the border) gave a superior prediction (Dice 0.90 versus 0.69). This is of particular concern because it shows that a smaller patch itself–without any overlapping tiles– can produce inferior predictions to the whole image.

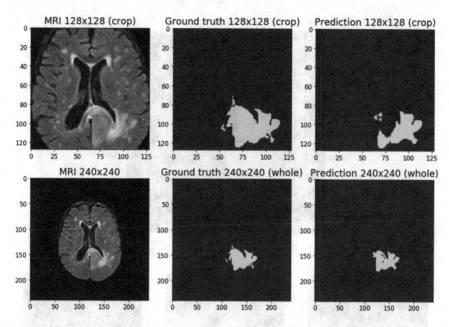

Fig. 3. The prediction from the 128×128 center crop of the MRI image (*Top*) is poorer (Dice = 0.69) than one based on the full 240×240 MRI image (Dice = 0.90) (*Bottom*) even though the model was trained on 128×128 image patches.

3.2   Multiple, Overlapping Tiles

Figure 4 shows a false positive prediction. In this case, the prediction based on using the whole image is correct (*i.e.* no tumor, *Right*), but the prediction based on overlapping tiles shows a high confidence (yellow) of tumor present in the MRI (*Left*).

In Fig. 5 we experiment with smaller tile dimensions and greater overlap between tiles. In most cases, a larger tile size and a larger overlap between successive tiles improved the prediction, but did not completely resolve the false positives and negatives.

For the 8,215 images in the test dataset, we found that using the whole image gave a 0.045 average increase in the Dice coefficient when compared to the tiling approach (Dice = 0.874 for whole image versus 0.829 for tiled approach, Fig. 6). In 7,146 of the cases (87%) using the whole image gave a better than or equal to Dice metric than the tiled approach. In 617 of the cases (7.5%), using the whole image gave more than a 0.1 increase in the Dice coefficient. In 63 of the cases (0.7%) using the tiled approach gave more than a 0.1 increase in the Dice coefficient. However, many of these cases involved very small ground truth masks and poor predictions by both models which may have led to the result (Fig. 7).

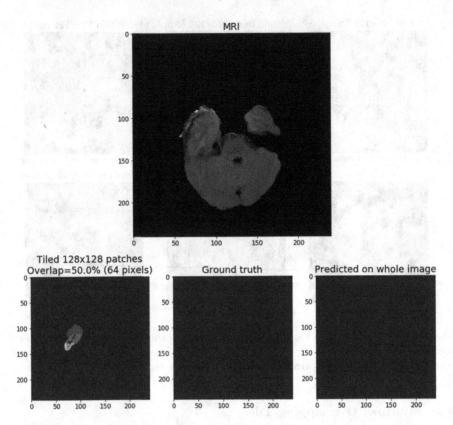

Fig. 4. 128 × 128 windows were slid over the whole image (*Top*) with a 50% window overlap in the horizontal and vertical directions. These 9 tiles were averaged to produce the tiled prediction (*Bottom left*). Although the tiled prediction does capture most of the true tumor, there are many false positive predictions in the tiled result. (*Bottom right*) The prediction using the whole image is far superior. (*Bottom center*) The radiologist's ground truth for the tumor. (Color figure online)

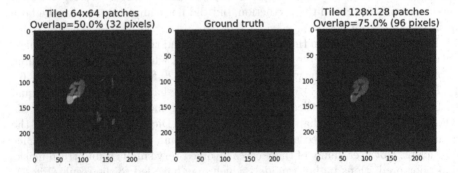

Fig. 5. Using a smaller patch size of 64 × 64 produces more false positives with the tiling method (*Left*). Using a greater overlap between patches does not completely reduce the false positive predictions (*Right*)

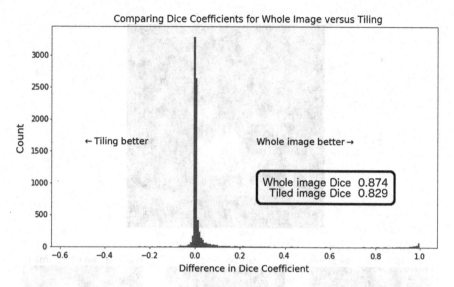

Fig. 6. The Dice metric between the prediction and the ground truth masks were compared for each image in the test dataset. The histogram shows the pairwise difference between using the whole image size and using the sliding tile method. The Dice metric using the whole image was on average 0.045 larger than using the tiling approach. Predictions based on the whole image were better or equal to tiling in 87% of the test dataset.

In Figs. 8, 9, and 10 we show three examples of where tiling leads to significant false negative predictions. In Fig. 8 tiling produces a low confidence prediction of the tumor mask (green pixels) whereas prediction on the whole image produces a high confidence prediction (yellow pixels). If a simple thresholding were used in this case, the tiling approach would result in a completely false negative prediction. In Fig. 9 we show that using the whole image was able to accurately predict even a very small tumor mask, but the tiling method on the same image again fails to detect the tumor with a significant confidence. In Fig. 10 we should that the tiling method produces a false negative prediction on the superior half of the tumor (green pixels).

4   Discussion

The overlapping tile approach is commonly used by researchers to apply fully convolutional models on large 2D and 3D images that would ordinarily not fit into available memory. Although this method works, we have demonstrated that clearly better predictions can be obtained by applying the convolutional model to the whole image.

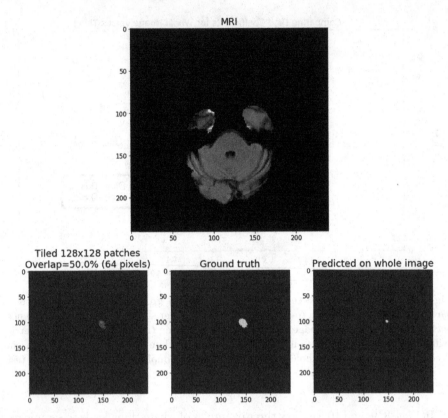

Fig. 7. One of the 0.7% of the cases where the tiled prediction gave a larger Dice coefficient (0.32) than the whole image prediction (0.21). In most of these cases, there were small ground truth masks and poor predictions generated by both methods.

Our results show that the tiling method often produces false positive and false negative predictions (Figs. 4, 8, 9, and 10). These false predictions can be reduced through a combination of increasing the tile dimensions and increasing the tile overlap. However, both of these corrections greatly increase the computational complexity of the model. For example, in our experiment, we used a tile dimension that was about one quarter size of the whole image (128×128 versus 240×240). With a 50% overlap between tiles, this cost 3 model predictions for each dimension– a total of 9 forward passes of the model in order to predict the whole image. If a 75% overlap were used instead, this would increase to 16 predictions (4×4). Therefore, the tiling approach can often lead to an $O(n^2)$ number of calculations for 2D images and a $O(n^3)$ for 3D images. These additional computations may easily negate any speed advantage attained by the tiling approach over the whole image approach– and at an increased cost in accuracy.

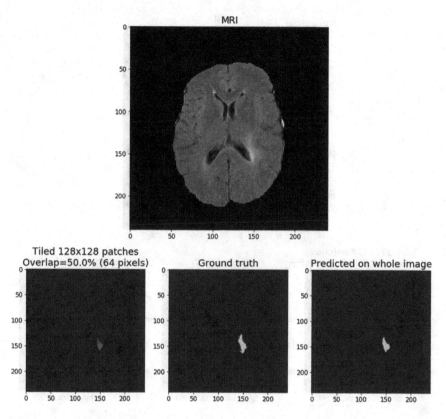

Fig. 8. The tiling approach produced a large false negative prediction (*Left*). If the prediction had been thresholded (*i.e.* $<0.5 = 0$; $\geq 0.5 = 1$) then tiling would have completely failed to predict the tumor within this slice (*Left*). (Color figure online)

Table 1. Memory and accuracy variations with training and inference methods

	Inf: FULL	Inf: TILED
Train: FULL	Inf Mem Req: High	Inf Mem Req: Low
	Inf Accuracy: High	Inf Accuracy: Low
Train: TILED	Inf Mem Req: High	Inf Mem Req: Low
	Inf Accuracy: High?	Inf Accuracy: High?

We hypothesize that the difference in predictions between the whole image and a tile may be due to the combination of max pooling and non-linear activation functions of the network. Note that this difference occurs even without averaging overlapping tile predictions (cf. Fig. 3). The larger field of view provided by the whole image approach may give a richer set of features at both training and inference time necessary to "push" the information past the ReLU activations and into the larger receptive fields created by the max pooling layer.

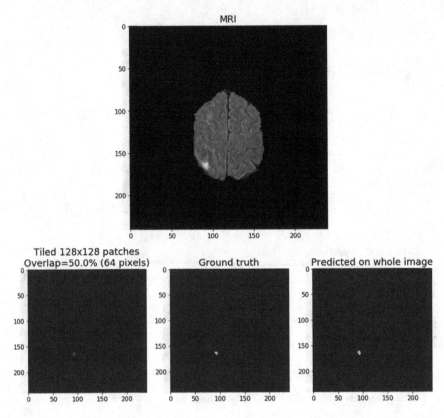

Fig. 9. Prediction on the whole image (*Right*) typically outperforms prediction using the tiling approach (*Left*) even when the ground truth masks are very small.

For example, a tumor might be too small to reliably detect in a narrowly defined field of view, but– over a larger field of view– the mass effect it produces on the surrounding tissue may provide enough additional feature information to increase the model's confidence. We liken this to a small pebble in a pond. The pebble may be too small to see, but its ripple effect on the surrounding water may still allow it to be detected.

We believe that researchers are artificially simplifying their models in order to fit into the memory limitations of current hardware. We suggest that instead researchers should be working with hardware and software manufacturers to more easily allow for greater memory capacity (Table 1). We believe that such efforts would help these models to be wider, deeper, and allow for larger inputs so that they can move to a new level in state of the art accuracy.

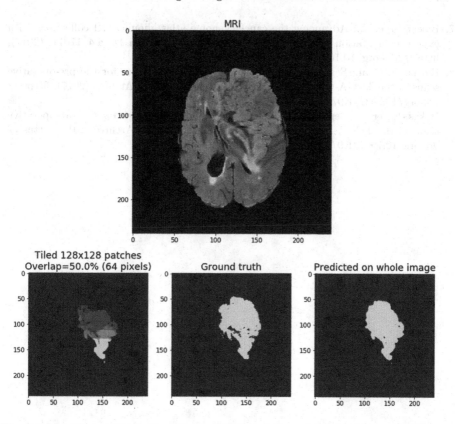

Fig. 10. The tiling methods the tiling method produces a false negative prediction on the superior half of the tumor (*Left*) but the whole image prediction correctly predicts the entire mass (*Right*). (Color figure online)

References

1. Greenspan, H., Van Ginneken, B., Summers, R.M.: Deep learning in medical imaging: overview and future promise of an exciting new technique. IEEE Trans. Med. Imaging **35**(5), 1153–1159 (2016)
2. Esteva, A., et al.: Dermatologist-level classification of skin cancer with deep neural networks. Nature **542**(7639), 115–118 (2017). https://doi.org/10.1038/nature21056
3. Tianqi, C., Bing, X., Chiyuan, Z., Guestrin, C.: Training deep nets with sublinear memory cost. arXiv:1604.06174v2 [cs.LG] 22 April 2016
4. Pinckaers, J.H.F.M., Litjens, G.J.S.: Training convolutional neural networks with megapixel images. arXiv:1804.05712v1 [cs.CV] 16 April 2018
5. Ronneberger, O., Fischer, P., Brox, T.: U-Net: convolutional networks for biomedical image segmentation, 18 May 2015. https://arxiv.org/abs/1505.04597
6. Menze, B.H., et al.: The multimodal brain tumor image segmentation benchmark (BRATS). IEEE Trans. Med. Imaging **34**(10), 1993–2024 (2015). https://doi.org/10.1109/TMI.2014.2377694

7. Bakas, S., et al.: Advancing the cancer genome atlas glioma MRI collections with expert segmentation labels and radiomic features. Nature Sci. Data **4**, 170117 (2017). https://doi.org/10.1038/sdata.2017.117
8. Bakas, S., et al.: Segmentation labels and radiomic features for the pre-operative scans of the TCGA-GBM collection. The Cancer Imaging Archive (2017). https://doi.org/10.7937/K9/TCIA.2017.KLXWJJ1Q
9. Bakas, S., et al.: Segmentation labels and radiomic features for the pre-operative scans of the TCGA-LGG collection. The Cancer Imaging Archive (2017). https://doi.org/10.7937/K9/TCIA.2017.GJQ7R0EF

An Update on Machine Learning in Neuro-Oncology Diagnostics

Thomas C. Booth[1,2](✉) [ID]

[1] School of Biomedical Engineering and Imaging Sciences,
King's College London, St. Thomas' Hospital, London SE1 7EH, UK
tombooth@doctors.org.uk
[2] Department of Neuroradiology, King's College Hospital NHS Foundation
Trust, London SE5 9RS, UK

Abstract. Imaging biomarkers in neuro-oncology are used for diagnosis, prognosis and treatment response monitoring. Magnetic resonance imaging is typically used throughout the patient pathway because routine structural imaging provides detailed anatomical and pathological information and advanced techniques provide additional physiological detail.

Following image feature extraction, machine learning allows accurate classification in a variety of scenarios. Machine learning also enables image feature extraction *de novo* although the low prevalence of brain tumours makes such approaches challenging.

Much research is applied to determining molecular profiles, histological tumour grade and prognosis at the time that patients first present with a brain tumour. Following treatment, differentiating a treatment response from a post-treatment related effect is clinically important and also an area of study. Most of the evidence is low level having been obtained retrospectively and in single centres.

Keywords: Neuro-oncology · Machine learning · Diagnostic

1 Introduction

1.1 Imaging Biomarkers

A biomarker can be defined as a characteristic that is measured as an indicator of normal biological processes, pathogenic processes, or responses to an exposure or intervention, including therapeutic interventions [1]. Molecular, histologic, imaging, or physiologic characteristics are types of biomarkers. In neuro-oncology, imaging biomarkers are used for diagnosis, prognosis and treatment response monitoring.

Magnetic resonance imaging is typically used throughout the patient pathway because routine structural imaging provides detailed anatomical and pathological information and advanced techniques provide additional physiological detail. Qualitative analysis of a new intracranial mass aides diagnosis and in routine clinical practice can determine whether or not to proceed to confirmatory biopsy or resection. For example, with some basic demographic information such as the age of the patient and with some basic clinical information, such as knowledge that the mass was found

© Springer Nature Switzerland AG 2019
A. Crimi et al. (Eds.): BrainLes 2018, LNCS 11383, pp. 37–44, 2019.
https://doi.org/10.1007/978-3-030-11723-8_4

incidentally whilst imaging for an unrelated condition, the qualitative routine structural imaging features of a grade 1 meningioma allow diagnosis with a high positive predictive value without the need for confirmatory biopsy. Advanced techniques allow quantitative analysis of masses which can also change management. For example, cerebral blood volume values obtained using dynamic susceptibility-weighted imaging within an area of tumour contrast enhancement, or 1H-magetic resonance spectroscopic ratios acquired from a tumour, may help determine whether a mass is of high histological grade (grade III or IV) in certain scenarios.

Some image analysis recommendations, which determine treatment response of high histological grade gliomas, have become common in the research setting and rely on simple linear metrics of simple image features, namely the product of the maximal perpendicular cross-sectional dimensions of contrast enhancing tumour [2, 3].

Unlike the above biomarkers where simple imaging features are apparent to the reporting clinician, much image analysis research aims to extract underlying information from the imaging dataset to develop biomarkers that may not be readily visible. Machine learning can be applied to different phases of image analysis research which sequentially consists of pre-processing images, feature estimation (quantifying or characterizing the image), feature selection (remove noise and random error in the underlying data), classification (decision or discriminant analysis) and evaluation [4].

1.2 Clinical Validity

Evaluation in image analysis research initially consists of analytical validation, where accuracy and reliability of the biomarker are assessed [5]. Accuracy determines how often a test is correct in a given population (the number of true positives and true negatives divided by the number of overall tests). Clinical validation is the testing of biomarker performance in a clinical trial. Biomarkers in neuro-oncology may not be rigorously proven to be analytically or clinically valid [5]. Validation instead may attempt to use a common biomarker thereby reducing the clinical validity. For example, an attempt to validate a new imaging biomarker for treatment response monitoring may involve comparing it to a common biomarker for treatment response, such as the product of the maximal perpendicular cross-sectional dimensions of contrast enhancing tumour. However, the common biomarker itself may not be rigorously proven to be clinically valid.

This update describes several illustrative research studies with a variety of designs aimed at developing imaging biomarkers for diagnosis, prognosis and treatment response monitoring using machine learning. Different machine learning strategies used in classification in particular, as well as feature estimation and selection, are demonstrated. The extent of analytical and clinical validation is highlighted. As with the illustrative studies described here, most research studies pertaining to machine learning and neuro-oncology are pioneering but the level of evidence is low [6]. Afterall, most studies are retrospective and performed in single centres.

2 Diagnostic Biomarkers

2.1 Pre-diagnostic Biomarkers

Pre-diagnostic or risk or susceptibility biomarkers are typically clinical or molecular and occur in the absence of overt neuro-oncological disease. An example could be the discovery of a patient with Li-Fraumeni syndrome. This is a hereditary cancer syndrome due to mutations in the tumour suppressor gene p53 where patients have a susceptibility for the development of glioma. Other examples include DNA repair gene polymorphisms, single-nucleotide polymorphisms and a history of ionizing radiation [5]. Imaging has had a negligible contribution to neuro-oncological pre-diagnosis.

2.2 Diagnostic Biomarkers

Diagnostic biomarkers are used to detect or confirm the presence of a disease or a subtype of the disease [1]. Both histology and molecular features are now frequently combined and 1p/19q chromosome arm co-deletion status and isocitrate dehydrogenase (IDH) mutation status are routinely acquired after biopsy in accordance with the 2016 World Health Organization Classification of Tumors of the Central Nervous System [7]. There has been much research using machine learning to extract molecular information from imaging, known as radiomics. The results have been promising but prospective clinical validation is required [5].

Example 1. The aim of this retrospective study was to use a machine-learning algorithm to generate a model predictive of IDH mutant status in high-grade gliomas based on clinical variables and multimodal features extracted from pre-operative routine MRI [8]. True IDH mutant status was determined following biopsy using a combination of immunohistochemistry, spectrometry and sequencing. The authors suggest that knowing the pre-operative IDH mutant status might counter the limited sensitivity of immunohistochemistry and might influence the extent of tumour resection, although there is limited evidence for these assertions. Pre- and post-contrast T1-weighted, T2-weighted, and apparent diffusion coefficient map images were obtained. Whole tumour, enhancing and non-enhancing tumour volumes as well as a tumour border region were segmented. Subregions delimited by apparent diffusion coefficient thresholds within the three volumes were also segmented. Imaging descriptors including location, first and second order (textural) statistics gave 2970 extracted features. Feature selection was performed using area under the receiving-operator characteristic curve (AUC) threshold and correlation. The remaining 386 features were used to build a model predictive of IDH mutant status by applying random forest to a 90 patient training dataset. The tree depth was set to 64 with a 4096 tree upper bound limit and bootstrapping applied. Tenfold cross validation was used. This gave 86% accuracy with an AUC of 0.88. The model was tested on a 30 patient in test dataset giving 89% accuracy and 0.92 AUC.

Heterogeneity metrics associated with ADC-delineated segmentation were the imaging features that contributed most in predicting IDH mutant status. Despite the multiple complex imaging features such as these, patient age gave the highest predictive value of IDH mutant status demonstrating the importance of including simple, accessible information as features in radiomic analyses. Unfortunately, other simple data such as Karnofsky Performance Status, which is known to be an important covariate in multivariate analyses of glioma survival, was not included. Nonetheless, the overall approach shows that machine learning allows combinations of features to be combined to give higher accuracy than single features alone including age.

A strength of the study is that routine imaging alone was used which makes translation to the clinic more feasible than if advanced imaging algorithms were also included. This is due to a frequent lack of standardization in many advanced imaging algorithms.

Common to most studies of diagnostic biomarkers, a limitation is that the findings relate to a single institution therefore the findings cannot be generalized elsewhere. Secondary high grade glioma were excluded, which presumably relates to exclusion of low grade gliomas that were followed up and then transformed. It is also noted that only enhancing tumours were included. Within the institution, the model can only be used within these constraints.

Example 2. In a similar retrospective study, a machine-learning algorithm was also applied to multimodal features extracted from pre-operative routine MRI to generate a model predictive of IDH mutant status (84 patients) [9]. In this example, grade II and III gliomas were studied and 1p/19q chromosome arm co-deletion status (67 patients) was also predicted as was grade (84 patients). Pre- and post-contrast T1-weighted, T2-weighted/FLAIR images were obtained from The Cancer Genome Atlas (TCGA)/The Cancer Imaging Archive (TCIA) dataset. Imaging descriptors with similarities to the previous study such as location, derived from Visually Accessible Rembrandt Images (VASARI), as well as second order statistics were determined.

Second order (textural) statistics and VASARI features were independently applied to raw images that had undergone a variety of manipulations such as down-sampling or grey-scale thresholding, using different sequences to give 3360 extracted features. Feature selection was performed using logistic regression and bootstrapping was performed to maximize the area under the receiving-operator characteristic curve giving models with <10 features. Using this methodology alone, second order statistic models performed better than VASARI models predicting IDH1 mutation status, 1p/19q co-deletion status and histological grade with AUCs of 0.86, 0.96, and 0.86, respectively. Random forest using 500 trees was then applied to combinations of clinical features and the two models of selected imaging features. IDH mutation status, 1p/19q co-deletion and histological grade were predicted with AUCs of 0.86, 0.89 and 0.78. Overall, texture played a dominant role in prediction. It is noteworthy that prediction of 1p/19q co-deletion status and grade was more accurate with logistic regression and bootstrapping methodology alone than when used as an input for random forest.

Analytical validation with a separate test dataset is required to improve analytical validity and make the findings more meaningful. However, even with further analytical validation the findings are unlikely to be translatable to the clinic as the fundamental constraint for clinical validation is that there was *a priori* knowledge that there were no grade IV gliomas in the dataset.

Example 3. In this small retrospective study a voxel-based unsupervised clustering method used a batch-learning self-organizing map (SOM) followed by *k*-means to determine regional histological grade from pre-operative routine MRI [10]. SOM is a neural network which can simplify features and remove outliers. k-means can identify features with similar patterns. Pre- and post-contrast T1-weighted and T2-weighted/FLAIR images from 36 patients with grade II-IV gliomas were processed and 161,157 extracted features underwent this two-level clustering to give clustered image maps. Segmented clustered image map regions corresponding to enhancing tumour tissue, non-enhancing tumour tissue, and oedematous tissue were described as class ratios which were used as inputs for supervised analysis. Classification was by a linear kernel support vector machine (SVM) using leave-one-out cross validation to distinguish low and high grade gliomas. The clustered image map with the optimal number of cluster classes gave an accuracy of 0.86 with 0.93 AUC. It was noted that a phenotype for high grade gliomas included high intensity of post-contrast T1-weighted and FLAIR images in contrast enhancing regions whereas a low grade phenotype showed high intensity of T2 images in these regions. Information from contrast enhancing regions alone made a large contribution to grade prediction with an accuracy of 0.82.

The method was applied prospectively to 4 patients with analysis of targeted biopsy tissue from representative classes which gave some limited evidence that the clusters gave meaningful information. It is noteworthy that no clinical parameters were used. Although this is a single centre study with a small number of patients, and without robust clinical validation, the approach to diagnostic biomarker development is an exemplar for how to minimise *a priori* knowledge.

3 Monitoring Biomarkers

Monitoring biomarkers are measured serially and may detect change in extent of disease, provide evidence of treatment exposure or assess safety [1]. There is an overlap with safety biomarkers which specifically determine any treatment toxicity. Monitoring blood or cerebral spinal fluid for circulating tumor cells, exosomes, and microRNAs shows promise [5]. However, imaging is particularly useful as it is non-invasive and captures the entire tumour volume and adjacent tissues and has led to recommendations to determine treatment response in trials [2, 3]. Clinical validation is typically not proven. Common biomarkers are frequently used in an attempt to indirectly validate the monitoring biomarker under development.

Example 1. The aim of this small glioblastoma study was to use a machine-learning algorithm to differentiate progression from pseudoprogression, at the earliest time point when an enlarging MRI-enhancing lesion is seen, using T2-weighted images alone [11]. Unsupervised feature estimation was performed using principal component analysis to investigate topological descriptors of image heterogeneity called Minkowski functionals. After confounders were identified (MRI field strength) and sensitivity to field strength demonstrated, a supervised analysis was performed. Feature selection reduced Minkowski functional, first order statistical and clinical features from 32 to 7. A radial basis function kernel support vector machine gave an accuracy of 0.88 in a retrospective training dataset of 17 patients and 0.86 in a prospective test dataset of 7 patients. Although not apparent to the reporting radiologist, the T2-weighted hyper-intensity phenotype of those patients with progression was heterogeneous, large and frond-like when compared to those with pseudoprogression. The pseudoprogression phenotype on T2-weighted images was shown to be a distinct entity and different from vasogenic oedema and radiation necrosis.

Additional analytical validation was performed firstly in the form of reliability testing which showed that a different operator performing segmentation achieved 100% classification concordance. Secondly, the same results using a different software package and a different operator were also obtained. Thirdly, a different feature selection method (random forest) and classifier (lasso) were used and also gave the same accuracy with 6 similar selected features.

A strength of the study is that T2-weighted images alone were used increasing the chance of translation. However, the study was performed in a single centre and, as the authors point out, the biomarker requires clinical validation in a larger multicentre test dataset.

Example 2. The aim of this small high grade glioma study was to use a machine-learning algorithm to differentiate progression from pseudoprogression at the earliest time point when an enlarging MRI-enhancing lesion is seen, using [18F]-fluoroethyl-L-tyrosine positron emission tomography. First and second order statistics were obtained from the images of 14 patients and underwent unsupervised consensus clustering. The cumulative distribution function then determined the optimal class size. Feature selection by predictive analysis of microarrays methodology using 10-fold cross validation reduced the features from 19 to 10. One of the 3 class PET-based clusters could differentiate progression and pseudoprogression, however the results were similar to the standard analysis method using maximal tracer uptake in the tumor divided by that in normally appearing brain tissue. The small, single centre study will require more analytical and clinical validation as the authors acknowledge.

4 Prognostic Biomarkers

Prognostic biomarkers identify the likelihood of a clinical event, recurrence, or progression based on the natural history of the disease [1]. They are generally associated with specific outcome such as overall survival or progression-free survival. Some molecular markers are prognostic biomarkers therefore there is some overlap with diagnostic biomarkers used to predict molecular markers (including IDH mutation status and 1p/19q co-deletion status).

Example 1. The aim of this retrospective study was to use a machine-learning algorithm to determine overall survival using imaging features from pre-operative routine MRI in patients with glioblastoma [13]. Pre- and post-contrast T1-weighted, FLAIR, DSC and diffusion tensor imaging (DTI) images were obtained from a retrospective training dataset of 105 patients. Enhancing tumour tissue, non-enhancing tumour tissue, and oedematous tissue regions were segmented with the glioma image segmentation and registration (GLISTR) segmentation algorithm which produced imaging descriptors including location and first order statistics and limited demographic information. From >150 features, 60 features with the best survival prediction following 10-fold cross validation were feature selected. Two linear kernel SVMs were used to classify patients as survivors or not at 6 and 18 months respectively and a combined prediction index calculated. Tenfold cross validation was used to determine the generalization accuracy of the predictive models to give an accuracy of 77% for the prediction of short/medium/long survivors. A prospective test dataset of 29 patients gave an accuracy of 79%.

Simple data such as Karnofsky Performance Status, which is known to be an important co-variate in multivariate analyses of glioma survival, were not included. An insightful aspect of this study is that histograms were produced in order to understand the predictive features: greater age, large tumour size, increased tumour diffusivity, larger regions of T2 hypointensity and highest perfusion peak heights, were all predictive of short survival. Although the findings have a plausible biological basis, translation is limited as this was performed in a single centre.

Example 2. The aim of this retrospective study was to use a machine-learning algorithm to determine overall survival of patients with high grade glioma using brain tumor segmentation (BRaTS) data [14]. Pre- and post-contrast T1-weighted, T2-weighted and FLAIR images were obtained from a retrospective training dataset of 163 patients. Segmented regions including enhancing tumour tissue, non-enhancing tumour tissue, and oedematous tissue regions were manually segmented. Features were selected by simple features such as location; discrete wavelet transform first and second order statistics; histograms alone; and a convolutional neural network (CNN) which gave over 4000 deep features. The CNN, AlexNet, used in transfer learning context consisted of five convolutional layers followed by three fully connected layers, with maximum pooling layers used in between the convolution and fully connected layers.

Patients were then classified as survivors or not at 10 and 15 months respectively. SVM, k-nearest neighbors (KNN), linear discriminant, tree, ensemble, and logistic regression were all independently applied to each set of features. A combination of CNN deep features and a linear discriminant classifier with 5-fold cross validation gave the best predictive result with a train dataset of 91% accuracy and a test dataset of 55% accuracy. Although interesting approaches to developing a prognostic biomarker were employed including using a CNN to generate features, the low test accuracy is suggestive of overfitting.

5 Conclusion

Machine learning and neuro-oncology are at an early stage of development and are not ready to be incorporated into the clinic as the level of evidence is low. Integration of data in addition to imaging, including demographic, clinical and molecular markers, may lead to increasingly accurate biomarkers. Development and validation of machine learning models applied to neuro-oncology require large, well-annotated datasets, and therefore multidisciplinary and multicentre collaborations are necessary.

Acknowledgments. This work was supported by the Wellcome/EPSRC Centre for Medical Engineering [WT 203148/Z/16/Z].

References

1. FDA-NIH Biomarker Working Group: BEST (Biomarkers, EndpointS, and other Tools) Resource. Food and Drug Administration (US), Silver Spring. Co-published by National Institutes of Health (US), Bethesda (2016)
2. MacDonald, D., Cascino, T.L., Schold, S.C., Cairncross, J.G.: Response criteria for phase II studies of supratentorial malignant glioma. J. Clin. Oncol. **8**, 1277–1280 (2010). https://doi.org/10.1200/JCO.1990.8.7.1277
3. Wen, P.Y., Macdonald, D.R., Reardon, D.A., Cloughesy, T.F., Sorensen, A.G., Galanis, E.: Updated response assessment criteria for high-grade gliomas: response assessment in neuro-oncology working group. J. Clin. Oncol. **28**, 1963–1972 (2010). https://doi.org/10.1200/JCO.2009.26.3541
4. Kassner, A., Thornhill, R.E.: Texture analysis: a review of neurologic MR imaging applications. Am. J. Neuroradiol. **31**(5), 809–816 (2010). https://doi.org/10.3174/ajnr.A2061
5. Cagney, D.N., Sul, J., Huang, R.Y., Ligon, K.L., Wen, P.Y., Alexander, B.M.: The FDA NIH biomarkers, endpoints, and other tools (BEST) resource in neuro-oncology. Neuro. Oncol. **20**(9), 1162–1172 (2017). https://doi.org/10.1093/neuonc/nox242
6. Howick, J., et al.: OCEBM Table of Evidence Working Group: The Oxford 2011 Levels of Evidence (2011). http://www.cebm.net/index.aspx?o=5653. Oxford Centre for Evidence-Based Medicine, Oxford (2016)
7. Louis, D.N., et al.: The 2016 world health organization classification of tumors of the central nervous system: a summary. Acta Neuropathol. **131**(6), 803–820 (2016). https://doi.org/10.1007/s00401-016-1545-1
8. Zhang, B., et al.: Multimodel MRI features predict isocitrate dehydrogenase genotype in high grade gliomas. Neuro. Oncol. **19**, 109–117 (2017)
9. Zhou, H., et al.: MRI features predict survival and molecular markers in diffuse lower-grade gliomas. Neuro. Oncol. **19**, 862–870 (2017)
10. Inano, R., et al.: Visualization of heterogeneity and regional grading of gliomas by multiple features using magneteic resonance-based clustered images. Sci. Rep. **6**, 30344 (2016)
11. Booth, T.C., et al.: Analysis of heterogeneity in T2-weighted MR images can differentiate pseudoprogression from progression in glioblastoma. PLoS One **12**(5), e0176528 (2017). https://doi.org/10.1371/journal.pone.0176528
12. Kebir, S., et al.: Unsupervised consensus cluster analysis of [18F]-fluoroethyl-L-tyrosine positron emission tomography identified textural features for the diagnosis of pseudoprogression in high grade glioma. Oncotarget **8**(5), 8294–8304 (2016)
13. Macyszyn, L., et al.: Imaging patterns predict patient survival and molecular subtype in glioblastoma via machine learning techniques. Neuro. Oncol. **18**, 417–425 (2016)
14. Chato, L., Latifi, S.: Machine learning and deep learning techniques to predict overall survival of brain tumor patients using MRI images. In: 17th IEEE International Conference on Bioinformatics and Engineering. IEEE Press, New York (2017). https://doi.org/10.1109/bibe.2017.00009

Brain Lesion Image Analysis

MIMoSA: An Approach to Automatically Segment T2 Hyperintense and T1 Hypointense Lesions in Multiple Sclerosis

Alessandra M. Valcarcel[1(✉)], Kristin A. Linn[1], Fariha Khalid[2],
Simon N. Vandekar[1], Shahamat Tauhid[2], Theodore D. Satterthwaite[3],
John Muschelli[4], Rohit Bakshi[2], and Russell T. Shinohara[1]

[1] Department of Biostatistics, Epidemiology, and Informatics,
University of Pennyslvania, Philadelphia, USA
alval@pennmedicine.upenn.edu
[2] Department of Neurology and Radiology, Brigham and Women's Hospital,
Boston, USA
[3] Department of Psychiatry, University of Pennsylvania, Philadelphia, USA
[4] Department of Biostatistics, Johns Hopkins University, Baltimore, USA

Abstract. Magnetic resonance imaging (MRI) is crucial for *in vivo* detection and characterization of white matter lesions (WML) in multiple sclerosis (MS). The most widely established MRI outcome measure is the volume of hyperintense lesions on T2-weighted images (T2L). Unfortunately, T2L are non-specific for the level of tissue destruction and show a weak relationship to clinical status. Interest in lesions appearing hypointense on T1-weighted images (T1L) ("black holes"), which provide more specificity for axonal loss and a closer link to neurologic disability, has thus grown. The technical difficulty of T1L segmentation has led investigators to rely on time-consuming manual assessments prone to inter- and intra-rater variability. We implement MIMoSA, a current T2L automatic segmentation approach, to delineate T1L. Using cross-validation, MIMoSA proved robust for segmenting both T2L and T1L. For T2L, a Sørensen-Dice coefficient (DSC) of 0.6 and partial AUC (pAUC) up to 1% false positive rate of 0.69 were achieved. For T1L, 0.48 DSC and 0.63 pAUC were achieved. The correlation between EDSS and manual versus automatic volumes were similar for T1L (0.32 manual vs. 0.34 MIMoSA) and T2L (0.34 vs. 0.34).

Keywords: Logistic regression · Inter-modal coupling · Multiple sclerosis

1 Introduction

Multiple sclerosis (MS) is a life-long chronic disease of the central nervous system with no known cure. MS is the most common autoimmune disorder globally with about 2.3 million people affected worldwide [1, 2]. The pathophysiology of MS includes development of lesions which occur in the white matter (WML) and exhibit inflammation, destruction of myelin sheaths, and axonal loss. The accumulation of WML is associated with long-term morbidity and disability and is visible on structural magnetic resonance imaging (MRI).

© Springer Nature Switzerland AG 2019
A. Crimi et al. (Eds.): BrainLes 2018, LNCS 11383, pp. 47–56, 2019.
https://doi.org/10.1007/978-3-030-11723-8_5

Common quantitative MRI metrics in MS include lesion volume and count which rely on accurate segmentation of WML. Lesion count and volume are often derived from three related pathological presentations of WML: (1) contrast-enhancing lesions (EL), which are thought to represent acute perivascular inflammatory activity following focal break-down of the blood brain barrier, (2) T2 hyperintense lesions (T2L), which detect the process of demyelination and axonal loss and non-specific damage unrelated to MS, and (3) persisting T1 hypointense lesions (T1L), which are the most demyelinated and damaged regions [3–5]. Figure 1 displays axial slices of FLAIR and T1-weighted images with manual delineations overlaid.

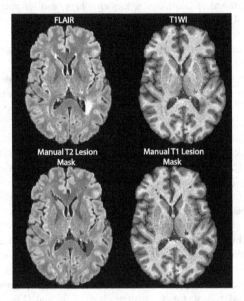

Fig. 1. Axial slices from an inhomogeneity corrected, registered, and intensity normalized MRI of a single subject are displayed in the top row. In the bottom row, manual lesion segmentation masks are overlaid on T1WI and FLAIR volumes.

Despite the existence of a number of automatic lesion segmentation methods [6–10], the majority delineate T2L solely. The sparsity of prior research is in part due to technical challenges: since T1L and their boundaries appear similar to gray matter [11] and are subtler than T2L, they are much more difficult to segment by manual and automatic methods. The simplest method was proposed by Filippi et al. using an expert-driven semi-automated thresholding approach to estimate lesion volumes [12]. Molyneux et al. similarly propose a semi-automated technique to delineate T1L in a multi-center study where they showed that T1L volume is a consistent and reproducible metric that can be applied to MRI data from various scanners [13]. Following these results, Datta et al. recently developed fully automated methods using fuzzy connectivity modeling [14]. Other methods proposed an algorithm to detect EL, T1L, and T2L using intensity-based statistical k-nearest neighbor classification combined with template-driven segmentation and partial volume artifact correction [15]. To automatically segment T1L, others proposed an approach that used a standard classification algorithm to partition T1-weighted images into gray matter, white matter, and cerebrospinal fluid and then finding T1L in the white matter by spatially voxel-wise testing using healthy controls as a reference [16]. Unfortunately, no approach has released publicly available software and these studies were based on relatively small MRI data sets with uniform patient demographics and lesion load [12–16].

In our previous work, a Method for Inter-Modal Segmentation Analysis (MIMoSA) was developed and validated as an automatic T2L segmentation method in people with MS [10]. In the present study, we extended the MIMoSA method to automatically segment T1L. Since no publicly available software for automatic detection of T1L exists, we automatically segment T2L using MIMoSA and used these measures as a reference for T1L performance. This was motivated by our findings that MIMoSA is a competitive T2L segmentation approach [10], and all T1L are also seen as T2L (but not vice-versa). Moreover, since the data acquired in this study were acquired under different protocol than data in the original development of MIMoSA, through the application of MIMoSA to segment T2L we validate and assess robustness of MIMoSA's accuracy across scanner platforms and protocols. For further comparison, OASIS, another validated T2L lesion segmentation algorithm [8], was used to automatically segment T1L. Finally, we examined correlations between lesion volume with clinical status measurements in order to determine if the reduction in noise associated with automatic lesion segmentation revealed stronger associations with disability.

2 Methods

2.1 Data and Preprocessing

Data were collected at the Brigham and Women's Hospital in Boston, Massachusetts. Forty patients, all with a clinical diagnosis of MS, were consecutively obtained from MRI scans at the center. Subjects had an examination by an MS specialist neurologist to assess the type of MS, the level of physical disability on the Expanded Disability Status Scale (EDSS), and ambulatory function on the timed 25-foot walk (T25FW). High-resolution 3D T1-weighted (T1WI), T2-weighted (T2WI), and fluid-attenuated inversion recovery (FLAIR) volumes of the brain were collected on a Siemens 3T Skyra instrument using a consistent scan protocol among subjects. In addition to the imaging sequences, T1L and T2L were manually segmented by an experienced reading panel of two observers under the supervision of an experienced observer. The observers determined the presence or absence of T1L together to form a single consensus segmentation between the two raters. In the event of a disagreement, a senior experienced observer was consulted. This procedure was repeated to segment T2L so that T1L and T2L were segmented by the two raters using a consensus approach but the lesion types are obtained independently. These T1L and T2L manual annotations were acquired manually and without the use of any automatic method.

All images were preprocessed prior to implementing the MIMoSA model, using the R (version 3.1.0, R Foundation for Statistical Computing, Vienna, Austria) packages extrantsr [17] and WhiteStripe [18] as well as Multi-Atlas Skull-Stripping (MASS) [19, 20]. After N4 inhomogeneity correction [21], volumes were co-registered across sequences for each subject using a rigid-body transformation with a Lanczos windowed sinc interpolator. To remove extracerebral voxels, MASS was implemented [19, 20]. As conventional MRI volumes are acquired in arbitrary units, statistical intensity normalization using WhiteStripe [18] was applied in order to model intensities across subjects.

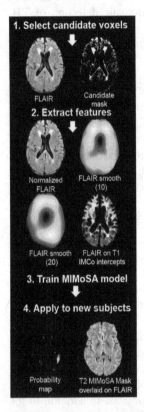

Fig. 2. The MIMoSA procedure is outlined with relevant images. Only features derived from FLAIR volumes are shown for simplicity.

2.2 MIMoSA: Method for Inter-modal Segmentation Analysis

MIMoSA was originally built to automatically segment T2L and extended in this pilot study to automatically segment T1L. As feature extraction is known to be pivotal for a segmentation algorithm's accuracy and generalizability, the MIMoSA method relies on features that capture the mean structure of each imaging modality as well as the covariance across modalities. The method requires FLAIR and T1 as image inputs but can also utilize T2 and PD if they were collected. The MIMoSA procedure is outlined in Fig. 2 and described below. First, MIMoSA identifies all candidate voxels for T2L presence defined as 85th percentile or above on the FLAIR. The algorithm then creates the features to be implemented in the model. The features included in the model are normalized MRI volumes, smoothed volumes with kernel parameters 10 and 20 mm, and inter-modal coupling (IMCo) intercept and slope coefficients for each combination of images as outcome and predictors [22]. With all relevant features calculated, the procedure then fits a local logistic regression based on gold standard manual segmentations to obtain coefficients which we then use to produce maps of the probability of lesion.

In the model below, $P\{L_i(v) = 1\}$ represents the probability that a voxel is part of a lesion where $L_i(v)$ is a random variable denoting voxel-level lesion presence. If there is a lesion in voxel v for subject i, then $L_i(v) = 1$, otherwise $L_i(v) = 0$. We model the probability that a voxel v contains lesion incidence with the following logistic regression model:

$$logit[P\{L_i(v) = 1\}] = \beta_0 + X_i^T(v)\beta + \mathfrak{G}X_i^T(v, 10)\{\beta_{10} + X_i(v) \otimes \beta_{10}^*\} - \mathfrak{G}X_i^T(v, 20)\{\beta_{20} + X_i(v) \otimes \beta_{20}^*\} + CX_{i,I}^T(v)\beta_I + CX_{i,S}^T(v)\beta_S, \tag{1}$$

where we denote the normalized images $X_i(v) = \left[T_{1,i}(v), FLAIR_i(v), T_{2,i}(v), PD_i(v)\right]^T$ and we use $\mathfrak{G}$ to denote the smoothing operator with parameter $\delta \in \{10\,\text{mm}, 20\,\text{mm}\}$, which takes a weighted average within each neighborhood $N(v, \delta)$ around v. We express the smoothed images in vector form by $\mathfrak{G}X_i(v, \delta) = \left[\mathfrak{G}\left(T_{1,i}(v); N(v, \delta)\right), \ldots, \mathfrak{G}(PD_i(v); N(v, \delta))\right]^T$, and we denote all combination of intercept and slope IMCo parameters respectively by $CX_{i,I}^T(v)$ and $CX_{i,S}^T(v)$. We use $\otimes$ to represent the Hadamard product. The interaction terms between the normalized volumes and the smoothed volumes, denoted by β_{j0}^*, contribute to the model by capturing differences between voxel intensities and their local mean intensities. These aid in mitigating artifacts due to residual field inhomogeneity in some cases, and generally improve lesion detection performance. We use a logistic regression because it is simple, easy to interpret, and computationally quick [8]. In the past, studies have compared classification methods and shown that simple methods often yield performance equivalent to more sophisticated methods so long as relevant biological features are included [23].

After fitting, the MIMoSA method can then be applied to new subjects, namely subjects not included in the training set, in order generate probability maps which we then threshold to create binary lesion segmentation masks. We select the threshold by an optimal thresholding algorithm that optimizes similarity of predicted segmentation masks in the training set with gold standard segmentations based on DSC. To automatically segment T2L and T1L separate models must be fit based on manual segmentations. We simply apply the MIMoSA procedure, built for T2L to T1L.

2.3 Statistical Analysis

Training and testing of MIMoSA methods was conducted using a cross-validation. In addition to implementing MIMoSA, a competitive T2L segmentation algorithm, OASIS, was also applied [8]. OASIS was specifically chosen for the present study as it can be easily trained using publicly available software and there are no publicly available data for benchmarking T1L automatic lesion segmentation. To fit the models and measure performance, 20 subjects were allocated to the training set and 20 subjects to the test set. MIMoSA and OASIS were then trained for T1L and T2L separately using subjects in the training set. After models were fit, the estimated coefficients were applied to the test set in order to generate probability maps. To generate lesion masks, the threshold produced from the optimal threshold algorithm described above was applied.

This procedure was iterated 100 times. In each fold, subject-level DSC and partial AUC (pAUC, up to 1% false positive rate) were recorded [24]. pAUC was estimated rather than traditional AUC since it only considers regions of the ROC space which correspond to clinically relevant values of specificity [25]. After calculation at the subject level, performance measures were averaged across subjects and cross-validation folds. Figure 3 shows the full cross-validation pipeline.

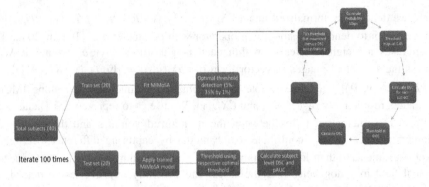

Fig. 3. Bootstrap cross-validation scheme to assess MIMoSA performance on T1 lesion (T1L) and T2 lesion (T2L). To identify the optimal threshold, probability maps for subjects in the training set were generated using the model previously fit. These maps were thresholded along a grid and Sørensen-Dice coefficient (DSC) was calculated. The threshold resulting in the maximum DSC across subjects in the training set was applied to the threshold in the test set.

3 Results

The DSC and pAUC false positive rates up to 1% are shown in Fig. 4. False positive rates above 1% are not clinically useful in MS lesion segmentation. For example, consider 1% of the volume of a healthy control subject is on the order of 10 cm^3, which is equal to the average lesion load of an MS subject. Results in Fig. 4 indicate competitive lesion segmentation performance of both T1L and T2L. The method accurately delineates T1L and T2L as exemplified by high DSC and pAUC. MIMoSA performance measures are all higher than OASIS results indicating superior automatic segmentation.

In practice, common applications of lesion segmentation metrics are for association studies with clinical status and evaluating therapeutic efficacy [5, 26]. In Table 1, we report the relationship between both manual and MIMoSA lesion segmentation metrics and clinical

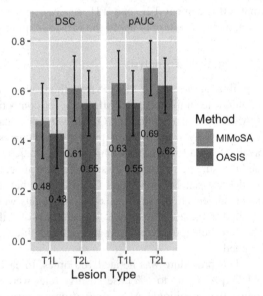

Fig. 4. Results from the cross-validation are presented. T1 lesion (T1L) and T2 lesion (T2L) average measures for Sørensen-Dice coefficient (DSC) and partial AUC (pAUC) with up to 1% false positive rate were averaged within each testing set and then across folds. Error bars are overlaid where standard deviation was calculated within cross-validation folds and averaged iterations.

measures. Volumetric measures were correlated with EDSS score and T25FW for T1L and T2L. The correlations displayed in this table show that $\hat{\rho}(MIMoSA)$ is equal to or larger than $\hat{\rho}(Manual)$.

Table 1. Clinical-MRI relationships using either manual lesion volume denoted as $\hat{\rho}(Manual)$ or MIMoSA lesion volume denoted as $\hat{\rho}(MIMoSA)$ averaged across folds are shown. Lesion volumes using T1 lesions (T1L) and T2 lesions (T2L) were correlated separately with Expanded Disability Status Scale (EDSS) score, timed 25-foot walk (T25FW), and disease duration.

Variable	Method	$\hat{\rho}(Manual)$	$\hat{\rho}(MIMoSA)$
EDSS	T1L	0.32	0.34
	T2L	0.33	0.34
T25FW	T1L	0.06	0.14
	T2L	0.06	0.08
Disease duration	T1L	0.12	0.30
	T2L	0.15	0.26

Lesion volume and count are important metrics for diagnosis and in the evaluation of therapeutic effectiveness. Thus, their accurate estimation from an automatic method is of the utmost importance. Figure 5 provides subject-level measures of volume and count using MIMoSA compared with manually acquired metrics, averaged across cross validation iterations. MIMoSA's estimation of lesion volume is extremely accurate as the T1L and T2L points all lie close to the y = x line. Additionally, the correlations presented overlaid on the graph are very close to 1. Lesion count is similarly very accurate for subjects with less than 25 lesions. As lesion count increases beyond this though, MIMoSA tends to undercount lesions.

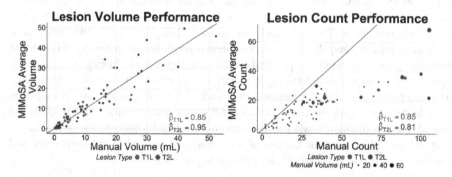

Fig. 5. Lesion volume and count are presented to compare manual segmentation with MIMoSA segmentation metrics. Volume and count for MIMoSA were obtained by averaging volume or count for each test subject across cross-validation folds (100). The solid line depicts the y = x line.

4 Discussion

MIMoSA is a fully automated segmentation method that utilizes changes in inter-modal covariance structure occurring in white matter pathology, and can be used to delineate T1L and T2L accurately, reliably, and efficiently in people with MS. The use of IMCo likely drives improvements in accuracy as IMCo measures appear to be robust to even small changes in intensity across imaging modalities. These measures are especially useful for detecting T1L, a challenging task since these lesions appear similar to gray matter. MIMoSA does not require human input, providing stability and consistency into delineation. The model can easily be adapted and trained for cases with more or fewer imaging sequences [8, 10]. The full modeling procedure can be easily and quickly implemented using software and documentation provided through Neuroconductor [27, 28].

MIMoSA provides accurate and reliable automatic segmentations of both T1L and T2L. Though T2L DSC and pAUC measures are slightly larger, indicating more similarity with manual segmentations, T1L performance was competitive. The automatic segmentation of T1L and T2L using the same procedure allows for a simple and consistent framework to obtain both metrics. Simultaneous delineation of T1L and T2L will lead to a better understanding of overall patient status. The correlation between manual volumes and MIMoSA volumes of WML indicates precision for total volume prediction. Since total lesion volume is commonly used in the assessment of new therapies in clinical trials [5], MIMoSA provides a promising alternative to manual segmentation in these settings. This may be especially useful for multi-center studies with a large number of patients or longitudinal studies with sequences collected over time.

Often lesion volumes are correlated with clinical covariates and disease status in patient management and clinical trials that evaluate therapy effectiveness. Therefore, automatic segmentation approaches should be as sensitive as manual measures. Correlations were provided to compare manual and MIMoSA segmentations with clinically relevant variables. Our results indicate that the relationship between MIMoSA volumetric assessments showed as close or better correlations compared with using manual segmentations. This was likely due to the stability and consistency introduced by an automatic method that requires no operator input. Segmentation of T1L can be challenging since the intensity profile is often indistinguishable from gray matter [5], especially with respect to delineating boundaries; thus, reliability in these areas may be the cause for the stronger correlation with covariates. For T2L evaluation, correlations seem to be approximately equal between MIMoSA and manual segmentations. In general, the measurements, whether obtained from manual segmentation or MIMoSA, were similar, advocating for the use of the automated method to cut cost, time, and introduce stability without sacrificing relation to patient status.

Acknowledgements. This project was supported in part by a pilot grant from the Center for Biomedical Computing and Analytics at the University of Pennsylvania as well as R01NS 085211, R21NS093349, R01NS060910, and R01MH112847 from the National Institutes of Health. The content is solely the responsibility of the authors and does not necessarily represent the official views of the National Institutes of Health.

References

1. Berer, K., Krishnamoorthy, G.: Microbial view of central nervous system autoimmunity. FEBS Lett. **588**, 4207–4213 (2014)
2. Feigin, V.L., et al.: Global, regional, and national burden of neurological disorders during 1990–2015: a systematic analysis for the global burden of disease study 2015. Lancet Neurol. **16**, 877–897 (2017)
3. Rovira, À., León, A.: MR in the diagnosis and monitoring of multiple sclerosis: an overview. Eur. J. Radiol. **67**, 409–414 (2008)
4. Katdare, A., Ursekar, M.: Systematic imaging review: multiple sclerosis. Ann. Indian Acad. Neurol. **18**, S24–S29 (2015)
5. Bakshi, R., Minagar, A., Jaisani, Z., Wolinsky, J.S.: Imaging of multiple sclerosis: role in neurotherapeutics. NeuroRx **2**, 277–303 (2005)
6. García-Lorenzo, D., Francis, S., Narayanan, S., Arnold, D.L., Collins, D.L.: Review of automatic segmentation methods of multiple sclerosis white matter lesions on conventional magnetic resonance imaging. Med. Image Anal. **17**, 1–18 (2013)
7. Meier, D.S., et al.: Dual-sensitivity multiple sclerosis lesion and CSF segmentation for multichannel 3T brain MRI. J. Neuroimaging Off. J. Am. Soc. Neuroimaging **28**, 36–47 (2017)
8. Sweeney, E.M., et al.: OASIS is automated statistical inference for segmentation, with applications to multiple sclerosis lesion segmentation in MRI. NeuroImage Clin. **2**, 402–413 (2013)
9. Shiee, N., Bazin, P.-L., Ozturk, A., Reich, D.S., Calabresi, P.A., Pham, D.L.: A topology-preserving approach to the segmentation of brain images with multiple sclerosis lesions. NeuroImage **49**, 1524–1535 (2010)
10. Valcarcel, A.M., et al.: MIMoSA: an automated method for intermodal segmentation analysis of multiple sclerosis brain lesions. J. Neuroimaging **28**, 389–398 (2018)
11. Ceccarelli, A., et al.: The impact of lesion in-painting and registration methods on voxel-based morphometry in detecting regional cerebral gray matter atrophy in multiple sclerosis. AJNR Am. J. Neuroradiol. **33**, 1579–1585 (2012). PubMed – NCBI. https://www.ncbi.nlm.nih.gov/pubmed/22460341
12. Filippi, M., Rovaris, M., Campi, A., Pereira, C., Comi, G.: Semi-automated thresholding technique for measuring lesion volumes in multiple sclerosis: effects of the change of the threshold on the computed lesion loads. Acta Neurol. Scand. **93**, 30–34 (1996)
13. Molyneux, P.D., et al.: The precision of T1 hypointense lesion volume quantification in multiple sclerosis treatment trials: a multicenter study. Mult. Scler. J. **6**, 237–240 (2000)
14. Datta, S., Sajja, B.R., He, R., Wolinsky, J.S., Gupta, R.K., Narayana, P.A.: Segmentation and quantification of black holes in multiple sclerosis. NeuroImage **29**, 467–474 (2006)
15. Wu, Y., et al.: Automated segmentation of multiple sclerosis lesion subtypes with multichannel MRI. NeuroImage **32**, 1205–1215 (2006)
16. Spies, L., et al.: Fully automatic detection of deep white matter T1 hypointense lesions in multiple sclerosis. Phys. Med. Biol. **58**, 8323–8337 (2013)
17. Muschelli, J.: extrantsr: extra functions to build on the ANTsR package (2014)
18. Shinohara, R.T., Muschelli, J.: WhiteStripe: white matter normalization for magnetic resonance images using WhiteStripe (2017)
19. Doshi, J., Erus, G., Ou, Y., Gaonkar, B., Davatzikos, C.: Multi-Atlas Skull-Stripping. Acad. Radiol. **20**, 1566–1576 (2013)
20. NITRC: CBICA: Multi Atlas Skull Stripping (MASS): Tool/Resource Info. https://www.nitrc.org/projects/cbica_mass/

21. Tustison, N.J., et al.: N4ITK: improved N3 bias correction. IEEE Trans. Med. Imaging **29**, 1310–1320 (2010)
22. Vandekar, S.N.: Subject-level measurement of local cortical coupling. NeuroImage **133**, 88–97 (2016)
23. Hand, D.J., et al.: Classifier technology and the illusion of progress. Stat. Sci. **21**, 1–14 (2006)
24. Sing, T., Sander, O., Beerenwinkel, N., Lengauer, T.: ROCR: visualizing classifier performance in R. Bioinformatics **21**, 3940–3941 (2005)
25. Walter, S.D.: The partial area under the summary ROC curve. Stat. Med. **24**, 2025–2040 (2005)
26. Zivadinov, R., Bakshi, R.: Role of MRI in multiple sclerosis I: inflammation and lesions. Front. Biosci. J. Virtual Libr. **9**, 665–683 (2004)
27. Neuroconductor. https://neuroconductor.org/
28. Mimosa—Neuroconductor. https://neuroconductor.org/package/details/mimosa

CNN Prediction of Future Disease Activity for Multiple Sclerosis Patients from Baseline MRI and Lesion Labels

Nazanin Mohammadi Sepahvand[1]($\boxtimes$), Tal Hassner[2], Douglas L. Arnold[3,4], and Tal Arbel[1]

[1] Centre for Intelligent Machines, McGill University, Montreal, Canada
nazsepah@cim.mcgill.ca
[2] The Open University of Israel, Ra'anana, Israel
[3] Montreal Neurological Institute, McGill University, Montréal, Canada
[4] NeuroRx Research, Montréal, Canada

Abstract. New T2w and gadolineum-enhancing lesions in Magnetic Resonance Images (MRI) are indicators of new disease activity in Multiple Sclerosis (MS) patients. Predicting future disease activity could help predict the progression of the disease as well as efficacy of treatment. We introduce a convolutional neural network (CNN) framework for future MRI disease activity prediction in relapsing-remitting MS (RRMS) patients from multi-modal MR images at baseline and illustrate how the inclusion of T2w lesion labels at baseline can significantly improve prediction accuracy by drawing the attention of the network to the location of lesions. Next, we develop a segmentation network to automatically infer lesion labels when semi-manual expert lesion labels are unavailable. Both prediction and segmentation networks are trained and tested on a large, proprietary, multi-center, multi-modal, clinical trial dataset consisting of 1068 patients. Testing based on a dataset of 95 patients shows that our framework reaches very high performance levels (sensitivities of 80.11% and specificities of 79.16%) when semi-manual expert labels are included as input at baseline in addition to multi-modal MRI. Even with inferred lesion labels replacing semi-manual labels, the method significantly outperforms an identical end-to-end CNN which only includes baseline multi-modal MRI.

Keywords: Multiple sclerosis · Magnetic resonance imaging ·
Disease activity · Deep learning

1 Introduction

Multiple sclerosis is traditionally known as a chronic inflammatory demyelinating disease of the central nervous system [7]. The presence of lesions in MRI is one of the hallmarks of MS. As a result, MRI has been used for diagnosis and to monitor disease progression and treatment response. The number of new or

A. Crimi et al. (Eds.): BrainLes 2018, LNCS 11383, pp. 57–69, 2019.
https://doi.org/10.1007/978-3-030-11723-8_6

enlarging T2w lesions as well as gadolinium-enhancing lesions have been used as markers of disease activity [14,16,20] which in turn is used as a clinical outcome to monitor the progression of disease and also the efficacy of new treatments in clinical trials for RRMS [10,21]. Hence, developing an automatic method to predict future disease activity from MRI could lead to better understanding of disease progression and help identify patients that can benefit from treatment. However, given the variability of lesion distribution in MRI, complexity of the evolution of lesions over time and the heterogeneity of the disease across the population in terms of clinical disease course, there are currently no established MR biomarkers that reliably predict the future disease activity.

We define future MRI disease activity as the *presence of any new/enlarging T2w lesions or gadolinium lesions at any period within two years of the trial*. Examples of active and inactive patients are shown in Fig. 1. Traditional biomarkers such as lesion counts and lesion volumes at baseline are not reliable predictors of future MRI disease activity due to the complexity of the disease. Furthermore, the presence of lesions at baseline does not necessary guarantee future MRI activity. Figure 2 illustrates how the absence of lesions at baseline does not guarantee the lack of future MRI disease activity, and a high lesion load at baseline does not guarantee the appearance of new/enlarging lesions two years ahead. As such, more sophisticated biomarkers are needed to reliably predict future MRI disease activity.

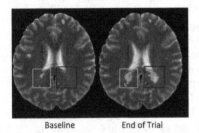 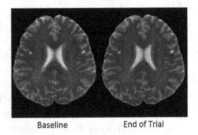

Baseline   End of Trial   Baseline   End of Trial

Fig. 1. Examples of active (left) and inactive patients (right). Lesions are highlighted in red. In each panel, the left image depicts T2w MRI at baseline and the right image is the same patient at the end of the second year. The patient in the left panel shows one enlarging (yellow bounding box) and one new lesion (blue bounding box) near the ventricles at year two. In the right panel, no new or enlarging lesions are present at year two. (Color figure online)

Although several automatic prediction methods predict the conversion of patients with preliminary symptoms to MS [1,3,23], only recently has the first machine learning approach been proposed for the prediction of future MS disease activity, in terms of future new/enlarging T2w lesions, based on baseline MRI [5]. This approach—a random forest classifier based on a *Bag-of-Lesion* representation—led to promising results on a proprietary dataset (sensitivity at 68.0% and specificity at 57.0%).

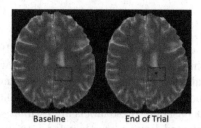 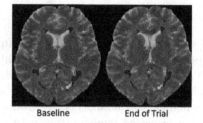

Baseline   End of Trial   Baseline   End of Trial

Fig. 2. Examples of patients whose baseline lesion loads are not good indicators of future disease activity: (left) active; (right) inactive. Lesions are highlighted in red. In each panel, the left image represents T2w MRI at baseline and the right image represents the same patient at the end of the second year. The patient in the left panel shows no baseline lesions, yet develops a new lesion (blue bounding box) at the end of the second year. The patient on the right, however, shows no new or enlarging lesions at the end of second year, despite having high lesion load at baseline. (Color figure online)

In recent years, deep learning has provided a wide range of powerful alternative frameworks with impressive results in both computer vision [2,12] and medical imaging fields [4,13]. In this work, we present the *first automatic, deep learning framework, a 3D CNN, for predicting future disease activity* from baseline MRI of patients with RRMS. We show how activity prediction accuracy is further enhanced through the inclusion of binary T2w lesion labels at baseline as inputs. These lesion labels help the network focus on areas of the brains with lesions, thereby aiding in network training. As expert lesion labels are expensive, time consuming, and hard to obtain, we further evaluate the performance of our prediction network in settings where semi-manual expert T2w lesion labels are not available. To this end, we develop a 3D Unet segmentation network [17] for automatic inference of T2w lesion labels. These labels are then used as an input to the prediction network.

Both prediction and segmentation networks are trained on a proprietary, multi-scanner, multi-center, clinical trial dataset of 1068 patients with RRMS. The performance of both networks is examined at operating point of interest on Receiver Operating Characteristic (ROC) curves. Our results indicate that using only baseline MRI and semi-manual expert lesion labels at baseline leads to very high prediction accuracies for future MRI disease activity over the next two years (accuracies of 80.21% and precision of 91.82%). This leads to the possibility of future development of precision medicine in RRMS. We further show that inclusion of baseline T2w lesion labels inferred from a Unet segmentation network provides good results, while still significantly outperforming a CNN based only on baseline MRI as inputs.

2 Proposed Framework

We develop a 3D CNN to predict future disease activity from baseline MR sequences. The prediction network takes three baseline MRI sequences (T1w, T2w and FLAIR) as well as T2w lesion labels as input for each patient and produces future MRI disease activity as binary labels (*active/inactive*). Should semi-manual lesion labels not be available at baseline, these labels are estimated using a proposed segmentation network. Inferred lesions are then fed into our prediction network along with baseline MR modalities (see Fig. 4(a)). Our segmentation network is a modified Unet, trained to segment lesions from baseline MR sequences. This network takes as input four acquired MR sequences (T1w, T2w, FLAIR, and Proton Density) and generates a lesion label with the same dimensions as the input brain volume (Fig. 4(b)).

2.1 Activity Prediction Network

The prediction network is a 3D CNN network with five convolutional layers followed by two dense layers. The architecture of the prediction network is illustrated in Fig. 3. As evident from the figure, each convolutional layer consists of two consecutive $3 \times 3 \times 3$ convolutions, each followed by a rectified linear unit (ReLu). In each convolutional layer, barring the first, a $2 \times 2 \times 2$ max-pooling with strides of two follows the two convolution units. The initial number of filters (feature maps) is set to four and this number is doubled after each max-pooling. Two fully-connected layers—with 16 and one neuron(s), respectively—are appended to the output of last convolutional layer. At the end of each convolutional layer, batch normalization [9] is applied. The network is trained with a dropout probability of 0.5 applied to the layers before both dense layers.

2.2 Segmentation Network

Should manual labels not be available, we wish to infer the lesion labels through an automatic framework. To this end, we develop a segmentation network which is a modified Unet (see Fig. 4(b)). Similar to the standard Unet, our network consists of contracting (encoding) layers followed by expanding (decoding) layers. The structure of the encoding path is very similar to the prediction network. Five convolutional layers each containing two $3 \times 3 \times 3$ convolutions where each convolution unit is followed by a ReLu. Every layer, except the first, is followed by a $2 \times 2 \times 2$ max-pooling layer with strides of two for down-sampling.

In the decoding path, each layer consists of a deconvolution of $3 \times 3 \times 3$ with strides of two for upsampling, a concatenation with the correspondingly feature map from the contracting path, and two $3 \times 3 \times 3$ convolutions each followed by a ReLu. The number of feature maps is halved after each upsampling. Similar to the prediction network, batch normalization is applied to the output of each convolutional layer in the encoding path. In the decoding path, batch normalization is applied to the output of deconvolution unit in each layer.

Semi-manual expert T2 lesion labels

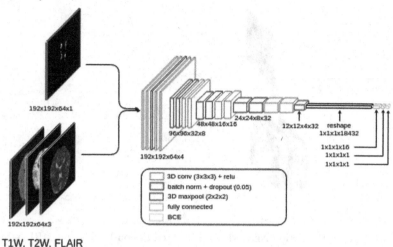

T1W, T2W, FLAIR

Fig. 3. Architecture of the activity prediction network. The network takes as input three MR sequences as well as T2w lesion labels at baseline and predicts future disease activity. All operations including convolution, max-pooling, and up-sampling are applied to 3D volumes.

2.3 Network Training

Training of both segmentation and prediction networks is performed by the Adam optimizer [11] using standard cross entropy loss [8]. To deal with the class imbalance in the segmentation task, the two classes (*lesion/non-lesion*) are weighted in a manner which is inversely proportional to their frequencies. Specifically, the weight for each class (w_{c_i}) is defined as total number of voxels (vox_{tot}) divided by the total number of class voxels ($n_{vox_{c_i}}$):

$$w_{c_i} = n_{vox_{tot}}/n_{vox_{c_i}} \quad c_i = 0 \text{ (inactive)}, 1 \text{ (active)}. \tag{1}$$

This produces a rough ratio of 1 (lesion) to 800 (non-lesion). The prediction task suffers from a similar imbalance problem: 75% of the patients in our trial are labeled as *active*. We address this imbalance by oversampling from the minority (inactive) class so that each batch has equal number of active and inactive samples. Since the ratio of active to inactive patients is 3 to 1, oversampling can be achieved by replicating inactive samples three times. This way, the total number of active and inactive samples are equal and therefore each batch can contain equal number of samples from both classes.

Inferred T2 lesion labels

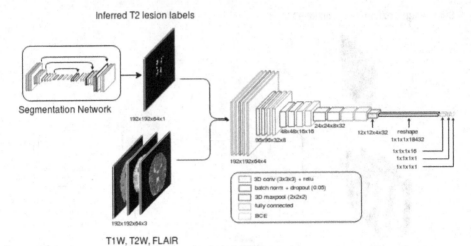

Segmentation Network

192x192x64x1

192x192x64x4

192x192x64x3

T1W, T2W, FLAIR

(a) Activity Prediction Network with Inferred Lesion Labels and MRI as Inputs

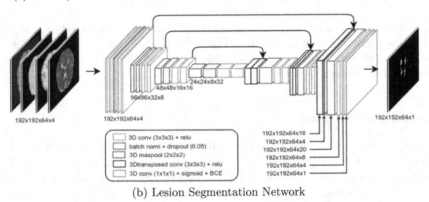

(b) Lesion Segmentation Network

Fig. 4. (a) Activity prediction network with automatically inferred labels as inputs, (b) lesion segmentation network. (a) The prediction network takes as input three MR sequences as well as T2w lesion labels generated by the segmentation network depicted in (b) and predicts future MRI disease activity. (b) The segmentation network takes as input four MR sequences and estimates T2w lesion labels at the baseline which are then fed into the prediction network demonstrated in (a). All operations including convolution, max-pooling and up-sampling are applied to 3D volumes.

3   Experiments and Results

3.1   Data Sets

To validate our framework for predicting future activity from baseline MRI, we conducted experiments using a proprietary dataset consisting of 1068 MS patient brain images, acquired during a large, multi-center, multi-scanner clinical trial. The trial was two years long, and MR scans were obtained at the beginning of the trial (baseline) and at the end of the first and second years. While samples from all time points were used for training, validation, and testing of the segmentation network, only baseline samples were used for the prediction network. T1-weighted (T1w), T2-weighted (T2w), Proton Density-weighted (PDw) and Fluid-attenuated inversion (FLAIR) are among MR sequences available for each subject and each timepoint. The dimensions of each volume are $192 \times 192 \times 64$, providing a resolution of $1\,mm \times 1\,mm \times 3\,mm$. Pre-processing included brain extraction [19], bias field inhomogeneity correction using N3 [18], Nyul image intensity normalization [15], and registration of all images to MNI-space.

In addition to four MR modalities, semi-manual expert T2w lesion labels are included in this dataset. These are comprised of 3D volumes with binary labels for lesions/non-lesions at each voxel. These labels are available for each patient at each time point and used for training the segmentation network. They were obtained through a semi-manual process where lesion labels were first generated by an in-house automated segmentation algorithm and then corrected by a trained expert reader.

Clinical data provided for this trial includes gadolinium and new/enlarging T2w lesion counts at the end of both first and second year. Due to missing clinical data for some patients, the total number of patient data available for the prediction task was 937. To calculate new/enlarging T2w lesion counts, lesion labels were generated through expert validation of an automatic longitudinal MS lesion segmentation framework [6]. Gadolinium counts were also provided for each patient in this dataset. These were estimated from post contrast T1w MRIs obtained after administration of contrast media (gadolinium). Gadolinium lesion segmentation was performed manually by trained experts. Binary MRI disease activity labels were defined based on the provided lesion counts. A patient was defined as being *active* if they had one or more new/enlarging T2 or gadolinium lesions.

3.2   Results

The dataset is divided into a training (80%), a validation (10%), and a test set (10%) in such a way that the ratio of active/inactive is the same (3/1) for all three splits. In this section, we first report the results of the prediction network trained to predict future disease activity from baseline multi-modal MRI as well as semi-manual expert lesion labels. This will be compared against using multi-modal MRI alone as inputs to the network. Next, the results of the lesion segmentation network will be shown. This leads to a quantitative analysis of the performance of the same network but with the semi-manual expert lesion labels replaced with the inferred lesions generated automatically by the segmentation network. Finally, a full comparison of the activity prediction results for all cases will be provided.

(I) Prediction with Semi-manual Expert T2w Lesion Labels. A 3D CNN network that takes as input three MR sequences (T1w, T2w and FLAIR) as well as T2w lesion labels at baseline is trained on a proprietary dataset.

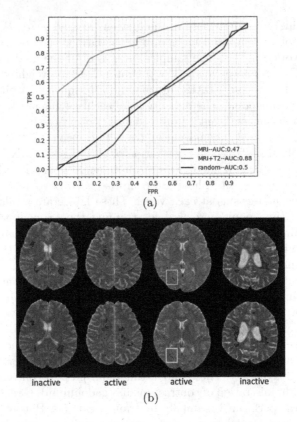

Fig. 5. Quantitative and qualitative results of the prediction network. (a) ROC curve with different inputs: Orange line: three MR sequences plus semi-manual, expert T2w lesion labels; blue: baseline MRI sequences alone; red: random performance. (b) Examples of brains for which the network prediction is successful: (top) Baseline MRI and lesion labels (red); (bottom) Year 2 MRI and lesion labels. Blue box depicts example of a new lesion. Yellow box depicts an enlarging lesion. (Color figure online)

Testing the performance of this network on a test set of 95 patients results in a sensitivity of 80.11% and a specificity of 79.16% (accuracies of 80.21%, precision of 91.82%) suggesting that our network is able to reliably predict the future MRI disease activity. To further evaluate the performance of our prediction network, an identical CNN network with only baseline MR sequences as input is trained. ROC curves, defined as True Positive Rate (TPR) vs False Positive Rate (FPR), for these two experiments are shown in Fig. 5(a). As is evident from the ROC curves, training an identical network with the same parameters and hyper-parameters values and using only baseline MR images leads to a performance barely better than random (with sensitivity of 8.45% and specificity of 79.16%). The low sensitivity in the baseline network is due to the fact that the network predicts the majority of the patients as inactive. This result shows that including lesion labels can significantly improve the performance of the prediction network over MRI alone. Qualitative examples of images for which network prediction is successful are also depicted in Fig. 5(b).

(II) T2w Lesion Segmentation and Detection. We examine the case where semi-manual lesion labels are not available, and are instead generated automatically. To this end, we train a Unet segmentation network with four MR sequences (T1w, T2w, FLAIR, and PDw) as input to estimate the lesion labels. Later, the inferred lesion labels will be fed, along with three MR sequences (T1w, T2w, FLAIR) at baseline, to the 3D CNN activity prediction network.

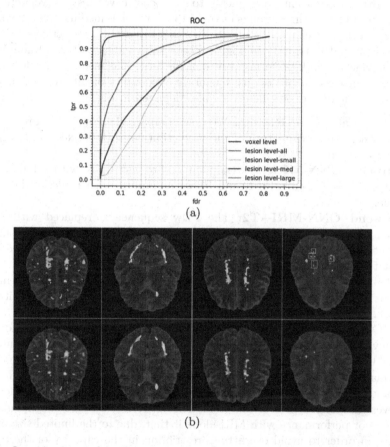

(a)

(b)

Fig. 6. Quantitative and qualitative results of the segmentation network. (a) ROC curves for both voxel level segmentation (blue) and lesion level detection split into three groups: small (red curve), medium (green) and large (magenta) lesions. (b) Examples of semi-manual expert segmentation (top) vs. output of segmentation network (bottom). While the segmentation network performs well for the three first brain images, a few small lesions are missed in the last case (inside red bounding boxes). (Color figure online)

The segmentation network is evaluated by reporting ROC curves for the test data. ROC curves, defined as TPR vs. False Detection Rate (FDR), are reported in Fig. 6 for both voxel-level segmentation and lesion-level detection. To obtain lesion-level detection statistics, TPR and FDR, from voxel segmentations, semi-manual, expert lesions smaller than three voxels are removed, in accordance with clinical protocol [22]. Candidate detected lesions are inferred using a simple connected component labeling method: a lesion is labeled as a true positive if the segmentation and its 18-connected neighborhood overlaps with at least three, or more than 50%, of the expert lesion voxels. Otherwise, it is labeled as a false positive. Insufficient overlap leads to a false negative. Lesion-level detection results are split into three groups of small (3–10 voxels), medium (11–50 voxels) and large (51+ voxels) according to their sizes and the ROC curves for each group is plotted separately in Fig. 6. The results indicate that, although the segmentation network performs very well for large and medium size lesions, it performs worse for small lesions. Qualitative examples of inferred lesion label slices shown against expert labels are also depicted in Fig. 6(b).

(III) Comparison of Results. We now compare the results of three experiments, where the architecture of the prediction network is fixed, and network inputs are varied:

End-to-end CNN-MRI: We train an end-to-end CNN with all four MR sequences (T1w, T2w, FLAIR, and PDw) as inputs, with no lesion labels at baseline.

End-to-end CNN-MRI+T2: The PDw sequence is replaced with *semi-manual, expert* lesion labels at baseline. PDw sequence is selected as it contains information that is available in the remaining sequences and is thus the least informative modality among all available MR sequences.

End-to-end CNN-MRI+segm T2: The semi-manual, lesion labels in the previous experiment are replaced with inferred labels from the segmentation network.

ROC curves for these three experiments are provided in Fig. 7. In addition, accuracy, precision, sensitivity, and specificity at an operation point of interest (FPR = 0.2) for each experiment are reported in Table 1. It is evident that the performance of the network with MRI alone as inputs is barely better than random, and that adding the semi-manual, expert lesion labels as input considerably improves the performance of the prediction network. One possible explanation for the poor performance with MRI alone is that, due to the limited size of the dataset, in order to avoid overfitting, restriction in the capacity of the model results in insufficient ability to learn from only baseline MR images. Our solution is to add lesion labels as an extra input modality to facilitate the training process by helping the network focus on the areas of interest.

The results also suggest that, although adding inferred lesion labels did not improve performance over MRI alone as much as adding the semi-manual expert labels, the gain in performance as compared with the baseline MRI alone is significant.

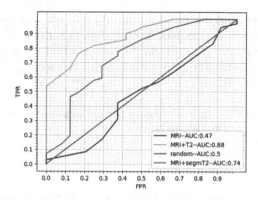

Fig. 7. ROC curves from three experiments on the prediction network with varying inputs. Orange line: three MR sequences plus semi-manual, expert T2w lesion labels, green line: three MR sequences plus inferred T2w lesion labels, blue: four MR sequences, red: random performance. (Color figure online)

Table 1. Quantitative comparison of results of three experiments on the same test data, reported at $FP = 0.2$.

Method	Accuracy	Precision	Specificity	Sensitivity
End-to-end CNN-MRI	26.31%	50.45%	79.16%	8.45%
End-to-end-MRI+segmT2	58.95%	88.09%	79.16%	54.12%
End-to-end-MRI+T2	80.21%	91.82%	79.16%	80.11%

The degradation in prediction performance when using inferred lesion labels over semi-manual labels is mainly due to the segmentation's drop in performance for small lesions, which in this case, make up approximately 40% of all lesions. This suggests that improving the accuracy of the segmentation network (not the focus of this work) would lead to significant accuracy improvements in the automatic prediction of future MRI disease activity.

4   Conclusions

We present the first deep learning framework for predicting future MRI disease activity in RRMS patients from baseline MR sequences. We show that prediction accuracy based on MRI alone does not perform much better than chance but improves substantially when baseline lesion labels are provided as additional inputs to the network. These results suggest the possibility of early prediction of future disease activity for RRMS patients, paving the way for the possibility of precision medicine in RRMS. Including semi-manual expert T2w lesion labels offers a remarkable boost in prediction accuracy showing accuracies of 80.21%, precision of 91.82%, specificity of 79.16% and sensitivity of 80.11%. When these are not available, we offer a deep learning framework for the automatic estimation of lesion labels. Our tests demonstrate that future MRI disease prediction using machine-generated lesion labels is not as accurate as prediction using semi-manual expert labels, but offers significant improvement over MRI sequences alone. Future work on improving lesion segmentation, particularly for small lesions, should significantly increase the accuracy of future disease prediction.

Acknowledgments. This work was supported by an award from the International Progressive MS Alliance (PA-1603-08175).

References

1. Barkhof, F., et al.: Comparison of MRI criteria at first presentation to predict conversion to clinically definite multiple sclerosis. Brain **120**(11), 2059–2069 (1997)
2. Bengio, Y., Courville, A., Vincent, P.: Representation learning: a review and new perspectives. TPAMI **35**(8), 1798–1828 (2013)
3. Brosch, T., Yoo, Y., Li, D.K.B., Traboulsee, A., Tam, R.: Modeling the variability in brain morphology and lesion distribution in multiple sclerosis by deep learning. In: Golland, P., Hata, N., Barillot, C., Hornegger, J., Howe, R. (eds.) MICCAI 2014. LNCS, vol. 8674, pp. 462–469. Springer, Cham (2014). https://doi.org/10. 1007/978-3-319-10470-6_58
4. Carass, A., et al.: Longitudinal multiple sclerosis lesion segmentation: resource and challenge. Neuroimage **148**, 77–102 (2017)
5. Doyle, A., Precup, D., Arnold, D.L., Arbel, T.: Predicting future disease activity and treatment responders for multiple sclerosis patients using a bag-of-lesions brain representation. In: Descoteaux, M., Maier-Hein, L., Franz, A., Jannin, P., Collins, D.L., Duchesne, S. (eds.) MICCAI 2017. LNCS, vol. 10435, pp. 186–194. Springer, Cham (2017). https://doi.org/10.1007/978-3-319-66179-7_22
6. Elliott, C., et al.: Temporally consistent probabilistic detection of new multiple sclerosis lesions in brain MRI. IEEE TMI **32**(8), 1490–1503 (2013)
7. Gold, R., et al.: Placebo-controlled phase 3 study of oral BG-12 for relapsing multiple sclerosis. N. Engl. J. Med. **367**(12), 1098–1107 (2012)
8. Goodfellow, I., Bengio, Y., Courville, A.: Deep Learning. MIT Press, Cambridge (2016)
9. Ioffe, S., Szegedy, C.: Batch normalization: accelerating deep network training by reducing internal covariate shift. In: ICML, pp. 448–456 (2015)
10. Kaunzner, U., Gauthier, S.: MRI in the assessment and monitoring of multiple sclerosis: an update on best practice. Ther. Adv. Neurol. Disord. **10**(6), 247–261 (2017)
11. Kingma, D., Ba, J.: Adam: a method for stochastic optimization. In: ICLR (2014)
12. Krizhevsky, A., Sutskever, I., Hinton, G.: Imagenet classification with deep convolutional neural networks. In: NIPS, pp. 1097–1105 (2012)
13. Menze, B., et al.: The multimodal brain tumor image segmentation benchmark (BRATS). IEEE TMI **34**(10), 1993 (2015)
14. Moccia, M., de Stefano, N., Barkhof, F.: Imaging outcome measures for progressive multiple sclerosis trials. Mult. Scler. J. **23**(12), 1614–1626 (2017)
15. Nyúl, L., Udupa, J.: On standardizing the MR image intensity scale. Magn. Reson. Med.: Off. J. Int. Soc. Magn. Reson. Med. **42**(6), 1072–1081 (1999)
16. Río, J., et al.: MR imaging in monitoring and predicting treatment response in multiple sclerosis. Neuroimaging Clin. **27**(2), 277–287 (2017)

17. Ronneberger, O., Fischer, P., Brox, T.: U-Net: convolutional networks for biomedical image segmentation. In: Navab, N., Hornegger, J., Wells, W.M., Frangi, A.F. (eds.) MICCAI 2015. LNCS, vol. 9351, pp. 234–241. Springer, Cham (2015). https://doi.org/10.1007/978-3-319-24574-4_28

18. Sled, J., Zijdenbos, A., Evans, A.: A nonparametric method for automatic correction of intensity nonuniformity in MRI data. IEEE TMI **17**(1), 87–97 (1998)

19. Smith, S.: Fast robust automated brain extraction. Hum. Brain Mapp. **17**(3), 143–155 (2002)

20. Sormani, M.P., Bruzzi, P.: MRI lesions as a surrogate for relapses in multiple sclerosis: a meta-analysis of randomised trials. Lancet Neurol. **12**(7), 669–676 (2013)

21. Stangel, M., et al.: Towards the implementation of 'no evidence of disease activity' in multiple sclerosis treatment: the multiple sclerosis decision model. Ther. Adv. Neurol. Disord. **8**(1), 3–13 (2015)

22. Windham, B., et al.: Small brain lesions and incident stroke and mortality: a cohort study. Ann. Intern. Med. **163**(1), 22–31 (2015)

23. Yoo, Y., et al.: Deep learning of brain lesion patterns for predicting future disease activity in patients with early symptoms of multiple sclerosis. In: Carneiro, G., et al. (eds.) LABELS/DLMIA-2016. LNCS, vol. 10008, pp. 86–94. Springer, Cham (2016). https://doi.org/10.1007/978-3-319-46976-8_10

Learning Data Augmentation for Brain Tumor Segmentation with Coarse-to-Fine Generative Adversarial Networks

Tony C. W. Mok[✉] and Albert C. S. Chung

Lo Kwee-Seong Medical Image Analysis Laboratory,
Department of Computer Science and Engineering,
The Hong Kong University of Science and Technology, Kowloon, Hong Kong
cwmokab@connect.ust.hk, achung@cse.ust.hk

Abstract. There is a common belief that the successful training of deep neural networks requires many annotated training samples, which are often expensive and difficult to obtain especially in the biomedical imaging field. While it is often easy for researchers to use data augmentation to expand the size of training sets, constructing and generating generic augmented data that is able to teach the network the desired invariance and robustness properties using traditional data augmentation techniques is challenging in practice. In this paper, we propose a novel automatic data augmentation method that uses generative adversarial networks to learn augmentations that enable machine learning based method to learn the available annotated samples more efficiently. The architecture consists of a coarse-to-fine generator to capture the manifold of the training sets and generate generic augmented data. In our experiments, we show the efficacy of our approach on a Magnetic Resonance Imaging (MRI) image, achieving improvements of 3.5% Dice coefficient on the BRATS15 Challenge dataset as compared to traditional augmentation approaches. Also, our proposed method successfully boosts a common segmentation network to reach the state-of-the-art performance on the BRATS15 Challenge.

1   Introduction

Accurate segmentation of a brain tumor from medical images is a crucial step for clinical diagnosis, evaluation, and follow-up treatment. Currently, the automatic segmentation methods which achieve state-of-the-art results are often using a deep learning approach. Modern deep learning models often consist of millions of parameters and learning these parameters requires massive annotated datasets to avoid overfitting to the training set. However, the problem is made challenging by the number of annotated training datasets often being limited in the medical imaging domain due to a couple of reasons. First, it is time-consuming and expensive for experts to accurately delineate the pixel-wise brain tumor region. Second, manual labeling also suffers from considerable intra-rater and inter-rater inconsistencies [5]. Third, there are various modalities and imaging protocols,

A. Crimi et al. (Eds.): BrainLes 2018, LNCS 11383, pp. 70–80, 2019.
https://doi.org/10.1007/978-3-030-11723-8_7

therefore a training set generated for one study is difficult to transfer to another study in practice.

To address these problems, we propose an automatic data augmentation approach for network-based brain tumor segmentation. Specifically, we present and evaluate a method for augmenting multimodal brain MRI images of high-grade (HG) and low-grade (LG) glioma patients, in which the generic augmented data enable the network-based method to learn the available annotated datasets more efficiently. Experimental results demonstrate that the proposed method effectively improves the segmentation accuracy of the network-based method, compared to the traditional data augmentation approach. It achieves improvements of 3.5% dice coefficient on the BRATS15 Challenge dataset as compared to traditional augmentation approaches.

2 Related Work

Data Augmentation. Data augmentation is essential to teach the network the desired invariance and robustness properties when only a few training samples are available. For medical image segmentation, different combinations of affine transformations are commonly used as data augmentation to teach the network the desired invariance and robustness properties. Ronneberger et al. [7] applied shift, rotation and elastic deformations to the microscopical images during training, while Milletari et al. [6] applied the random deformation to prostate MRI volumes using dense deformation field with B-spline interpolation. For brain tumor segmentation, scaling, rotation and flipping have also been applied to multimodal brain MR images for data augmentation [9]. Typical data augmentation approaches fail to increase the diversity of the training data, i.e., different parameters for MR imaging protocol, tumor size, shape, location, and appearance. The contribution of this work is that we have developed an automatic way to learn a more generic augmentation so that not only the rotational and scaling invariance, high-level information such as the shape of tumor and contextual information can also be augmented.

Generative Adversarial Networks. In the domain of computer vision, Generative adversarial networks (GANs) [2] have elicited considerable attention. GANs aim to model the data distribution by forcing the generated sample to be indistinguishable from the data. They have also proven successful in a wide variety of applications such as image generation [1,8], image manipulation [13] and image inpainting [11]. Recently, various coarse-to-fine frameworks of GANs have been proposed [4,10] to generate high-quality and high-resolution images, e.g., 1024×1024 pixels. Inspired by their successes, we propose a new coarse-to-fine boundary-aware GANs suitable to generate generic augmented MR images for brain tumor segmentation.

3 Methods

3.1 Preliminaries of Generative Adversarial Networks

Typical Generative Adversarial Networks (GANs), comprise a generator G and a discriminator D that are trained to compete with each other alternatively. The generator G is optimized to generate the data distribution p_{data} by generating the images that are indistinguishable for the discriminator D to differentiate from real images. While D is optimized to distinguish real images and synthetic images generated by G. The training objective is similar to a two-player min-max game as follows:

$$\min_{G} \max_{D} \mathcal{L}_{GAN}(D, G) = \mathbb{E}_{x \sim p_{data}}[\log D(x)] + \mathbb{E}_{z \sim p_z}[\log(1 - D(G(z)))]. \quad (1)$$

where x is a real sample from the target data distribution p_{data}, and z is a noise vector sampled from distribution p_z.

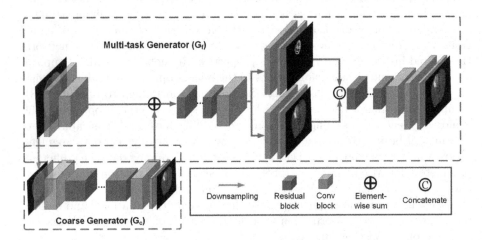

Fig. 1. Network architecture of proposed generator.

3.2 Coarse-to-Fine Boundary-Aware Generator

To generate high-resolution MR images of brains with realistic detail, we propose a Coarse-to-fine Boundary-aware Generative Adversarial Networks (CB-GANs). In our proposed method, the noise vector in traditional GANs is replaced by a label map of 2D axial slice from a 3D MR volume as a conditioning variable. We explain how we diversify the generated data without using a noise vector as input in Sect. 3.4.

Coarse-to-Fine Generator. Our generator is decomposed into two different sub-generators: G_c and G_f. G_c is the coarse generator while G_f is the fine generator. The generator G is then given by the tuple $G = \{G_c, G_f\}$. The coarse generator G_c aims to sketch the primitive shape and texture of multimodal brain MR images from a label map at a lower resolution and the fine generator G_f aims to correct the defects and completes the details of the low-resolution MR images from the coarse generator G_c.

The G_c and G_f consist of three components, namely, a convolutional down-sampling block, a set of residual blocks, and a transposed convolutional block. The resolution of the input label map to G_f is the same as the training data, while the resolution of the input label map to G_c is 4× smaller than the training data (2x smaller along each axis). Different from the residual block in coarse generator G_c, the residual block in fine generator G_f takes the element-wise sum of the output of G_c and the input feature maps of previous layers from G_f as the input. The element-wise sum operation helps integrate the global and local information from G_c and G_f.

Boundary-Aware Generator. Although the above coarse-to-fine framework can already produce high resolution natural images, it remains a challenge to produce a high quality synthetic MR image of a brain tumor that serves the purpose of data augmentation, given the corresponding label map. Because the size of the tumor core in MR images is often small compared to the other encephalic regions, in which the networks may fail to notice that the details of the tumor core are important. Preserving accurate tumor boundaries is important for augmented data to teach the network the desired invariance and robustness properties. To address this problem, we propose a multi-task generator G_f to replace the original fine generator.

The structure of the proposed generator is illustrated in Fig. 1. Instead of treating the image generation task as a single problem, we formulate it as a multi-task problem by exploring auxiliary information, which can simultaneously infer the location and boundary of the complete tumor. Specifically, two different branches are added to the final layer of G_f in order to output the MR image of a brain with a tumor and the boundaries of the complete tumor. After that, the outputs from the two new branches are concatenated and fed into a residual block followed by a non-linear activation layer. Therefore, the boundary and texture information from the new branches are fused together to output the final image. The mean-square-error loss is used for the boundary extraction task, as shown in the following:

$$\mathcal{L}_b(x, y) = \frac{1}{n_i} \sum_n \sum_i (P(x_{n,i}; \theta) - y_{n,i})^2, \tag{2}$$

where θ is the weight parameters in the generator. $\mathcal{L}_b$ refers to the mean-square-error loss for the boundary extraction task. $x_{n,i}$ and $y_{n,i}$ are the i-th pixel and ground truth in the n-th image used for training, respectively. P refers to the predicted probability for the pixel $x_{n,i}$.

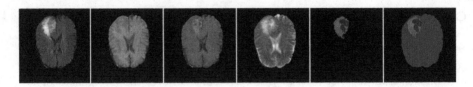

Fig. 2. Example of a synthetic high-grade glioma tumor. Left to right: FLAIR, T1, T1c, T2, expert delineation (Ground truth) and semantic label map (Red: necrosis, Green: edema, Yellow: non-enhancing tumor, Blue: enhancing tumor). (Color figure online)

3.3   Adversarial Training

Multi-discriminators. As the resolution of the synthetic image increases, the difficulty for the discriminator to differentiate real and synthetic images also increase. When there is only a single discriminator, the discriminator needs to have a large receptive field that is able to capture both global, i.e., tumor location, and local, i.e., tumor texture and shape, information from the input image. However, this may not be a good idea as it implies we will need a discriminator which has either a deep network or large convolutional kernels. Both options require a large memory and may easily suffer from overfitting to the training data.

To address this challenge, we adopt multi-discriminators with different scales of input as our discriminator $D = \{D_1, D_2, D_3, D_4\}$. The four discriminators have identical architectures but operate at different image scales, which is similar to [10]. Specifically, real MR images and synthesized MR images are downsampled by factors 2, 4 and 8 using the bilinear interpolation to create input for D of 4 scales. Throughout the experiments, we find that using four discriminators can achieve the optimal performance and further increasing the number of discriminators cannot improve the quality of synthetic image.

Perceptual Loss. We further improve the GAN's loss by incorporating a modified perceptual loss. The main idea of the perceptual loss function is that if the synthetic image is similar to the real image, the ith-layer feature maps of the discriminator should be also similar when the synthetic and real images pass through it. The modified perceptual loss $\mathcal{L}_P(G, D_k)$ is calculated as:

$$\mathcal{L}_P(G, D_k) = \mathbb{E}_{(x,c)} \sum_{i=1}^{L} \frac{1}{N_i} [||D_k^{(i)}(x, c) - D_k^{(i)}(G(z), z)||_2^2], \tag{3}$$

where $D_k^{(i)}$ represents the ith-layer feature maps of discriminator D_k, L is the total number of layers, N_i denotes the number of elements in each layer, x denotes the real MRI image, c denotes the label map and z denotes the deformed label map.

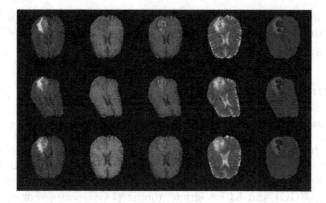

Fig. 3. Comparison of traditional augmentation and our proposed method. First row: Original image. Second row: Augmented image using elastic deformation. Third row: synthetic image generated by our approach. Left to right: FLAIR, T1, T1c, T2 and semantic label map

Therefore, our full training objective combines both GANs loss and modified perceptual loss as:

$$\min_{G} \left(\left(\max_{D_{k \in \{1,2,3,4\}}} \sum_{k=1,2,3,4} \mathcal{L}_{GAN}(D_k, G) \right) + \lambda_1 \mathcal{L}_b(x, y) + \lambda_2 \sum_{k=1,2,3,4} \mathcal{L}_P(G, D_k) \right).$$

$$(4)$$

3.4 Diversity of Augmented Data

Using Deformed Semantic Label Maps. The traditional augmentation approach for object segmentation often uses different combinations of affine transformations, such as shifting, rotation, and zoom, to leverage the knowledge of invariances in a task. However, such knowledge implied by these affine transformations is limited. For example, the shape, location and appearance of a complete tumor in a multi-modal MR image can vary significantly in the testing data, but the augmented image produced by the typical data augmentation fails to "simulate" such changes. Although some interpolation-based techniques such as elastic deformation can cause a slight variation in the shape of the augmented image, it may bring about damage and noise to the training data, as shown in Fig. 3, if the deformation field varies a lot.

Instead, we propose applying the elastic deformation to the label map. After that, we create a set of semantic labels from the deformed label maps. Specifically, we label 1 to 5 for necrosis, edema, non-enhancing tumor, enhancing tumor, non-tumor brain regions and 0 for everything else in the semantic labels. We use the semantic labels instead of the label maps as input for our proposed CB-GANs. By providing the information of the contour of the brain to the generator, it further diversifies the synthetic brain MR image with different shapes and prevents model collapse, i.e., prevents the model from generating a set of realistic MR images with the identical shape and context of the brain. Figures 2 and 3 show the example of the synthetic image generated by CB-GANs with corresponding semantic label map.

4   Experiments and Results

4.1   Data and Pre-processing

Experiments have been performed using brain MRI sequences from BRATS15 datasets. The dimensions of each MR volume are $240 \times 240 \times 155$ pixels. BRATS15 provides both the training and test sets. The training set consists of 220 high-grade glioma (HGG) and 54 low-grade glioma (LGG) cases. For each case, it includes 4 modalities (Flair, T1, T1-contrast (T1c) and T2) that were skull-stripped and co-registered. Pixel-wise ground truths that annotate the complete tumor, which are verified by radiologists, are provided in the training set. For the testing set, it consists of 110 cases each with 4 modalities. Unlike the training set, the ground truth labels of the test cases are hidden from the public and evaluation is carried out via an online system. Extensive evaluation has been carried out on three tasks: (1) the complete tumor (necrosis, edema, non-enhancing and enhancing tumor) (2) the tumor core (necrosis, non-enhancing and enhancing tumor) (3) enhancing tumor region. For each MR image, we normalize the intensities of each modality to have zero-mean and unit variance.

4.2   Network Architectures

Generator Architectures. For generator networks, we adopt our backbone architectures from Wang et al. [10] with some modifications. Below, we follow the naming convention used in the Wang et al. [10]. Let $c7s1$-k denote a 7×7 Convolution-BatchNorm-ReLU layer with k filters and stride 1. dk denotes a 3×3 Convolution-BatchNorm-ReLU layer with k filters and stride 2. Rk denotes a residual block that contains two 3×3 convolutional layers with the same number of filters on both layers. uk denotes a 3×3 fractional-strided-Convolution-BatchNorm-ReLU layer with k filters and stride $\frac{1}{2}$. Note that we will replace the activation layer from ReLU to Tanh for the final layer of each generator.

Our coarse generator G_c:
$c7s1$-64, d128, d256, d512, d1024, R1024, R1024, R1024, R1024, u512, u256, u128, u64, $c7s1$-4

Our fine generator G_f:
$c7s1$-32, d64, R64, R64, R64, u32, $c7s1$-4, $concat\{c7s1$-2, $c7s1$-4$\}$, R64, R64, $c7s1$-4

Discriminator Architectures. For discriminator networks, we use 4 Convolution-BatchNorm-LeakyReLU blocks for each discriminator. Let Ck denote 4×4 Convolution-BatchNorm-LeakyReLU blocks with k filters and stride

2. At the last layer, we add a sigmoid activation layer at the end to produce a 1-dimensional output. We use leaky ReLUs with default slope 0.2. All our four discriminators share the identical architecture as follows:
$C64, C128, C256, C512$

4.3 Network Configuration and Training

During the experiments, we employ two sets of convolution neural networks (CNNs). The first set of CNNs is the proposed CB-GANs as shown in Fig. 1, which is used for generic data augmentation. While the second set of CNNs is the U-Net [7], which is used for the segmentation task. We first trained the CB-GANs with back-propagation using the Adam optimizer with initial learning rate 0.0002 and momentum 0.5 for both generators and discriminators. We use CB-GANs to augment the training data during the training phase of U-Net. U-Net is trained with the same learning rate as CB-GANs. All the network are trained from scratch. The method is implemented using Pytorch.

In terms of computation time, it takes about 4 days to train the CB-GANs and 20 h to train the segmentation network for 100 epochs on a Nvidia GTX1080 Ti GPU. Moreover, we define the typical augmentation to be a combination of rotation (-10 to $10°$), zoom (0.98x to 1.02x) and random horizontal flip (50%) that apply to the training data.

4.4 Evaluation

We validate our approach by using it to augment the annotated training sets for the segmentation tasks and show that we have achieved strong gains, in terms of the Dice overlap metric between the automated segmentation and the radiologist annotation label map, over traditional augmentation baselines. We randomly split the training set in BRATS15 into two subsets, resulting in 234 training and 40 validation multimodal volumes. The full test set in BRATS15 is used as our test set, which includes 110 patients. First, we conduct the component testing on the test set to evaluate the impact of coarse-to-fine framework and proposed boundary loss function. Table 1 compares the segmentation performance between a baseline GANs, a coarse-to-fine GANs and the proposed CB-GANs. It shows that if the coarse-to-fine framework and boundary loss function were added, there is an improvement in Dice values for the tumor core task, giving an average 3.6% improvement in Dice. This is probably because the coarse-to-fine framework GANs and boundary loss function can correct defects and generate a clear boundary for small tumor regions in synthetic images.

We also compare the performance of the proposed method to the traditional augmentation method as listed in Table 2. Both data augmentation methods are able to improve the segmentation performance by a significant Dice value. Our proposed method further improves the performance over traditional data augmentation methods on average by 3.5% of Dice values and achieves a significant improvement in Dice for the complete tumor task.

Table 1. Segmentation performance on the BRATS15 testing set. GANs: proposed architecture without coarse-to-fine framework and boundary loss function. C-GANs: coarse-to-fine GANs. CB-GANs: our proposed method.

Method	Dice			Precision			Sensitivity		
	Complete	Core	Enh.	Complete	Core	Enh.	Complete	Core	Enh.
GANs	0.80	0.58	0.55	0.84	0.80	0.62	0.80	0.55	0.51
C-GANs	0.82	0.60	0.55	**0.87**	0.80	**0.66**	0.81	0.55	0.52
CB-GANs (ours)	**0.84**	**0.63**	**0.57**	**0.87**	**0.82**	0.65	**0.84**	**0.57**	**0.54**

Table 2. Performance on the BRATS15 testing set. w/o DA: without any data augmentation. w/ DA: with typical data augmentation. w/ Proposed: with proposed generic data augmentation method.

Method	Dice			Precision			Sensitivity		
	Complete	Core	Enh.	Complete	Core	Enh.	Complete	Core	Enh.
w/o DA	0.79	0.54	0.43	0.85	0.79	**0.66**	0.78	0.47	0.37
w/ DA	0.81	0.61	0.55	0.85	**0.82**	0.64	0.80	0.54	**0.54**
w/ Proposed	**0.84**	**0.63**	**0.57**	**0.87**	**0.82**	0.65	**0.84**	**0.57**	**0.54**

Table 3. Comparison to the state-of-the-art results on the BRATS15 testing set.

Method	Dice			Precision			Sensitivity		
	Complete	Core	Enh.	Complete	Core	Enh.	Complete	Core	Enh.
Kamnitsasa17 [3]	**0.85**	0.67	**0.63**	0.85	**0.84**	0.63	**0.87**	0.60	0.66
Zhao17 [12]	0.84	**0.73**	0.62	**0.89**	0.76	0.63	0.82	**0.76**	**0.67**
2D U-net w/ proposed	0.84	0.63	0.57	0.87	0.82	**0.65**	0.84	0.57	0.54

Finally, we compare our proposed method with two state-of-the-art methods as listed in Table 3. Kamnitsasa et al. [3] method, achieving a top ranking in both BRATS15 and ISLES15 Challenge, using a dual pathway deep 3D CNNs to segment the tumor region and 3D fully connected Conditional Random Field to reduce the false positive, while Zhao [12] joins three segmentation models which uses 2D image patches from different views. Our results are competitive with both methods and give better result for the enhancing tumor task in terms of Dice precision.

Also, one advantage of our proposed model is its relatively low computational cost in both the training and testing phases as we only leverage simple 2D CNNs with no post-processing method. Kamnitsasa et al. [3] reported a running time of 3 min using a 3 GB GPU to segment one case, while 6–12 min was reported by

Zhao [12]. With our proposed method, we achieve 2.1 s for one case in inference time as the architecture of U-Net has much fewer learning parameters.

5   Conclusion

In this paper, we propose a novel, automatic and network-based data augmentation method for brain tumor MR image segmentation. The main contribution is that we propose a generic way to augment training data that is able to teach network-based methods the desired invariance and robustness properties for segmentation tasks. We have shown that the proposed coarse-to-fine framework and boundary loss function in GANs lead to improved augmented data and segmentation quality. We have also shown that our method can boost a common segmentation network to reach the state-of-the-art multi-scale deep networks' performance with the relatively low computational cost at inference time and outperforms the traditional augmentation methods.

References

1. Arjovsky, M., Chintala, S., Bottou, L.: Wasserstein generative adversarial networks. In: International Conference on Machine Learning, pp. 214–223 (2017)
2. Goodfellow, I., Pouget-Abadie, J., Mirza, M., et al.: Generative adversarial nets. In: Advances in Neural Information Processing Systems, pp. 2672–2680 (2014)
3. Kamnitsas, K., Ledig, C., Newcombe, V.F.J., et al.: Efficient multi-scale 3D CNN with fully connected CRF for accurate brain lesion segmentation. MedIA **36**, 61–78 (2017)
4. Karras, T., Aila, T., Laine, S., Lehtinen, J.: Progressive growing of GANs for improved quality, stability, and variation. arXiv preprint arXiv:1710.10196 (2017)
5. Menze, B., Jakab, A., Bauer, S., et al.: The multimodal brain tumorimage segmentation benchmark (BRATS). IEEE Trans. Med. Imaging **34**(10), 1993–2024 (2015)
6. Milletari, F., Navab, N., Ahmadi, S.A.: V-net: fully convolutional neural networks for volumetric medical image segmentation. In: 2016 Fourth International Conference on 3D Vision (3DV), pp. 565–571. IEEE (2016)
7. Ronneberger, O., Fischer, P., Brox, T.: U-Net: convolutional networks for biomedical image segmentation. In: Navab, N., Hornegger, J., Wells, W.M., Frangi, A.F. (eds.) MICCAI 2015. LNCS, vol. 9351, pp. 234–241. Springer, Cham (2015). https://doi.org/10.1007/978-3-319-24574-4_28
8. Rosca, M., Lakshminarayanan, B., Warde-Farley, D., Mohamed, S.: Variational approaches for auto-encoding generative adversarial networks. arXiv preprint arXiv:1706.04987 (2017)
9. Shen, H., Wang, R., Zhang, J., McKenna, S.J.: Boundary-aware fully convolutional network for brain tumor segmentation. In: Descoteaux, M., Maier-Hein, L., Franz, A., Jannin, P., Collins, D.L., Duchesne, S. (eds.) MICCAI 2017. LNCS, vol. 10434, pp. 433–441. Springer, Cham (2017). https://doi.org/10.1007/978-3-319-66185-8_49
10. Wang, T.C., Liu, M.Y., et al.: High-resolution image synthesis and semantic manipulation with conditional GANs. arXiv preprint arXiv:1711.11585 (2017)

11. Yeh, R.A., Chen, C., Lim, T.Y., et al.: Semantic image inpainting with deep generative models. In: CVPR, pp. 5485–5493 (2017)
12. Zhao, X., Wu, Y., Song, G., et al.: A deep learning model integrating FCNNs and CRFs for brain tumor segmentation. MedIA **43**, 98–111 (2018). https://doi.org/10.1016/j.media.2017.10.002
13. Zhu, J.-Y., Krähenbühl, P., Shechtman, E., Efros, A.A.: Generative visual manipulation on the natural image manifold. In: Leibe, B., Matas, J., Sebe, N., Welling, M. (eds.) ECCV 2016. LNCS, vol. 9909, pp. 597–613. Springer, Cham (2016). https://doi.org/10.1007/978-3-319-46454-1_36

Multipath Densely Connected Convolutional Neural Network for Brain Tumor Segmentation

Cong Liu[1], Weixin Si[2,3(✉)], Yinling Qian[2], Xiangyun Liao[2], Qiong Wang[2], Yong Guo[1], and Pheng-Ann Heng[2,3]

[1] College of Mechanical and Electrical Engineering, Central South University, Changsha, China
[2] Guangdong Provincial Key Laboratory of Computer Vision and Virtual Reality Technology, Shenzhen Institutes of Advanced Technology, Chinese Academy of Sciences, Shenzhen, China
wxsics@gmail.com
[3] Department of Computer Science and Engineering, The Chinese University of Hong Kong, Hong Kong, China

Abstract. This paper presents a novel Multipath Densely Connected Convolutional Neural Network (MDCNN) for automatically segmenting glioma with unknown sizes, shapes and positions. Our network architecture is based on the Multipath Convolutional Neural Network [21], which considers both local and contextual patches of segmentation information, including original MRI images, symmetry information and spatial information. Motivated to reduce the feature loss induced by under-utility of feature maps, we propose to fuse feature maps from original local and contextual paths at three different units and introduce three more densely connected paths. Consequently, three auxiliary segmentation paths together with original local and contextural paths forms the complete segmentation network. The model's training and validation are performed on the BraTS2017 dataset. Experimental results demonstrate that the proposed network is capable to effectively extract more accurate tumor locations and contours with improved stability.

1 Introduction

Glioma is a type of tumor that occurs in the brain and spinal cord, and it accounts for almost 80% of primary malignant brain tumors [20]. Statistics show that brain and other central nervous system (CNS) tumors are the second most common cancers in children and adolescents [6]. What's worse is that brain tumor in children easily leads to related diseases, such as visual impairment and hemiplegia.

Last few decades have witnessed significant improvement in clinical neuro-oncology treatment, including neurosurgery surgery, chemotherapy, radiotherapy, physiotherapy and *etc.* [23]. Medical images, such as computed tomography (CT), ultrasound and magnetic resonance imaging (MRI), act as an irreplaceable role in these techniques, since they can provide indispensable information for diagnosis and treatment. Comparing with other medical imaging modalities, MRI can form clear and complete pictures of the lesions from various image sequences [3]. This advantage makes MRI being widely used in hospitals and clinics for medical diagnosis and treatment of many diseases [7].

Whereas MRI brings great conveniences for tumor diagnosis, it's not straightforward to accurately segment MRI images for extracting the patient-specific important clinical information, and their diagnostic features [15, 24]. Moreover, due to the large amount of data, it is time consuming and tedious for the radiologists to manually annotate and segment the data, resulting in limited usage for objective quantitative analysis [3]. To tackle this issue, great efforts are put for automatic brain tumor segmentation. Considerable research has been conducted for glioma segmentation in MRI images, it's still a challenge task to achieve satisfied results due to several inherent difficulties:

- It requires large dataset to learn stable characteristics feature map because of the uncertainty in shape and characteristics of glioma.
- It's difficult to balance between large context features and local detail features because of various region sizes of different tissue distributions.
- It easily causes class imbalance due to inhomogeneity of label classification.

Abundant of solutions are proposed aiming at solving this problem from different perspectives. To reduce continuous resolution reduction, Long *et al.* [14] propose to segment images using Fully Convolutional Networks (FCN) and achieve pixel level segmentation on images with various scales. However, the results has limited accuracy because max-pooling and sub-sampling would reduce the resolution of feature maps. To solve this problem, SegNet [1] and U-Net [19] add up-sampling layers (decoder network) after down-sampling layers (encoder network) to map low resolution feature maps to full input resolution feature maps. This architecture improves segmentation accuracy due to more features retained from training data. U-Net has been successfully adopted for real brain tumor segmentation [4].

With the depth of the network deepens, SegNet and U-Net would "get lazy" in feature extraction, *i.e.*, it would easily lose abstract features. ResNet [9] are proposed to improve the learning capability of deeper features with deeper models. It reformulates the layers as learning residual functions with reference to the layer inputs, rather than learning unreferenced functions.

By introducing direct connections between any two layers with the same feature-map size, DenseNet [11] can naturally scale to hundreds of layers without exhibiting optimization difficulties. Moreover, DenseNet requires substantially fewer parameters and less computation because of the adoption of hyperparameter settings optimized for residual networks.

Meanwhile, dual path cascade architecture also has achieved good results on brain tumor segmentation task [8]. [12] simultaneously exploits local features and more global contextual features than traditionally CNN architecture and obtains competitive performance with much higher efficiency. [21] further improves the network of [12] by considering spatial information and high asymmetry property caused by the contrast between healthy brain tissue and tumor tissue.

Some other researchers design segmentation networks by integrating advantages from multiple different networks. [18] proposes a natural image segmentation algorithm by introducing a special residual function with a path of multiple stream. [11] employed this method on brain tumor segmentation and achieved appealing results. [16] and [17] integrate several components into CNNs, including residual structure, up-sampling, down-sampling and symmetric/asymmetric convolution structure. Therein, E-Net [16], significantly improve efficiency with competitive results while [17] performs better in boundary segmentation by enlarging perception of different convolution sum. In addition to residual structure, [5] continuously fuses feature information from different convolutional layers to retain detail features and reduce in-balance problem.

In this paper, we propose a Multi-path Densely Connected Convolutional Neural Network (MDCNN) which fuses information in multiple stages to improve the stability and reduce feature loss, and define a loss function to support multi-path segmentation.

2   Database

Data from BarTS Challenge 2017 [2], which is composed of training, validation and test datasets, is employed for our experiments. Since the date of competition has expired and unable to get the test datasets, we only use the training and validation dataset. The training dataset contains MRI data of 285 subjects, 210 with high grade glioma (HGG) and 75 with low grade glioma (LGG). Each training subject provides four MRI modalities, namely T-weighted(T1), post-contract T1-weighted(T1ce), T2-weighted(T2) and T2 Fluid Attenuated Inversion Recovery(FLAIR). In addition, four ground truth classes are given by images with resolution $240 \times 240 \times 155$. The four classes represent necrotic and non-enhancing tumor (NCR and NET, label 1), peritumoral edema (ED, label 2), enhancing tumor (ET, label4) and everything else (label 0), respectively.

80% of the training dataset (168 HGG and 60 LGG subjects) is used for training and the remaining 20% (42 HGG and 15 LGG subjects) is used for validation. At last, the validation dataset without GT from 46 patients is used to test the proposed network.

3   Methods

Our method is motivated to make full usage of features for better segmentation performance by integrating advantages from multi-path Convolutional Neural Network [21] and densely connected multi-scale feature maps [11]. The work of [21] uses multi-path to encode both contextual and local information and proposes specific pre-processing and post-processing procedures for better performance. However, it hasn't taken full advantage from feature maps. Multi-scale feature maps

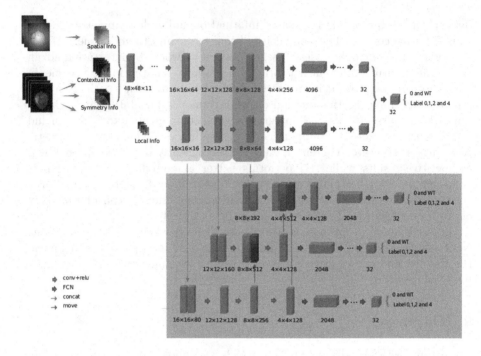

Fig. 1. Illustration of multipath densely connected CNN. (Color figure online)

and dense connectivity are introduced to further improve the segmentation accuracy. This shares similar idea as MSDNET [10], which is originally proposed to classify images with strong randomness and large variation range. So, the proposed method not only fuse the feature maps at the final stage of the two paths, but also introduce fusion in multiple stages and add dense connection between the fused feature maps. Even local and contextural paths may lose some features, the segmentation network of fusion feature can compensate with different features and thus can improve the performance. This architecture can continuously strengthen the transmission between feature maps, thus making more efficient usage of feature maps and reducing vanishing-gradient [10, 11]. Even with more parameters, the network is more stable and has stronger learning ability.

Figure 1 illustrates the network structure. The network extracts both large contextual patches and local patches as multi-path CNN does, so it can well retain contextual information. On the basis work of [21], symmetry information is used to differentiate healthy tissue and tumor tissue. To facilitate re-usage of feature maps for fusion in the same scale, we change the patch sizes on the contextual region path and on local region path in multi-path CNN to $48 \times 48 \times 11$ and $20 \times 20 \times 4$, respectively. In addition to convolution operation in each path, we fuse feature maps from large region path and local region path. The fusion operation can make better usage of feature maps. The first fusion of feature maps from two paths is conducted when the size of feature map is 16×16 (Illustrated in the red part in Fig. 1). The fusion generates the first auxiliary densely connected segmentation path. In the same manner, the second and third auxiliary densely connected segmentation path are generated by fusing the feature maps with size 12×12 and 8×8 (Illustrated in the blue and yellow regions respectively). Further, first auxiliary layer with size 16×16 is proceeded to layer with size 8×8 (See the yellow rectangle in Fig. 1) and fused to the second auxiliary densely connected segmentation path. Similarly, the second auxiliary layer is proceeded to layer with size 4×4 and fused to the third auxiliary layer. Each fused feature map will generate an auxiliary densely connected segmentation path with normal convolution and FCN. We also tried to introduce fusion on the layer with patch size 4×4 and found that it can't significantly improve the performance but introduces more parameters. So we finally chose to use three fusion segmentation paths for the best performance. Except for the parameters shown in Fig. 1, the layers between $48 \times 48 \times 11$ to $16 \times 16 \times 64$ and the FCN parameters on the main path and branches are given in Table 1.

After several convolution operation and dropout regularization [22], contextural and local feature vectors are concatenated and put forward for another FC layer. Finally, softmax function is adopted for classification. In the training process, we employ DisturbLabel regularization, which randomly set GT to 0 or 1, to improve the stability.

Table 1. Detail MDCNN parameters.

Region	Block	Output size
Large region	$\{5 \times 5 \text{ Conv+ReLU}\} \times 2$	$40 \times 40 \times 32$
	2×2 Max-pooling	$20 \times 20 \times 32$
	$\{5 \times 5 \text{ Conv+ReLU}\} \times 2$	$12 \times 12 \times 128$
Main path	FCN $\times 4$	256
		128
		32
		4/2 classes
Auxiliary path	FCN $\times 3$	128
		32
		4/2 classes

Besides, considering that brain tumor segmentation has relatively lower randomness, we introduce the idea of dense connection [10] to our net. In addition, the segmentation results from four paths, *i.e.*, main path and three densely connected paths, are weighted by 0.25 summed up for the final loss. The loss function is:

$$L(X; \omega) = \sum_{s \in S} \lambda_1 L_s(X; \omega_s) + \sum_{s \in S} \lambda_2 L_s(X; \omega_s), \tag{1}$$

in which, S is the number of branches, X is the input with size $[N, h * w * d]$, ω_s is the weight, λ_1 and λ_2 are the weights correlating to the ratio of the two cases. L_s is the loss function defined for each branch:

$$L_s = \frac{1}{N} \sum_{i=1}^{N} \sum_{c=1}^{C} Z_{i,c}^{gt} \cdot \log q_{i,c} \tag{2}$$

in which, N is the number of samples in the batches, $q_{i,c}$ means the segmentation results with respect to label c, and $Z_{i,c}^{gt}$ is the corresponding label. In this paper, we simultaneously consider two classification cases, namely two classes (with label 0 or 1) and four classes (with label 0,1,2,4), to further enhances the learning capability.

4   Results

The method is implemented with Python and trained with TensorFlow. The hardware platform is PC with NVIDIA Titan X (12 GB) GPU and Intel® Xeon(R) CPU E3-1231 v3 @ 3.40 GHz x8 32 GB). The detailed parameters in network are given in Fig. 1 and Table 1.

In pre-processing, all data are clipped into range $[-2.0, 2.0]$. The sizes of contextual patches and local patches are 48×48 and 20×20, respectively. Dropout probability used during the entire training process is 0.5. Weights for the two classification cases (λ_1 and λ_2) are 0.2 and 0.4. These parameters are empirically chosen based on [21]. In the training process, Adama optimization algorithm [13] is used to update weights. Different parameters are used in Adama optimizers in different phases.

Two phases are defined in training, rather than 3 phases use by [21]. The two phases correlate to pre-training and fine training. During the first phase (iteration ≤ 9000), we randomly chose data of 5 subjects and extract 50 samples for each class in each iteration. During the second phase ($9000 < $ iteration ≤ 13000), 10 subjects and 25 samples for each class are randomly chosen in each iteration. 10^{-4} and 10^{-5} are chosen as the learning rate for these two phases.

In post-processing, thresholding is first applied to convert the output, which represents the probability of being different regions, to integer values matching with the GT data type. The thresholding values for being tumorous or being enhancing tumor are 0.905 and 0.45, respectively. In addition, after thresholding, connected components with less than 3000 voxels will be re-classified as healthy.

Our experiment is conducted on various label combination, including whole tumor region with labels 1, 2, 4 (WT), core tumor region with labels 1 and 4 (TC), and enhancing tumor region with label 4 (ET). We adopt 20% of the training dataset (Data of 57 subjects) and validation dataset (Data of 46 subjects) in validation and testing, respectively. Besides, Dice score and 95% Hausdorff are used for measurement. The results all we show in the tables are received after uploading our segmentation results to the official website. Since BraTS2017 testing dataset is not available, we use the validation dataset for testing. The

Table 2. Comparison with [21] after post-processing in terms of average Dice score and 95% Hausdorff distance for the BraTS 2017 training subset and validation datasets.

	Dataset	Dice			Hausdorff		
		ET	WT	TC	ET	WT	TC
Results of [21]	Training (57 cases)	0.671	0.859	0.725	8.882	11.279	10.989
	Validation	0.678	0.877	0.713	12.749	11.013	14.000
Our results	Training (57 cases)	0.685	0.896	0.822	6.515	6.614	7.816
	Validation	0.705	0.889	0.756	6.016	6.161	10.355

Table 3. Comparison on results without post-processing in terms of average Dice score and 95% Hausdorff distance for the BraTS 2017 training subset.

	Dataset	Dice			Hausdorff		
		ET	WT	TC	ET	WT	TC
Results of [21]	Training (57 cases)	0.626	0.751	0.687	27.336	36.910	30.001
Our results	Training (57 cases)	0.652	0.820	0.793	14.505	26.186	17.170

Table 4. Segmentation results on LGG training subjects from BraTS 2017 training datasets.

	Dataset	Dice			Hausdorff		
	Training (LGG, 15cases)	ET	WT	TC	ET	WT	TC
Results of [21]	Without post-processing	0.334	0.800	0.532	39.820	38.987	23.124
	With post-processing	0.358	0.890	0.544	19.962	11.841	12.352
Our results	Without post-processing	0.334	0.873	0.670	30.291	28.567	16.009
	With post-processing	0.372	0.895	0.676	18.516	5.530	9.367

evaluation results of validation dataset are also received after uploading our segmentation results to the official website, because we have no the Ground Truth.

Table 2 shows the results of our method and that of [21] both with post-processing. The thresholding of being tumorous or being enhancing tumor used by [21] are 0.916 and 0.4. We can find that our method achieve better performance for all segmentation tasks (WT, TC and ET) than [21]. On the training dataset, the presented method gains nearly 10% improvement, and on the validation dataset, we achieve 3% improvement in average.

Table 3 compares the results without post-processing from our method and that from [21]. The comparison is conducted on the testing dataset from training data. It shows our method achieves 6.7% improvements in average.

Tables 4 and 5 are specified segmentation results on LGG and HGG training dataset with or without post-processing. It can be observed that our results with or without post-processing are prior to the result from [21]. Besides, ET segmentation results on LGG dataset are significantly poorer than that on HGG

Table 5. Segmentation results on HGG training subjects from BraTS 2017 training datasets.

	Dataset	Dice			Hausdorff		
	Training (HGG, 42cases)	ET	WT	TC	ET	WT	TC
Results of [21]	Without post-processing	0.730	0.734	0.742	24.364	36.169	36.471
	With post-processing	0.783	0.847	0.780	6.244	11.078	10.502
Our results	Without post-processing	0.765	0.814	0.837	10.371	28.567	17.585
	With post-processing	0.797	0.896	0.874	3.372	7.001	7.262

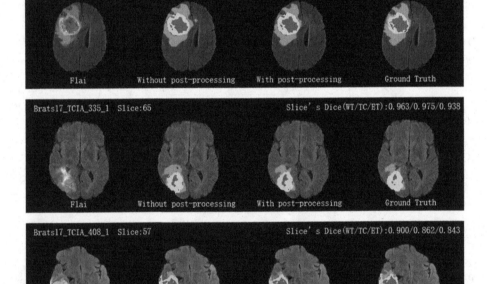

Fig. 2. Segmentation results with MDCNN. The first two row are segmentation results on two HGG subjects, and the last row shows results of one LGG subject. Regions with label 1, 2 and 4 are colored in red, green and yellow respectively. (Color figure online)

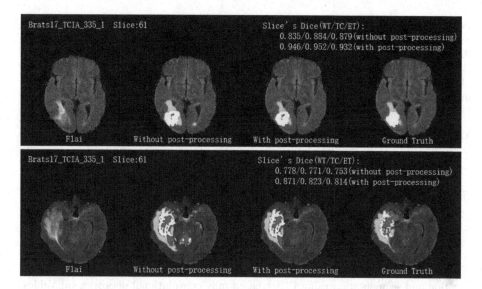

Fig. 3. Comparison on results with and without post-processing. Regions with labels 1, 2 and 4 are colored in red, green and yellow respectively. (Color figure online)

dataset. The main reason might be that LGG dataset is much smaller than HGG dataset. Thus, collecting more LGG training subjects would improve the performance.

Several segmentation results are illustrated in Figs. 2 and 3. It can be noticed that, post-processing can clearly improve the segmentation results on those subjects with more smooth boundaries. In this case, post-processing can effectively filter out redundant false details, thus improving the performance. Meanwhile, segmentation for those subjects with bumping region boundaries rely more on the segmentation capability of the network, and the promotion from post-processing would be small.

5   Conclusions and Future Work

This paper presents a novel Multi-path Densely Connected Convolutional Neural Network for glioma segmentation. By introducing densely connections on fused feature maps, the proposed model can well retain both abstract features and specific features for deep network. Thus, the network becomes stable and with higher learning capability. Experiments indicate that our method achieves the highest performance for various segmentation tasks. In future, we intend to reduce the memory cost of the network and extend the framework to more clinical applications.

Acknowledgment. This work was supported in part by the Shenzhen Science and Technology Program (No. JCYJ20160429190300857 and No. JCYJ2017041 3162617606), Research Grants Council of the Hong Kong Special Administrative Region (Project No. GRF 14203115) and the National Natural Science Foundation of China (Grant No. 61802385). Yinling Qian and Weixin Si are the corresponding authors.

References

1. Badrinarayanan, V., Kendall, A., Cipolla, R.: SegNet: a deep convolutional encoder-decoder architecture for image segmentation. IEEE Trans. Pattern Anal. Mach. Intell. **39**(12), 2481–2495 (2017)
2. Bakas, S., et al.: Advancing The Cancer Genome atlas glioma MRI collections with expert segmentation labels and radiomic features. Sci. Data **4**, 170117 (2017)
3. Bauer, S., Wiest, R., Nolte, L.-P., Reyes, M.: A survey of MRI-based medical image analysis for brain tumor studies. Phys. Med. Biol. **58**(13), R97 (2013)
4. Dong, H., Yang, G., Liu, F., Mo, Y., Guo, Y.: Automatic brain tumor detection and segmentation using U-Net based fully convolutional networks. In: Valdés Hernández, M., González-Castro, V. (eds.) MIUA 2017. CCIS, vol. 723, pp. 506–517. Springer, Cham (2017). https://doi.org/10.1007/978-3-319-60964-5_44
5. Fidon, L., et al.: Generalised Wasserstein dice score for imbalanced multi-class segmentation using holistic convolutional networks. arXiv preprint arXiv:1707.00478 (2017)
6. Gittleman, H.R., et al.: Trends in central nervous system tumor incidence relative to other common cancers in adults, adolescents, and children in the United States, 2000 to 2010. Cancer **121**(1), 102–112 (2015)
7. Gordillo, N., Montseny, E., Sobrevilla, P.: State of the art survey on MRI brain tumor segmentation. Magn. Reson. Imaging **31**(8), 1426–1438 (2013)
8. Havaei, M., et al.: Brain tumor segmentation with deep neural networks. Med. Image Anal. **35**, 18–31 (2017)
9. He, K., Zhang, X., Ren, S., Sun, J.: Deep residual learning for image recognition. In: Proceedings of the IEEE Conference on Computer Vision and Pattern Recognition, pp. 770–778 (2016)
10. Huang, G., Chen, D., Li, T., Wu, F., van der Maaten, L., Weinberger, K.Q.: Multi-scale dense networks for resource efficient image classification. arXiv preprint arXiv:1703.09844 (2017)
11. Huang, G., Liu, Z., Van Der Maaten, L., Weinberger, K.Q.: Densely connected convolutional networks. In: Proceedings of the IEEE Conference on Computer Vision and Pattern Recognition (2017)
12. Kamnitsas, K., et al.: Efficient multi-scale 3D CNN with fully connected CRF for accurate brain lesion segmentation. Med. Image Anal. **36**, 61–78 (2017)
13. Kingma, D.P., Ba, J.: Adam: a method for stochastic optimization. arXiv preprint arXiv:1412.6980 (2014)
14. Long, J., Shelhamer, E., Darrell, T.: Fully convolutional networks for semantic segmentation. In: Proceedings of the IEEE Conference on Computer Vision and Pattern Recognition, pp. 3431–3440 (2015)
15. Menze, B.H., et al.: The multimodal brain tumor image segmentation benchmark (BRATS). IEEE Trans. Med. Imaging **34**(10), 1993–2024 (2015)

16. Paszke, A., Chaurasia, A., Kim, S., Culurciello, E.: ENet: a deep neural network architecture for real-time semantic segmentation. arXiv preprint arXiv:1606.02147 (2016)

17. Peng, C., Zhang, X., Yu, G., Luo, G., Sun, J.: Large kernel matters-improve semantic segmentation by global convolutional network. arXiv preprint arXiv:1703.02719 (2017)

18. Pohlen, T., Hermans, A., Mathias, M., Leibe, B.: Full-resolution residual networks for semantic segmentation in street scenes. arXiv preprint (2017)

19. Ronneberger, O., Fischer, P., Brox, T.: U-Net: convolutional networks for biomedical image segmentation. In: Navab, N., Hornegger, J., Wells, W.M., Frangi, A.F. (eds.) MICCAI 2015. LNCS, vol. 9351, pp. 234–241. Springer, Cham (2015). https://doi.org/10.1007/978-3-319-24574-4_28

20. Schwartzbaum, J.A., Fisher, J.L., Aldape, K.D., Wrensch, M.: Epidemiology and molecular pathology of glioma. Nat. Rev. Neurol. **2**(9), 494 (2006)

21. Sedlar, S.: Brain tumor segmentation using a multi-path CNN based method. In: Crimi, A., Bakas, S., Kuijf, H., Menze, B., Reyes, M. (eds.) BrainLes 2017. LNCS, vol. 10670, pp. 403–422. Springer, Cham (2018). https://doi.org/10.1007/978-3-319-75238-9_35

22. Srivastava, N., Hinton, G., Krizhevsky, A., Sutskever, I., Salakhutdinov, R.: Dropout: a simple way to prevent neural networks from overfitting. J. Mach. Learn. Res. **15**(1), 1929–1958 (2014)

23. Tabatabai, G., et al.: Molecular diagnostics of gliomas: the clinical perspective. Acta Neuropathol. **120**(5), 585–592 (2010)

24. Yamahara, T., et al.: Morphological and flow cytometric analysis of cell infiltration in glioblastoma: a comparison of autopsy brain and neuroimaging. Brain Tumor Pathol. **27**(2), 81–87 (2010)

Multi-institutional Deep Learning Modeling Without Sharing Patient Data: A Feasibility Study on Brain Tumor Segmentation

Micah J. Sheller[1(✉)], G. Anthony Reina[1], Brandon Edwards[1], Jason Martin[1], and Spyridon Bakas[2(✉)] (iD)

[1] Intel Corporation, Santa Clara, CA 95052, USA
micah.j.sheller@intel.com
[2] Center for Biomedical Image Computing and Analytics (CBICA),
University of Pennsylvania, Philadelphia, PA 19104, USA
sbakas@upenn.edu

Abstract. Deep learning models for semantic segmentation of images require large amounts of data. In the medical imaging domain, acquiring sufficient data is a significant challenge. Labeling medical image data requires expert knowledge. Collaboration between institutions could address this challenge, but sharing medical data to a centralized location faces various legal, privacy, technical, and data-ownership challenges, especially among international institutions. In this study, we introduce the first use of federated learning for multi-institutional collaboration, enabling deep learning modeling without sharing patient data. Our quantitative results demonstrate that the performance of federated semantic segmentation models (Dice = 0.852) on multimodal brain scans is similar to that of models trained by sharing data (Dice = 0.862). We compare federated learning with two alternative collaborative learning methods and find that they fail to match the performance of federated learning.

Keywords: Machine learning · Deep learning ·
Glioma · Segmentation · Federated · Incremental · BraTS

1 Introduction

Gliomas describe tumors of the central nervous system with vastly heterogeneous radiographic, histologic, and molecular landscape. There is mounting evidence that tumor subregions apparent at the radiographic level reflect various histologically distinct tumor subregions with different biological properties. Accurate segmentation of these subregions is considered the basis for extracting quantitative imaging features corresponding to specific anatomical regions, which when integrated using advanced computational approaches, can enable assessment of this radiographic heterogeneity, evaluation of disease properties, and correlation with treatment response, patient prognosis, and molecular characteristics [1–6].

© Springer Nature Switzerland AG 2019
A. Crimi et al. (Eds.): BrainLes 2018, LNCS 11383, pp. 92–104, 2019.
https://doi.org/10.1007/978-3-030-11723-8_9

The brain tumor segmentation (BraTS) challenge describes a successful effort to create a publicly available multi-institutional dataset for benchmarking and quantitatively evaluating the performance of computer-aided segmentation algorithms [7–10]. However, such centralization of data, notwithstanding multi-institutional collaborations, is challenging because (a) data availability is more limited when compared with real-world/photographic imagery, and (b) sharing data to a centralized location may be cumbersome, especially in international configurations, due to various legal, privacy, technical, and data ownership challenges [11,12].

Considering the difficulty of creating public centralized medical imaging datasets, this paper introduces the first use of *federated learning* (FL) [13] for medical imaging. Specifically, we apply FL on the BraTS data to build an effective segmentation model that learns the variation across multiple institutions, without sharing any patient data, by iteratively aggregating locally-trained models at a centralized server. Although we applied FL to supervised semantic segmentation using a deep convolutional neural network (CNN) architecture, namely U-Net, FL works with any supervised machine learning (ML) architecture. Furthermore, we compare FL against two alternative collaborating learning techniques: *institutional incremental learning* (IIL), where each institution trains the shared model in turn, and *cyclic institutional incremental learning* (CIIL), which is IIL done in rounds with prescribed numbers of epochs [14]. We find that IIL performs poorly compared to FL and CIIL, while CIIL is less stable and harder to validate than FL, resulting in an inferior model.

Prior demonstrations of FL have focused on either toy problems or end-user tasks [13,15–17]. Our work is the first demonstration of FL in the medical domain, for institution-level tasks, specifically applied on clinically-acquired data.

2   Materials and Methods

2.1   Federated Learning (FL) Overview

In traditional ML solutions, all collaborating data owners (i.e., institutions) upload their data to a central server for training. Contrarily, in FL, the owners do not share their data, but they train the shared model locally instead, and only send model updates to the central server. The server then aggregates the updates and sends the new shared parameters to the data owners for further training (as often as desired), or application. Specifically, the aggregation is performed as a weighted average of institutional updates, with the weighting at a particular institution given as the fraction of total data instances that reside at that institution. Each iteration of this process: local training, update aggregation, and distribution of new parameters, is called a *federated round* (Fig. 1).

Hyper-parameters and Convergence. Along with the typical hyper-parameters of deep learning architectures (e.g., batch size, optimizer, learning

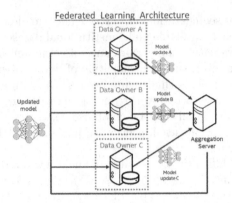

Fig. 1. System architecture of federated learning.

rate), FL also includes: (a) epochs per round (EpR), (b) number of participants in each round, and (c) model update compression/pruning methods [13]. EpR influences convergence, as learning rate and batch size do in traditional training: i.e., more EpR can speed up convergence, but there are diminishing returns, especially when the institutions' datasets are not independent and identically distributed (IID) [18], and more EpR requires more compute per institution. We study this issue by experimenting with different sizes of simulated federations, from 4 up to 32 institutions, and training for various numbers of EpR.

Hyper-parameters of number of participants per round and update compression/pruning methods are not assessed in this study, as they are generally used to mitigate participant limitations that are less relevant in the medical domain (e.g., limited networks).

Differential Privacy. Though FL ensures raw data is never shared between collaborators, additional measures may be desired to prevent certain information from being obtained through model updates. For example, noise can be added before sending an update, to obscure the presence of any collection of samples in the institution's dataset, and an accounting can be made as to the likelihood that such a determination can be made from the resulting model. The model is then said to have a degree of 'differential privacy'.

Differentially private training has been studied for ML use cases other than semantic segmentation [23,24]. However, applying noise to updates generally slows training, and there is a point at which training must stop to prevent increasing the likelihood of information leakage beyond an acceptable level. It is possible that the model has not reached the desired utility at that point. Differentially private ML models in the medical domain could be very desirable given the privacy issues surrounding medical data. However, we leave the study of differentially private training for segmentation models to future work.

2.2   Institutional Incremental Learning (IIL)

IIL is a simple collaborative learning approach, where institutions train a shared model in succession. Each institution trains the model only once, and may train the model however it chooses. Compared to FL, IIL requires less bandwidth as each institution needs to transmit the model once and receive it twice (once to train and once to receive the final version). The major disadvantages of IIL are (a) the drop in performance as the number of institutions increases, and (b) the problem of catastrophic forgetting [19, 20], where previously learned patterns are forgotten when new training data replace the previous data.

2.3   Cyclic Institutional Incremental Learning (CIIL)

CIIL changes IIL by repeating the IIL process, i.e., cycling through the institutions, and by fixing the number of epochs at each institution to reduce forgetting. In a CIIL cycle, each institution trains the model in series for a specific number of epochs before passing the updated model to the next participant. In contrast, during a federated round, each institution trains the model in parallel for a specific number of epochs, after which the institutions' model updates are aggregated to form the updated model. CIIL and FL share most of their software and infrastructure requirements, except for the FL aggregator. Although existing literature [14] reports parallel collaborative training (e.g., FL) as more logistically complex, the results in the present study show that for CIIL to achieve comparable results to FL, CIIL requires additional validation overhead that makes it more complex and less efficient than FL.

2.4   U-Net Topology

For our analysis, we implemented[1] a U-Net topology of a deep CNN [21] (Fig. 2). The model takes as input a single channel image and outputs an equivalently-sized, binary mask in which each pixel is assigned a class label. The network mimics the architecture of an autoencoder, with a contracting path that captures context (via max pooling) and an expanding path that enables localization (via upsampling). Unlike the standard autoencoder, each feature map in the expanding path is concatenated with a corresponding feature map from the contracting path, augmenting downstream feature maps with spatial information acquired using smaller receptive fields. Intuitively, this allows the network to consider features at various spatial scales. Since its introduction in 2015, U-Net has quickly become one of the standard deep learning topologies for image segmentation and has been instrumental in creating prediction models for segmenting nerves in ultrasound images, lungs in CT scans, and even interference in radio telescopes. All of our collaborative learning experiments in this study used this model with a dropout parameter of 0.2 and upsampling set to true.

---

[1] https://github.com/NervanaSystems/topologies/tree/master/distributed_unet.

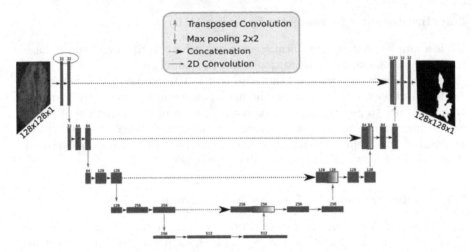

Fig. 2. U-Net network diagram. Numbers above each layer indicate the number of channels in that layer. Note that the channel count (purple circle) differs from the original design [21] by a factor of 2. (Color figure online)

2.5 BraTS Dataset

To quantitatively evaluate FL in a medical imaging context, we used the BraTS 2018 training dataset [7–10], which contains multi-institutional multi-modal magnetic resonance imaging (MRI) brain scans from patients diagnosed with gliomas. The radiographically abnormal regions of each brain scan have been manually annotated using 3 distinct labels corresponding to (i) peritumoral edematous/invaded tissue, (ii) non-enhancing/solid and necrotic/cystic tumor core, and (iii) enhancing tumor.

Since the objective of this study was to assess the performance of FL on a clinically-relevant task and not to develop a new segmentation method, we have only focused on the whole tumor volume, defined as the union of all three labels, only for the patients diagnosed with a high-grade glioma and we only considered the FLAIR modality for the input channel to the model.

3   Experimental Results

Our experiments compare traditional ML using data-sharing with collaborative configurations of FL, IIL and CIIL. For our collaboration experiments, we distribute the data among institutions in two different ways: (1) the actual BraTS distribution, i.e. the real-world data distribution, and (2) simulated distributions of 4 to 32 institutions, in steps of powers of two. The simulated distributions were created by randomly and proportionally splitting the subjects among the collaborating institutions, ensuring that each patient's data is assigned to only one institution. Table 1 shows the average number of subjects per distribution for the simulated distributions, as well as the actual distribution of subjects across institutions. Note that the actual distribution is quite imbalanced, with a single institution contributing nearly half the data.

Table 1. Distribution of data for all experimental configurations, simulated and real.

Type of distribution	Institutions	Average subjects per institution
Data-sharing	1	178
Simulated	4	44.5
Simulated	8	22.2
Simulated	16	11.1
Simulated	32	5.6
Real (BraTS distribution)	10	70, 27, 17, 12, 11, 9, 6, 6, 4, 3

For the 'data-sharing' and 'simulated' distributions, we randomly chose 32 subjects to hold out for validation on unseen data, prior to distributing the data among institutions. For the real BraTS distribution, we increased the unseen set to 45 subjects. This is because for institutions with only 4 and 5 subjects, contributing just 1 patient represents 20–25% of their data, so we increased the unseen set proportion per institution to better balance their contributions. This does slightly penalize the results of the real distribution experiments compared to the data-sharing and simulated experiments.

Tables 2 and 3 compare the data-sharing and three collaborative methods for the real and simulated distributions. For the data-sharing experiments, we show the best result from multiple model initializations. This matches normal practice for centralized training. Testing multiple model initializations may not be considered reasonable for collaborative methods, so we show the mean and standard deviation across multiple runs. For CIIL, we show results for all cycles over multiple runs, since CIIL does not provide a validation method for choosing the best cycle from a series of cycles. We discuss this further in Sect. 3.3.

Table 2. Data-sharing, FL, IIL and CIIL experiment results for the real BraTS data distribution.

Method	Validation DC	Percent of data-sharing DC
Data-sharing	0.862	100%
FL	0.852 ± 0.002	98.7%
CIIL	0.82 ± 0.04	95%
IIL	0.803 ± 0.042	93%

Table 3. Comparing FL, CIIL and IIL for collaborations of 4–32 institutions. Val.DC : Validation DC. D.S.DC : Percent of Data-Sharing DC.

Institutions	FL		CIIL		IIL	
	Val.DC	D.S.DC	Val.DC	D.S.DC	Val.DC	D.S.DC
4	0.862 ± 0.003	99.9%	0.843 ± 0.011	97.7%	0.841 ± 0.004	97.4%
8	0.865 ± 0.002	100.2%	0.839 ± 0.016	97.3%	0.823 ± 0.014	95.4%
16	0.863 ± 0.002	99.9%	0.82 ± 0.032	95%	0.82 ± 0.018	95%
32	0.857 ± 0.001	99.3%	0.809 ± 0.023	93.7%	0.701 ± 0.058	81.20%

3.1 Benchmarking Metric

The quantitative performance evaluation metric for the BraTS challenge has always been the Dice Coefficient (DC), a similarity measure in the range $[0, 1]$ that reflects a ratio of the intersection over the union of the predictions and ground truth, defined as:

$$DC = \frac{2|P \cap T|}{|P| + |T|} \quad (1)$$

where P and T are the prediction and ground truth masks, respectively.

The inter-rater agreement for expert neuro-radiologists measured by the DC was reported in the original BraTS benchmark paper [7] equal to 0.85±0.08 (mean±std for the whole tumor segmentation). Furthermore, state-of-the-art models for this dataset have DC of greater than or equal to 0.85 [22].

We used the Adam optimizer (learning rate 0.0005) to minimize the negative log of DC. To further increase numerical stability, we added a Laplace smoothing of 1 and we algebraically rearranged the final loss function to replace division with log subtraction:

$$loss = \log(|P| + |T| + 1) - \log(2|P \cap T| + 1) \quad (2)$$

3.2 Baseline U-Net Results

The model trained to state-of-the-art accuracy within 3 epochs and reached a peak validation DC of 0.862 (Fig. 3A) (15% holdout data). A qualitative assessment of two MRI slices from the validation dataset shows a good agreement between the model predictions and the manually annotated boundaries (Fig. 3B).

3.3 BraTS Distribution Results

Figure 4 shows comparative results across data-sharing, FL, IIL and CIIL for our U-Net implementation over the BraTS 2018 training dataset. In the collaborative experiments, the patient data was divided among the institutions exactly as it was collected by the BraTS data contributors. Figure 4A shows how the scores

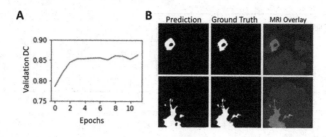

Fig. 3. (A) Validation DC scores over training epochs. The model peaked at 12 epochs and achieves a validation DC score of greater than 0.86. (B) Model performance on two images from the test set MRI. The model predicted mask closely matches the ground truth labels. An overlay of the ground truth with the original MRI slice.

vary across multiple runs, while Fig. 4B shows the validation score after each pass over the full training data. For the FL and CIIL experiments, the number of EpR was one. Note that in FL, each institution trains in parallel and the updates are averaged, such that the effective learning rate is less than that of the other methods. Furthermore, because FL trains in parallel, FL rounds and CIIL cycles are not wall-clock-equivalent.

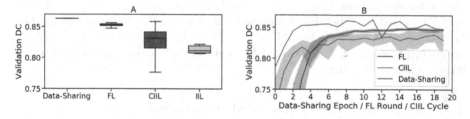

Fig. 4. Comparing centralized learning, FL, IIL and CIIL for the actual BraTS data distribution. The x-axis in (B) shows passes over the full dataset (epochs). Epochs are not equivalent in wall-clock time. The shading in (B) is min/max.

For our CIIL results in Fig. 4A, we show the distribution of scores after every cycle, rather than the best scores of a given run, to emphasize the importance of validation during training: CIIL does not support proper validation during training, as the model is not synchronized across the institutions until training is completed.

Despite the significant imbalance in numbers of subjects per institution (Table 1), FL achieves 98.7% of the centralized validation DC, as do the best CIIL results. However, CIIL is less stable (Fig. 4), with a wide range of scores after each cycle. The instability of CIIL means that to learn a good model with CIIL, the model must be evaluated after each cycle. However, to evaluate the model, each institution must receive a copy of the model to test against its validation data. This adds additional communication overhead (i.e., an institution

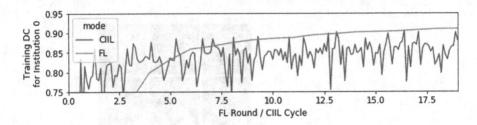

Fig. 5. CIIL catastrophic forgetting: first institution's training DC.

must receive the model 1 extra time per cycle) and requires a method to aggregate the results, at which point CIIL becomes arguably more complex than FL and with greater communication cost.

IIL learns a relatively poor model, averaging only 93% of the validation DC, and suffers similar instability as CIIL. For our IIL experiments, each institution trained until there was no improvement in validation DC for eight epochs (as measured by its own validation data), passing the best-performing model to the next institution.

Evaluation of the training data DC scores for institution 0 during CIIL and FL training reveals that the CIIL models suffer from some amount of catastrophic forgetting [19,20], i.e., the model "forgets" some of what it learned from the earlier institutions (Fig. 5). We verified that the peaks in the training DC for CIIL indeed correspond to immediately after institution 0 trained the model. Forgetting could cause the instability we see in CIIL. We leave further investigation to future work. By comparison, FL maintains the training DC for institution 0 throughout its training.

3.4   Results for Random Simulated Distributions

Figure 6 shows comparative results across FL, IIL and CIIL for the simulated data distributions, where, each institution was assigned roughly the same number of subjects (no subject's data was split across institutions), so they were far more balanced than the actual distribution. Note that in Fig. 6, the y-axis scale is different for 32 institutions, as the CIIL and IIL results were quite poor.

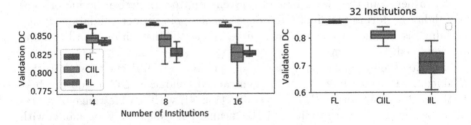

Fig. 6. Comparing FL, CIIL and IIL for collaborations of 4–32 institutions. Note that 32 simulations have a different y-axis range.

Our FL results show remarkable consistency on the simulated distributions, achieving 99+% of the data-sharing results in all simulations, and FL also achieves superior results when compared to the best CIIL models. Even with 32 institutions, where each institution averaged fewer than 6 subjects, FL trains efficiently. In contrast, CIIL and IIL show considerable instability, with standard deviations 10x that of FL for 16 and 32 institutions.

Figure 7 shows that for FL, while different numbers of institutions converge to similar model quality, they do not converge at the same rate. Note the different x-axis scales for 16 and 32 institutions. The causes for this are two-fold. First, with less data per institution, the model deltas are smaller at each round. Second, though the data is randomly distributed, the per-institution datasets become small enough that the individual institutions' datasets are less similar.

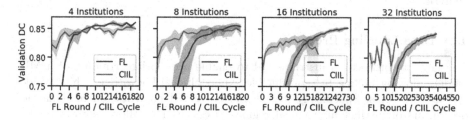

Fig. 7. FL and CIIL over round/cycle (1 epoch per). Confidence intervals are 0–100%. Note that 16 and 32 institutions are shown to 30 and 50 rounds, respectively.

By comparing FL experiments for 16 and 32 institutions at various EpR (Fig. 8), we note a convergence slowdown caused by smaller model deltas. The convergence slowdown is not quite proportional to the decrease in epochs, especially for 16 institutions.

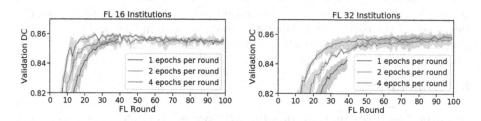

Fig. 8. FL over rounds for various EpR (16 and 32 institutions). Min/max shading.

3.5   Hyper-parameters

All our experiments used a batch size of 64 and learning rate 5e-4 (Adam optimizer). We developed several solutions for adapting Adam for FL, all of which worked equivalently well in this domain. We leave investigating these options in a broader set of domains for future work.

4   Practical Considerations

4.1   Data Pre-processing

Although the data is not centrally shared in FL, sources of variation across equipment configurations and acquisition protocols need to be considered. The uncontrolled varying acquisition environment of standard clinical practice, where the highest throughput of medical images is produced, make such data of limited use and significance in large-scale analytical studies, whereas data from more controlled environments (such as clinical trials) are more suitable. Since standardization of the acquisition protocols cannot be controlled, the pre-processing approaches should account for harmonization of heterogeneous data, allowing for integration and facilitating easier multi-institutional collaboration for large-scale analytics.

4.2   Data Labeling Protocol

The labeling protocol is instrumental to enable appropriate training of a ML model, allowing linking to reproducible expert clinical knowledge, while avoiding operator bias. Specifically, the definition and documentation of semantic descriptors of distinct anatomical regions is essential to allow reproducibility across institutions.

4.3   Addition/Removal of Collaborators

Institutions could be added, or removed, after some time of training, in any of the above collaborative learning configurations (FL, IIL, CIIL). In such cases, the model resulting from further collaborative training is expected to be qualitatively similar (after a transition period) to one obtained by training from scratch with the new set of collaborators. Eventually, any missing data would be forgotten. New data patterns will be learned subject to the limitations observed in this paper of the particular collaborative configuration, in face of the new data distribution. We leave such studies, however, to future work.

5   Conclusions

Our experiments demonstrate that the collaborating clinical institutions could train a model without sharing their data, using federated learning (FL). Our FL experiments achieve 99% of the model performance of a data-sharing model even with imbalanced datasets, such as the actual BraTS institutional distribution, or relatively few samples per participant, such as our simulation of 32 institutions with 6 subjects per institution. While CIIL may seem a simpler alternative, in order to select a good model, full validation must be run often, such as at the end of each cycle. These validations would require the same synchronization and aggregation steps as FL, and would even add communication costs above FL.

Finally, IIL and CIIL do not scale well to large number of institutions with small amounts of data.

Translation and adoption of such a FL system in a clinical configuration for multi-institutional collaboration, towards producing computer-aided analytics and assistive diagnostics, is expected to have a catalytic impact towards precision medicine, especially since introducing knowledge from another institution would improve the performance of the trained models without the need to share patient data, thereby overcoming potential privacy or data ownership concerns.

Acknowledgements. Research reported in this publication was partly supported by the National Institutes of Health (NIH) under award numbers NIH/NINDS:R01NS042645 and NIH/NCI:U24CA189523. The content of this publication is solely the responsibility of the authors and does not necessarily represent the official views of the NIH.

References

1. Bakas, S., et al.: In vivo detection of EGFRvIII in glioblastoma via perfusion magnetic resonance imaging signature consistent with deep peritumoral infiltration: the ϕ-index. Clin. Cancer Res. **23**(16), 4724–4734 (2017). https://doi.org/10.1158/1078-0432.CCR-16-1871

2. Chang, K., et al.: Residual convolutional neural network for the determination of IDH status in low- and high-grade gliomas from MR imaging. Clin. Cancer Res. **24**(5), 1073–1081 (2018). https://doi.org/10.1158/1078-0432.CCR-17-2236

3. Korfiatis, P., Kline, T.L., Lachance, D.H., Parney, I.F., Buckner, J.C., Erickson, B.J.: Residual deep convolutional neural network predicts MGMT methylation status. J. Digit. Imaging **30**(5), 622–628 (2017). https://doi.org/10.1007/s10278-017-0009-z

4. Binder, Z.A., et al.: Epidermal growth factor receptor extracellular domain mutations in glioblastoma present opportunities for clinical imaging and therapeutic development. Cancer Cell **34**(1), 163–177 (2018). https://doi.org/10.1016/j.ccell.2018.06.006

5. Akbari, H., et al.: Imaging Surrogates of infiltration obtained via multiparametric imaging pattern analysis predict subsequent location of recurrence of glioblastoma. Neurosurgery **78**(4), 572–580 (2016). https://doi.org/10.1227/NEU.0000000000001202

6. Macyszyn, L., et al.: Imaging patterns predict patient survival and molecular subtype in glioblastoma via machine learning techniques. Neuro-Oncology **18**(3), 417–425 (2016). https://doi.org/10.1093/neuonc/nov127

7. Menze, B.H., et al.: The multimodal brain tumor image segmentation Benchmark (BRATS). IEEE Trans. Med. Imaging **34**(10), 1993–2024 (2015). https://doi.org/10.1109/TMI.2014.2377694

8. Bakas, S., et al.: Advancing The Cancer Genome Atlas glioma MRI collections with expert segmentation labels and radiomic features. Nat. Sci. Data **4**, 170117 (2017). https://doi.org/10.1038/sdata.2017.117

9. Bakas, S., et al.: Segmentation labels and radiomic features for the pre-operative scans of the TCGA-GBM collection. In: The Cancer Imaging Archive (2017). https://doi.org/10.7937/K9/TCIA.2017.KLXWJJ1Q

10. Bakas, S., et al.: Segmentation labels and radiomic features for the pre-operative scans of the TCGA-LGG collection. In: The Cancer Imaging Archive (2017). https://doi.org/10.7937/K9/TCIA.2017.GJQ7R0EF
11. Tresp, V., Overhage, J.M., Bundschus, M., Rabizadeh, S., Fasching, P.A., Yu, S.: Going digital: a survey on digitalization and large-scale data analytics in healthcare. Proc. IEEE **104**, 2180–2206 (2016). https://doi.org/10.1109/JPROC.2016.2615052
12. Chen, M., Qian, Y., Chen, J., Hwang, K., Mao, S., Hu, L.: Privacy protection and intrusion avoidance for cloudlet-based medical data sharing. IEEE Trans. Cloud Comput. 1 (2017). https://doi.org/10.1109/TCC.2016.2617382
13. Brendan McMahan, H., Moore, E., Ramage, D., Hampson, S., Agera y Arcas, B.: Communication-efficient learning of deep networks from decentralized data. ArXiv e-prints (2016)
14. Chang, K., et al.: Distributed deep learning networks among institutions for medical imaging. J. Am. Med. Inform. Assoc. **25**(8), 945–954 (2018). https://doi.org/10.1093/jamia/ocy017
15. Geyer, R.C., Klein, T., Nabi, M.: Differentially Private Federated Learning: A Client Level Perspective. ArXiv e-prints (2017)
16. Bagdasaryan, E., Veit, A., Hua, Y., Estrin, D., Shmatikov, V.: How To Backdoor Federated Learning. ArXiv e-prints (2018)
17. Brendan McMahan, H., Ramage, D., Talwar, K., Zhang, L.: Learning Differentially Private Recurrent Language Models. ArXiv e-prints (2017)
18. Zhao, Y., Li, M., Lai, L., Suda, N., Civin, D., Chandra, V.: Federated Learning with Non-IID Data. ArXiv e-prints (2018)
19. French, R.M.: Catastrophic forgetting in connectionist networks. Trends Cogn. Sci. **3**, 128–135 (1999). https://doi.org/10.1016/S1364-6613(99)01294-2
20. Kirkpatrick, J., et al.: Overcoming catastrophic forgetting in neural networks. Proc. Nat. Acad. Sci. **114**, 3521–3526 (2017). https://doi.org/10.1073/pnas.1611835114
21. Ronneberger, O., Fischer, P., Brox, T.: U-Net: Convolutional Networks for Biomedical Image Segmentation. ArXiv e-prints (2015)
22. Zhao, X., Wu, Y., Song, G., Li, Z., Zhang, Y., Fan, Y.: A deep learning model integrating FCNNs and CRFs for brain tumor segmentation. Med. Image Anal. **43**, 98–111 (2018). https://doi.org/10.1016/j.media.2017.10.002
23. Shokri, R., Smatikov, V.: Privacy-preserving deep learning. In: CCS 2015 Proceedings of the 22nd ACM SIGSAC Conference on Computer and Communications Security, pp. 1310–1321 (2015). https://doi.org/10.1145/2810103.2813687
24. Abadi, M., et al.: Deep learning with differential privacy. In: CCS 2016 Proceedings of the 2016 ACM SIGSAC Conference on Computer and Communications Security, pp. 308–318 (2016). https://doi.org/10.1145/2976749.2978318

Patient-Specific Registration of Pre-operative and Post-recurrence Brain Tumor MRI Scans

Xu Han[1]([✉]), Spyridon Bakas[2], Roland Kwitt[3], Stephen Aylward[4], Hamed Akbari[2], Michel Bilello[2], Christos Davatzikos[2], and Marc Niethammer[1]

[1] Department of Computer Science, UNC Chapel Hill, Chapel Hill, NC, USA
`xhs400@cs.unc.edu`
[2] Center for Biomedical Image Computing and Analytics,
Perelman School of Medicine, University of Pennsylvania, Philadelphia, PA, USA
[3] University of Salzburg, Salzburg, Austria
[4] Kitware Inc., New York, USA

Abstract. Registering brain magnetic resonance imaging (MRI) scans containing pathologies is challenging primarily due to large deformations caused by the pathologies, leading to missing correspondences between scans. However, the registration task is important and directly related to personalized medicine, as registering between *baseline pre-operative* and *post-recurrence* scans may allow the evaluation of tumor infiltration and recurrence. While many registration methods exist, most of them do not specifically account for pathologies. Here, we propose a framework for the registration of longitudinal image-pairs of individual patients diagnosed with glioblastoma. Specifically, we present a combined image registration/reconstruction approach, which makes use of a patient-specific principal component analysis (PCA) model of image appearance to register baseline pre-operative and post-recurrence brain tumor scans. Our approach uses the post-recurrence scan to construct a patient-specific model, which then guides the registration of the pre-operative scan. Quantitative and qualitative evaluations of our framework on 10 patient image-pairs indicate that it provides excellent registration performance without requiring (1) any human intervention or (2) prior knowledge of tumor location, growth or appearance.

1 Introduction

Glioblastoma is the most common and aggressive malignant brain tumor that heavily and heterogeneously infiltrates surrounding tissue. This infiltration complicates treatment [1], as it is difficult to precisely localize the extent of infiltration. Considering that more than 80% of patients have a local tumor recurrence close to the initial resection cavity [2] (hence to the infiltrated brain tissue), we identify the need to accurately map correspondences between pre-operative (*pre*) and post-recurrence (*post*) brain tumor scans. Such registrations would

© Springer Nature Switzerland AG 2019
A. Crimi et al. (Eds.): BrainLes 2018, LNCS 11383, pp. 105–114, 2019.
https://doi.org/10.1007/978-3-030-11723-8_10

support research into the early detection of tumor recurrence, e.g., enable the identification of subtle imaging phenotypic characteristics of tumor recurrence. Even though correspondences are established between longitudinal image-pairs, where scans are expected to be comparable (as they are of the same patient), registration is challenging due to there being two sources of image appearance changes: *first*, the *pre* scans contain tumors and mass effect deformations; *second*, the *post* scans typically contain tumor resection cavities (where the tumor used to be in the *pre* scan) and show signs of tumor infiltration and recurrence.

Registration in the presence of pathologies may employ cost function masking [3] to exclude regions without clear correspondences and hence avoid influencing the registration's image similarity term. Alternatively, one could combine cost function masking with a model of infiltration and mass effect [4]. A joint segmentation-registration method has also been proposed [5,6], that incorporates a tumor growth model and that estimates a patient-specific atlas to guide the segmentation and registration, while using 4 MRI modalities. Building upon this and considering that both *pre* and *post* scans are of the same patient, Kwon et al. [7] proposed a framework to jointly segment and register the *post* to the *pre* scans. However, it may be challenging to estimate the deformations of *pre* scans with large tumors and *post* scans with large mass effect relaxation. In addition, these segmentation-registration approaches require manual interaction in the form of seed-points to initialize the growth model and to model the intensity distribution of each brain tissue across modalities. This complicates the use for large-scale studies and hampers the clinical translatability of these methods.

To account for missing correspondences in pathological regions, an alternative strategy is to estimate a *quasi-normal image* by learning from population data and use it for registration. Quasi-normal images can, for example, be estimated by a low-rank/sparse (LRS) decomposition [8] or via deep variational encoder-decoder architectures [9]. However, these methods either blur the normal tissue appearance and compromise the registration results or require a large number of training images. Inspired by the LRS decomposition, Han et al. [10] proposed a joint PCA/image-reconstruction model, which also decomposes the pathological image into two parts: (1) normal tissue appearance is captured by a statistical (PCA) model; and (2) large pathologies are captured via a total-variation (TV) term, which avoids blurring of the normal tissue and retains fine details in the quasi-normal image. The reconstructed quasi-normal image is then used for atlas registration. One could directly apply this method independently to the *pre* and the *post* scans, and then register the resulting quasi-normal images. However, this strategy would ignore the fact that these scans come from the same patient and the statistical model in the atlas space may not adequately capture the normal appearance for a specific patient; consequently, the registration quality may be impaired. Similarly, Kwon et al. [11] extended their framework [7] by incorporating an inpainting strategy to account for pathological regions. However, prior knowledge about the tumor of each scan, comprising of seed-points with associated radii and initial intensity modeling of each brain tissue type, is required for the algorithm. This manual interactive step, in addition to introducing an extra burden to the method's usability and increasing the time footprint of the method, also affects the objectivity and repeatability of the obtained results. All these together have a direct impact on the consideration of the method for potential clinical translation, as well as for large research studies. Therefore, a method combining the benefits of pathology modeling with patient-specificity, while eliminating manual interactions, is highly desirable.

Contributions. In this work, we present an automatic, repeatable, patient-specific registration approach for *pre* and *post* brain MRI scans that requires only a single modality. This is accomplished through careful adjustment of the PCA/image-reconstruction model [10]. In particular, we show how to (1) model each patient separately to improve the registration results and (2) leverage the decomposition's TV term to intrinsically exclude the estimated pathology in case the image is not well-aligned to the target space.

Organization. Section 2 describes our patient-specific registration framework, Sect. 3 presents the qualitative and quantitative evaluations of our approach, compared against other state-of-the-art methods, and Sect. 4 concludes the paper with a discussion and an outlook on future work.

2 Methodology

We first present an overview of the low-rank/sparse approach [8] and the PCA-TV model [10]. We then propose modifications for patient-specific registration.

2.1 Low-Rank/Sparse (LRS) Decomposition

In the LRS approach, images $\{I_i : i = 1, 2, ..., n\}$ are first arranged as columns of a matrix $I = [I_1, ..., I_n]$, where n describes the number of images. This matrix is then decomposed into a low-rank matrix $L = [L_1, ..., L_n]$ and a sparse matrix $S = [S_1, ..., S_n]$ by solving the problem:

$$\{L, S\} = \arg\min_{L,S}(\|L\|_* + \lambda\|S\|_1), \quad \text{s.t.} \quad I = L + S, \tag{1}$$

where $\| \cdot \|_*$ is the nuclear norm (i.e., a convex approximation of the rank), $\| \cdot \|_1$ denotes the ℓ_1 norm, and λ weighs the penalty on sparse term. Liu et al. [8] proposed a low-rank-based registration method by alternating the LRS decomposition and registering the low-rank image to an atlas. Upon convergence, the low-rank matrix contains the normal information from all images, while the sparse matrix obtains the estimated pathology. The low-rank images are then used for registration. While effective, the approach suffers two shortcomings: *First*, it requires optimization over the *entire* population, which is ineffective and computationally expensive. *Second*, while it recovers normal appearance in pathological regions, normal tissue areas are blurred which may negatively affect registration results.

2.2 PCA-TV Model

Inspired by the LRS framework and the Rudin-Osher-Fatemi (ROF) image denoising model [12], Han et al. proposed a PCA-TV registration framework [10]. It registers all the "normal" images, i.e., images from healthy controls, to an atlas space only *once*, followed by a PCA on the warped normal images. The PCA basis is kept fixed, and the PCA-TV model decomposes the image by solving the following problem:

$$\{\hat{L}, T, \alpha\} = \underset{\hat{L}, T, \alpha}{\arg\min} \frac{1}{2}\|\hat{L} - B\alpha\|_2^2 + \gamma\|\nabla T\|_{2,1}, \quad \text{s.t.} \quad \hat{I} = \hat{L} + T, \quad (2)$$

where $\hat{I}$ denotes the "pathological image", i.e., an image with at least a pathology/tumor, after we subtract the mean. $\|\nabla T\|_{2,1} = \sum_i \|\nabla T_i\|_2$, i is the spatial location, $\{\alpha\}$ are the PCA coefficients and B is the PCA basis. The model consists of (1) a quasi-low-rank part $\hat{L}$ that is close to the PCA space and retains image detail, and (2) a TV term, which captures pathologies that are large, spatially contiguous, and not expressed by the PCA basis. The quasi-normal image is obtained by adding the mean image to the quasi-low-rank image. Overall, this model is more effective than the LRS decomposition, as it works on just one image and explicitly leverages spatial information.

Additionally, an iterative regularization strategy can be used after the decomposition, just as for the ROF model [13]. In particular, after solving (2) and obtaining $\tilde{L}_0 = \hat{L}$ and α_0, for $k \geq 1$, one can iteratively solve

$$\{\tilde{L}_k, T_k, \alpha_k\} = \underset{\tilde{L}_k, T_k, \alpha_k}{\arg\min} \frac{1}{2}\|\tilde{L}_k - B\alpha_k\|_2^2 + \gamma\|\nabla T_k\|_{2,1} \quad \text{s.t.} \quad \hat{I}_k = \tilde{L}_k + T_k, \quad (3)$$

where $\hat{I}_k = \hat{I} + \tilde{L}_{k-1} - B\alpha_{k-1}$. After N regularization steps, the TV term T_N captures the pathology and the quasi-low-rank term can be obtained by subtracting the TV term from the input image, i.e., $\hat{L}_N = \hat{I} - \hat{T}_N$.

The entire framework iteratively alternates between the image decomposition and atlas registration. Each iteration includes the registration of the quasi-normal image to the atlas, the transformation of the input image to the atlas space, and the decomposition of the warped image in the atlas space. In addition, to avoid accumulating deformation errors, the quasi-normal image is always transformed back to the original image space prior to registration.

2.3 Patient-Specific Registration

When registering the *pre* to the *post* scan, one could simply apply the PCA-TV model (Sect. 2.2) independently on each scan and then register the corresponding quasi-normal images. However, this ignores that both scans are of the same patient. In addition, both the LRS and the PCA-TV approaches register quasi-normal images to the atlas during each iteration, but never use the sparse/TV information. In case an image contains tumors with large mass effect, it is drastically misaligned with the population images. Hence, the decomposition may not work sufficiently, unless the image is well-aligned with the atlas. This is especially true during the first iteration of registration and decomposition. To overcome these shortcomings and improve the registration of *pre* and *post* scans, we propose the following key adjustments to the PCA-TV model.

PCA-TV-Mask Model. When we compute the decomposition in the first iteration, the image is only affinely aligned to the atlas. We apply Otsu thresholding to the TV image, to obtain a coarse mask of the pathological region (TV-mask). This mask is then used during the registration, i.e., we register the quasi-normal image to the atlas, but use the TV-mask for cost-function masking of the tumor. Once the image is better aligned to the atlas via a deformable registration, we remove the TV-mask and use the entire quasi-normal image for registration. We refer to this improvement as the *PCA-TV-mask model*.

Patient-Specific PCA. Considering (i) that the *post* scan is relatively free from mass effects (e.g., except for scarring) and (ii) that the tumor resection cavity is easily modeled via the TV term, we propose the following two-step strategy. In the *first step*, we apply the PCA-TV-mask model to the *post* scan, resulting in a quasi-normal reconstructed image, in addition to registering the *post* scan to the atlas space. In the *second step*, we use the inverse transformation of the first step to map all normal images into the *post* scan space and then construct a new PCA basis from this data. Importantly, we can now use this new PCA basis together with the quasi-normal *post* image (now warped back to the patient space and used as atlas) to run the PCA-TV-mask model on the *pre* scan. Overall, this strategy allows *direct* registration between the *pre* and the *post* scans. Another advantage of using this patient-specific strategy is that by running PCA in the patient-specific space, the normal space spanned by the PCA basis is expected to be more consistent with the *pre* scan, which in turn improves the decomposition and registration results, when compared with the original framework.

3   Experiments

We evaluate our framework on 10 pairs of *pre* and *post* clinically-acquired scans of patients diagnosed with de novo (primary) glioblastoma. Each timepoint contains native (T1) and contrast-enhanced T1-weighted (T1-CE), T2-weighted and FLAIR MRI. All modalities of each patient are skull-stripped, bias-field corrected, and affinely co-registered to the *pre* T1-CE scan of this patient that describes a $192 \times 256 \times 192$ volume with voxel size of $0.977 \times 0.977 \times 1.0$ [mm^3]. For quantitative evaluation, we use manually seeded landmarks from two clinical experts. The first expert placed 20 landmarks within 30[mm] from the tumor region and 30 landmarks outside the 30[mm] region in each *pre* scan. Then, both experts independently placed matching landmarks in the *post* scans. The landmarks placed by the first expert are considered the gold-standard and the ones placed by the second expert serve as a baseline comparison, referred to as RATER. In our experiments, we only use the T1 volumes from each patient and run 6 iterations of registration and decomposition. The remaining 3 modalities were only

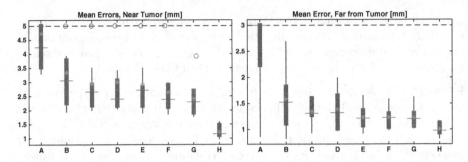

Fig. 1. Boxplots of the mean landmark errors. For each method, the landmark errors are computed against the gold-standard. On each box, the red line is the median and the green star is the mean. The bottom and top edges of the box denote the 25th and 75th percentiles, respectively, the whiskers extend to the most extreme datas that are not considered outliers and the outliers are plotted in circle. (A) `AFFINE`; (B) `GREEDY`; (C) `DRAMMS`; (D) `ANTs`; (E) `NiftyReg`; (F) `PCA-TV`; (G) `PCA-PS`; (H) `RATER`. Our result is plotted in red. (Color figure online)

used by the experts for seeding the landmarks. We pick 100 normal images from OASIS [14] and select 50 as PCA basis. For registration, we use `NiftyReg` [15] as B-spline registration with the default settings and local normalized cross correlation as similarity measure (`--lncc 40`). The TV-mask is used in the first iteration when the image is only affinely aligned to the target image. After B-spline registration, we remove the TV-mask for subsequent iterations. We also apply the regularization steps in the last three iterations. γ in (2) and (3) for the decomposition model is chosen as 1 if no regularization step is used and 2 if regularization steps are used. We compare with `AFFINE` [16], `GREEDY` [17], `DRAMMS` [18], `ANTS` [19], `NiftyReg` [15] and `PCA-TV` [10]. Although `PORTR` [7] was specifically designed for this task, we did not include it in our analysis, as our intent is to compare methods that do not require multiple modalities or manual interaction and hence more easily translate to clinical use.

We compute the mean landmark error for each region of each patient (Fig. 1) and we note that all deformable methods are better than affine registration, but worse than `RATER`. Compared to other deformable methods, our patient-specific approach improves the results in the close-to-tumor region. We also improve results in the region far away from the tumor, except when comparing with `NiftyReg` and the original `PCA-TV` model. In fact, as shown in Table 2, the improvements in the close-to-tumor region are statistically significant, assessed via a one-tailed paired Wilcoxon signed-rank test with a Benjamini-Hochberg procedure to control the false discovery rate at level $\alpha = 0.05$. For far-from-tumor regions, the results are only significant when compared to `AFFINE` and `GREEDY`. We also calculate the effect sizes with each paired rank test. Most of the tests result in large or medium effect sizes.

We also evaluate the statistics of the paired landmark errors in both regions (Table 1). For each landmark, we calculate the differences of the errors between

Table 1. Statistic results for all paired landmark errors in both regions. For each landmark, we calculate the paired error; i.e., we subtract the landmark error of compared method from the landmark error of our method. This is to calculate the improvement obtained by our method. For each compared method, we rank the paired landmark errors and show the statistics in the table. The green boxes indicate results where errors from our framework are smaller.

	Near Tumor[mm]						Far from Tumor[mm]					
	5%	25%	50%	75%	95%	Mean	5%	25%	50%	75%	95%	Mean
AFFINE	-1.03	0.22	1.65	3.71	7.58	2.32	-0.60	0.79	1.35	3.64	6.90	2.11
GREEDY	-1.01	-0.28	0.31	1.18	6.03	0.94	-0.80	-0.17	0.05	0.43	2.77	0.36
DRAMMS	-1.45	-0.55	0.18	0.79	4.68	0.52	-1.15	-0.28	0.14	0.52	1.16	0.13
ANTs	-1.44	-0.31	0.17	0.80	6.10	0.59	-0.68	-0.18	0.08	0.37	1.32	0.17
NiftyReg	-1.21	-0.19	0.12	0.60	3.35	0.51	-0.50	-0.14	-0.02	0.12	0.55	0.01
PCA_TV	-1.06	-0.29	0.11	0.57	2.08	0.23	-0.45	-0.14	0.00	0.15	0.55	0.01
RATER	-4.67	-1.86	-0.79	0.07	1.06	-1.18	-1.99	-0.74	-0.14	0.44	1.18	-0.21

our framework and competing methods. Compared to RATER, our method shows worse performance on more than 50% of the landmarks. However, when comparing to other automatic registration methods, although at some landmarks our method performs worse than others by less than 1.5[mm] near the tumor and 1[mm] far away from the tumor, as shown at 5% statistics, it shows better performance on more than 50% of the landmarks. In fact, the improvement, especially near the tumor, can be larger than 5[mm], as shown at 95% statistics in the table. Furthermore, on average, we perform better than other registration methods by 0.5[mm] near the tumor and by less than 0.2[mm] far away from the tumor. This is consistent with the green stars shown in Fig. 1. Our patient-specific method also improves over the PCA-TV model near the tumor which illustrates its utility and the benefit of the patient-specific model.

Table 2. *p*-values and effect sizes for one-tailed paired Wilcoxon signed-rank test. We compare all methods (except for RATER) with our patient specific framework. Green boxes indicate statistically significant results after false discovery rate correction or effect sizes that are at least medium (>0.3).

		AFFINE	GREEDY	DRAMMS	ANTs	NiftyReg	PCA-TV
p-values	Near	9.77e-4	4.90e-3	1.37e-2	1.86e-2	4.90e-3	3.22e-2
	Far	2.00e-3	1.37e-2	4.20e-2	0.116	0.423	0.385
effect sizes	Near	0.6268	0.5584	0.4900	0.4672	0.5584	0.4217
	Far	0.6040	0.4900	0.3989	0.2849	0.0570	0.0798

Finally, Fig. 2 shows example results from three patients, where we register the *pre* to the *post* scans. For the PCA-TV model and our patient-specific PCA-PS model, we reconstruct the quasi-normal images from each patient which are used to guide the registrations. Although the visual differences between our method and the PCA-TV model are subtle, other results show that by modeling the pathologies registrations are qualitatively more accurate. Note that, Fig. 2(c) illustrates the T2-FLAIR scans for the *post* images, only for visualization purposes, to better depict the surgically-imposed cavities of these illustrated examples. All the applied registration methods use only the T1 volumes.

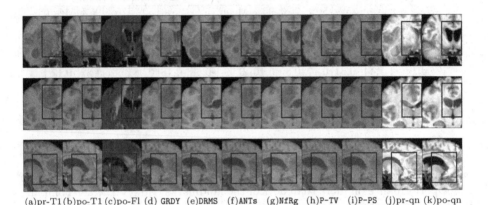

(a)pr-T1 (b)po-T1 (c)po-Fl (d) GRDY (e)DRMS   (f)ANTs   (g)NfRg (h)P-TV   (i)P-PS (j)pr-qn (k)po-qn

Fig. 2. Example registration results from three patients. (a) and (b) show the *pre* and the *post* T1 scans. (c) shows the *post* T2-FLAIR scans, only for visualization purposes. (d)-(i) show registration results of *pre* to *post* from GREEDY, DRAMMS, ANTs, NiftyReg, PCA-TV, and our patient-specific model, PCA-PS. In addition, (j) and (k) show the quasi-normal reconstructions of the *pre* and *post* scans, respectively. The red box highlights major differences. (Color figure online)

4   Conclusion

We proposed a patient-specific registration framework based on a PCA-TV-mask model, which registers pre-operative and post-recurrence scans of the same patient. The framework uses the *post* scan, which is relatively free from mass effects, to build a patient-specific PCA basis, and directly registers the *pre* scan to the patient space. The validation results show that our framework is more effective than the PCA-TV model, as well as other registration methods that do not explicitly model pathologies. In addition, our framework does not require any manual interaction, neither in the form of segmentation nor as tumor seeding, and only requires a single modality. In future work we will explore our method for different diseases, for example, to register acute and chronic image pairs from patients with traumatic brain injuries.

Acknowledgements. Research reported in this publication was supported by the National Institutes of Health (NIH) and the National Science Foundation (NSF), under award numbers NIH:2R44NS081792, NIH/NINDS:R01NS042645,

NIH/NCI:U24CA189523, NSF/ECCS-1148870, and NSF/EECS-1711776. The content of this publication is solely the responsibility of the authors and does not necessarily represent the official views of the NIH, or the NSF.

References

1. Price, S.J., Jena, R., Burnet, N.G., Carpenter, T.A., Pickard, J.D., Gillard, J.H.: Predicting patterns of glioma recurrence using diffusion tensor imaging. Eur. Radiol. **17**(7), 1675–1684 (2007)
2. Milano, M.T., et al.: Patterns and timing of recurrence after temozolomide-based chemoradiation for glioblastoma. Int. J. Rad. Oncol. Biol. Phys. **78**(4), 1147–1155 (2010)
3. Brett, M., Leff, A.P., Rorden, C., Ashburner, J.: Spatial normalization of brain images with focal lesions using cost function masking. NeuroImage **14**(2), 486–500 (2001)
4. Niethammer, M., et al.: Geometric metamorphosis. In: Fichtinger, G., Martel, A., Peters, T. (eds.) MICCAI 2011. LNCS, vol. 6892, pp. 639–646. Springer, Heidelberg (2011). https://doi.org/10.1007/978-3-642-23629-7_78
5. Gooya, A., et al.: GLISTR: glioma image segmentation and registration. IEEE TMI **31**(10), 1941–1954 (2012)
6. Bakas, S., et al.: GLISTRboost: combining multimodal MRI segmentation, registration, and biophysical tumor growth modeling with gradient boosting machines for glioma segmentation. In: Crimi, A., Menze, B., Maier, O., Reyes, M., Handels, H. (eds.) BrainLes 2015. LNCS, vol. 9556, pp. 144–155. Springer, Cham (2016). https://doi.org/10.1007/978-3-319-30858-6_13
7. Kwon, D., Niethammer, M., Akbari, H., Bilello, M., Davatzikos, C., Pohl, K.M.: PORTR: pre-operative and post-recurrence brain tumor registration. IEEE TMI **33**(3), 651–667 (2014)
8. Liu, X., Niethammer, M., Kwitt, R., Singh, N., McCormick, M., Aylward, S.: Low-rank atlas image analyses in the presence of pathologies. IEEE TMI **34**(12), 2583–2591 (2015)
9. Yang, X., Han, X., Park, E., Aylward, S., Kwitt, R., Niethammer, M.: Registration of pathological images. In: Tsaftaris, S.A., Gooya, A., Frangi, A.F., Prince, J.L. (eds.) SASHIMI 2016. LNCS, vol. 9968, pp. 97–107. Springer, Cham (2016). https://doi.org/10.1007/978-3-319-46630-9_10
10. Han, X., Yang, X., Aylward, S., Kwitt, R., Niethammer, M.: Efficient registration of pathological images: a joint PCA/image-reconstruction approach. In: ISBI (2017)
11. Kwon, D., Zeng, K., Bilello, M., Davatzikos, C.: Estimating patient specific templates for pre-operative and follow-up brain tumor registration. In: Navab, N., Hornegger, J., Wells, W.M., Frangi, A.F. (eds.) MICCAI 2015. LNCS, vol. 9350, pp. 222–229. Springer, Cham (2015). https://doi.org/10.1007/978-3-319-24571-3_27
12. Rudin, L.I., Osher, S., Fatemi, E.: Nonlinear total variation based noise removal algorithms. Phys. D: Nonlinear Phenom. **60**(1–4), 259–268 (1992)
13. Osher, S., Burger, M., Goldfarb, D., Xu, J., Yin, W.: An iterative regularization method for total variation-based image restoration. Multiscale Model. Simul. **4**(2), 460–489 (2005)
14. Marcus, D.S., Wang, T.H., Parker, J., Csernansky, J.G., Morris, J.C., Buckner, R.L.: Open access series of imaging studies (OASIS): cross-sectional MRI data in young, middle aged, nondemented, and demented older adults. J. Cogn. Neurosci. **19**(9), 1498–1507 (2007)

15. Modat, M., et al.: Fast free-form deformation using graphics processing units. Comput. Methods Programs Biomed. **98**(3), 278–284 (2010)

16. Modat, M., Cash, D.M., Daga, P., Winston, G.P., Duncan, J.S., Ourselin, S.: Global image registration using a symmetric block-matching approach. J. Med. Imaging **1**(2), 024003 (2014)

17. Avants, B.B., Epstein, C.L., Grossman, M., Gee, J.C.: Symmetric diffeomorphic image registration with cross-correlation: evaluating automated labeling of elderly and neurodegenerative brain. Med. Image Anal. **12**(1), 26–41 (2008)

18. Ou, Y., Sotiras, A., Paragios, N., Davatzikos, C.: Dramms: deformable registration via attribute matching and mutual-saliency weighting. Med. Image Anal. **15**(4), 622–639 (2011)

19. Avants, B.B., Tustison, N., Song, G.: Advanced normalization tools (ANTS). Insight J. **2**, 1–35 (2009)

Segmentation of Post-operative Glioblastoma in MRI by U-Net with Patient-Specific Interactive Refinement

Ashis Kumar Dhara[1]([✉]), Kalyan Ram Ayyalasomayajula[1], Erik Arvids[2],
Markus Fahlström[2], Johan Wikström[2], Elna-Marie Larsson[2],
and Robin Strand[1,2]

[1] Centre for Image Analysis, Uppsala University, Uppsala, Sweden
`ashis.kumar.dhara@it.uu.se`
[2] Department of Surgical Sciences, Radiology, Uppsala University, Uppsala, Sweden

Abstract. Accurate volumetric change estimation of glioblastoma is very important for post-surgical treatment follow-up. In this paper, an interactive segmentation method was developed and evaluated with the aim to guide volumetric estimation of glioblastoma. U-Net based fully convolutional network is used for initial segmentation of glioblastoma from post contrast MR images. The max flow algorithm is applied on the probability map of U-Net to update the initial segmentation and the result is displayed to the user for interactive refinement. Network update is performed based on the corrected contour by considering patient specific learning to deal with large context variations among different images. The proposed method is evaluated on a clinical MR image database of 15 glioblastoma patients with longitudinal scan data. The experimental results depict an improvement of segmentation performance due to patient specific fine-tuning. The proposed method is computationally fast and efficient as compared to state-of-the-art interactive segmentation tools. This tool could be useful for post-surgical treatment follow-up with minimal user intervention.

1 Introduction

Glioblastoma is the most aggressive form of brain tumor and accounts for more than 80% of all primary brain malignancies [1]. Volumetric change estimation is very important for post-surgical treatment follow-up of patients with glioblastoma. Fully-automated segmentation of glioblastoma is difficult due to presence of necrosis and post-operative haemorrhages. There are different types of inter-active tools existing in the literature based on the mode of outlining such as click-based [2], contour-based [3] and bounding box-based [4] and scribbling. Manual tumor segmentation is time consuming and the results are subject to inter- and intra-observer errors. In clinical practice, the diameter of glioblastoma in the biggest representative slice is often considered as the representative of the volume. However, the diameter is not a proper estimation of volumetric change as it is affected by the presence of post-operative cavities.

© Springer Nature Switzerland AG 2019
A. Crimi et al. (Eds.): BrainLes 2018, LNCS 11383, pp. 115–122, 2019.
https://doi.org/10.1007/978-3-030-11723-8_11

Several automatic and semi-automatic methods have been developed using support vector machines [5], random forests [6], markov random field [7] for segmentation of brain tumor. The majority of reported work on brain tumor segmentation use 2D filters for convolution [8,9]. Structured predictions is implemented through convolution neural networks for better segmentation of brain tumor [9]. The U-Net [10] architecture considers the high-level features and the local appearance of tumor for better segmentation. Interactive segmentation methods have been implemented in several medical software [11,12]. These methods require a relatively large amount of user interaction for segmentation of images with ambiguous boundaries.

In this paper, to reduce the time of user interaction for segmentation of glioblastoma, initial segmentation obtained from U-Net is refined by graph cut [13] algorithm and Smart paint [14]. The proposed method is novel as it includes an interactive step that levies a minimal effort on the expert who has to provide manual intervention at regions that have been mislabelled. It saves manual annotation of a few thousand pixels by the expert per patient. The work introduced in the paper are preliminary results along the direction of integrating energy minimization based post-processing in a semantic segmentation on a dataset with no available ground truth except for an expert's judgement. Therefore, an attempt has been made to pick a fully convolution neural network based encoder-decoder architecture accepted by the medical community, such as U-Net and apply well understood post processing algorithm such as graph-cut to establish a baseline. The contributions of this paper are (i) design of a deep learning based interactive segmentation framework and (ii) implementation of patient-specific interactive refinement to deal with large content variations among images.

2 Materials and Methods

2.1 Database

A clinical MR image database (prospective research project, approved by local ethical committee) of brain tumors with world health organization (WHO) grade gliomas IV (glioblastoma) is used in this study. The database consists of longitudinal MR scans (post-operative) of 15 patients. The total number of MRI volumes of 15 subjects is 85. Contrast enhanced T1-weighted MR images were obtained from those 15 patients at defined time intervals. The boundaries of contrast enhanced tumour tissue is annotated by an expert radiologist using FsLeyes package [15] through slice by slice interaction. The average time spent for boundary annotation of each MRI volume is 12 min.

2.2 Interactive Segmentation Framework

The block diagram of the proposed interactive segmentation framework is provided in Fig. 1. The availability of a sufficient number of annotated images is often a bottleneck in medical image analysis and U-Net [10] fills this gap by producing good results with a lower number of images. Therefore, the U-Net is used in the proposed framework. A set of cropped ROIs is used for training the U-Net and the trained model is named T_i. Segmentation is performed for images of any MR instance of a patient based on T_i and the results are shown to user for interactive refinement by scribbling in foreground and background region. The pre trained model is updated based on the corrected mask created by interactive correction.

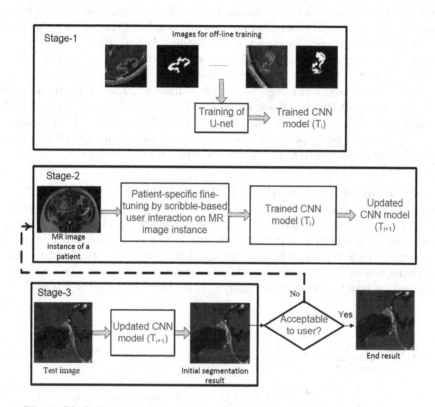

Fig. 1. Block diagram of the proposed interactive segmentation framework

Architecture of U-Net. The proposed framework used U-Net [10] as it considers the high-level information and local appearance information of an object for better segmentation. It consists of an *encoder* and *decoder* paths. The architecture makes the use of strong data augmentation to use the available annotated samples more efficiently. The encoder path has five convolutional blocks and each block has two convolutional layers. The convolution is performed with a filter size of 3×3 and a stride of 1. Max pooling is performed with stride 2×2 to the end of every block except the last block. The size of the input images is 64×64 and it decreases from 64×64 to 4×4. In the decoder path, transpose convolution is applied in each layer with a filter size of 3×3 and stride of 2×2. In each block, the size of the feature maps is doubled but the width of the feature maps is decreased by factor of 2. Therefore, the size of feature maps increases from 4×4 to 64×64. High-level information is represented at up-sampling blocks, and low-level features are transferred through skip connections.

Off-Line Training of U-Net. Patches of size 64×64 are extracted from axial slices of brain MR images for the training of U-Net. Rotation and flip are used to increases the number of training images. In training of U-Net, a stochastic gradient-based optimization [16] (Adam) is applied to minimize the cross-entropy based cost function. The cross-Entropy loss function has been used for this purpose due to its efficiency in binary classification. Adam utilizes the first and second moments of gradients for updating and correcting moving average of the current gradients. The learning rate for the Adam optimizer is set to 0.0001, weights of background and foreground are initialized as $1 : 10$, and training were performed up to $60,000$ iterations. Dropout [17] is used to reduce over-fitting. A training and validation dataset pair (I_r, I_{r_g}); (I_v, I_{v_g}), consisting of MRI images from patients with post-operative glioblastoma is used to train a CNN based on vanilla U-Net architecture from scratch, to obtain a preliminary CNN model T_i (stage-1 of Fig. 1).

Update of Initial Segmentation Obtained by U-Net. The output of U-Net is updated using a graph cut algorithm [13]. In case of interpreting an images as a graph, its pixels are considered as vertices and image edges are considered as graph edges. The optimal solution results in a cut on the graph that minimizes the energy, which corresponds to maximizing the flow along those edges. The underlying idea is to formulate the refined segmentation of image I as an energy minimization problem as given in the Eq. (1) resulting in a binary image B

$$
\begin{aligned}
E_I(B) = &\sum_{i=0}^{m}\sum_{j=0}^{n}[L_{ij}^t(1 - B_{ij}) + L_{ij}^b B_{ij}] && \text{(Data fidelity term)} \\
&+ \sum_{i=0}^{m}\sum_{j=0}^{n} C_{ij}^h(B_{ij} \neq B_{i+1,j}) && \text{(Horizontal continuity term)} \quad (1) \\
&+ \sum_{i=0}^{m}\sum_{j=0}^{n} C_{ij}^v(B_{ij} \neq B_{i,j+1}) && \text{(Vertical continuity term)}
\end{aligned}
$$

where $L_{ij}^t, L_{ij}^b, C_{ij}^h, C_{ij}^v$ are costs associated with labeling a pixel indexed (i, j) as belonging to tumor, brain, smoothness along horizontal and vertical directions, respectively. Labeling costs are governed by Eq. (2)

$$
L_{ij}^t = P(t_{ij}|I_{ij}; W) = \frac{e^{l_{t_{ij}}}}{e^{l_{t_{ij}}} + e^{l_{b_{ij}}}}, L_{ij}^b = P(b_{ij}|I_{ij}; W) = 1 - P(t_{ij}|I_{ij}; W), \quad (2)
$$

where the labeling cost L_{ij}^t, L_{ij}^b are interpreted as a probabilities $P(t_{ij}|I_{ij}; W)$, $P(b_{ij}|I_{ij}; W)$ of the pixel I_{ij} at (i, j) being labeled as tumor or brain respectively.

The probabilities are obtained by applying a *Softmax function* over the class scores $l_{t_{ij}}, l_{b_{ij}}$ derived from the U-Net model with weights W for tumor and brain classes respectively.

Smoothness cost is governed by Eq. (3)

$$C_{ij}^h = \begin{cases} 0, & E_{ij} \wedge (I_{ij} < I_{i+1,j}) \\ 0, & E_{i+1,j} \wedge (I_{ij} \geq I_{i+1,j}), \\ \phi, & otherwise \end{cases} \quad C_{ij}^v = \begin{cases} 0, & E_{ij} \wedge (I_{ij} < I_{i,j+1}) \\ 0, & E_{i,j+1} \wedge (I_{ij} \geq I_{i,j+1}) \\ \phi, & otherwise \end{cases}$$

$$(3)$$

which encourages label consistency on either sides of the edge E_{ij} for an associated cost ϕ. Details on graph-cut based semantic segmentation improvement can be found in [18].

Patient-Specific Refinement of Pre-trained Model. The procedure of interactive refinement is described in stage-2 and stage 3 of Fig. 1. A test dataset I_{s_1} for which no ground truth exists, is segmented using T_i to obtain segmented output I_o. A graph-cut is applied to the class probability maps from T_i to get an output I_g.

- Though I_g can be directly used for interactive segmentation, it is empirical found that using $I_g \odot I_o$ (*Hadamard product* of the two images) helps reduce false positives.
- $I_g \odot I_o$ is used in an interactive tool by an expert to obtain the ground truth segmentation for I_{s_1} as I_{sg_1}.

The pre-trained U-Net model T_i from Stage-1 is re-trained on (I_{s_1}, I_{sg_1}) to get a refined model T_{i+1}. Stages 2, 3 can be iterated to get better models using T_{i+1}. Images of a particular instance of longitudinal data is used for patient specific fine-tuning and another instance of that patient (I_t, I_{t_g}) is used for testing to report the dice similarity coefficient (DSC).

3 Experiments, Results and Discussion

Training of U-Net has been performed in Linux environment using a 11 GB GPU (Zotac GeForce GTX 1080 Ti) on a system with Core-i7 processor and 32 GB RAM. The network architecture is implemented in Python using the PyTorch library. DSC is considered as performance metric for evaluation of segmentation results, as defined below:

$$DSC = \frac{2N_{tp}}{2N_{tp} + N_{fp} + N_{fn}} \qquad (4)$$

where, N_{tp}, N_{tn}, N_{fp} and N_{fn} is number of true positive, true negative, false positive and false negative pixels, respectively.

The segmentation results is evaluated by considering 15-fold cross-validation. A trained medical student, who is familiar with the glioblastomas database, has corrected the initial segmentation using Smart paint [14] for network update. Segmentation was also performed using several interactive segmentation tools such as 3D paint and brush tool of ITK-Snap [19] and 3D-Slicer [20] for comparative study. Box plots of DSC achieved by several methods are shown in Fig. 2. The result shows substantial improvement over existing methods. The proposed approach is straight forward and simple. It is an intentional choice to explore the improvement of segmentation based on subject specific refinement. The max-flow algorithm plays an important role in improving segmentation accuracy in the boundary region. The improvement of mean DSC is around 3% due to the inclusion of max-flow algorithm. Moreover, there is reduction of computational time if max-flow algorithm as it is applied on the result of U-Net. Interactive refinement is required in more than 90% of the cases. Therefore, max-flow can not be avoided. Comparative study of interactive segmentation time required by the expert is presented in Table 1. Time of interactive segmentation to achieve a sufficiently high accuracy is noted for all methods. The proposed method takes less time to achieve acceptable segmentation result.

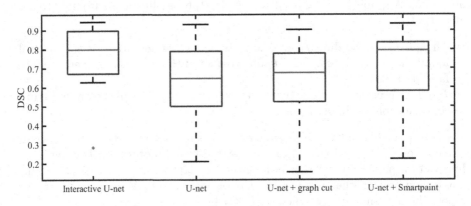

Fig. 2. Box plot of DSC of a single MRI volume of each subjects using U-Net, interactive U-Net, U-Net followed by graph cut and the expert's refinement of U-Net using Smartpaint. In each box, the central mark is the median, the edges of the box are the 25th and 75th percentiles.

Table 1. Time of interaction of the proposed and competing techniques.

Methods	Time (mean ± std) in sec.
ITK-snap from scratch	185.8 ± 68.46
3D-slicer from scratch	223.53 ± 66.06
U-Net followed by Smartpaint	83.66 ± 27.09
Proposed interactive U-Net	80.26 ± 24.19

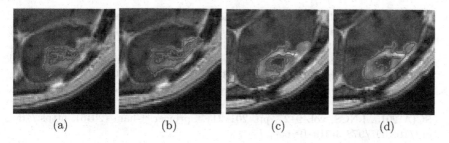

(a)   (b)   (c)   (d)

Fig. 3. Segmentation evaluation examples. (a), (c) initial segmentation by U-Net and (b), (d) improved segmentation after patient-specific fine-tuning. Ground truth is indicated by red and segmented image is indicated by blue. (Color figure online)

The initial segmentation results of U-Net and result after patient-specific interactive refinement is shown in Fig. 3. Initial segmentation by U-Net is poor due to the presence of a necrosis and post-operative haemorrhage. It is observed that incorporating certain image features related to the patient, improves the segmentation results.

4 Conclusion

A patient-specific interactive refinement method is implemented using U-Net based fully convolutional network. The proposed interactive segmentation framework can provide both efficient and robust segmentation compared to other interactive tools. The network is able to learn patient-specific features from the feedback provided by an expert user and results better segmentation for rest of the longitudinal MR images of the same subject. This method will be helpful in treatment follow-up analysis in neuroradiology.

References

1. Ostrom, Q.T., et al.: CBTRUS statistical report: primary brain and central nervous system tumors diagnosed in the United States in 2008-2012. Neuro-oncology **17**(Suppl. 4) (2015)
2. Haider, S.A., et al.: Single-click, semi-automatic lung nodule contouring using hierarchical conditional random fields. In: 2015 IEEE 12th International Symposium on Biomedical Imaging (ISBI), pp. 1139–1142. IEEE (2015)
3. Xu, C., Prince, J.L.: Snakes, shapes, and gradient vector flow. IEEE Trans. Image Process. **7**(3), 359–369 (1998)
4. Rother, C., Kolmogorov, V., Blake, A.: GrabCut: interactive foreground extraction using iterated graph cuts. In: ACM Transactions on Graphics (TOG), vol. 23, pp. 309–314. ACM (2004)
5. Lee, C.-H., Wang, S., Murtha, A., Brown, M.R.G., Greiner, R.: Segmenting brain tumors using pseudo–conditional random fields. In: Metaxas, D., Axel, L., Fichtinger, G., Székely, G. (eds.) MICCAI 2008. LNCS, vol. 5241, pp. 359–366. Springer, Heidelberg (2008). https://doi.org/10.1007/978-3-540-85988-8_43

6. Tustison, N.J., et al.: Optimal symmetric multimodal templates and concatenated random forests for supervised brain tumor segmentation (simplified) with ANTsR. Neuroinformatics **13**(2), 209–225 (2015)

7. Menze, B.H., et al.: The multimodal brain tumor image segmentation benchmark (BRATS). IEEE Trans. Med. Imaging **34**(10), 1993–2024 (2015)

8. Lyksborg, M., Puonti, O., Agn, M., Larsen, R.: An ensemble of 2D convolutional neural networks for tumor segmentation. In: Paulsen, R.R., Pedersen, K.S. (eds.) SCIA 2015. LNCS, vol. 9127, pp. 201–211. Springer, Cham (2015). https://doi.org/10.1007/978-3-319-19665-7_17

9. Dvorak, P., Menze, B.: Structured prediction with convolutional neural networks for multimodal brain tumor segmentation. In: Proceedings of the Multimodal Brain Tumor Image Segmentation Challenge, pp. 13–24 (2015)

10. Ronneberger, O., Fischer, P., Brox, T.: U-Net: convolutional networks for biomedical image segmentation. In: Navab, N., Hornegger, J., Wells, W.M., Frangi, A.F. (eds.) MICCAI 2015. LNCS, vol. 9351, pp. 234–241. Springer, Cham (2015). https://doi.org/10.1007/978-3-319-24574-4_28

11. Armstrong, C.J., Price, B.L., Barrett, W.A.: Interactive segmentation of image volumes with live surface. Comput. Graph. **31**(2), 212–229 (2007)

12. Cates, J.E., Lefohn, A.E., Whitaker, R.T.: GIST an interactive, GPU-based level set segmentation tool for 3D medical images. Med. Image Anal. **8**(3), 217–231 (2004)

13. Boykov, Y., Kolmogorov, V.: An experimental comparison of min-cut/max-flow algorithms for energy minimization in vision. IEEE Trans. Pattern Anal. Mach. Intell. **26**(9), 1124–1137 (2004)

14. Malmberg, F., Strand, R., Kullberg, J., Nordenskjöld, R., Bengtsson, E.: Smart paint a new interactive segmentation method applied to MR prostate segmentation. In: MICCAI Grand Challenge: Prostate MR Image Segmentation 2012 (2012)

15. Paul, M.: FSLeyes. https://fsl.fmrib.ox.ac.uk/fsl/fslwiki/FSLeyes

16. Kingma, D., Ba, J.: Adam: a method for stochastic optimization. arXiv preprint arXiv:1412.6980 (2014)

17. Hinton, G.E., Srivastava, N., Krizhevsky, A., Sutskever, I., Salakhutdinov, R.R.: Improving neural networks by preventing co-adaptation of feature detectors. preprint arXiv:1207.0580 (2012)

18. Ayyalasomayajula, K.R., Brun, A.: Historical document binarization combining semantic labeling and graph cuts. In: Sharma, P., Bianchi, F.M. (eds.) SCIA 2017. LNCS, vol. 10269, pp. 386–396. Springer, Cham (2017). https://doi.org/10.1007/978-3-319-59126-1_32

19. Yushkevich, P.A., Gao, Y., Gerig, G.: ITK-SNAP: an interactive tool for semi-automatic segmentation of multi-modality biomedical images. In: 2016 IEEE 38th Annual International Conference of the Engineering in Medicine and Biology Society (EMBC), pp. 3342–3345. IEEE (2016)

20. Fedorov, A., et al.: 3D slicer as an image computing platform for the quantitative imaging network. Magn. Reson. Imaging **30**(9), 1323–1341 (2012)

Characterizing Peritumoral Tissue Using DTI-Based Free Water Elimination

Abdol Aziz Ould Ismail[1]($\boxtimes$), Drew Parker[1],
Moises Hernandez-Fernandez[1], Steven Brem[2], Simon Alexander[3],
Ofer Pasternak[4], Emmanuel Caruyer[5], and Ragini Verma[1]

[1] Center for Biomedical Image Computing and Analytics (CBICA), Radiology,
University of Pennsylvania, Philadelphia, PA, USA
AbdolAziz.OuldIsmail@uphs.upenn.edu,
Ragini@pennmedicine.upenn.edu
[2] Neurosurgery, University of Pennsylvania, Philadelphia, PA, USA
[3] Synaptive Medical Inc., Toronto, ON, Canada
[4] Psychiatry and Radiology, Harvard Medical School, Boston, MA, USA
[5] CNRS, IRISA, Britanny, France

Abstract. Finding an accurate microstructural characterization of the peritumoral region is essential to distinguish between edema and infiltration, enabling the distinction between tumor types, and to improve tractography in this region. Characterization of healthy versus pathological tissue is a key concern when modeling tissue microstructure in the peritumoral area, which is muddled by the presence of free water (e.g., edema). Although diffusion MRI (dMRI) is being used to obtain the microstructural characterization of tissue, most methods are based on advanced dMRI acquisition schemes that are infeasible in the clinical environment, which predominantly uses diffusion tensor imaging (DTI), and are mostly for healthy tissue. In this paper, we propose a novel approach for microstructural characterization of peritumoral tissue, that involves multi-compartment modeling and a robust free water elimination (FWE) method to improve the estimation of free water in both healthy and pathological tissue. As FWE requires the fitting of two compartments, it is an ill-posed problem in DTI acquisitions. Solving this problem requires an optimization routine, which in turn relies on an initialization step for finding a solution, which we optimally choose to model the presence of edema and infiltration unlike existing schemes. We have validated the method extensively on simulated data, and applied it to data from brain tumor patients to demonstrate the improvement in tractography in the peritumoral region, which is important for surgical planning.

Keywords: Diffusion tensor imaging (DTI) · Fractional anisotropy (FA) · Mean diffusivity (MD) · White matter (WM) · Grey matter (GM) · Peritumoral region · Free water (FW) · Volume fraction (VF) · Non-linear optimization · Tractography

© Springer Nature Switzerland AG 2019
A. Crimi et al. (Eds.): BrainLes 2018, LNCS 11383, pp. 123–131, 2019.
https://doi.org/10.1007/978-3-030-11723-8_12

1   Introduction

The characterization of tissue microstructure in the peritumoral region is crucial for accurate surgical planning (e.g., using tractography), to identify infiltration for targeted treatment, and to provide new radiomic features for distinguishing tumor subtypes. Accurate white matter (WM) microstructural modeling using diffusion magnetic resonance imaging (dMRI) has the potential to provide a characterization of the peritumoral tissue. Such microstructure modeling is expected to have a substantial impact on future clinical studies in pathologies.

A key aspect for exploring the microstructural properties is to estimate the free water component within a tissue of interest. Not accounting for free water can cause the fitted diffusion tensors to have an erroneously low fractional anisotropy (FA), causing some tracking algorithms to stop prematurely in the contaminated regions. Hence, a correct estimation of free water is crucial.

In one of the earliest attempts for estimating free water, a bi-tensor model was proposed [1], separating the derived diffusion signal into isotropic (free water) and tissue-based components. This approach acknowledged the fact that the bi-tensor model fitting has infinitely many solutions for single-shell data [2], and posited that the ill-posed nature of the problem could be addressed by appropriate initialization of the model parameters and by spatial regularization that stabilizes the fit. The regularized approach significantly contributed to the field of free water elimination (FWE), allowing a better reconstruction of healthy fornix tracts, as well as improving tractography in the peritumoral region. However, the initialization that was originally proposed is acquisition dependent [3]. In addition, there are various pathological conditions in which the initialization in [1] may lead to an inaccurate estimation of the free water compartment and diffusion indices (FA, MD) [4].

In this work, we propose a novel initialization approach based on the estimation of free water maps from simulated ground truth data. This initialization relies on prior and heuristic constraints that were validated across a range of simulated free water fractions, anisotropy levels, and underlying diffusivities. Such an approach allows for optimizing the FWE in the entire brain, including regions affected by pathologies such as edema and infiltration. Therefore, the initialization method that we propose is applicable to standard clinical acquisitions, enabling retrospective investigations and providing novel insight into peritumoral regions. As the method is applicable on any clinically acquired data, the free water map is expected to become a crucial feature in radiomics.

2   Methods

2.1   Free Water Elimination

The proposed FWE initialization is based on a bi-compartment model: a tensor for modeling the underlying tissue, and an isotropic free water compartment. Specifically, we fit [1]

$$A_i = fe^{-bq_i^T D q_i} + (1-f)e^{-bd} \tag{1}$$

where the first term models the tissue and the second term represents the free water compartment. f is the tissue volume fraction, A_i is the signal attenuation of the diffusion weighted image acquired along the ith gradient direction, b is the amount of diffusion weighting, q_i is i^{th} gradient direction, D represents the diffusion tensor used for modeling the tissue compartment, and d is the diffusivity in the isotropic compartment, which is fixed at 3.0×10^{-3} mm^2/s. Fitting this model using a single-shell dMRI acquisition is a problem with infinitely many solutions. Previously, the fitting was stabilized by using a regularizer post-initialization [1]. There, Pasternak *et al.* proposed an initial estimate of the tissue compartment f based on scaling the unweighted images (S_0) with respect to the mean signal of representative WM voxels (S_t) and CSF voxels (S_w) using:

$$f_{t=0} = 1 - \frac{\log(S_0/S_t)}{\log(S_w/S_t)} \tag{2}$$

In [1], f was additionally constrained within bounds that were enforced during the non-linear fit stage via

$$\frac{A_{min} - e^{-bd}}{e^{-b\lambda_{max}} - e^{-bd}} < f_{\lambda_{min,max}} < \frac{A_{max} - e^{-bd}}{e^{-b\lambda_{min}} - e^{-bd}} \tag{3}$$

where λ_{max} and λ_{min} are the expected maximal and minimal diffusivity of the tensor, A_{max} and A_{min} are the maximum and minimum measured signal attenuations within a region of interest. This initialization approach suppressed the free water in healthy tissue, yet in the final estimation in this approach as well as other FWE approaches it is clear that free water content in healthy WM tissue is low but not zero [1, 5]. Moreover, the initialization proposed in Eq. (2) is solely based on T2 contrast observed in b0 images. While this T2 contrast gives us information about the amount of free water, it may miss additional information that we can only infer using the diffusion-weighted attenuation.

An alternative approach is to constrain the corrected tissue signal attenuation by assuming that the mean diffusivity (MD) in the healthy tissue is around a fixed diffusivity, e.g., $MD_{tissue} = 0.60 \times 10^{-3}$ mm^2/s [5, 6], and thus the free water compartment explains the increase in MD in voxels affected by partial voluming. As such:

$$f_{MD} = \frac{e^{-bMD} - e^{-bd}}{e^{-bMD_{tissue}} - e^{-bd}} \tag{4}$$

where MD is the mean diffusivity from the standard tensor fit in a voxel of interest.

Although such an approach improves the estimation in the healthy tissue, it struggles to characterize regions that are highly contaminated by free water or have tumor cells that restrict the diffusivity further in the peritumoral area, i.e., where MD_{tissue} is not a good initial value. The underlying problem in all these approaches, is that the initialization is designed to either address the healthy tissue or tissue with pathology, but not both.

To alleviate these issues, we propose an initialization to the FWE problem that is a hybrid between $f_{\lambda_{min,max}}$ and f_{MD}. Aiming to obtain a better estimation of the free water compartment in both healthy tissue as well as voxels contaminated by edema or partial voluming, we introduce an initialization that is a log-linear interpolation between $f_{\lambda_{min,max}}$ and f_{MD}. We propose:

$$f_{initial} = f_{MD}^{\alpha} \times f_{\lambda_{min,max}}^{1-\alpha} \tag{5}$$

where α is set to $\alpha = f_{t=0}$ (see Eq. (2)). That is, for tissue with healthy appearing T2, we constrain the free water using f_{MD} (Eq. (4)), and in regions that appear like CSF, or edematous in the T2 contrast, we estimate the free water compartment based on Eq. (3).

2.2   Evaluation

We will evaluate the different FWE initialization approaches on simulated data (where the ground truth is known) as well as on brain tumor patients. On the patients, we will evaluate the fit in the healthy tissue, defined as white matter outside the peritumoral region, as well as in peritumoral region, and its effect on tracking for the purposes of surgical planning.

2.2.1   Simulated Data

To have a sample with ground truth, simulated data was generated with varying ground truth mean diffusivities, anisotropy, and free water volume fractions (Table 1). These simulated datasets follow a bi-tensor model where one of the compartments represents tissue and the second is isotropic with a fixed diffusivity (3.0×10^{-3} mm^2/s) [7]. For the unweighted images, we calculate the transverse magnetization as a linear sum of the contribution of each compartment:

$$M_T = VF_{wm}\rho_{WM}\left(1 - e^{-TR/T1_{WM}}\right)e^{-TE/T2_{WM}} + VF_{CSF}\rho_{CSF}\left(1 - e^{-TR/T1_{CSF}}\right)e^{-TE/T2_{CSF}} \tag{6}$$

where VF is the volume fraction, ρ is the proton density, and TR/TE are repetition and echo times. For every experiment, we perform 10 noise realizations by adding Rician noise equivalent to SNR = 40.

Table 1. Parameters selected for simulating data to validate the proposed method

	Values	Number of experiments
MD values	0.4–1.0×10^{-3} mm^2/s	4
FA values	0–1	11
Free water volume fractions	0–1	11
Rotation of the tensor directions	–	10
Noise realizations	–	10
Total		48,400

2.2.2 Brain Tumor Datasets

138 brain tumor patients were included (88 glioblastoma/50 metastasis). The data was acquired on Siemens 3T Verio scanner with TR/TE = 5000/86 ms, b = 1000 s/mm^2, 3 b0, and 30 gradient directions. Automatic tumor and peritumoral region segmentation was performed on coregistered T1, T2, T2-FLAIR and T1-contrast enhanced images using GLISTR [8].

2.3 Tractography

Tractography was performed, using Diffusion Toolkit [9], on the FWE tensor field as well as the standard tensor field, in voxels with FA values exceeding 0.2, using the 2nd order Runge-Kutta algorithm with an angle threshold of 45° and a step size of 1 mm. Five bundles of interest (corticospinal tract, inferior frontal, inferior fronto-occipital, uncinate and arcuate fasciculi) were extracted from each tractogram in each hemisphere, using the RecoBundles algorithm [10], with a pruning parameter of 7 mm. Finally, a "*coverage measure*" was defined and calculated for each patient and each of the two tractograms. "Coverage Measure" is the percentage of voxels in the peritumoral edema region containing a streamline from any of the ten bundles of interest.

3 Results

3.1 Evaluation on Simulated Data

We compare the proposed method to previous approaches [1, 4] that fit a bi-compartment model. Our results show that the FW compartment is estimated to be nearly 0 in the healthy WM, in the previous approaches (Fig. 1C). In simulated edema scenarios (volume fraction (VF) = 0.6), The proposed initialization of free water and FA is closer to the ground truth in simulated edematous/infiltrated regions (Fig. 1A–F), as compared to the other initialization approaches.

FWE in Simulated Data

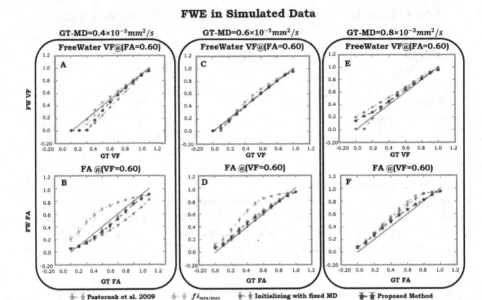

Fig. 1. Three different ground truth (GT) MD values (GT-MD) (1st, 2nd, and 3rd columns) were investigated using different GT FW and GT FA values. For details on the simulation parameters, see Table 1. Four FWE initializations were performed (Pasternak *et al.* [1], $f_{\lambda_{min,max}}$ (i.e., $\alpha = 0$), f_{MD}, (i.e., $\alpha = 1$) and the proposed method with varying alpha. The free water (FW VF) and FA (FW FA) outputs of each method are compared to the corresponding ground truth values. The 1st row represents the investigation of the FW maps in a simulated WM across different FW fractions. The 2nd row is for FW FA maps for a simulated edema scenario (VF = 0.60).

3.2 *In-Vivo* Data: Application on Brain Tumor Datasets

Figure 2 shows the FW and FA maps, pre- and post-FWE, using the proposed approach, f_{MD} initialization, as well as $\alpha = 0$ and $\alpha = 1$. Several things are observed: (i) FA maps from the corrected tensors show that various approaches were successful in increasing FA in the edema region (Fig. 2A); (ii) FW is estimated to be nearly zero in many regions of healthy WM by previous approaches, as compared to our proposed method; (iii) a large number of physiologically implausible voxels (defined as MD < 0.40×10^{-3} mm^2/s [5, 6]) were observed when using the method in [1] (Fig. 2C). Although this was alleviated by use of $\alpha = 0$, this approach underestimated the free water [5, 6] in almost the entire healthy WM (Fig. 2C). The proposed method, as well as the one that used $\alpha = 1$, were able to eliminate these implausible voxels, without underestimating the FW in the healthy tissue (Fig. 2C).

3.3 Tractography

Figure 3A shows the comparative results of tracking in the peritumoral regions, pre- and post-FWE (using the proposed initialization) showing that the streamlines were able to travel through the peritumoral region more after FWE. Additionally, in Fig. 3B, a quantitative evaluation using our *coverage measure* shows that the tracking coverage in the peritumoral region is significantly improved with FWE.

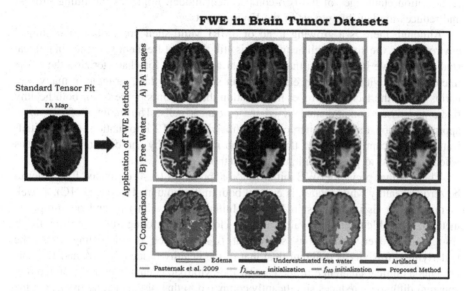

Fig. 2. (**A**) FA maps from the various FWE initializations. (**B**) Free water maps using different initialization strategies. (**C**) Location of implausible and voxels with underestimated FW (FW volume fraction < 0.02) in the different investigated methods.

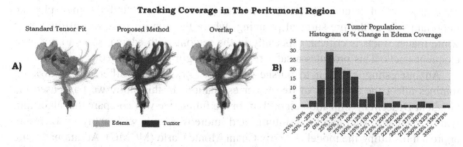

Fig. 3. (**A**) Comparison of tracking in the peri-tumoral region (gray blob), pre- and post FWE in the left arcuate and CST reconstructed in a tumor patient using the proposed initialization. In the overlap of the approaches, we see the significant increase of streamlines coverage in the edema. (**B**) Tracking coverage in the edema was computed before and after FWE. A paired t-test analysis of tractography showed a significant increase in peritumoral region coverage using FWE tractography (t = 6.91, p < 1 $\times$ 10^{-9}). (Color figure online)

4 Discussion

In this manuscript, we proposed a novel initialization method for estimating the free water compartment from single-shell acquisitions. Our proposed method addresses initialization challenges of the two-compartment model, improves the fitting process and reduces implausible voxels in the FWE output.

Although there is a growing trend of dMRI Multi-shell protocols in the clinical environments, the single-shell acquisition is still the most frequent sequence in clinical studies. Here we have shown that traditional approaches for characterizing the tissue microstructure with free-water elimination in single shell data, struggle in many scenarios due to the lack of a proper initialization that accounts properly for both healthy tissue and tissue contaminated by edema or cancer cells. The findings of our novel approach demonstrate that interpolating diffusivity and T2 information, in the initialization, significantly improved the estimation in the healthy and contaminated tissues when compared to previous initialization methods [1, 4].

The fact that previous FWE initialization approaches [1, 4] estimated free water to be zero in healthy white matter, is evident both in the simulated data (Fig. 1C), as well as the healthy tissue regions in Fig. 2C. Initializations with f_{MD} and our proposed method, were able to produce values that are closer to the ground truth in healthy tissue (Fig. 2). However, simulations show that f_{MD} approach is misleading when the underlying diffusivity is different than that of the healthy tissue (Fig. 1A and B). This presents a clinical challenge in the case of the cytotoxic edema, where the underlying apparent diffusivity reduces significantly compared to that of the healthy tissue. On the other hand, the proposed approach yields a robust solution across different diffusivities in the simulated peritumoral region.

Furthermore, the proposed strategy for initializing the parameters of a bi-compartment model leads to a better characterization of the underlying white matter in areas affected by edema. This affects the results of tractography, allowing the reconstruction of tracts impacted by edema (Fig. 3). These findings may play an important role in the pre-surgical planning, where the non-invasive reconstruction of white matter tracts is critical, especially in the regions with edema, so as to not miss eloquent tracts. This provides a safer resection margin for the surgeon.

An interesting question to pursue in future work is how different optimization routines affect the accuracy of free water estimation. In this study we have used the Gradient Descent optimization algorithm. In the future, we will compare this algorithm with Levenberg-Marquardt algorithm, and moreover, we will apply a Bayesian approach for fitting the model, Markov Chain Monte Carlo (MCMC). Additionally, the differences that are evident in the peritumoral tissue in the metastatic tumors and the GBMs, suggest the potential of free water measures in differentiating tumor subtypes. This will be undertaken as future work.

References

1. Pasternak, O., Sochen, N., Gur, Y., Intrator, N., Assaf, Y.: Free water elimination and mapping from diffusion MRI. Magn. Reson. Med. **62**, 717–730 (2009)
2. Scherrer, B., Warfield, S.K.: Why multiple b-values are required for multi-tensor models: evaluation with a constrained log-Euclidean model. In: ISBI 2010, pp. 1389–1392 (2010)
3. Parker, D., Ould Ismail, A., Alexander, S., Verma, R.: The role of bias field correction in the free water elimination problem. In: ISMRM, June 2018
4. Ould Ismail, A., Parker, D., Alexander, S., Caruyer, E., Pasternak, O., Verma, R.: Free watER iNvariant Estimation of Tensor (FERNET): addressing the issue of edema in clinically feasible acquisitions. In: ISMRM, June 2018
5. Hoy, A.R., Koay, C.G., Kecskemeti, S.R., Alexander, A.L.: Optimization of a free water elimination two-compartment model for diffusion tensor imaging. NeuroImage **31**(103), 323–333 (2014)
6. Kodiweera, C., Wu, Y.C.: Data of NODDI diffusion metrics in the brain and computer simulation of hybrid diffusion imaging (HYDI) acquisition scheme. Data Brief **1**(7), 1131–1138 (2016)
7. Garyfallidis, E., et al.: Dipy, a library for the analysis of diffusion MRI data. Front. Neuroinform. **21**(8), 8 (2014)
8. Gooya, A., et al.: GLISTR: glioma image segmentation and registration. IEEE Trans. Med. Imaging **31**(10), 1941–1954 (2012)
9. Wang, R., Benner, T., Sorensen, A.G., Wedeen, V.J.: Diffusion toolkit: a software package for diffusion imaging data processing and tractography. In: Proceedings of the International Society for Magnetic Resonance in Medicine 2007, vol. 15, no. 3720 (2007)
10. Garyfallidis, E., et al.: Recognition of white matter bundles using local and global streamline-based registration and clustering. NeuroImage **170**, 283–295 (2017)

Deep 2D Encoder-Decoder Convolutional Neural Network for Multiple Sclerosis Lesion Segmentation in Brain MRI

Shahab Aslani[1,2(✉)], Michael Dayan[1], Vittorio Murino[1,3], and Diego Sona[1,4]

[1] Pattern Analysis and Computer Vision (PAVIS),
Istituto Italiano di Tecnologia (IIT), Genoa, Italy
{shahab.aslani,michael.dayan,vittorio.murino,diego.sona}@iit.it
[2] Science and Technology for Electronic and Telecommunication Engineering,
University of Genoa, Genoa, Italy
[3] Dipartimento di Informatica, University of Verona, Verona, Italy
[4] NeuroInformatics Laboratory, Fondazione Bruno Kessler, Trento, Italy

Abstract. In this paper, we propose an automated segmentation approach based on a deep two-dimensional fully convolutional neural network to segment brain multiple sclerosis lesions from multimodal magnetic resonance images. The proposed model is made as a combination of two deep subnetworks. An encoding network extracts different feature maps at various resolutions. A decoding part upconvolves the feature maps combining them through shortcut connections during an upsampling procedure. To the best of our knowledge, the proposed model is the first slice-based fully convolutional neural network for the purpose of multiple sclerosis lesion segmentation. We evaluated our network on a freely available dataset from ISBI MS challenge with encouraging results from a clinical perspective.

Keywords: Segmentation · Multiple sclerosis · Convolutional neural network

1 Introduction

Multiple Sclerosis (MS) is one of the most common demyelination diseases having direct effects on the central nervous system, especially on white matter (WM), which can be visualized through magnetic resonance imaging (MRI) scans. The detection of all MS lesions is an important task as it can help characterizing the progression of the disease and monitoring the efficacy of a candidate treatment [14].

In literature, there are both manual and automatic methods for MS lesion segmentation. Manual segmentation usually provides accurate results with the drawbacks of being time-consuming, affected by expert skills and biased towards a given expert. This highlights the importance of automatic segmentation methods, which can be faster, not affected by the expertise variability and unbiased [4].

© Springer Nature Switzerland AG 2019
A. Crimi et al. (Eds.): BrainLes 2018, LNCS 11383, pp. 132–141, 2019.
https://doi.org/10.1007/978-3-030-11723-8_13

Methods of automated MS lesion segmentation can be arbitrarily classified in two main types: empirical approaches typically based on a heuristic series of image-processing operations, and machine learning approaches.

Image-processing based methods are faster but generally depend on the manual set-up of specific parameters, for example, the choice of thresholds, as in He et al. [7], where an adaptive procedure segments unhealthy regions with a multistep pipeline of morphological operations.

On the contrary, machine learning based approaches particularly supervised methods can be slower but learn automatically from a training dataset previously labeled by an expert. For example, Jesson et al. [8] proposed a three-stage pipeline to discriminate healthy tissues from lesions, where intensity distributions were used to train a random forest classifier.

Recently, deep learning methods, in particular, convolutional neural networks (CNNs), have shown excellent performance with various applications [9]. One of the most important advantages of these methods over other supervised algorithms is that they can learn themselves how to design features directly from data during the training procedure. It is important to mention that over the last years, CNNs have also been used in biomedical image analysis with state-of-the-art results in different problems [13].

Regarding the literature, there exist a few proposed methods based on CNNs for segmenting MS lesions. In [1], a three-dimensional (3D) CNN is designed to use shortcut connections between layers of the network, which allow concatenating the features from deep layers to shallow layers. Recently, Valverde et al. [15] proposed a patch-based method relying on a cascade of two 3D CNNs. In this approach, the extracted volumetric patches are used to train the first network. Then, a second network is used to refine the training on samples misclassified by the first network.

In this paper, we present a pipeline for automatic MS lesion segmentation based on two-dimensional (2D) CNNs. In this work, we concentrated on whole-brain segmentation in order to avoid some common problems like the neglect of global information of patch-based approaches, and the overfitting of 3D segmentation due to the small sample set issue. The CNN architecture used in this approach is a modified version of Residual Network (ResNet) [6] which has been proposed for image classification. To the best of our knowledge, this is the first slice-based (whole-brain) fully convolutional end-to-end encoder-decoder network proposed for MS lesion segmentation. The robustness of the method is improved by exploiting the volumetric slicing in all three possible imaging planes (axial, coronal and sagittal). Indeed, we used different imaging axes of each 3D input MRI in an ensemble framework to exploit the contextual information in all three anatomical planes. Moreover, this model can be used as a multi-modal network to make use of all of the information available within each within each MRI modality available, typically fluid-attenuated inversion-recovery (FLAIR), T1-weighted (T1w), and T2-weighted (T2w).

2   Method

2.1   Input Data Preparation

From each original volumetric MRI modality, axial, coronal and sagittal planes
are considered by extracting 2D slices along the x, y, z axes of the 3D image.
Since the size of the imaging planes differed according to the imaging axes, we
zero padded each slice (while centering the brain), so that to obtain the same
consistent size irrespective of the imaging plane. Further, the same consistent size
was applied across modalities. Then, slices belonging to each plane orientation
and each modality were stacked together to create a single multi-channel input
stack. Since three modalities were used in our experiments, the obtained multi-
channel slices included three channels which can be represented as RGB images.
Figure 1 illustrates the described procedure using three modalities, FLAIR, T1w,
and T2w.

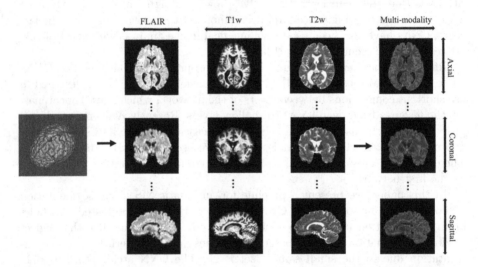

Fig. 1. Feature extraction pipeline. From each original 3D MRI image, axial, coronal
and sagittal planes were extracted for each modality. Last column: in our specific
application which 3 modalities were used (FLAIR, T1w, T2w), multi-channel slices
(represented here as RGB images) were created by grouping together the corresponding
slices of each modality.

2.2   Network Architecture

Recently very deep CNNs showed outstanding performance in computer vision
problems. In particular, ResNet [6] based on residual connections, gave significant improvement in image recognition tasks. Deep networks are hard to train
because of the vanishing gradient problem during the back-propagation procedure. Therefore, when the network goes deeper, its performance gets saturated.

The authors in [6] addressed the mentioned problem by proposing the network called ResNet. The main idea of the ResNet is to use identity shortcut connection between layers of the network which have some benefits like preventing of vanishing gradient and also not adding computational complexity to the network. In this work, we modified ResNet50 (version with 50 layers) for a pixel-wise segmentation task inspired by the idea of Fully Convolutional Networks (FCNs) [10]. The easiest way to convert a ResNet to a segmentation network is to replace the last prediction layer with a dense pixel-wise prediction layer as described in FCNs. Since the output of the last convolutional layer of ResNet is very coarse compared with the input image resolution (32 times smaller than the original image) upsampling such high level feature maps with a simple operation like bilinear interpolation as described in FCNs is not an effective solution. Therefore, in order to address this problem, we propose a multi-pass upsampling network using the advantages of multi-level feature maps with skip connections.

In deep networks, features from deep layers include high-level semantic information. On the contrary, features from early layers contain low-level spatial information. It was shown that features from middle layers also provide information which can be effective to increase the performance of the segmentation [13]. Therefore, combining multi-level features from different stages of the network makes the feature map richer than just using single scale feature maps. The intuition behind this work is to use these multi-level feature maps by adding multiple upsampling network with skip connections [13] to the ResNet output of all intermediate layers. The diagram of the proposed network for segmentation can be seen in Fig. 2.

We divided the ResNet50 into 5 blocks in the downsampling part according to the resolution of feature maps. In the upsampling subnetwork, the encoded features from different scales are decoded step by step using upsampling fused features (UFF) blocks. Each UFF block includes one upconvolutional layer with kernel size 2×2 and stride 2, one concatenation or fusion layer and two convolution layers with kernel sizes 3×3. After each layer, a rectifier linear activation function (ReLU) is applied [12]. The upconvolutional layer is used to transform low-resolution feature maps into the higher resolution maps. Then a simple concatenation layer is used for combining the two sets of input feature maps. Two convolution layers are further used for adaptation as described in [13], and the output goes to the next block. The number of feature maps after each UFF block is halved. At the end of the network, a soft-max layer of size 2 is used to get output probability maps, identifying pixel-wise positive (lesion) or negative (non-lesion) classes.

3   Experiments

3.1   Data

To evaluate the proposed model, we used the dataset from ISBI 2015 Longitudinal MS Lesion Segmentation Challenge; which includes 19 subjects divided into two sets, 5 subjects for training and 14 subjects for testing. All training and

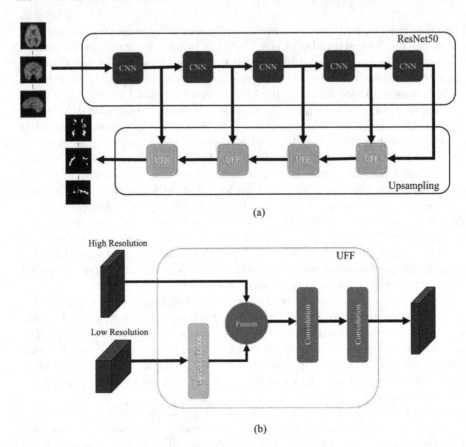

(a)

(b)

Fig. 2. (a) General framework of the proposed network for MS segmentation. The first sub-network (ResNet50) encodes the input 2D slices into different resolutions. This sub-network was divided into 5 blocks with respect to the resolution of the representations during the encoding. The second sub-network (Upsampling) decodes the representations provided by the encoder network. This sub-network gradually converts low-resolution representations back to the original resolution of the input image using UFF blocks. (b) Details of the proposed UFF block. Each UFF block has two set of input representations with different resolutions. This block is responsible to upsampling the low-resolution representations and combines them with high-resolution representations.

testing data have the same 1 mm-isotropic resolution. Each subject has MRI data with a different number of time-points, normally ranging between 4 to 6. Moreover, for each time-point, T1w, T2w, proton density-weighted (PDw), and FLAIR image modalities were provided. All training images have been segmented manually by two different raters and the segmented images are publicly available. For the test set, there is no public available ground truth. In order to evaluate the performance of the proposed method over the test dataset, the associated lesion binary mask must be submitted to the challenge website for evaluation [2].

3.2   Training and Testing

To train the proposed CNN, firstly, a training dataset was created using the pipeline mentioned in the previous section. In order to remove uninformative samples from the whole training set, a subset was determined by selecting only slices with at least one lesion pixel. This meant that 2D slices without lesions were omitted from the training set. In order to optimize network weights and early stopping criteria, we split the training dataset into different training and validation sets depending on the experiments as described in the following section. According to the network initialization, in the first subnetwork, the pre-trained ResNet50 on ImageNet was used and the weights from the second sub-network (Upsampling) were randomly initialized. Adaptive learning rate method (ADADELTA) [16] was used to tune the learning rate and a binary cross-entropy was used as loss function. The maximum number of training epochs was fixed to 500, and the best model was selected according to the validation set.

We evaluated then the proposed network with unseen test data with respect to the corresponding experiments. For each subject, we first extracted all the slices from the test set, following the approach described in the previous section. Feeding each 2D slice to the network, we got as output the associated 2D binary lesion classification map. Since the original data was duplicated three times in the input, once for each slice orientation (coronal, axial, sagittal), concatenating the binary lesion maps belonging to the same orientation resulted in three 3D lesion classification maps. These three lesion maps were combined via majority voting (the most frequent lesion classification was selected).

We implemented our proposed model in Keras [3] using a Nvidia GTX Titan X GPU.

3.3   Data Augmentation

As suggested in [5], simple off-line data augmentation was applied to the training dataset in order to increase training samples. Increasing training samples has been shown to increase the performance of the network. Therefore, we increased the number of the samples by a factor of 5 simply by either rotating each extracted slice by 4 possible angles ($5°$, $10°$, $-5°$, $-10°$) and flipping (right to left) of the images with their original rotation (no combination of flipping and rotation were included in the data augmentation procedure).

3.4   Evaluation

For evaluation purposes, two different experiments were implemented according to the availability of ground truth. In the first experiment, we ignored the official ISBI test set so that to only consider data with the available ground truth. In

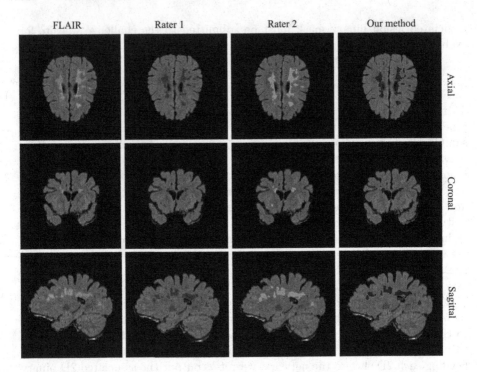

Fig. 3. An example of our network results in the axial, coronal and sagittal planes. First column: original FLAIR modality from different views, second column: ground truth related to the rater 1, third column: ground truth related to the rater 2, last column: segmentation output from the proposed method.

order to get a fair result, we did a leave-one-out cross-validation training (at subject level: 3 subjects for training, 1 subject for validation and 1 subject for testing). In this experiment, Dice Similarity Coefficient (DSC), Lesion-wise True Positive Rate ($LTPR$), and Lesion-wise False Positive Rate ($LFPR$) measures were used for evaluation.

The DSC is computed as;

$$DSC = (2 \times TP)/(FN + FP + 2 \times TP) \tag{1}$$

Where TP, FN and FP indicate the true positive, false negative and false positive voxels respectively. $LTPR$ denotes the number of lesions in the reference segmentation that overlap with a lesion in the output segmentation, divided by the number of lesions in the reference segmentation (lesion recall). $LFPR$ denotes the number of lesions in the output segmentation that do not overlap with a lesion in the reference segmentation, over the total number of lesions in the produced segmentation (lesion precision).

For the second experiment, the official ISBI test set was used as our test set so the ground truth was not available. We trained the network using leave-one-out cross-validation over all 5 subjects in the training set (4 subjects for training and 1 subject for validation). We evaluated the ensemble of 5 trained models on the test set and then for a final prediction, we did majority voting over all classifiers. The 3D output binary lesion maps were submitted to the website of ISBI for evaluation purposes. In this experiment, a score is measured online (using the challenge website) according to the results on that test set. As described in [2], the mentioned score is a weighted average of different metrics including DSC, $LTPR$, $LFPR$, positive prediction value (PPV) and absolute volume difference (AVD). PPV is the ratio between the number of true positive voxels and the total number of positive voxels. AVD is the absolute difference of volumes divided by the true volumes.

3.5 Results

In the first experiment, as described previously, we evaluate the performance of our network on the training set. Table 1 shows the performance of our method in comparison with other previously proposed methods. As can be seen, our method has the highest performance regarding LTPR metric while having a high DSC which means that the proposed method can identify lesion areas with higher precision than other methods while having a good overlap in terms of lesion volume overall. Figure 3 shows an example of the output of our network in comparison to the corresponding ground truth.

Table 1. Comparison of our method with the other state-of-the-art methods. GT1 and GT2 show that the corresponding model was trained using annotation provided by rater 1 and rater 2 as the ground truth respectively.

Method	Rater 1			Rater 2		
	DSC	LTPR	LFPR	DSC	LTPR	LFPR
Rater 1	-	-	-	0.7320	0.6450	0.1740
Rater 2	0.7320	0.8260	0.3550	-	-	-
Jesson et al. [8]	**0.7040**	0.6111	**0.1355**	**0.6810**	0.5010	**0.1270**
Maier et al. [11] (GT1)	0.7000	0.5333	0.4888	0.6555	0.3777	0.4444
Maier et al. [11] (GT2)	0.7000	0.5555	0.4888	0.6555	0.3888	0.4333
Brosch et al. [1] (GT1)	0.6844	0.7455	0.5455	0.6444	0.6333	0.5288
Brosch et al. [1] (GT2)	0.6833	0.7833	0.6455	0.6588	0.6933	0.6199
Ours (GT1)	0.6980	**0.7460**	0.4820	0.6510	**0.6410**	0.4506
Ours (GT2)	0.6940	**0.7840**	0.4970	0.6640	**0.6950**	0.4420

In the second experiment, the performance of the proposed method was also evaluated on the official ISBI test set using the challenge web service[1]. At the time we submitted the results, we obtained a score of 89.85 which is comparable

---

[1] http://iacl.ece.jhu.edu/index.php/MSChallenge.

to the ISBI inter-rater score scaled to 90. The detailed result for each subject is available online on the ISBI MS lesion segmentation challenge website.

4   Discussion and Conclusion

We have proposed a supervised method for the brain MS lesion segmentation. The presented approach is a deep end-to-end CNN including two pathways, a contracting path which extracts multi-resolution representations by encoding the input image and an expanding path which decodes the provided representations gradually by upsampling and fusing them. Our CNN has been trained using whole-brain slices as inputs to take advantage of the spacial information about the location and shape of MS lesions. Moreover, it has been designed for multi-modality (FLAIR, T1w, T2w) and multi-planes (axial, coronal and sagittal) analysis of MRI images.

The proposed method has been evaluated using the publicly available dataset (ISBI 2015 challenge). Comparing with other state-of-the-art methods, our experiments demonstrated that the proposed architecture performed better which has high capability to effectively identify unhealthy regions (LTPR = 0.7840) while having overall a good overlap with the ground truth in terms of overall lesion volume (DSC = 0.6980). This can be particularly important in clinical settings where detecting all potential lesions is prioritized over discarding easily identifiable false negatives.

Unlike previously proposed 3D-based CNN approach by Brosch et al. [1] which used a single short-cut connection between the deepest and the shallowest layers, our proposed architecture includes multiple short-cut connections between several layers of the network combining multi-level features from different stages of the network. In our opinion, the obtained results suggest that the combination of multi-level features during the upsampling procedure helps network to exploit more contextual information of the shape of the lesions. This could explain why the segmentation performance of our proposed network (DSC = 69.80) improved compared with the method proposed by Brosch et al. [1] (DSC = 0.6844).

The proposed method also has some limitations. Our network cannot use four-dimensional (4D) modalities such as functional MRI or diffusion MRI. Moreover, the maximum number of MRI modalities that can be used in our architecture is three. This results from the fact that we used pre-trained ResNet as the encoder part in our network, which can only handle an input with three channels. Therefore in the case of the more modalities available, one would be restricted to choose three amongst all. Another limitation is that CNN based approaches in MS segmentation highly depend on the training which is costly to acquire due to the time consuming manual segmentation by experts it requires.

Acknowledgments. We respectfully acknowledge NVIDIA for GPU donation.

References

1. Brosch, T., Tang, L.Y., Yoo, Y., Li, D.K., Traboulsee, A., Tam, R.: Deep 3D convolutional encoder networks with shortcuts for multiscale feature integration applied to multiple sclerosis lesion segmentation. IEEE Trans. Med. Imaging **35**(5), 1229–1239 (2016)
2. Carass, A., et al.: Longitudinal multiple sclerosis lesion segmentation: resource and challenge. NeuroImage **148**, 77–102 (2017)
3. Chollet, F., et al.: Keras (2015). https://github.com/fchollet/keras
4. García-Lorenzo, D., Francis, S., Narayanan, S., Arnold, D.L., Collins, D.L.: Review of automatic segmentation methods of multiple sclerosis white matter lesions on conventional magnetic resonance imaging. Med. Image Anal. **17**(1), 1–18 (2013)
5. Havaei, M., et al.: Brain tumor segmentation with deep neural networks. Med. Image Anal. **35**, 18–31 (2017)
6. He, K., Zhang, X., Ren, S., Sun, J.: Deep residual learning for image recognition. In: Proceedings of the IEEE Conference on Computer Vision and Pattern Recognition, pp. 770–778 (2016)
7. He, R., Narayana, P.A.: Automatic delineation of Gd enhancements on magnetic resonance images in multiple sclerosis. Med. Phys. **29**(7), 1536–1546 (2002)
8. Jesson, A., Arbel, T.: Hierarchical MRF and random forest segmentation of ms lesions and healthy tissues in brain MRI. In: The Longitudinal MS Lesion Segmentation Challenge (2015)
9. Krizhevsky, A., Sutskever, I., Hinton, G.E.: Imagenet classification with deep convolutional neural networks. In: Advances in Neural Information Processing Systems, pp. 1097–1105 (2012)
10. Long, J., Shelhamer, E., Darrell, T.: Fully convolutional networks for semantic segmentation. In: Proceedings of the IEEE Conference on Computer Vision and Pattern Recognition, pp. 3431–3440 (2015)
11. Maier, O., Handels, H.: MS lesion segmentation in MRI with random forests. In: Proceedings of the 2015 Longitudinal Multiple Sclerosis Lesion Segmentation Challenge, pp. 1–2 (2015)
12. Nair, V., Hinton, G.E.: Rectified linear units improve restricted Boltzmann machines. In: Proceedings of the 27th International Conference on Machine Learning (ICML 2010), pp. 807–814 (2010)
13. Ronneberger, O., Fischer, P., Brox, T.: U-Net: convolutional networks for biomedical image segmentation. In: Navab, N., Hornegger, J., Wells, W.M., Frangi, A.F. (eds.) MICCAI 2015. LNCS, vol. 9351, pp. 234–241. Springer, Cham (2015). https://doi.org/10.1007/978-3-319-24574-4_28
14. Steinman, L.: Multiple sclerosis: a coordinated immunological attack against myelin in the central nervous system. Cell **85**(3), 299–302 (1996)
15. Valverde, S., et al.: Improving automated multiple sclerosis lesion segmentation with a cascaded 3D convolutional neural network approach. NeuroImage **155**, 159–168 (2017)
16. Zeiler, M.D.: ADADELTA: an adaptive learning rate method. arXiv preprint arXiv:1212.5701 (2012)

Shallow vs Deep Learning Architectures for White Matter Lesion Segmentation in the Early Stages of Multiple Sclerosis

Francesco La Rosa[1,3(✉)], Mário João Fartaria[1,2,4], Tobias Kober[1,2,4],
Jonas Richiardi[2,4], Cristina Granziera[5,6], Jean-Philippe Thiran[1,4],
and Meritxell Bach Cuadra[1,3,4]

[1] LTS5, Ecole Polytechnique Fédérale de Lausanne, Lausanne, Switzerland
francesco.larosa@epfl.ch
[2] Siemens Healthcare AG, Lausanne, Switzerland
[3] Medical Image Analysis Laboratory, CIBM, University of Lausanne,
Lausanne, Switzerland
[4] Radiology Department, Lausanne University Hospital, Lausanne, Switzerland
[5] Translational Imaging in Neurology Basel,
Department of Medicine and Biomedical Engineering,
University Hospital Basel and University of Basel, Basel, Switzerland
[6] Neurologic Clinic and Policlinic, Departments of Medicine,
Clinical Research and Biomedical Engineering,
University Hospital Basel and University of Basel, Basel, Switzerland

Abstract. In this work, we present a comparison of a shallow and a deep learning architecture for the automated segmentation of white matter lesions in MR images of multiple sclerosis patients. In particular, we train and test both methods on early stage disease patients, to verify their performance in challenging conditions, more similar to a clinical setting than what is typically provided in multiple sclerosis segmentation challenges. Furthermore, we evaluate a prototype naive combination of the two methods, which refines the final segmentation. All methods were trained on 32 patients, and the evaluation was performed on a pure test set of 73 cases. Results show low lesion-wise false positives (30%) for the deep learning architecture, whereas the shallow architecture yields the best Dice coefficient (63%) and volume difference (19%). Combining both shallow and deep architectures further improves the lesion-wise metrics (69% and 26% lesion-wise true and false positive rate, respectively).

1 Introduction

Multiple Sclerosis (MS) is a demyelinating disease that affects the central nervous system. Demyelination results in focal lesions that appear with higher frequency in the white matter (WM). Magnetic resonance imaging (MRI) is a fundamental tool for MS diagnosis and monitoring of disease evolution as well as response to therapy. Currently, expert's manual annotations are considered the clinical gold standard for MS lesion identification. However, as this task is time-consuming

ⓒ Springer Nature Switzerland AG 2019
A. Crimi et al. (Eds.): BrainLes 2018, LNCS 11383, pp. 142–151, 2019.
https://doi.org/10.1007/978-3-030-11723-8_14

and prone to inter and intra-observer variations, many automated methods for MS lesion detection and segmentation have been proposed in the literature [1]. In this context, supervised techniques that learn and train from manually annotated examples have been proven to be the most successful in detection of MS WM lesions [2–4]. In the last years, deep learning architectures have achieved remarkable successes and have recently proven good performance in MS lesion segmentation as well [5–7].

In order to compare automated lesion segmentation methods, several computational imaging challenges have been proposed at international conferences [2–4], providing very valuable benchmark datasets for validation. However, these evaluation scenarios are based on patients with relatively high lesion load, and reported results are often computed on scans exhibiting relative large lesion sizes. Thus, the performance of deep supervised techniques on early stages of MS and at small lesion sizes remains to be proven.

In this work, we aim at comparing shallow with novel deep learning architectures using data from early stages of the disease in challenging conditions, i.e. exploring minimum lesion sizes as given by neuroradiological conventions [8] and even pushing the limit below.

To this end, we have selected two recently published MS segmentation methods. First, we have applied a supervised k-NN method combined with partial volume (PV) modeling [9,10], specifically developed on subjects with a low disease burden and small lesions. Second, we have used a recently and publicly available deep learning approach based on a cascade of two 3D patch-wise convolutional neural networks (CNNs) [5]. At the time of the writing of this work, this CNNs method achieved the best result on the MICCAI2008 and MSSEG2016 challenges [2,4] and competitive performance on other clinical datasets. Furthermore, we explore a straightforward prototype combination of these two methods. Both methods and their combination are trained on the same clinical dataset and validated on a pure test set. The results are analyzed considering different minimum lesion volume and total lesion load, as these are important evidences for early stage disease patients with low disabilities.

2   Methodology

2.1   Datasets

The study was approved by the Ethics Committee of our institution, and all patients gave written informed consent prior to participation. The training dataset was composed of 32 patients, 18 female/14 male, mean age 34 ± 10 years, with Expanded Disability Status Scale (EDSS) scores ranged from 1 to 2 (mean 1.6 ± 0.3). Mean lesion volume is 0.11 ± 0.40 ml (range 0.001–7.03 ml). Mean lesion load per case was 6.0 ± 7.2 ml (range 0.3–37.2 ml). MRI acquisitions were performed on a 3T MRI scanner (Magnetom Trio, Siemens Healthcare, Erlangen, Germany). Both 3D MPRAGE and 3D FLAIR were acquired with a resolution of $1 \times 1 \times 1.2$ mm^3.

The test dataset was made up of 73 patients, 50 females and 23 males (mean age 38 ± 10 years). EDSS scores ranged from 1 to 7.5 (mean 2.6 ± 1.5). Mean lesion volume was 0.25 ± 3.29 ml (range 0.002–159.827 ml). Mean lesion load per case was 14.3 ± 27.9 ml (range 0.2–162.9 ml). Both 3D MPRAGE and 3D FLAIR were acquired at $1 \times 1 \times 1 \, mm^3$ but with different Siemens scanners: 5 subjects at 1.5T with MAGNETOM Aera, and the other patients at 3T with either Prisma_fit, TrioTim, or Skyra systems.

Manual Segmentation: In the training set, MS lesions were detected by consensus by one radiologist and one neurologist, with respectively 6 and 11 years of experience. The lesion volumes were then delineated in each image by a trained technician. Testing set lesions segmentation was performed by the Medical Image Analysis Center-MIAC [11] based on a standardized semi-automated method and furhter experts quality check, which has been extensively applied to phase II and III clinical trials.

2.2   Pre-processing

The same pre-processing steps were applied to the training and testing datasets. First, the two image contrasts were rigidly registered to the same space (MPRAGE) using the ELASTIX C++ library [12]. Second, all cases were skull-stripped using BET [13] and bias-corrected using N4 [14,15].

2.3   LeMan-PV

LeMan-PV is a Bayesian PV estimation (PVE) algorithm, where spatial constraints for GM and lesions are included to drive the segmentation [10]. The spatial constraint for GM is an atlas-based probability map, and spatial constraints for lesions are derived from a kNN-supervised-based approach [16]. LeMan-PV has proven its good performance, and improvements as compared to state-of-the-art methods, in a leave-one-out experiments with MS patients with low lesion loads and small lesions. As in [9], initial mean tissue intensities and hyperparameters (symmetric penalty matrix A, and amount of spatial smoothness β) were set and a patient with relatively high lesion load chosen as a reference to train the PV estimation algorithm. Specifically, A coefficients were $a_1 = 11.25$, $a_4 = 14.33$, $a_5 = 0.47$, $a_6 = 12.21$, $a_7 = 1.33$, $a_8 = 16.93$, and $\beta = 0.5$. Patient mean intensities were set determined beforehand by histogram matching with the same reference patient used for hyperparameter setting [17].

2.4   CNNs

A novel MS segmentation method based on a cascade of two 3D patch-wise CNNs has recently been proposed [5]. The two networks have the same architecture and number of parameters, but don't share the same weights. Added to the above pre-processing steps, additional intensity normalization was performed, applying a histogram matching technique [17]. Afterwards, the first CNN receives as input patches of size $11 \times 11 \times 11$ from different MRI modalities, centered around a voxel of interest. Only voxels with a FLAIR intensity over a threshold optimized in the validation phase are considered. Lesion candidates from the first CNN are then given as input to the second one, which mainly has the task of reducing the false positives. In order to overcome the problem of data imbalance, before each CNN the negative class is undersampled, and the same number of positive and negative patches are obtained. Binary output masks are computed by linearly thresholding the probabilistic lesion masks given as output by the second network.

Table 1. Network architecture. c indicates the number of MRI modalities.

Layer	Type	Output size	Feature maps
0	Input	$c \times 11 \times 11 \times 11$	-
1	Convolutional	$32 \times 11 \times 11 \times 11$	32
2	Convolutional	$32 \times 11 \times 11 \times 11$	32
3	Max-pooling	$32 \times 5 \times 5 \times 5$	-
4	Convolutional	$64 \times 5 \times 5 \times 5$	64
5	Convolutional	$64 \times 5 \times 5 \times 5$	64
6	Max-pooling	$64 \times 2 \times 2 \times 2$	-
7	FC	256	256
8	Softmax	2	2

We have applied the same architecture [5] publicly available at [18] (see Table 1). Each convolutional layer is followed by a ReLU activation function and a batch normalization regularization. Dropout ($p = 0.5$) is applied before the first fully-connected layer. The networks were trained with the adaptive learning rate method (ADADELTA) [19], a batch size of 128, and early stopping as in the original paper. From the training dataset 7 cases were kept for validation, leaving 25 cases for training. With these the binarisation threshold was optimized considering equally the dice coefficient and the lesion false positive rate. In the original work [5] the CNNs were trained with 20 to 35 cases. Therefore, having a comparable number of patients for training, we hypothesize that this method should not perform worse in our study.

2.5   Combination of LeMan-PV with CNNs

It has been shown that for segmentation tasks, CNNs can benefit from prior probability maps fed in as an additional input channel [20,21]. Moreover, combining different classifiers has also been a successful technique for improving the final results in supervised learning in several works [22–24]. Here, we propose a naive prototype combination (PV-CNNs) of both approaches described above.

The concentration lesion maps generated by LeMan-PV are included as an additional input channel of the first CNN during training and testing. In this way, additional prior information on lesions was given to the network with the aim of improving the final segmentation.

3   Results

We compared the results of LeMan-PV, CNNs, and PV-CNNs strategies (see Fig. 1). In line with three MS lesion segmentation challenges [2–4], we computed the following evaluation metrics: overlap Dice coefficient (Dice), lesion-wise false positive (LFPR) and lesion-wise true positive (LTPR) rates, voxel-wise true positives (TP), and volume difference (VD), according to [3,25]. Rather than a leave-one-out analysis [5,10], we present our results on a pure testing set of 73 patients cases acquired with different scanners. These two factors allow us to evaluate the generalization of the proposed methods in a setting close to the clinical scenario (shown in Table 3).

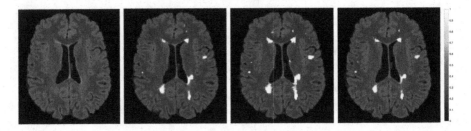

Fig. 1. Segmentation results (lesion probability (CNNs, PV-CNNs) and concentration (LeMan-PV)), from left to right: ground truth, LeMan-PV, CNNs, PV-CNNs. Reduction of LFPR is observed in PV-CNNs.

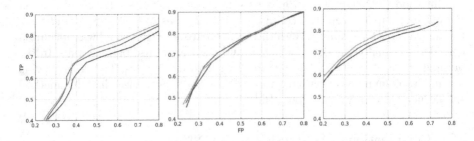

Fig. 2. ROC curves for different minimum lesion size: 5 (blue), 10 (orange), 15 (yellow) mm^3. From left to right: LeMan-PV, CNNs, PV-CNNs. (Color figure online)

Quantitative evaluation at different lesion sizes (5, 10, 15 mm^3) is given by ROC curves in Fig. 2. Both LeManPV and PV-CNNs performed better at bigger minimum lesion size. However, CNNs did not show this behavior in our cohort, presenting similar ROC curves for all minimum lesion sizes.

As in the original studies [5,9], in what follows, a minimum lesion size of $5\,mm^3$ is considered. Median values for the whole test dataset are reported in Table 2. LeManPV achieved the best Dice coefficient and volume difference. However, in terms of LFPR and LTPR, the CNNs performed better. The combination of the two methods outperformed them singularly in these lesion-wise metrics.

Table 2. Median values of the evaluation metrics for each method considered.

Method	Dice	LFPR	LTPR	TP	VD
LeMan-PV	**0.63**	0.37	0.57	**0.66**	**0.19**
CNNs	0.57	0.30	0.66	0.56	0.26
PV-CNNs	0.60	**0.26**	**0.69**	**0.66**	0.40

Segmentation results by Dice coefficient, TP, LFPR, and LTPR are given in the boxplots of Fig. 3. Results are split in groups of patients according to their total lesion volume (TLV). In agreement with [16], we considered a low (TLV < 5 ml), moderate (5 ml ≤ TLV ≤ 15 ml) and high (TLV > 15 ml) total lesion burden. Statistically significant differences between the methods are computed with Wilcoxon signed-rank test ($p < 0.05$ uncorrected). Interestingly, PV-CNNs achieved the best Dice coefficient for low and medium TLV, but its performance drops for high lesion load. We hypothesize that the lower number of cases in this category (only 15 patients) downgrades the classification results for CNNs weakness to statistics. Overall, besides the presence of some outliers, LeMan-PV and CNNs showed a similar behavior at low and medium lesion loads. Regarding the TP, there are not significant differences between the three TLV. On the other hand, the LFPR decreases for all methods as the TLV increases. This represents an understandable behavior, as higher lesion load cases are expected to be better segmented. Curiously, and opposite to TP, the LTPR follows a similar trend. However, as stated above, the low number of patients at highest lesion load prevents us from drawing conclusions.

Volume differences are given (top row, for low and medium TLV patients only, bottom row: all dataset) by Bland-Altman plots (Fig. 4). Slightly better results were obtained when combining both architectures, with a mean volume difference of −133.21 ml. However, a different behavior is shown when including the high TLV patients, with an increase of the mean volume difference to 3250, 6410 and 7483 ml for LeMan-PV, CNNs and PV-CNNs respectively.

Finally, the effect of the scanner type is briefly investigated. Table 3 shows the mean Dice coefficient for the four different scanner types used to acquire the testing cases. For all segmentation methods, the highest Dice coefficient is achieved for the cases acquired with the TrioTim scanner. However, in this work the number of cases for each scanner is highly unbalanced. Therefore, further studies, with enlarged datasets, will be needed to quantify accuracy versus scanner type.

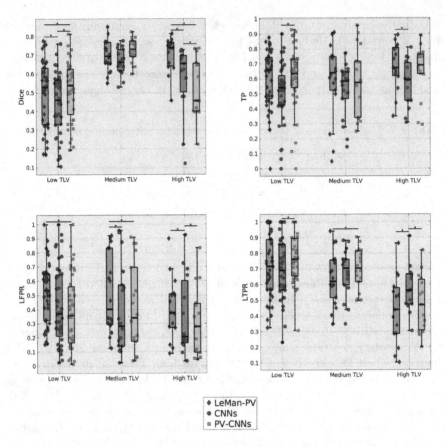

Fig. 3. Box plots of the Dice coefficient, TP, LFPR, and LTPR for the three methods considering the different TLV of the testing cases ($p < 0.05$ is indicated by *).

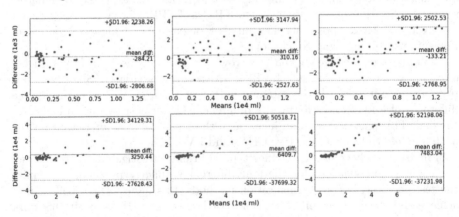

Fig. 4. Bland-Atlmann plots of low and medium TLV cases (top row) and of the whole dataset (bottom row) showing the volume differences of the three methods analyzed. From left to right: LeMan-PV, CNNs, PV-CNNs.

Table 3. Mean Dice coefficient of the testing cases for the different scanners they were acquired with.

Scanner	N. cases	Dice (range)		
		LeMan-PV	CNNs	PV-CNNs
Aera	5	0.59 (0.47–0.64)	0.47 (0.12–0.75)	0.52 (0.29–0.63)
TrioTim	6	0.63 (0.50–0.81)	0.61 (0.44–0.75)	0.65 (0.51–0.80)
Prisma_fit	11	0.54 (0.31–0.74)	0.48 (0.26–0.64)	0.53 (0.34–0.72)
Skyra	51	0.59 (0.16–0.84)	0.53 (0.11–0.78)	0.56 (0.19–0.80)

4   Conclusion

In this work, we presented the comparison of two of the most recent automated methods for WM lesions segmentation published in literature. In particular, we have tested a Bayesian partial volume estimation algorithm (LeMan-PV) [9] and a novel deep learning architecture based on a cascade of CNNs [5]. Both methods were tested on a pure test dataset composed of 73 cases, mainly belonging to early stage disease patients. The CNNs achieved the lowest LFPR of 30%. This confirms, as claimed in the original paper [5], that they are an effective method for reducing false positives. However, LeMan-PV showed the best segmentation results with the highest Dice coefficient (63%) and smallest volume difference (19%), indicating that PV might be still an asset for good delineation. Further analysis indicates a slight dependence of LeMan-PV performance on the minimum lesion size considered, whereas the CNNs didn't show this behavior. Furthermore, a combination of the two methods (PV-CNNs) was implemented. Providing the CNNs with the probability maps of the LeMan-PV improved the LFPR (26%) and LTPR (69%) but did not perform well in terms of VD. Those results confirm that the hybrid of the two methods is also effective for WM lesion segmentation of early stages disease cases. However, further improvements are needed to increase the segmentation accuracy of low lesion burden cases, in which these automated methods achieved the worst performance (median Dice around 0.5). These cases are indeed of great importance for detecting MS lesions in the early stages of the disease. Future work will include experimenting with advanced combinations of these methods, training and testing on different datasets, and verifying if the results depend on the scanner used.

Acknowledgements. The work is supported by the Centre d'Imagerie BioMédicale (CIBM) of the University of Lausanne (UNIL), the Swiss Federal Institute of Technology Lausanne (EPFL), the University of Geneva (UniGe), the Centre Hospitalier Universitaire Vaudois (CHUV), the Hôpitaux Universitaires de Genève (HUG), and the Leenaards and Jeantet Foundations. This project is also supported by the European Union's Horizon 2020 research and innovation program under the Marie Sklodowska-Curie project TRABIT (agreement No. 765148). CG is supported by the Swiss National Science Foundation grant SNSF Professorship PP00P3-176984.

References

1. Garcia-Lorenzo, D., et al.: Review of automatic segmentation methods of multiple sclerosis white matter lesions on conventional magnetic resonance imaging. Med. Image Anal. **17**(1), 1–18 (2013)
2. Styner, M., et al.: 3D segmentation in the clinic: a grand challenge II: MS lesion segmentation. MIDAS J. **2008**, 1–6 (2008)
3. Carass, A., et al.: Longitudinal multiple sclerosis lesion segmentation: resource and challenge. NeuroImage **148**, 77–102 (2017)
4. Commowick, O., Cervenansky, F., Ameli, R.: MSSEG challenge proceedings: multiple sclerosis lesions segmentation challenge using a data management and processing infrastructure. In: MICCAI 2016 (2016)
5. Valverde, S., et al.: Improving automated multiple sclerosis lesion segmentation with a cascaded 3D convolutional neural network approach. NeuroImage **155**, 159–168 (2017)
6. Brosch, T., et al.: Deep 3D convolutional encoder networks with shortcuts for multiscale feature integration applied to multiple sclerosis lesion segmentation. Trans. Med. Imaging **35**(5), 1229–1239 (2016)
7. Roy, S., et al.: Multiple sclerosis lesion segmentation from brain MRI via fully convolutional neural networks. arXiv:1803.09172 (2018)
8. Grahl, S., et al.: Defining a minimal meaningful lesion size in multiple sclerosis. Mult. Scler. J. **23**, P538-237 (2017)
9. Fartaria, M.J., et al.: Partial volume-aware assessment of multiple sclerosis lesions. NeuroImage: Clin. **18**, 245–253 (2018)
10. Fartaria, M.J., Roche, A., Meuli, R., Granziera, C., Kober, T., Bach Cuadra, M.: Segmentation of cortical and subcortical multiple sclerosis lesions based on constrained partial volume modeling. In: Descoteaux, M., Maier-Hein, L., Franz, A., Jannin, P., Collins, D.L., Duchesne, S. (eds.) MICCAI 2017. LNCS, vol. 10435, pp. 142–149. Springer, Cham (2017). https://doi.org/10.1007/978-3-319-66179-7_17
11. https://miac.swiss/en/
12. Klein, S., et al.: Elastix: a toolbox for intensity-based medical image registration. IEEE Trans. Med. Imaging **29**(1), 196–205 (2010)
13. Smith, S.M.: Fast robust automated brain extraction. Hum. Brain Mapp. **17**(3), 143–155 (2002)
14. Tustison, N.J., et al.: N4ITK: improved N3 bias correction. IEEE Trans. Med. Imaging **29**(6), 1310–1320 (2010)
15. Kikinis, R., Pieper, S.D., Vosburgh, K.G.: 3D slicer: a platform for subject-specific image analysis, visualization, and clinical support. In: Jolesz, F.A. (ed.) Intraoperative Imaging and Image-Guided Therapy, pp. 277–289. Springer, New York (2014). https://doi.org/10.1007/978-1-4614-7657-3_19. ISBN 978-1-4614-7656-6
16. Fartaria, M.J., et al.: Automated detection of white matter and cortical lesions in early stages of multiple sclerosis. J. Magn. Reson. Imaging **43**(6), 1445–1454 (2016)
17. Crimi, A., Commowick, O., Ferré, J.C., Maarouf, A., Edan, G., Barillot, C.: Semi-automatic classification of lesion patterns in patients with clinically isolated syndrome. In: 2013 IEEE 10th International Symposium on Biomedical Imaging (ISBI), pp. 1102–1105. IEEE, April 2013
18. https://github.com/sergivalverde/cnn-ms-lesion-segmentation
19. Zeiler, M.D.: ADADELTA: an adaptive learning rate method. arXiv:1212.5701 (2012)

20. Luo, K., et al.: A CNN-based segmentation model for segmenting foreground by a probability map. In: Intelligent Signal Processing and Communication Systems (ISPACS), IEEE ISBI 2017 (2017)
21. Zotti, C., et al.: GridNet with automatic shape prior registration for automatic MRI cardiac segmentation. arXiv:1705.08943 (2017)
22. Kotsiantis, S.B., Zaharakis, I.D., Pintelas, P.E.: Machine learning: a review of classification and combining techniques. Artif. Intell. Rev. 26(3), 159–190 (2006)
23. Fartaria, M.J., et al.: An ensemble of 3D convolutional neural networks for central vein detection in white matter lesions. In: MIDL 2018 Abstract Submission (2018)
24. Kamnitsas, K., et al.: Ensembles of multiple models and architectures for robust brain tumour segmentation. arXiv:1711.01468 (2017)
25. Geremia, E., Clatz, O., Menze, B.H., Konukoglu, E., Criminisi, A., Ayache, N.: Spatial decision forests for MS lesion segmentation in multi-channel magnetic resonance images. NeuroImage 57, 378–390 (2011)

Detection of Midline Brain Abnormalities Using Convolutional Neural Networks

Aleix Solanes[1,2]([envelope]), Joaquim Radua[1], and Laura Igual[3]

[1] FIDMAG Research Foundation, Barcelona, Spain
al.solanes@gmail.com
[2] Department of Psychiatry and Forensic Medicine,
Autonomous University of Barcelona, Barcelona, Spain
[3] Department of Mathematics and Computer Science,
University of Barcelona, Barcelona, Spain

Abstract. Patients with mental diseases have an increased prevalence of abnormalities in midline brain structures. One of these abnormalities is the cavum septum pellucidum (CSP), which occurs when the septum pellucidum fails to fuse. The detection and study of these brain abnormalities in Magnetic Resonance Imaging requires a tedious and time-consuming process of manual image analysis. It is also problematic when the same abnormality is analyzed manually by different experts because different criteria can be applied. In this context, it would be useful to develop an automatic method for locating the abnormality and give the measure of its depth. In this work, we explore, for the first time in the literature, an automated detection method based on CNNs. In particular, we compare different CNN models and classical machine learning classification algorithms to face this problem on a dataset of 861 subjects (639 patients with mood or psychotic disorders and 223 healthy controls) and obtain very promising results, reaching over 99% of accuracy, sensitivity and specificity.

Keywords: CSP · CNN · Deep learning · Brain · MRI

1 Introduction

Different studies have demonstrated that the prevalence of abnormalities in midline brain structures in patients with schizophrenia, as well as other mood and psychotic disorders, is increased with respect to healthy patients [1]. Concretely, in this work, we focus on a variant of cerebrospinal fluid (CSF) space formed between the leaflets of the septum pellucidum. This zone is called the cavum septum pellucidum (CSP) (see Fig. 1). During fetal development both laminae tend to fuse in an anterior to posterior manner between the third and the sixth month of life, when these laminae do not fuse completely, the space in-between both laminae is the CSP. The importance of this finding can help to diagnose and study these disorders in terms of neurodevelopmental etiology. In other

© Springer Nature Switzerland AG 2019
A. Crimi et al. (Eds.): BrainLes 2018, LNCS 11383, pp. 152–160, 2019.
https://doi.org/10.1007/978-3-030-11723-8_15

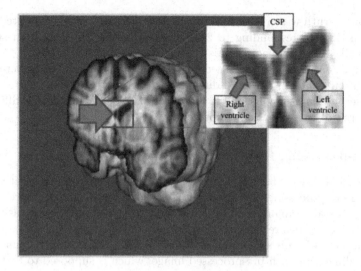

Fig. 1. Cavum Septum Pellucidum (CSP) seen from the anterior section.

words, the study of this abnormalities could help to investigate the causation or origination of these pathologies, like schizophrenia or bipolar disorders.

The presence of a CSP with a length from anterior to posterior part of the brain of around 1–1.4 mm is considered normal anatomy of the brain, with an incidence of 60–80% [2], so a larger cavity is considered as an abnormality.

Manually detecting this kind of abnormalities is a tedious task, and defining accurately the depth of this cavity can be quite controversial. Thus, a fully-automated method would help researchers and clinical professionals to study these abnormalities in detail. Up to our knowledge, there is no work in the literature regarding this issue. In this work, we propose different deep and shallow machine learning methods to detect CSP slide-by-slide in Magnetic Resonance Images (MRI). In particular, we implement two Convolutional Neural Networks (CNNs) models with different architectures and a variant adding contextual information in the input to the network. We also use different classical shallow machine learning classification algorithms with different dimensionality reduction techniques, in order to compare the performances.

2 Data and Pre-processing

2.1 Data

In this study, the data consists in 861 subjects, from which 639 are patients with schizophrenia, bipolar disorder and other psychotic or mood disorders, and 223 are healthy controls. We use T1-weighted MRI. All subjects had been scanned with a 1.5 Tesla GE Signa scanner (General Electric Medical Systems) located at Sant Joan de Déu Hospital in Barcelona.

A ground truth was created specifically for this study which consisted in 888 slices with CSP corresponding to 213 subjects and 26510 slices from subjects without CSP. As can be noted, this is a highly unbalanced problem.

A second dataset has been used to test the methods in a completely new dataset, from a different site and different scanner.

This second dataset consisted in 500 slices of 10 subjects from OASIS (Open Access Series of Imaging Studies) project[1].

2.2 Pre-processing

Given the small size of the CSP with respect to the whole MRI image, it is useful to define a pre-processing pipeline in order to extract a Region of Interest (ROI) containing the abnormality and discard the areas not useful for the task.

All MRI volumes are pre-processed using the same pipeline automatically. The main goal is to obtain an image that will have the posterior genus at the same coordinate for each pre-processed image, which is supposed to be the first slice of the CSP. The pipeline contains the following steps:

- **Skull-stripping:** We remove the skull using FSL-BET [7] (Brain Extraction Tool) with a fractional intensity threshold of 0.4. This is a necessary preparation step for the registration and segmentation steps using FSL guidelines[2].
- **Registration:** We transform all images in order that they share a common coordinate system using FLIRT [8] (FMRIB's Linear Image Registration Tool).
- **Segmentation of brain tissues and spatial intensity correction:** We segment the different tissues (white matter, grey matter and CSF). We first use white matter segmentation to find the posterior genus. This structure indicates the beginning of the CSP, and according to our clinical team, should not be deeper than 50 slices. Finally, we apply FSL-FAST[3] which uses a hidden Markov random field and an associated expectation-maximization algorithm to correct for spatial intensity variations caused by RF inhomogeneities.
- **Region of Interest definition:** Given the first slide, we define a volume of $50 \times 30 \times 30$ voxels beginning at the posterior genus and defining 50 slices from the anterior to the posterior part of the brain, where the CSP should appear.
- **Intensity normalization:** The standardization of a dataset is a common requirement for many machine learning estimators. Many estimators expect individual features to look like standard normally distributed data. Thus we normalize the intensities per patient to be all within the same range.
- **Slicing to 2D images:** We extract 2D 50×30 coronal slices.

Finally, to increase the number of samples of this CSP class (underrepresented), we flip the image in both horizontal and vertical axes. This *data*

---

[1] https://www.oasis-brains.org/.

[2] https://fsl.fmrib.ox.ac.uk/fslcourse/lectures/reg.pdf.

[3] https://fsl.fmrib.ox.ac.uk/fsl/fslwiki/FAST.

augmentation procedure was consensuated with clinical experts from FIDMAG Research Foundation[4].

3   Methods

Given the 2D ROI images we propose to classify them as containing CSP or not using different machine learning methods.

3.1   Shallow Machine Learning Methods

Shallow machine learning methods generally use a first step to define features from the images, and before applying the classification algorithm. Next, we detail the dimensionality reduction and classification algorithms used in this work.

In order to reduce the dimensionality of the data, we consider two different approaches, feature selection and feature extraction. The former, selects the more important features of the image, and the latter builds derived features to better classify data.

We apply *Extremely Randomized Trees* (extra-trees) classifier [6] as a feature selection algorithm. In order to extract meaningful features we use *Principal Component Analysis* (PCA) and *Histograms of Oriented Gradients* (HOG).

We compare the performance of the following four shallow classification algorithms, by using the features extracted with both PCA and HOG: *K Nearest Neighbors* (KNN), *Linear Support Vector Classifier* (Linear SVC), *Random Forests* and *Linear Discriminant Analysis* (LDA).

We implement these methods using scikit-learn[5] (Python), opencv[6] and nilearn[7].

3.2   Convolutional Neural Networks Models

Deep machine learning methods directly learn the features from the images. In this study, we consider two different 2D CNN models which receive RAW ROIs as input instead of features extracted from the images [5].

We explore two different approaches, illustrated on Figs. 2 and 3. The first model stacks two convolution layers with a 3×3 kernel at the beginning (see Fig. 2), while the second one only implements one convolutional layer (see Fig. 3). Both models differ also in an extra dropout layer implemented before the last fully connected layer on the first CNN.

Note that the definition of the field of view in the ROI can be decisive for the classifier performance because of the large size of the brain, when comparing to the small size of the region we are interested in. Thus, the definition of a ROI

---

[4] https://fidmag.com.
[5] http://scikit-learn.org.
[6] https://opencv.org/.
[7] http://nilearn.github.io/.

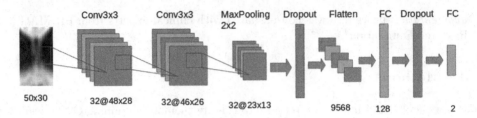

Fig. 2. First CNN model

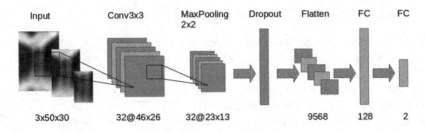

Fig. 3. The second CNN model with 3 levels of zoom as input.

that reduces the quantity of unnecessary information could benefit our method performance [9].

In order to learn from contextual information of the CSP, we propose to change the input of the CNN model by adding three different channels, containing three different levels of zooming of the ROI, see Fig. 4.

We expect that this information can help to better detect the CSP.

In addition, we test to add these three different levels of zooming of the ROI as three different layers, and join them in a *merge* layer before the first convolutional layer of the model. We only extend the second model, since in our experimental analysis, it was the best performing method and we pretended the contextual information to improve only the best CNN model. Figure 3 shows the architecture of the second model with 3 levels of zoom as input.

We implemented the different models using the framework Keras[8] (in Python).

4   Experimental Results

In this section we present the experimental results of the classification of the 2D images as containing CSP or not using the proposed machine learning methods.

In Table 1, we compare the results of the two CNN models, the variant of the second model, and different common machine learning algorithms. As can be seen, the results of the three CNN models are equivalent, although the last variant gives slightly better sensitivity and specificity (Table 2).

---

[8] https://github.com/fchollet/keras.

Fig. 4. Example of three levels of zoom used in second CNN model.

Table 1. Results on first dataset. Images from same scanner.

Method	Accuracy	Sensitivity	Specificity
EXTRA-TREES	0.965	0.94	0.99
LDA (PCA)	0.924	0.94	0.91
LDA (HOG)	0.957	0.96	0.95
KNN (PCA)	0.983	0.99	0.98
KNN (HOG)	0.991	0.99	0.99
LSVC (PCA)	0.920	0.92	0.92
LSVC (HOG)	0.955	0.96	0.95
ADABOOST (PCA)	0.937	0.94	0.93
ADABOOST (HOG)	0.925	0.93	0.92
RANDOM FORESTS (PCA)	0.968	0.96	0.98
RANDOM FORESTS (HOG)	0.920	0.93	0.91
CNN 1	0.976	0.98	0.97
CNN 2	0.981	0.98	0.98
CNN 2 (3 zoom levels as channels)	0.988	0.99	0.99
CNN 2 (3 zoom levels as layers)	0.974	0.99	0.96

Table 2. Results on second dataset. Images from different scanner.

Method	Accuracy	Sensitivity	Specificity
KNN (HOG)	0.801	0.809	0.794
CNN 2 (3 zoom levels as channels)	0.958	0.96	0.949

The best model is CNN 2 with 3 zoom levels as channels. This means that adding contextual information is important to improve the performance of the classifier.

In Fig. 5, we show qualitative results of True Positives (TP), True Negatives (TN), False Positives (FP) and False Negatives (FN) for the KNN method and

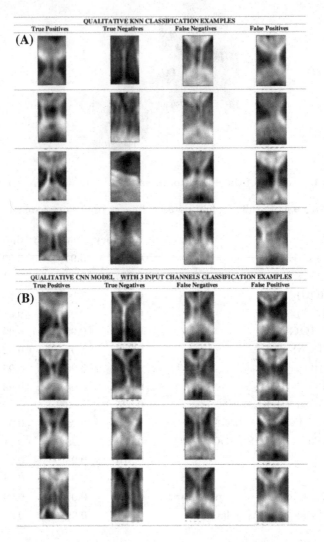

Fig. 5. (A) Top: Qualitative results of KNN. (B) Down: Qualitative results of best CNN

the best CNN method, respectively. As it can be seen in the examples, the definition of the CSP is not a simple task due to the fact that the boundaries are difficult to discriminate.

Our CNN method is able to detect the presence of CSP from front to rear, in axial projection of a MRI image. These approaches can classify almost perfectly slices where the CSP can be easily seen by an expert, like in the two first images in Fig. 5A. The model failed on certain small or thick CSP, concretely at first or last slices of the CSP, like the all images in column *false positives* in Fig. 5A.

Note that KNN and the best CNN approach reached 99% of accuracy on the first dataset, however they differ in the type of errors they make. CNNs tend to fail on images where the CSP is difficult to identify, like in the image in last column second row from Fig. 5B; whereas KNN seems to fail also on images easier to identify, like in third image in first and last rows in Fig. 5B. Despite having high accuracies, KNN method gives some false positives and false negatives in axial slices of the brain volume where it is anatomically impossible that CSP appears. Moreover, KNN, unlike CNNs, is highly dependent on the preprocessing (see Table 1, where KNN accuracy decreased 19% with respect to the results obtained on first dataset, whereas CNN only decreased 4%) and it is sensible to noise and local structure of the data.

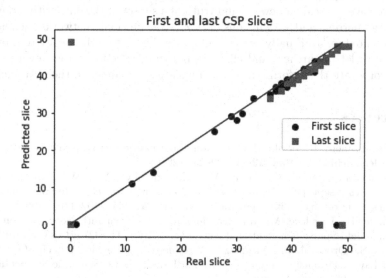

Fig. 6. First and last slice of CSP predicted vs. real in a regression plot. Variance of 0.8.

The validity of the model can be seen in Fig. 6, where we used 90 subjects to compare the results of our model with the results manually annotated by an expert. We annotated both the first and last slice of the CSP and placed then in a graphic comparing the predicted ones with the result expected. A perfect model prediction would place all slices over the diagonal line, and except from a few outliers (false positives or false negatives), according to Fig. 6, the predicted slices are close to the expert manual annotation.

5   Conclusions and Future Work

In this work, we studied, for the first time in the literature, the automatic detection of midline brain abnormalities in MRI slide-by-slide. For this purpose, we compared three different approaches based on CNNs with other classical shallow machine learning methods. KNN and the best CNN approach reached 99% of accuracy when training and validation images come from the same scanner, but KNN decrease the accuracy up to a 80.1% when classifying images from a different site, while CNN only decreased 4% in this scenario.

Up to our knowledge, there are no studies, in the literature, with such an amount of patients and studying the relation of this volume with mental disorders. In a further study, we want to face the problem by using 3D segmentation algorithms, such as the one in [3].

We expect the use of a 3D model will help to take into account spatial coherence information that can give robustness to results. We also plan to face the segmentation problem to obtain information of the volume of the CSP. For that, we have to create a new ground truth of a subset of the data with manually delineated borders and we will consider fully-convolutional networks such as the one presented in [4]. Finally, we will integrate the best methodology to a public software to let researchers and clinical professionals conduct their own studies and incorporate this information in a translational manner to their patients.

References

1. Landin-Romero, R., et al.: Midline brain abnormalities across psychotic and mood disorders. Schizophr. Bull. **42**(1), 229–238 (2016)
2. Nopoulos, P., Swayze, V., Flaum, M., Ehrhardt, J.C., Yuh, W.T.C., Andreasen, N.C.: Cavum septi pellucidi in normals and patients with schizophrenia as detected by magnetic resonance imaging. Biol. Psychiatry **41**(11), 1102–1108 (1997)
3. Graham, B., Engelcke, M., van der Maaten, L.: 3D Semantic Segmentation with Submanifold Sparse Convolutional Networks. arXiv:1711.10275v1 cs.CV (2017)
4. Jégou, S., Drozdzal, M., Vazquez, D., Romero, A., Bengio, Y.: The One Hundred Layers Tiramisu: Fully Convolutional DenseNets for Semantic Segmentation. arXiv:1611.09326 cs.CV (2016)
5. Havaei, M., Guizard, N., Larochelle, H., Jodoin, P.-M.: Deep learning trends for focal brain pathology segmentation in MRI, arXiv (2016)
6. Geurts, P., Ernst, D., Wehenkel, L.: Extremely randomized trees. Mach. Learn. **63**(1), 3–42 (2006)
7. Smith, S.M.: Fast robust automated brain extraction. Hum. Brain Mapp. **17**(3), 143–155 (2002)
8. Jenkinson, M., Bannister, P.R., Brady, J.M., Smith, S.M.: Improved optimisation for the robust and accurate linear registration and motion correction of brain images. NeuroImage **17**(2), 825–841 (2002)
9. Ghafoorian, M., et al.: Location sensitive deep convolutional neural networks for segmentation of white matter hyperintensities. Sci. Rep. **7**(1), 5110 (2017)

Deep Autoencoding Models for Unsupervised Anomaly Segmentation in Brain MR Images

Christoph Baur[1]([✉]), Benedikt Wiestler[2], Shadi Albarqouni[1],
and Nassir Navab[1,3]

[1] Computer Aided Medical Procedures (CAMP), TU Munich, Munich, Germany
c.baur@tum.de
[2] Department of Diagnostic and Interventional Neuroradiology,
Klinikum rechts der Isar, TU Munich, Munich, Germany
[3] Whiting School of Engineering, Johns Hopkins University, Baltimore, USA

Abstract. Reliably modeling normality and differentiating abnormal appearances from normal cases is a very appealing approach for detecting pathologies in medical images. A plethora of such unsupervised anomaly detection approaches has been made in the medical domain, based on statistical methods, content-based retrieval, clustering and recently also deep learning. Previous approaches towards deep unsupervised anomaly detection model local patches of normal anatomy with variants of Autoencoders or GANs, and detect anomalies either as outliers in the learned feature space or from large reconstruction errors. In contrast to these patch-based approaches, we show that deep spatial autoencoding models can be efficiently used to capture normal anatomical variability of entire 2D brain MR slices. A variety of experiments on real MR data containing MS lesions corroborates our hypothesis that we can detect and even delineate anomalies in brain MR images by simply comparing input images to their reconstruction. Results show that constraints on the latent space and adversarial training can further improve the segmentation performance over standard deep representation learning.

1 Introduction

Brain MR images are frequently acquired for detecting and diagnosing pathologies, monitoring disease progression and treatment planning. The manual identification and segmentation of pathologies in brain MR data is a tedious and time-consuming task. In an attempt to aid the detection and delineation of brain lesions arising from Multiple Sclerosis (MS), tumors or ischemias, the medical image analysis community has proposed a great variety of methods. Outstanding levels of performance have been achieved with recent supervised deep learning methods. However, their training requires vast amounts of labeled data which often is not available. Further, these approaches suffer from limited generalization since in general, training data rarely comprises the gamut of all possible

© Springer Nature Switzerland AG 2019
A. Crimi et al. (Eds.): BrainLes 2018, LNCS 11383, pp. 161–169, 2019.
https://doi.org/10.1007/978-3-030-11723-8_16

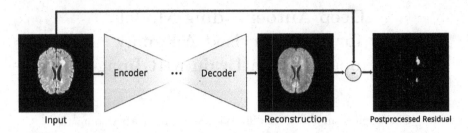

Fig. 1. The proposed anomaly detection concept at a glance. A simple subtraction of the reconstructed image from the input reveals lesions in the brain.

pathological appearances [17]. Given the constrained anatomical variability of the healthy brain, an alternative approach is to model the distribution of healthy brains, and both detect and delineate pathologies as deviations from the norm. Here, we formulate the problem of brain lesion detection and delineation as an unsupervised anomaly detection (UAD) task based on state-of-the-art deep representation learning, requiring only a set of normal data and no segmentation-labels at all. The detection and delineation of pathologies are thereby obtained from a pixel-wise reconstruction error (Fig. 1). To the best of our knowledge, this is the first application of deep convolutional representation learning for UAD in brain MR images which operates on entire MR slices at full resolution.

Related Work. In the medical field, many efforts have been made towards UAD, which can be grouped into methods based on statistical modeling, content-based retrieval or clustering and outlier detection [17]. Weiss et al. [19] employed Dictionary Learning and Sparse Coding to learn a representation of normal brain patches in order to detect MS lesions. Other unsupervised MS lesion segmentation methods rely on thresholding and 3D connected component analysis [6] or fuzzy c-means clustering with topology constraints [16]. Notably, only few approaches have been made towards deep learning based UAD. Vaidhya et al. [18] utilized unsupervised 3D Stacked Denoising Autoencoders for patch-based glioma detection and segmentation in brain MR images, however only as a pre-training step for a supervised model. Recently, Schlegl et al. [13] presented the AnoGAN framework, in which they create a rich generative model of normal retinal Optical Coherence Tomography (OCT) patches using a Generative Adversarial Network (GAN). Assuming that the model cannot properly reconstruct abnormal samples, they classify query patches as either anomalous or normal by trying to optimize the latent code of the GAN based on a novel mapping score, effectively also leading to a delineation of the anomalous region in the input data. In earlier work, Seeböck et al. [14] trained an Autoencoder and utilized a one-class SVM on the compressed latent space to distinguish between normal and anomalous OCT patches. A plethora of work in the field of deep learning based UAD has been devoted to videos primarily based on Autoencoders (AEs) due to their ability to express non-linear transformations and the ability to detect anomalies directly from poor reconstructions of input data [2,4,12].

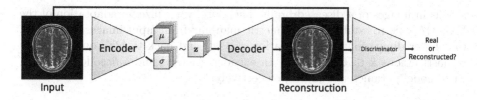

Fig. 2. An overview of our VAE-GAN for anomaly segmentation

Very recently, first attempts have also been made with deep generative models such as Variational Autoencoders [1,7] (VAEs), however limited to dense neural networks and 1D data. Noteworthy, most of this work focused on the detection rather than the delineation of anomalies.

A major advantage of AEs is their ability to reconstruct images with fairly high resolution thanks to a supervised training signal coming from the reconstruction objective. Unfortunately, they suffer from memorization and tend to produce blurry images. Unconditional GANs [3] have shown to produce very sharp images from random noise thanks to adversarial training, however the training is very unstable and the generative process is prone to mode collapse. VAEs have also shown that AEs can be turned into generative models, and both concepts have also been combined into the VAE-GAN [8] and α-GAN [11], yielding frameworks with the best of both worlds.

Contribution. Inarguably, AnoGAN is a great concept for UAD in patch-based and small resolution scenarios, but as our experiments show, unconditional GANs lack the capability to reliably synthesize complex, high resolution brain MR images. Further, the approach requires a time-consuming iterative optimization of the latent code. To overcome these issues, we propose to utilize deep convolutional autoencoders to build models that capture "global" normal anatomical appearance rather than a variety of local patches. In order to determine the benefits of mapping healthy anatomy to a well-structured, latent manifold, we also employ the VAE. In our experiments, we first compare dense and spatial variants of AEs and VAEs in the task of unsupervised MS lesion delineation and report significant improvements of spatial autoencoding models over traditional ones. In addition, we further augment the spatial variants with an adversarial network to improve realism of the reconstructed samples, ultimately turning the models into an AE-GAN [9] and a novel spatial VAE-GAN [8]. With the help of adversarial training, we notice additional minor, but insignificant improvements.

2 Methodology

As a novelty in this work, we employ deep generative representation learning to model the distribution of the healthy brain, which should enable the model to fully reconstruct healthy brain anatomy while failing to reconstruct anomalous lesions in images of a diseased brain. Therefore, we utilize an adaptation of the VAE-GAN [8] to establish a parametric mapping from input images $\mathbf{x} \in \mathbb{R}^{H \times W}$ to a lower dimensional representation $\mathbf{z} \in \mathbb{R}^d$ and back to high quality image reconstructions $\hat{\mathbf{x}} \in \mathbb{R}^{H \times W}$ using an encoder $\mathrm{Enc}(\cdot; \theta)$ and a decoder $\mathrm{Dec}(\cdot; \phi)$, with model parameters θ and ϕ, respectively:

$$\mathbf{z} \sim \mathrm{Enc}(\mathbf{x}; \theta), \quad \hat{\mathbf{x}} = \mathrm{Dec}(\mathbf{z}; \phi), \quad \text{s.t.} \quad \mathbf{z} \sim \mathcal{N}(0, I) \tag{1}$$

Like in [8], the latent space $\mathbf{z}$ is constrained to follow a multivariate normal distribution (MVN) $\mathcal{N}(0, I)$, which we leverage for encoding images of normal brain anatomy. Further, we employ a discriminator network $\mathrm{Dis}(\cdot; \psi)$ with model parameters ψ which classifies its input as either real or reconstructed.

Training. We optimize the framework using two loss functions in an alternating fashion. The parameters of the VAE component of the model are optimized using:

$$\begin{aligned} \mathcal{L}_{VAE} &= \lambda_1 \mathcal{L}_{rec} + \lambda_2 \mathcal{L}_{prior} + \lambda_3 \mathcal{L}_{adv} \\ &= \lambda_1 \|\mathbf{x} - \hat{\mathbf{x}}\|_1 + \lambda_2 \mathcal{D}_{KL}(\mathbf{z} \| \mathcal{N}(0, I)) - \lambda_3 \log(\mathrm{Dis}(\hat{\mathbf{x}})) \end{aligned} \tag{2}$$

The discriminator parameters are trained as commonly seen in the GAN framework [3]:

$$\mathcal{L}_{Dis} = -\log(\mathrm{Dis}(\mathbf{x})) - \log(1 - \mathrm{Dis}(\hat{\mathbf{x}})), \tag{3}$$

Originally, VAE-GAN used an abstract reconstruction loss on the latent space of the discriminator Dis rather than a pixelwise reconstruction objective $\mathcal{L}_{rec}$, which was not helpful for our purpose. For $\mathcal{L}_{rec}$, we thus used the pixelwise ℓ_1-distance between input image and reconstruction. $\mathcal{L}_{prior}$ is the KL-Divergence between the distribution of generated $\mathbf{z}$ and a MVN, which is only used to regularize the weights θ of the encoder. The third part $\mathcal{L}_{adv}$ is the adversarial loss which forces the decoder to generate images that are likely to fool the discriminator in its task to distinguish between real and reconstructed images. Both $\mathcal{L}_{VAE}$ and $\mathcal{L}_{Dis}$ are used for optimization in an alternating manner, i.e. for every training batch, first the discriminator is trained, then the encoder-decoder is updated to produce more realistic reconstructions.

A peculiarity of our approach is the fully convolutional encoder-decoder architecture which we use in order to preserve spatial information in the latent space, i.e. $\mathbf{z} \in \mathbb{R}^{h \times w \times c}$ is a multidimensional tensor. Figure 2 shows our VAE-GAN, and a depiction of different AE architectures which we also compare is given in Fig. 3.

Distinction from Other Autoencoding Models. Setting $\lambda_3 = 0$ in Eq. 2 ultimately turns the framework into a VAE. Further, setting $\lambda_2 = 0$ and replacing the stochastic bottleneck $\mathbf{z} \sim \mathrm{Enc}(\mathbf{x}; \theta)$ with a deterministic $\mathbf{z} = \mathrm{Enc}(\mathbf{x}; \theta)$ directly regressed by the encoder yields a normal AE. Note that for both the AE and VAE, no discriminator network is required.

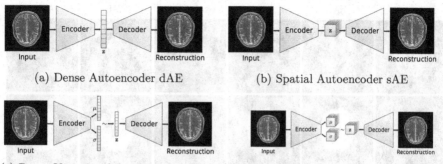

(a) Dense Autoencoder dAE   (b) Spatial Autoencoder sAE

(c) Dense Variational Autoencoder dVAE (d) Spatial Variational Autoencoder sVAE

Fig. 3. An overview of different Autoencoder frameworks

Anomaly Detection. Once a model is trained, anomalies are delineated by (1) computing the pixelwise ℓ_1-distance between an input image and its reconstruction, (2) applying a median filter to the resulting residual image to remove tiny structures and (3) thresholding the filtered image to obtain a binary segmentation.

3   Experiments and Results

Given the variants of AE and our proposed framework, we investigate (i) whether autoencoding deep networks can be utilized in general to learn to reconstruct complex brain MR images, (ii) how the dimensionality of $\mathbf{z}$ affects the reconstruction capabilities of a model, (iii) the effect of constraining $\mathbf{z}$ to be well structured and (iv) if adversarial training enhances the quality of reconstructed images. In the following paragraphs we first introduce the dataset, provide implementational details and then describe the experiments.

Datasets. For our experiments, we use an inhouse dataset which provides a rich variety of images of healthy brain anatomy - a necessity for our approach. The dataset consists of FLAIR images from 83 subjects with healthy brains (training) and 49 subjects with MS lesions (testing) acquired with a Philips Achieva 3T scanner. All images have been co-registered to the SRI24 ATLAS [10] to reduce appearance variability and skull-stripped with ROBEX [5]. The resulting images have been denoised using CurvatureFlow [15] and normalized into the range [0,1]. In order to obtain sufficient reconstruction quality when training the models, it was necessary to narrow the view on a region of the brain and thus, per subject, we focused on 20 consecutive axial slices (256×256px) around the midline.

Implementation. We build upon the basic architecture proposed in [8] and perform only minor modifications affecting the latent space (see Table 1). Across different architectures we keep the model complexity of the encoder-decoder part the same to allow for a valid comparison. All models have been trained for

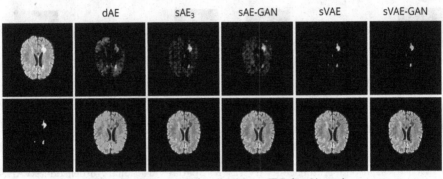

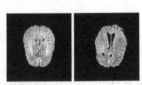

Fig. 4. 1st Column: a selected axial slice and its ground-truth segmentation; Succeeding columns show the filtered difference images (top row) and the resulting segmentation augmented to the input image (bottom row) for the following models defined in Table 1 (in order): dAE, sAE$_3$, sAE-GAN, sVAE and sVAE-GAN.

150 epochs in minibatches of size 8, using a learning rate of 0.001 for the reconstruction objective and 0.0001 for the adversarial training on a single nVidia 1080Ti GPU with 8 GB of memory. Thanks to the reconstruction objective, the training of both the AE-GAN and VAE-GAN was very stable and none of the models collapsed.

Evaluation Metrics. We measure the performance of the different models by the mean and standard deviation of the Dice-Score/F1-Score across different testing patients, the Area under the Precision-Recall Curve (AUPRC) as well as the average segmentation time per slice.

3.1 Anomaly Detection

We first trained normal convolutional AE & VAE with a dense latent space of dimensionality 512 and found that, besides not being capable of reconstructing brain lesions, they also lack the capability to reconstruct fine details such as the brain convolutions (Fig. 4). Similar to [2,4], we then make the architecture fully convolutional to ensure that spatial information is not lost in the bottleneck of the model. Notably, this heavily increases the dimensionality of $\mathbf{z}$. We thus vary the number of feature maps of the spatial AE to investigate the impact on reconstruction quality of normal and anomalous samples. We identify $\mathbf{z} = 16 \times 16 \times 64$ as a good parameterization and use it in further experiments on a spatial VAE, a spatial AE-GAN [9] and a spatial VAE-GAN. Further, we also trained an AnoGAN which we had to stop and evaluate after 82 epochs of training due to occuring instabilities (see

Fig. 5. Realistic (left) and unrealistic (right) samples generated with AnoGAN.

Fig. 5 for unrealistic samples produced by AnoGAN after further epochs). The required iterative reconstruction of testing samples was computed in 100 steps.

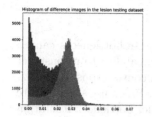

Fig. 6. The histogram of residuals for normal (blue) and anomalous (red) pixels using our VAE-GAN. (Color figure online)

Postprocessing. After reconstruction of all the slices, we apply some postprocessing steps to reduce the number of False Positives. For every patient, we apply a $5 \times 5 \times 5$ median filter to the reconstructed subvolumes to filter out small residuals, usually belonging to brain convolutions. Further, we multiply the residuals with slightly eroded brain masks to remove skull stripping artifacts, threshold the resulting volumes to obtain a binary segmentation mask and remove tiny 3D connected components with an area less than 6 voxels as they are unlikely to be lesions. The threshold is model specific and determined as the 98th percentile of the models reconstruction errors on the training dataset. We chose this percentile empirically from the histogram of residuals obtained from both normal and abnormal data (Fig. 6). The performance of each model is reported in Table 1. A comparison of processed residual images and final segmentations of various models can be seen in Fig. 4.

3.2   Results

The highest AUPRC has been obtained with the VAE-GAN, closely followed by the AE-GAN. The spatial VAEs and AEs which do not leverage adversarial training produce only slightly inferior scores, however. All spatial autoencoding models significantly outperform the ones with a dense bottleneck and, except for sAE_1, also the AnoGAN, though. Expectedly, the DICE-score is not necessarily

Table 1. Results of our experiments on unsupervised MS lesion segmentation. We report the Dice-Score (mean and std. deviation across patients) as well as the avg. reconstruction time per sample in seconds. Prefixes d or s stand for dense or spatial.

Modeltype	z	DICE ($\mu \pm \sigma$)	AUPRC	Avg. Reco.-time [s]
dAE	512	0.1276 ± 0.1461	0.3575	0.0128
sAE_1	$8 \times 8 \times 64$	0.1973 ± 0.1906	0.3227	0.0121
sAE_3	$16 \times 16 \times 64$	0.5855 ± 0.1984	0.6813	0.0118
sAE-GAN [9]	$16 \times 16 \times 64$	0.5263 ± 0.1978	0.6988	0.0144
dVAE	512	0.1661 ± 0.1779	0.3229	0.0108
sVAE	$16 \times 16 \times 64$	0.5922 ± 0.1958	0.6890	0.0129
sVAE-GAN	$16 \times 16 \times 64$	$\mathbf{0.6050 \pm 0.1927}$	**0.6906**	0.0154
AnoGAN [13]	64	0.3748 ± 0.2192	0.4178	19.8547

in line with the reported AUPRCs since the 98th percentile is not guaranteed to be a good threshold for every model.

4   Discussion and Conclusion

Our experiments show that AE & VAE models with dense bottlenecks cannot reconstruct anomalies, but at the same time lack the capability to reconstruct important fine details in brain MR images such as brain convolutions. By utilizing spatial AEs with sufficient bottleneck resolution, i.e. 16 × 16px sized feature maps, we can mitigate this problem. Noteworthy, a smaller bottleneck resolution of 8 × 8px seems to lead to a severe information loss and thus to large reconstruction errors in general. By further constraining the latent space to follow a MVN distribution and introducing adversarial training, we notice marginal improvements over the non-generative models, leaving us with the impression that adversarial training is not required in this particular setting. As expected, spatial autoencoding clearly outperforms the AnoGAN and is considerably faster. While AnoGAN requires an iterative optimization, which consumes ∼19 s for a single reconstruction, all of the AE models require only a fraction of a second. Interestingly, even though the models operate on 2D data, the segmentations seem very consistent among neighboring axial slices.

In summary, we presented a comparison of deep autoencoding models for fast UAD as well as a novel spatial VAE-GAN which encode the full context of brain MR slices. We believe that the approach does not only open up opportunities for unsupervised brain lesion segmentation, but can also act as prior information for supervised deep learning.

Acknowledgements. We thank our clinical partners from Klinikum Rechts der Isar for providing us with their dataset.

References

1. An, J., Cho, S.: Variational autoencoder based anomaly detection using reconstruction probability. In: Special Lecture on IE, vol. 2, pp. 1–18 (2015)
2. Chong, Y.S., Tay, Y.H.: Abnormal Event Detection in Videos using Spatiotemporal Autoencoder. CoRR (2017)
3. Goodfellow, I.J., et al.: Generative adversarial nets. In: NIPS (2014)
4. Hasan, M., Choi, J., Neumann, J., Roy-Chowdhury, A.K., Davis, L.S.: Learning temporal regularity in video sequences. In: 2016 IEEE Conference on Computer Vision and Pattern Recognition (CVPR), pp. 733–742. IEEE (2016)
5. Iglesias, J.E., Liu, C.Y., Thompson, P.M., Tu, Z.: Robust brain extraction across datasets and comparison with publicly available methods. IEEE Trans. Med. Imaging **30**(9), 1617–1634 (2011)
6. Iheme, L.O., et al.: Concordance between computer-based neuroimaging findings and expert assessments in dementia grading. In: SIU, pp. 1–4 (2013)
7. Kingma, D.P., Welling, M.: Auto-Encoding Variational Bayes. CoRR (2013)

8. Larsen, A.B.L., Sønderby, S.K., Winther, O.: Autoencoding beyond pixels using a learned similarity metric. CoRR cs.LG (2015)
9. Pathak, D., Krähenbühl, P., Donahue, J., Darrell, T., Efros, A.A.: Context encoders: feature learning by inpainting. In: 2016 IEEE Conference on Computer Vision and Pattern Recognition (CVPR), pp. 2536–2544. IEEE (2016)
10. Rohlfing, T., Zahr, N.M., Sullivan, E.V., Pfefferbaum, A.: The SRI24 multichannel atlas of normal adult human brain structure. Hum. Brain Mapp. **31**(5), 798–819 (2009)
11. Rosca, M., Lakshminarayanan, B., Warde-Farley, D., Mohamed, S.: Variational approaches for auto-encoding generative adversarial networks. arXiv preprint arXiv:1706.04987 (2017)
12. Sabokrou, M., Fathy, M., Hoseini, M.: Video anomaly detection and localisation based on the sparsity and reconstruction error of auto-encoder. Electron. Lett. **52**(13), 1122–1124 (2016)
13. Schlegl, T., Seeböck, P., Waldstein, S.M., Schmidt-Erfurth, U., Langs, G.: Unsupervised Anomaly Detection with Generative Adversarial Networks to Guide Marker Discovery. CoRR cs.CV (2017)
14. Seeböck, P., et al.: Identifying and Categorizing Anomalies in Retinal Imaging Data. CoRR cs.LG (2016)
15. Sethian, J.A., et al.: Level set methods and fast marching methods. J. Comput. Inf. Technol. **11**(1), 1–2 (2003)
16. Shiee, N., Bazin, P.L., Ozturk, A., Reich, D.S., Calabresi, P.A., Pham, D.L.: A topology-preserving approach to the segmentation of brain images with multiple sclerosis lesions. NeuroImage **49**(2), 1524–1535 (2010)
17. Taboada-Crispi, A., Sahli, H., Hernandez-Pacheco, D., Falcon-Ruiz, A.: Anomaly detection in medical image analysis. In: Handbook of Research on Advanced Techniques in Diagnostic Imaging and Biomedical Applications, pp. 426–446. IGI Global (2009)
18. Vaidhya, K., Thirunavukkarasu, S., Alex, V., Krishnamurthi, G.: Multi-modal brain tumor segmentation using stacked denoising autoencoders. In: Crimi, A., Menze, B., Maier, O., Reyes, M., Handels, H. (eds.) BrainLes 2015. LNCS, vol. 9556, pp. 181–194. Springer, Cham (2016). https://doi.org/10.1007/978-3-319-30858-6_16
19. Weiss, N., Rueckert, D., Rao, A.: Multiple sclerosis lesion segmentation using dictionary learning and sparse coding. In: Mori, K., Sakuma, I., Sato, Y., Barillot, C., Navab, N. (eds.) MICCAI 2013. LNCS, vol. 8149, pp. 735–742. Springer, Heidelberg (2013). https://doi.org/10.1007/978-3-642-40811-3_92

Brain Tumor Detection and Classification from Multi-sequence MRI: Study Using ConvNets

Subhashis Banerjee[1,2(✉)], Sushmita Mitra[1], Francesco Masulli[3], and Stefano Rovetta[3]

[1] Machine Intelligence Unit, Indian Statistical Institute, Kolkata, India
mail.sb88@gmail.com, sushmita@isical.ac.in
[2] Department of CSE, University of Calcutta, Kolkata, India
[3] DIBRIS, University of Genova, Genoa, Italy
{francesco.masulli,stefano.rovetta}@unige.it

Abstract. In this paper, we thoroughly investigate the power of Deep Convolutional Neural Networks (ConvNets) for classification of brain tumors using multi-sequence MR images. First we propose three ConvNets, which are trained from scratch, on MRI patches, slices, and multi-planar volumetric slices. The suitability of transfer learning for the task is next studied by applying two existing ConvNets models (VGGNet and ResNet) pre-trained on ImageNet dataset, through fine-tuning of the last few layers. Leave-one-patient-out (LOPO) testing scheme is used to evaluate the performance of the ConvNets. Results demonstrate that ConvNet achieves better accuracy in all cases where the model is trained on the multi-planar volumetric dataset. Unlike conventional models, it obtains a testing accuracy of 97% without any additional effort towards extraction and selection of features. We also study the properties of self-learned kernels/filters in different layers, through visualization of the intermediate layer outputs.

Keywords: Convolutional neural network · Deep learning · Brain tumor · Glioblastoma multiforme · Multi-sequence MRI · Transfer learning

1 Introduction

Glioblastoma Multiforme constitute 80% of all malignant brain tumors originating from the glial cells in the central nervous system. Based on the aggressiveness and infiltrative nature of the gliomas the World Health Organization (WHO) broadly classified them into two categories, viz. Low-grade gliomas (LGG), consisting of low-grade and intermediate-grade gliomas (WHO grades II and III), and high-grade gliomas (HGG) or glioblastoma multiforme (GBM) (WHO grade IV) [1]. Although most of the LGG tumors have slower growth rate compared to HGG and are responsive to treatment, there is a subgroup of LGG tumors

© Springer Nature Switzerland AG 2019
A. Crimi et al. (Eds.): BrainLes 2018, LNCS 11383, pp. 170–179, 2019.
https://doi.org/10.1007/978-3-030-11723-8_17

which (if not diagnosed earlier and left untreated) can lead to GBM. Histological grading, based on stereotactic biopsy test, is the gold standard for detecting the grade of brain tumors. The biopsy procedure requires the neurosurgeon to drill a small hole into the skull guided by MRI, from which a sample of the tissue is collected. There are many risk factors involving the biopsy test, including bleeding from the tumor and brain due to the biopsy needle, which can cause severe migraine, stroke, coma and even death. Other risks involve infection or seizures [2] and misleading histological grading [3]. In this context multi-sequence MRI plays a major role in the detection, diagnosis, and management of brain cancers in a non-invasive manner. Decoding of tumor phenotype using noninvasive imaging is a recent field of research, known as *Radiomics* [4], and involves the extraction of a large number of quantitative imaging features that may not be apparent to the human eye. Quantitative imaging features, extracted from MR images, have been investigated in literature for the assessment of brain tumors [5]. Ref. [6] presents an adaptive neuro-fuzzy classifier, based on linguistic hedges (ANFC-LH), for predicting the brain tumor grade using 56 3D quantitative MRI features extracted from the corresponding segmented tumor volume(s).

Although the techniques demonstrate good disease classification, their dependence on hand-crafted features requires extensive domain knowledge, involves human bias, and is problem-specific. Subsequently manual or semi-automatic localization and segmentation of the region of interest (ROI) or volume of interest (VOI) is also needed to extract the quantitative imaging features [7]. ConvNets offer state-of-the-art framework for image recognition or classification [8]. These networks automatically learn mid-level and high-level representations or abstractions from the input training data, in the form of convolution filters that are updated during the training process. It works directly on raw input (image) data, and learn the underlying representative features of the input which are hierarchically complex, thereby ruling out the need for specialized hand-crafted image features. However training a ConvNet from scratch is generally difficult because it essentially requires large training data. In medical applications data is typically scarce, and expert annotation is expensive. Transfer learning offers a promising alternative, in case of inadequate data, to fine tune a ConvNet pretrained on a large set of available labeled images from some other category [9].

In this paper we exhaustively investigate the performance of ConvNets, with and without transfer learning, for non-invasive brain tumor detection and grade prediction from multi-sequence MRI. Tumors are typically heterogeneous, depending on cancer subtypes, and contain a mixture of structural and patch-level variability. Prediction of the grade of a tumor may thus be based on either the image patch containing the tumor, or the 2D MRI slice containing the image of the whole brain including the tumor, or the 3D MRI volume encompassing the full image of the head enclosing the tumor. While in the first case only the tumor patch is necessary as input, the other two cases require the ConvNet to learn to localize the ROI (or VOI) followed by its classification. Therefore, the first case needs only classification while the other two cases additionally require detection or localization. Since the performance and complexity of ConvNets depend on the difficulty level of the problem and the type of input data representation, we prepare here three kinds viz. (i) Patch-based, (ii) Slice-based, and (iii) Volume-based data, from the original MRI dataset. Three ConvNet models are developed corresponding to each case, and trained from scratch. We also compare two state-of-the-art ConvNet architectures, viz. VGGNet [10] and ResNet [8], with parameters pre-trained on ImageNet using transfer learning (via fine-tuning).

The rest of the paper is organized as follows. Section 2 provides details about the data, its preparation in patch, slice and volumetric modes, along with some preliminaries of ConvNets and transfer learning. Section 3 introduces the proposed ConvNet architectures. Section 4 describes the experimental results, demonstrating the effectiveness in terms of both qualitative and quantitative. Finally conclusions are provided in Sect. 5.

2   Materials and Methods

In this section we provide a brief description of the data preparation at three levels of resolution, followed by an introduction to convolutional neural networks and transfer learning.

2.1   Brain Tumor Data

All experiments are performed on the TCGA-GBM [11] and TCGA-LGG [12] datasets, downloaded from The Cancer Imaging Archive (TCIA) [13]. The TCGA GBM and LGG dataset consists of 262 and 199 cases. We consider four MRI sequences for a patient MRI scan, encompassing native (T1) and post-contrast enhanced T1-weighted (T1C), T2-weighted (T2), and T2 Fluid-Attenuated Inversion Recovery (FLAIR). Since the available data is inadequate to train a 3D ConvNet model, here we formulate 2D ConvNet models based on the MRI patches (encompassing the tumor region) and slices, followed by a multi-planar slice-based ConvNet model that incorporates the volumetric information as well.

Patch-Based Dataset: The slice with the largest tumor region is first identified. Keeping this slice in the middle, a set of slices before and after it are considered for extracting 2D patches containing the tumor regions using a bounding-box. This bounding-box is marked, corresponding to each slice, based on the ground truth image. The enclosed image region is then extracted. We use a set of 20 slices for extracting the patches. In case of MRI volumes from HGG (LGG) patients, four (ten) 2D patches [with a skip over 5 (4) slices] are extracted for each of the MR sequences. Therefore a total of $262 \times 4 = 1048$ HGG and $199 \times 5 = 995$ LGG patches, with four channels each, constitute this dataset.

Slice-Based Dataset: Complete 2D slices, with visible tumor region, are extracted from the MRI volume. The slice with the largest tumor region, along with a set of 20 slices before and after it, are extracted from the MRI volume in a sequence similar to that of the patch-based approach. While for HGG patients 4 (with a skip over 5) slices are extracted, in the case of LGG patients 10 (with a skip of 2) slices are used.

Multi-planar Volumetric Dataset: Here 2D MRI slices are extracted along all three anatomical planes, viz. axial (X-Z axes), coronal (Y-X axes), and sagittal (Y-Z axes), in a manner similar to that described above.

2.2 Convolutional Neural Networks

The fundamental constituents of a ConvNet consist of the input, convolution, activation, pooling and fully-connected layers. Some additional layers include the dropout, and batch-normalization layers.

Input Layer: This serves as the entry point of the ConvNet, accepting the raw pixel value of the input image. Here input is a 4-channel brain MRI patch/slice denoted by $I \in \mathbb{R}^{4 \times w \times h}$, where w and h represent the resolution of the image.

Convolution Layer: It is the core building block of a ConvNet. Each convolution layer is composed of a filter bank (set of convolutional filters/kernels of same width and height). A convolutional layer takes an image or feature maps as input, and performs the convolution operation between the input and each of these filters by sliding (as stride) the filter over the image to generate a set of (same as the number of filters) activation maps or the feature map.

Activation Layer: Output responses of the convolution and fully connected layers pass through some nonlinear activation function, such as a Rectified Linear Unit (ReLU), for transforming the data. ReLU, is a popular activation function for deep neural networks due to its computational efficiency and reduced likelihood of vanishing gradient.

Pooling Layer: This follows each convolution layer to typically reduce computational complexity by downsampling of the convoluted response maps. It combines spatially close, possibly redundant, features in the feature maps; thereby, making the representation more compact and invariant to small changes in an image like the insignificant details.

Fully-Connected Layer: The features learned through a series of convolutional and pooling layers are eventually fed to a fully-connected layer, typically a Multilayer Perceptron. The term "fully-connected" implies that every neuron in a layer is connected to every neuron of the following layer. The purpose of the fully-connected layer is to use these features for categorizing the input image into different classes, based on the training dataset.

Additional layers like Batch-Normalization reduce initial covariate shift. The cost function for the ConvNets is chosen as binary cross-entropy (for a two-class problem).

2.3 Transfer Learning

Typically the early layers of a ConvNet learn low-level image features, which are applicable to most vision tasks. The later layers, on the other hand, learn high-level features which are more application-specific. Therefore, shallow fine-tuning of the last few layers is usually sufficient for transfer learning. A common practice is to replace the last fully-connected layer of the pre-trained ConvNet with a new fully-connected layer, having as many neurons as the number of classes in the new target application. The rest of the weights, in the remaining layers, of the pre-trained network are retained. However, when the distance between the source and target applications is significant than one may need to induce deeper fine-tuning. This is equivalent to training a shallow neural network with one or more hidden layers. An effective strategy is to initiate fine-tuning from the last layer, and then incrementally include deeper layers in the tuning process until the desired performance is achieved.

3   ConvNets for Brain Tumor Grading

The ConvNet architectures are illustrated in Fig. 1. PatchNet is trained on the patch-based dataset, and provides the probability of a patch belong to HGG or LGG. SliceNet gets trained on the slice-based dataset, and predicts the probability of a slice being from HGG or LGG. Finally VolumeNet is trained on the multi-planar volumetric dataset, and predicts the grade of a tumor from its 3D representation using the multi-planar 3D MRI data. We use filters of size (3×3) for our ConvNet architectures. A greater number of filters, involving deeper convolution layers, allows for more feature maps to be generated. This compensates for the decrease in size of each feature map caused by "valid" convolution and pooling layers. Due to the complexity of the problem and bigger size of the input image, the SliceNet and VolumeNet architectures are deeper as compared to the PatchNet.

Pre-trained VGGNet (16 layers), and ResNet (50 layers) architectures, trained on the ImageNet dataset, are employed for transfer learning. Even though ResNet is deeper than VGGNet, the model size of ResNet is substantially smaller due to the usage of global average pooling rather than fully-connected layers. Transferring from the non-medical to the medical image domain is achieved through fine-tuning of the last convolutional block of each model, along with the fully-connected layer (top-level classifier) of each model. Fine-tuning of a trained network is achieved by retraining on the new dataset, while involving very small weight updates.

Since the base models were trained on RGB images, and accept single input with three channels, we train and test them on the slice-based dataset involving three MR sequences ($T1C$, $T2$, $FLAIR$). The $T1C$ sequence was found to perform better than $T1$, when used in conjunction with $T2$ and $FLAIR$. The following section presents the results for the proposed three level ConvNet architectures, along with that of the fine-tuned models involving transfer learning.

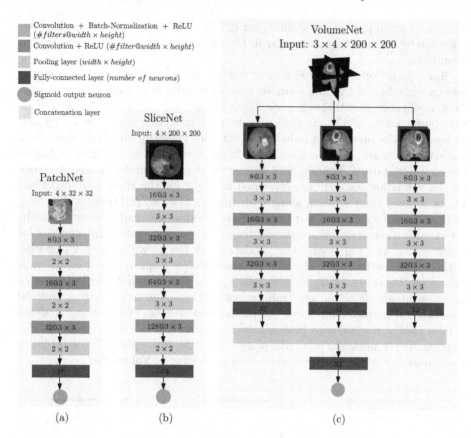

Fig. 1. Three level ConvNet architectures (a) PatchNet, (b) SliceNet, and (c) VolumeNet.

4   Experimental Results

The ConvNets were developed using TensorFlow, with Keras in Python. The experiments were performed on a desktop machine with Intel i7 CPU (clock speed 3.40 GHz), having 4 cores, 32 GB RAM, and NVIDIA GeForce GTX 1080 GPU with 8 GB VRAM. The operating system was Ubuntu 16.04. The quantitative and qualitative evaluation of the resuls are elaborated below.

We used leave-one-patient-out (LOPO) test scheme for quantitative evaluation. Although LOPO test scheme is computationally expensive, it allows availability of more data as required for ConvNets training. LOPO testing is robust and well-suited to our application, with results being generated for each individual patient. Therefore, in cases of misclassification, a patient sample may be further investigated. The ConvNet models PatchNet, SliceNet, VolumeNet, were trained on the corresponding datasets using Stochastic Gradient Descent (SGD) optimization algorithm with learning rate = 0.001 and momentum = 0.9, using mini-batches of size 32 samples generated from the corresponding training dataset. A small part of the training set (20%) was used for validating the ConvNet model after each training epoch, for parameter selection and detection of overfitting.

Since deep ConvNets entail a large number of free trainable parameters, the effective number of training samples were artificially enhanced using real-time data augmentation – through some linear transformation. Training and validation performance of the three ConvNets were measured using *Accuracy* and $F_1 Score$. In the presence of imbalanced data one typically prefers $F_1 Score$ over *Accuracy* because the former considers both false positives and false negatives during computation. Training and validation *Accuracy* and loss, and $F_1 Score$ on the validation dataset, are presented in Fig. 2 for the three proposed ConvNets, trained from scratch, along with that for the two pre-trained ConvNets (VGGNet, and ResNet). The plots demonstrate that VolumeNet gives the highest classification performance during training, reaching maximum accuracy on the training set (100%) and the validation set (98%) within just 20 epochs. Although the performance of PatchNet and SliceNet is quite similar on the validation set (PatchNet - 90%, SliceNet - 92%), it is observed that SliceNet achieves better accuracy (95%) on the training set (perhaps due to overfitting after 50 epochs). The performance of two the pre-trained models (VGGNet and ResNet) exhibit similar results, with both achieving around 85% accuracy on the validation set. All the networks reached a plateau after the 50th epoch. This establishes the superiority of the 3D volumetric level processing of VolumeNet.

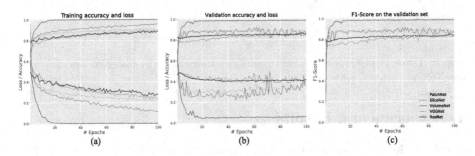

Fig. 2. Comparative performance of the networks.

After training, the networks were evaluated on the hold-out test set employing majority voting. Each patch or slice from the test dataset was from a single test patient in the LOPO framework, and was categorized as HGG or LGG. The class with maximum number of slices or patches correctly classified was indicative of the grade of the tumor. In case of equal votes the patient was marked as "ambiguous". The LOPO testing scores are displayed in Table 1. VolumeNet is observed to achieve the best LOPO test accuracy (97.19%), with zero "ambiguous" cases as compared to the other four networks. SliceNet is also found to provide good LOPO test accuracy (90.18%). Both the pre-trained models show

Table 1. Comparative LOPO test performance

ConvNets	Classified	Misclassified	Ambiguous	Accuracy
PatchNet	242	39	4	84.91%
SliceNet	257	26	2	90.18%
VolumeNet	**277**	**8**	**0**	**97.19%**
VGGNet	239	40	6	83.86%
ResNet	242	42	1	84.91%

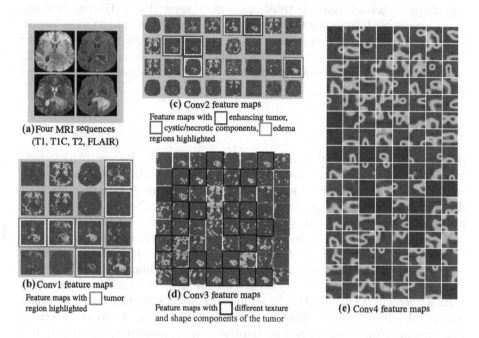

(a) Four MRI sequences (T1, T1C, T2, FLAIR)

(b) Conv1 feature maps
Feature maps with ☐ tumor region highlighted

(c) Conv2 feature maps
Feature maps with ☐ enhancing tumor, ☐ cystic/necrotic components, ☐ edema regions highlighted

(d) Conv3 feature maps
Feature maps with ☐ different texture and shape components of the tumor

(e) Conv4 feature maps

Fig. 3. (a) Four sequences of an MRI slice from a sample HGG patient. Intermediate layer outputs/feature maps, generated by SliceNet, at different levels by (b) Conv1, (c) Conv2, (d) Conv3 and (e) Conv4.

similar LOPO test accuracy as PatchNet. This is interesting because it demonstrates that with a little fine-tuning one can achieve a test accuracy similar to that by the patch-level ConvNet trained from scratch on a specific dataset.

The ConvNets were next investigated through visual analysis of their intermediate layers. Visualizing the output of any convolution layer can help determine the description of the learned kernels. Figure 3 illustrates the intermediate convolution layer outputs (after ReLU activation) of the proposed SliceNet (Fig. 1(b)) architecture on sample MRI slices from an HGG patient.

The visualization of the first convolution layer activations (or feature maps) (Fig. 3(b)) indicates that the ConvNet has learned a variety of filters to detect edges and distinguish between different brain tissues like white matter (WM), gray matter (GM), cerebrospinal fluid (CSF), skull and background. Most importantly, some of the filters could isolate the ROI (or the tumor); on the basis of which the whole MRI slice may be classified. Most of the feature maps generated by the second convolution layer (Fig. 3(c)) mainly highlight the tumor region and its subregions; like enhancing tumor structures, surrounding cystic/necrotic components and the edema region of the tumor. Thus the filters in the second convolution layer learn to extract deeper features from the tumor by focusing on the ROI (or tumor). The texture and shape of the tumor get enhanced in the feature maps generated from the third convolution layer (Fig. 3(d)). For example, small, distributed, irregular tumor cells get enhanced (one of the most important tumor grading criteria called "CE-Heterogeneity"). Finally the last layer (Fig. 3(e)) extracts detailed information about more discriminating features, by combining these to produce a clear distinction between images of different types of tumors.

5   Conclusion

An exhaustive study was made to demonstrate the effectiveness of ConvNets for non-invasive, automated detection and grading of brain tumors from multi-sequence MR images. Three novel ConvNet architectures were developed for distinguishing between HGG and LGG. Three level ConvNet architectures were designed to handle images at patch, slice and multi-planar modes. This was followed by exploring transfer learning for the same task, by fine-tuning two existing ConvNet models. The scheme for incorporating volumetric tumor information, using multi-planar MRI slices, achieved the best test accuracy of 97.19%. Visualization of the intermediate layer outputs/feature maps demonstrated the role of kernels/filters in the convolution layers in automatically learning to detect tumor features closely resembling different tumor grading criteria. It was also observed that existing ConvNets, trained on natural images, performed adequately just by fine-tuning their final convolution layer on the MRI dataset. This investigation allows us to conclude that deep ConvNets could be a feasible alternative to surgical biopsy for brain tumors.

References

1. Louis, D.N., et al.: The 2007 WHO classification of tumours of the central nervous system. Acta Neuropathol. **114**(2), 97–109 (2007)
2. McGirt, M.J., et al.: Independent predictors of morbidity after image-guided stereotactic brain biopsy: a risk assessment of 270 cases. J. Neurosurg. **102**(5), 897–901 (2005)
3. Chandrasoma, P.T., Smith, M.M., Apuzzo, M.L.J.: Stereotactic biopsy in the diagnosis of brain masses: comparison of results of biopsy and resected surgical specimen. Neurosurgery **24**(2), 160–165 (1989)
4. Mitra, S., Uma Shankar, B.: Medical image analysis for cancer management in natural computing framework. Inf. Sci. **306**, 111–131 (2015)

5. Zhou, M., et al.: Radiomics in brain tumor: image assessment, quantitative feature descriptors, and machine-learning approaches. Am. J. Neuroradiol. **39**(2), 208–216 (2017)
6. Banerjee, S., Mitra, S., Shankar, B.U.: Synergetic neuro-fuzzy feature selection and classification of brain tumors. In: Proceedings of IEEE International Conference on Fuzzy Systems (FUZZ-IEEE), pp. 1–6 (2017)
7. Banerjee, S., Mitra, S., Uma Shankar, B., Hayashi, Y.: A novel GBM saliency detection model using multi-channel MRI. PLoS ONE **11**(1), e0146388 (2016)
8. He, K., Zhang, X., Ren, S., Sun, J.: Deep residual learning for image recognition. In: Proceedings of the IEEE Conference on Computer Vision and Pattern Recognition, pp. 770–778 (2016)
9. Tajbakhsh, N., et al.: Convolutional neural networks for medical image analysis: full training or fine tuning? IEEE Trans. Med. Imaging **35**(5), 1299–1312 (2016)
10. Simonyan, K., Zisserman, A.: Very deep convolutional networks for large-scale image recognition, arXiv preprint arXiv:1409.1556 (2014)
11. Scarpace, L., et al.: Radiology data from the cancer genome atlas glioblastoma multiforme [TCGA-GBM] collection. Cancer Imaging Arch. (2016)
12. Pedano, N., Flanders, A., Scarpace, L., et al.: Radiology data from the cancer genome atlas low grade glioma [TCGA-LGG] collection. Cancer Imaging Arch. (2016). https://wiki.cancerimagingarchive.net/display/Public/TCGA-LGG#aab8e738e9b547979f782c25057bfbef
13. Clark, K., et al.: The cancer imaging archive (TCIA): maintaining and operating a public information repository. J. Digit. Imaging **26**, 1045–1057 (2013)

Voxel-Wise Comparison with *a-contrario* Analysis for Automated Segmentation of Multiple Sclerosis Lesions from Multimodal MRI

Francesca Galassi[1]([✉]), Olivier Commowick[1], Emmanuel Vallee[2], and Christian Barillot[1]

[1] Inria, CNRS, Inserm, IRISA, VisAGeS, Rennes, France
francesca.galassi@inria.fr
[2] FMRIB, NDCN, University of Oxford, Oxford, UK

Abstract. We introduce a new framework for the automated and unsupervised segmentation of Multiple Sclerosis lesions from multimodal Magnetic Resonance images. It relies on a voxel-wise approach to detect local white matter abnormalities, with an *a-contrario* analysis, which takes into account local information. First, a voxel-wise comparison of multimodal patient images to a set of controls is performed. Then, region-based probabilities are estimated using an *a-contrario* approach. Finally, correction for multiple testing is performed. Validation was undertaken on a multi-site clinical dataset of 53 MS patients with various number and volume of lesions. We showed that the proposed framework outperforms the widely used FDR-correction for this type of analysis, particularly for low lesion loads.

Keywords: Multiple Sclerosis · Voxel-wise comparison · a-contrario

1 Introduction

Multiple Sclerosis (MS) is a chronic inflammatory-demyelinating disease of the central nervous system [1]. Magnetic Resonance Imaging (MRI) is fundamental in MS to characterize and quantify MS lesions. The number and volume of lesions are used for MS diagnosis, to track its progression and to evaluate treatments [2]. Conventional MRI in MS usually consists in Fluid-Attenuated Inversion Recovery (FLAIR), T2-weighted (T2-w) and T1-weighted (T1-w) images. Accurate identification of MS lesions in MR images is extremely difficult due to variability in lesion location, size and shape, in addition to anatomical variability between subjects. Since manual segmentation requires expert knowledge, it is time consuming and prone to intra- and inter-expert variability, several methods have been proposed to automatically segment lesions [1,6]. In order to reduce false lesion detections, segmentation algorithms have to integrate complementary

A. Crimi et al. (Eds.): BrainLes 2018, LNCS 11383, pp. 180–188, 2019.
https://doi.org/10.1007/978-3-030-11723-8_18

information from multimodal data. Although many solutions have been proposed, e.g. 3-class tissue classification and Machine Learning (ML) approaches [1], the challenge remains to provide segmentation techniques that work regardless of the type of MS lesion or MRI protocol.

MS lesion segmentation algorithms are generally prone to detection of false positives, especially voxel-wise approaches, where inference is performed directly on the voxel-wise probabilities. We propose to tackle this problem by replacing classical methods for correction for multiple testings, e.g. Bonferroni and FDR-correction, with a locally multivariate inference: the *a-contrario* analysis [3].

We present a novel framework for the automated segmentation of MS lesions from multimodal MRI, based on a comparison at the voxel level between a patient and a model of healthy controls with an *a-contrario* approach. In Sect. 2, we present the steps of the proposed framework and the evaluation metrics. Then, in Sect. 3 we illustrate the experiments, performed on a multi-site clinical dataset. Finally, we discuss the results and conclude in Sect. 4.

2 Materials and Methods

2.1 MS Lesion Detection Framework

The *a-contrario* Approach. The *a-contrario* approach is a locally multivariate procedure which uses the size of a local excursion set as statistic [3]. An a-contrario framework was previously presented to extract patterns of abnormal perfusions in individual patients [4]. Its general steps can be summarized as follows: *(i)* a voxel-wise probability map is computed under a background model (i.e. the null-hypothesis in statistical decision theory [5]), *(ii)* a locally multivariate probability is estimated, and *(iii)* a correction for multiple testing is performed. We propose to apply the *a-contrario* approach to the segmentation of MS lesions from multimodal MRI as follows.

(i) Voxel-Wise Probability Map. In [7], a general methodology for the comparison, at a voxel level, of a patient model with a group of models was presented. We adopted a similar approach to compute the input voxel-wise probability map of the *a-contrario* analysis. Precisely, at a given voxel, we compared an intensity vector $V_P \in \mathbb{R}^h$, where h is the dimension and P indicates the patient, with a set of intensity vectors V_j from the control group, with $j = 1, ..., N$ controls. These intensity vectors were created from the image modalities (i.e. in our workflow we used FLAIR and T2-w modalities). The group of controls is assumed to follow a multivariate Normal distribution $\mathcal{N}(\overline{V}, \Sigma_V)$, where $\overline{V}$ and Σ_V denote respectively the average and covariance matrix of the control group. Thus, the difference statistic between V_P and $\overline{V}$ can be computed as a Mahalanobis distance $d^2(V_P) = (V_P - \overline{V})^T \Sigma_V^{-1}(V_P - \overline{V})$. $d^2(V_P)$ varies between zero and infinity, with smaller values if the patient vector more likely belongs to $\mathcal{N}(\overline{V}, \Sigma_V)$. The test p-value can be computed as:

$$p(V_P) = 1 - F_{h,N-h}(d^2(V_P)) \tag{1}$$

where $F_{h,N-h}$ is the cumulative distribution function of a Fisher distribution with parameters h and $N - h$. The obtained p-value map was employed as the input for the region-based probabilities estimation.

(ii) Region-Based Probabilities. The uncorrected p-value map was partitioned into regions, namely a grid of spheres of radius r centered at each voxel. A set of uncorrected p-value thresholds $p = \{p_1, ..., p_T\}$ was defined i.e. a set of decision thresholds. For a threshold p_i, the p-value map was thresholded to produce a binary map referred to as a *rare event* map. For each region s, the number of *rare events* occurring at a level p_i was computed and denoted as k_s. Hence, the probability π_i^s of having k_s or more *rare events* was calculated from the tails of the binomial distribution:

$$\pi_i^s = P(X \geq k_s), \quad \text{with} \quad X \sim B(n, p_i) \tag{2}$$

where n is the total number of voxels in the sphere s, i.e. the number of tests. The probability π_i^s associated to a region s was then assigned to its center voxel. Of all region-based probabilities, only the minimum probability over all p-value thresholds p_i, $min(\pi_i^s)$, was retained per voxel.

(iii) Correction for Multiple Testing. The probability map from step *(ii)* was then corrected for multiple testing. The probability map was converted to a *Number of False Alarms* (NFA) map, i.e. the number of false detections in the background, as:

$$\text{NFA}_s = N_s T min(\pi_i^s) \tag{3}$$

where N_s and T are the total number of regions and p-value thresholds, respectively. Last, the NFA map was thresholded so that regions with NFA $> \epsilon$ were discarded to obtain ϵ−significant regions, where ϵ is the detection threshold.

Post-processing. After the *a-contrario* analysis, the segmentation outcome may still include false positives due to e.g. registration errors, noise and artifacts. A few post-processing steps were therefore performed to reduce these false detections. A candidate lesion was discarded if one of the following conditions was verified: *(i)* it did not belong to an hyper-intensities mask, *(ii)* it was not sufficiently located in the white matter, *(iii)* its size was lower than $3\,\text{mm}^3$. The hyper-intensities mask was computed by performing Otsu's thresholding [8] on the product of the T2-w and FLAIR images of a subject [9]. The white matter probability map was calculated from the control subjects and then thresholded at 0.7 to obtain a mask.

2.2 Dataset and Pre-processing

MS Patients. We evaluated the proposed method on the MICCAI 2016 MS lesion segmentation challenge dataset [10]. It included 53 images of patients suffering from MS (15 training images and 38 testing images; evaluation on the testing images can be performed by submission to the evaluation platform[1]). They were acquired in four different sites (Siemens 3T Verio, Siemens Aera 1.5T, Philips 3T Ingenia, GE 3D Discovery). The MR imaging protocol included 3D T1-w, T2-w and 3D FLAIR anatomical images. More details on the imaging protocol are available on the challenge website[1]. For each subject, manual delineations of MS lesions from seven trained radiologists were provided; the ground truth was computed from the seven independent manual segmentations using LOP STAPLE [11].

Group of Controls. 20 MRI datasets of healthy subjects were acquired on a Siemens 3T Verio scanner. The MR imaging protocol included: 3D T1-w (matrix size: $256 \times 256 \times 160$, resolution: $1 \times 1 \times 1$ mm^3); T2-w (matrix size: $192 \times 256 \times 44$, resolution: $1 \times 1 \times 3$ mm^3); 3D FLAIR (matrix size: $256 \times 256 \times 160$, resolution: $1 \times 1 \times 1$ mm^3).

Pre-processing. MR images were denoised [12], rigidly registered towards T1-w images [13], skull-stripped [14] and bias corrected [15]. The proposed framework relies on a voxel-wise comparison of a patient to a set of controls. Hence, it requires that patient and controls images are in the same coordinates system, i.e. corresponding voxels describe the same spatial position, and corresponding anatomical tissues show the same intensity profile. A template image was generated from the set of controls images by applying a method derived from [16], which constructs an unbiased atlas representing the average intensity and shape of a number of images. Patient images were registered to the template image using a linear registration, based on a block-matching algorithm [13], followed by a dense non-linear registration [17]. In order to reduce inter-subject variability, intensities were normalized using k-means [18].

2.3 Evaluation of MS Lesion Detection

The quality of the proposed segmentation framework was assessed using three metrics:

(i) Dice Similarity Coefficient (DSC), i.e. the spatial overlap between the result R and the ground truth G:

$$DSC = 2\frac{|R \cap G|}{|R| + |G|} \tag{4}$$

(ii) Positive Predictive Value (PPV), i.e. the proportion of true positive lesions TP_R within the segmented N lesions:

$$PPV = 2\frac{TP_R}{N} \tag{5}$$

---

[1] https://portal.fli-iam.irisa.fr/msseg-challenge/overview.

(iii) $F1$ score, i.e. the weighted average of the lesion sensitivity Se_L and the positive predictive value PPV:

$$F1 = 2\frac{Se_L PPV}{Se_L + PPV} \tag{6}$$

These two last metrics evaluated the algorithm in terms of detection of individual lesions, independently of their contour quality i.e. at the lesion level and not at the voxel level.

Comparison with False Discovery Rate Correction. Inference in voxel-wise comparison approaches is generally performed directly on the p-value map by applying a False Discovery Rate (FDR) correction for multiple comparison [7]. The widely applied Benjamini-Hochberg procedure enables controlling the expected proportion of false positives when considering all tests, e.g. it ensures that no more than a ratio $q = 5\%$ of detections are false positives [19]. For comparison with our method, we replaced the *a-contrario* analysis with the FDR correction. Hence, we applied the method in [19] to the voxel-wise probability map as obtained from step *(i)*, followed by the same post-processing steps. We evaluated the outcomes using the three metrics presented above. We explored the significance of the differences in the scores obtained by the two approaches using the Wilcoxon test (a p-value < 5% was considered significant).

3   Results

3.1   Implementation and Computation Time

The framework was implemented in Python and employed in-house tools[2] for the pre-processing and post-processing steps. In the *a-contrario* framework, the radius r of a sphere was equal to two voxels, the set of p-values was $p = \{ 1.10^{-05}, 1.10^{-04}, 1.10^{-03} \}$, and $\epsilon = 1$. The computation time to process a subject on a laptop with an Intel Core i7 CPU 2.40 GHz (8 cores) was approximately 10 min.

3.2   Evaluation of MS Lesion Detection

Figure 1 shows a representative case of uncorrected p-value map from step *(i)* and detected MS lesions as obtained with the proposed framework. In Fig. 2, two segmentations outcomes as obtained with the two methods, i.e. the proposed method and the FDR-corrected voxel-wise probability map, are reported. From visual inspection, it appears that both methods are capable of detecting the true lesions; however, the FDR correction approach seems to be more prone to false positives than the proposed approach.

---

[2] Anima: Open source software for medical image processing from the INRIA VIS-AGES team.

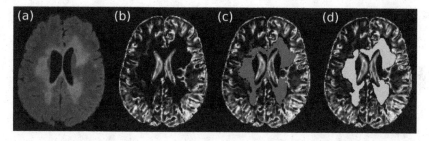

Fig. 1. (a) Original FLAIR image followed by (b) its uncorrected p-value map and superimposed MS lesion segmentations from (c) experts segmentation and (d) proposed framework.

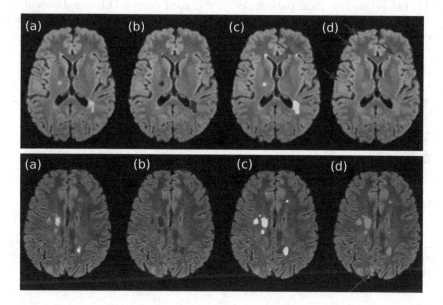

Fig. 2. (a) Original FLAIR image followed by FLAIR image and superimposed MS lesion segmentations from: (b) experts segmentation, (c) proposed framework, (d) FDR-correction. Arrow heads show some false detected lesions: green arrows for false positive on both (c) and (d), red arrows in (d) only. (Color figure online)

Table 1. Average scores per metric and p-value of the Wilcoxon test on corresponding sets of scores.

	DSC	PPV	F1 score
Proposed framework	0.51*	0.56*	0.32*
FDR-correction	0.48	0.45	0.25
Wilcoxon p-value	0.007	$1.86.10^{-8}$	0.03

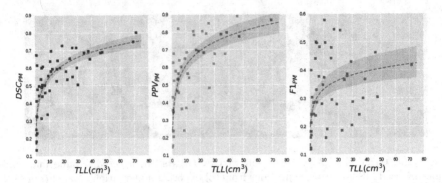

Fig. 3. Metrics as obtained with the proposed method (PM) for increasing Total Lesion Load (TLL) per patient. From the left: DSC, PPV, and F1 score. TLL varied from about $0.5\,cm^3$ to $70\,cm^3$. A log regression model is fitted to the data and a 95% confidence interval for that regression is shown.

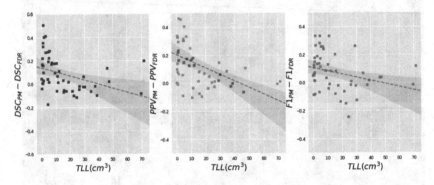

Fig. 4. The differences in scores as obtained with the two approaches for increasing Total Lesion Load. From the left: DSC, PPV, and F1 score. A linear regression model is fitted to the data and a 95% confidence interval for that regression is shown.

For each patient and for both the methods, we computed the three evaluation metrics. The average scores are reported in Table 1, together with the outcomes of the Wilcoxon test. In Fig. 3, the three scores for the proposed method are reported for increasing Total Lesion Load (TLL). Figure 4 shows the differences in scores per patient between the proposed framework and the FDR-correction approach for increasing TLL, where positive difference values indicate that the first outperforms the latter. The Wilcoxon test indicates that the scores are significantly different.

Generally, the proposed method outperforms the classical approach. This is particularly evident for low lesion loads, whereas the two performances tend to converge for high lesion loads. The highest improvements of the proposed method over the FDR correction approach were 36% in DSC (TTL $\approx 3\,cm^3$), 73% in PPV (TTL $\approx 3\,cm^3$), and 31% in F1 score (TTL $\approx 8\,cm^3$). The average improvements were about 10% in DSC, 20% in PPV, and 10% in F1 score.

Overall, we observed that all the scores tend to decrease with the total lesion load. This can be partially explained by the disagreement among the experts, which increases and hence becomes more relevant for a lower lesion load.

4   Conclusion

In this paper, we have proposed an automatic and unsupervised framework for the segmentation of MS lesions from multimodal MRI. It computes a voxel-wise probability map by comparing a patient with a group of controls, and it estimates locally multivariate probabilities using an *a-contrario* approach. Experiments have shown that the method outperforms the classical FDR-correction approach. Improvements increase with decreasing total lesion load, indicating that the proposed method is more specific and sensitive for patients with low lesion loads. The performance of the method relies on parameters, i.e. size of a region and set of thresholds, that must be accurately tuned on a set of cases.

Evaluation was performed on the MICCAI 2016 MS lesions segmentation challenge dataset, comprising clinical images acquired with different MR scanners and acquisition protocols [10]. This is an important aspect when developing techniques that are meant to be employed in the clinical practice. Compared to the results from the challenge results board (see footnote 1), the accuracy of the proposed framework was similar to that of the top rank strategies. Compared to other multivariate approaches, such as Machine Learning techniques, it has the clear advantage of being simple and not computationally intensive. These are important benefits, as the primary objective of the proposed framework is to assist radiologists in the clinical practice.

References

1. Garcia-Lorenzo, D., Francis, S., Narayanan, S., Arnold, D.L., Collins, D.L.: Review of automatic segmentation methods of multiple sclerosis white matter lesions on conventional magnetic resonance imaging. Med. Image Anal. **17**(1), 1–18 (2013)
2. Polman, C.H., et al.: Diagnostic criteria for multiple sclerosis: 2005 revisions to the Mc Donald criteria. Ann. Neurol. **58**, 840–846 (2005)
3. Robin, A., Moisan, L., Le Hgarat-Mascle, S.: An a-contrario approach for sub-pixel change detection in satellite imagery. IEEE Trans. Pattern Anal. Mach. Intell. **32**, 1977–93 (2010)
4. Maumet, C., Maurel, P., Ferré, J.C., Barillot, C.: An a contrario approach for the detection of patient-specific brain perfusion abnormalities with arterial spin labelling. NeuroImage **134**, 424–433 (2016)
5. Rousseau, F., et al.: An a contrario approach for change detection in 3D multi-modal images: application to multiple sclerosis in MRI. In: IEEE International Symposium on Biomedical Imaging (ISBI), pp. 2069–2072 (2007)
6. Crimi, A., Commowick, O., Ferr, J.C., Maarouf, A., Edan, G., Barillot, C.: Semi-automatic classification of lesion patterns in patients with clinically isolated syndrome. In: IEEE International Symposium on Biomedical Imaging (ISBI), pp. 1102–1105 (2013)

7. Commowick, O., Maarouf, A., Ferré, J.C., Ranjeva, J.P., Edan, G., Barillot, C.: Diffusion MRI abnormalities detection with orientation distribution functions: a multiple sclerosis longitudinal study. Med. Image Anal. **22**, 114–23 (2015)

8. Otsu, N.: A threshold selection method from gray-level histograms. IEEE Trans. Syst. Man Cybern. **9**, 62–66 (1979)

9. Gabr, R.E., Hasan, K.M., Haque, M.E., Nelson, F.M., Wolinsky, J.S., Narayana, P.A.: Optimal combination of FLAIR and T2 weighted MRI for improved lesion contrast in multiple sclerosis. J. Magn. Reson. Imaging **44**, 1293–1300 (2016)

10. Commowick, O., et al.: Objective Evaluation of Multiple Sclerosis Lesion Segmentation using a Data Management and Processing Infrastructure. bioRxiv 367557 (2018)

11. Akhondi-Asl, A., Hoyte, L., Lockhart, M.E., Warfield, S.K.: A logarithmic opinion pool based STAPLE algorithm for the fusion of segmentations with associated reliability weights. IEEE Trans. Med. Imaging **33**, 1997–2009 (2014)

12. Coupe, P., et al.: An optimized blockwise nonlocal means denoising filter for 3-D magnetic resonance images. IEEE Trans. Med. Imaging **27**, 425–41 (2008)

13. Commowick, O., Wiest-Daesslé, N., Prima, S.: Block matching strategies for rigid registration of multimodal medical images. In: IEEE International Symposium on Biomedical Imaging (ISBI), pp. 700–703 (2012)

14. Manjn, J.V., Coup, P.: volBrain: an online MRI brain volumetry system. Front. Neuroinformatics **10**, 30 (2016)

15. Tustison, N.J., et al.: N4ITK: improved N3 bias correction. IEEE Trans. Med. Imaging **29**, 1310–1320 (2010)

16. Guimond, A., Meunier, J., Thirion, J.P.: Average brain models. Comput. Vis. Image Underst. **77**, 192–210 (2000)

17. Commowick, O., Wiest-Daesslé, N., Prima, S.: Automated diffeomorphic registration of anatomical structures with rigid parts: application to dynamic cervical MRI. MICCAI **15**, 163–170 (2012)

18. Virmani, D., Taneja, S., Malhotra, G.: Normalization based K means Clustering Algorithm. arXiv:1503.00900 (2015)

19. Hochberg, Y., Tamhane, A.: Multiple Comparison Procedures. Wiley, Hoboken (1987)

A Graph Based Similarity Measure for Assessing Altered Connectivity in Traumatic Brain Injury

Yusuf Osmanlıoğlu[1](✉), Jacob A. Alappatt[1], Drew Parker[1], Junghoon Kim[2], and Ragini Verma[1]

[1] Center for Biomedical Image Computing and Analytics, Department of Radiology, University of Pennsylvania, Philadelphia, USA
yusuf.osmanlioglu@uphs.upenn.edu
[2] CUNY School of Medicine, The City College of New York, New York, USA

Abstract. Traumatic brain injury (TBI) arises from disruptions in the structural connectivity of brain, which further manifests itself as alterations in the functional connectivity, eventually leading to cognitive and behavioral deficits. Although patient-specific measures quantifying the severity of disease is crucial due to the heterogeneous character of the disease, neuroimaging based measures that can assess the level of injury in TBI using structural and functional connectivity is very scarce. Taking a graph theoretical approach, we propose a measure to quantify how dissimilar a TBI patient is relative to healthy subjects using their structural and functional connectomes. Over a TBI dataset with 39 moderate-to-severe TBI patients that are examined 3, 6, and 12 months post injury, and 35 healthy controls, we demonstrate that the dissimilarity scores obtained by the proposed measure distinguish patients from controls using both modalities. We also show that the dissimilarity scores significantly correlate with post-traumatic amnesia, processing speed, and executive function among TBI patients. Our results indicate the applicability of the proposed measure in quantitatively assessing the extent of injury. The measure is applicable to structural and functional connectivity, paving the way for a joint analysis in the future.

Keywords: Traumatic brain injury · Severity measure · Graph similarity · Connectome · DTI · fMRI

1 Introduction

Traumatic brain injury (TBI) is mainly considered as a white matter disorder arising from axonal injuries, which can result from causes such as fall or traffic accidents. Injury of axons cause disruptions in neural communication, eventually leading to psychological, cognitive, and emotional disturbances, making it a significant public health burden [1]. Although the relationship between trauma induced white matter injury and the neuropsychological outcomes in TBI has

© Springer Nature Switzerland AG 2019
A. Crimi et al. (Eds.): BrainLes 2018, LNCS 11383, pp. 189–198, 2019.
https://doi.org/10.1007/978-3-030-11723-8_19

been established for over two decades [2], neuroimaging measures that effectively quantify the injury burden based on connectivity disruptions started emerging recently and are still limited [3,4]. Heterogeneous character of TBI necessitates subject-specific investigations for providing patient-specific diagnostics and prognostics, which further highlights the need for imaging based measures.

Several studies in the literature have investigated the relationship between brain structure and behavioral outcomes in TBI on the basis of diffusion characteristics such as FA and MD [5]. Another major direction of research in elucidating this relationship has focused on changes in the topology of the structural brain network [6]. Recently, a summary measure of diffuse connectivity alterations to quantify overall injury burden was proposed in [4] which demonstrated high correlations with cognitive deficits. However, measures to quantify structural and functional disruptions in TBI together and investigating its relationship with cognitive outcomes of the disease has been sparsely studied.

Connectomes can naturally be represented as graphs, making a wide array of graph theoretical tools available for connectomic analysis. This, in turn, makes it possible to investigate structural and functional characteristics of the brain while preserving its organizational features [7]. Although fundamental graph theory measures such as centrality and small worldness are widely applied to connectomic analysis [8], application of advanced graph theory tools to brain data is still in its early stages. Graph similarity measures are prime examples of such tools that found very limited applications in brain analysis to date despite their great potentials.

Since the brain is commonly represented as connectomes, which are weighted graphs in essence, finding similarities between them can be considered as a graph matching problem. The main idea in here is to find a mapping between nodes and edges of two graphs while minimizing an overall similarity score. Although widely applied in pattern recognition and computer vision over several decades [9,10], use of graph matching as a similarity measure in neuroscience very scarce [11]. One of the early uses of graph similarity measures in connectomics is the application of graph edit distance in network classification of the epileptic brain [12]. Graph embedding [13] and graph kernels [14] are utilized in decoding of brain states in fMRI, which in turn is used for performing classification. Further utilization of graph similarity measures for quantifying neurological diseases and disorders is still a fertile area of investigation.

In this preliminary study, we explore graph similarity measures as a quantifier of injury on TBI using structural and functional connectomes, and investigate its relationship with emerging cognitive deficits. Specifically, over a longitudinal TBI dataset of 39 patients and 35 controls, we consider connectomes as weighted graphs and define dissimilarity between patients and controls by taking a graph matching approach to calculate similarity between two subjects. We demonstrate that there is significant group difference between patients and controls in both structural and functional connectomes, when graph based dissimilarity of subjects is used as the metric of injury. We further show that the proposed dissimilarity score also correlates well with the cognitive scores of the patients.

2   Materials and Methods

2.1   Dataset

We use a traumatic brain injury dataset consisting of 35 controls and 39 TBI patients [4]. For each subject, DTI and rs-fMRI data was acquired, once for healthy controls and three times for patients with TBI at approximately 3, 6, and 12 months post-injury, respectively. In addition to neuroimaging, TBI patients were also subjected to behavioral assessment at these time points.

Preprocessing of DTI Data: For each subject, DTI data was acquired on a Siemens 3T TrioTim scanner with a 8 channel head coil (single shot spin echo sequence, $TR/TE = 6500/84$ ms, $b = 1000$ s/mm^2, 30 gradient directions). 86 region of interests from the Desikan atlas [15] were extracted to represent the nodes of the structural network. A mask was defined using voxels with an FA of at least 0.1 for each subject. Deterministic tractography was performed to generate and select 1 million streamlines, seeded randomly within the mask. Angle curvature threshold of 60°, and a min and max length threshold of 5 mm and 400 mm were applied, resulting in an 86 × 86 adjacency matrix of weighted connectivity values, where each element represents the number of streamlines between regions.

Preprocessing of rs-fMRI Data: For each subject, resting state fMRI data was acquired on a Siemens 3T TrioTim scanner with a 8-channel head coil (single-shot, multi-slice, gradient-echo (GE) echoplanar (EPI) sequence, $TR/TE = 3000/3$ ms, interleaved acquisition, 3 mm isotropic voxel dimensions). The resting state fMRI data were pre-processed using preprocessing pipeline in [16]. The first 6 volumes of the BOLD 4-D time series data were discarded to allow signal stabilization. All functional time series were slice-time corrected, motion corrected to the median image using a rigid registration, and co-registered with the anatomical MPRAGE image. The DVARS method [17] was used to estimate the degree of image intensity change across volumes attributed to motion where volumes with excessive motion were flagged and not used for further analysis. Confound regression was performed to regress out the average BOLD signal from non-gray matter tissue compartments. Pairwise correlations were calculated across 86 nodes of the Desikan atlas [15] using Pearson's correlation, and finally an 86 × 86 resting state functional connectivity matrix was built for each subject.

Cognitive Measures: In our analysis, we included four clinical measures for assessing outcomes of injury that are widely used in TBI research: first three measures are used for quantifying cognitive function, and the fourth is used for measuring injury severity.

(i) **Processing speed (PS):** We used the Processing Speed Index from the Wechsler Adult Intelligence Scale-IV [18] to determine the speed of mental processing.

(ii) **Executive function (EF):** Following five tests were used to measure different aspects of executive function: Controlled Oral Word Association Test [19], Trail Making Test-Part B [20], Color-Word Interference Test, and Digits Backward and Letter-Number Sequencing subtests from the Wechsler Memory Scale IV [21]. In order to reduce type I error and increase signal-to-noise ratio, we constructed a composite score by identifying the rank of a participant on each individual measure, and then averaging the ranks across five measures.

(iii) **Verbal learning (VL):** The Rey Auditory-Verbal Learning Test [22] was used to evaluate verbal learning.

(iv) **Duration of post-traumatic amnesia (PTA):** PTA is a behavioral index of the severity of the neurological injury, which is calculated as the number of days between the TBI and the time within 72 h that the participant was fully oriented.

2.2   Graph Matching as a Similarity Measure

Having obtained structural and functional connectomes of patients and controls, our goal is to investigate connectivity based (dis)similarities across the two groups. Since graph theory provides a rich repertoire of tools that can be utilized to characterize the properties of brain networks and solve various network related problems [23], we considered connectomes as graphs. A graph is an ordered pair $G = (V, E)$ consisting of a set of nodes V, and a set of edges $E \subset V \times V$ that define a relation between node pairs. Connectomes can be effectively represented with weighted graphs where nodes correspond to brain regions and the weighted edges correspond to connectivity between brain regions with edge weights quantifying the strength of pairwise connectivity. Since the structural connectomes that we obtained from tractography and functional connectomes that we obtained via Pearson's correlation are symmetrical matrices indicating the nondirectional nature of relation between regions, we represented the connectomes with undirected weighted graphs.

Once connectomes are represented as graphs, an efficient way of calculating (dis)similarity across subjects is through *graph matching* [9], where the goal is to find a mapping between the nodes of two graphs along with an overall score quantifying their similarity. In this study, we formulated the graph matching as an instance of the linear assignment problem [24]. Specifically, given two graphs $\mathcal{P} = (V_\mathcal{P}, E_\mathcal{P})$ and $\mathcal{Q} = (V_\mathcal{Q}, E_\mathcal{Q})$, the aim is to find the optimal one-to-one mapping $f : V_\mathcal{P} \rightarrow V_\mathcal{Q}$ between their nodes while minimizing the following objective function:

$$\sum_{p \in V_\mathcal{P}} c(p, f(p)) \tag{1}$$

where $c(\cdot, \cdot)$ is a cost function determining the cost of assigning each node in $\mathcal{P}$ to a corresponding node in $\mathcal{Q}$. In order to define assignment cost between nodes of graphs, we annotated each node with an 86-dimensional feature vector which represented their connectivity with the rest of the nodes in the graph. We then considered Euclidean distance between feature vectors of nodes as the cost function. We used the Hungarian algorithm [25] for solving the matching problem, obtaining the optimal one-to-one mapping f between the nodes of the two graphs.

Solution to the graph matching can provide two similarity measures. First one is a summary measure obtained by summing the assignment costs of nodes in (1) based on the calculated mapping f. Hence, matching connectomes of a healthy control and a severe TBI patient, one would expect to get a high dissimilarity score. The second measure, denoted matching accuracy, leverages apriori knowledge of true matching between nodes by quantifying the similarity as the ratio of nodes that matched to its counterpart in the other graph over total number of nodes. Contrary to the previous measure, one would expect to get a lower score for comparing a healthy control with a severe TBI patient. In this study, we investigated the first similarity measure and left the analysis with the second measure to an extensive future study.

2.3   Experimental Setup

In our analysis, we considered the dissimilarity of individual subjects relative to the control population as a measure of TBI severity. For each patient at all three time points, we calculated its dissimilarity with each healthy control subject using graph matching. We then considered the average of these distances as the dissimilarity of the patient with respect to healthy control population. In order to analyze the distribution of dissimilarities across subject groups, we similarly calculated the average dissimilarity of each healthy subject with respect to the rest of the healthy control population. We finally calculated the z-score of each of the dissimilarity scores with respect to the healthy control population to obtain a standardized dissimilarity measure. In our analysis, we used weighted graph representation of structural and functional connectomes of subjects separately for calculating dissimilarities. While investigating dissimilarity using functional data, we separately evaluated the complete functional connectome having both positive and negative correlations, and two additional connectomes that consist of only positive and only negative functional edges.

3   Results and Discussion

Here, we demonstrate the efficacy of the proposed connectomic dissimilarity measure over the TBI dataset in distinguishing subject groups and its correlation with cognitive scores of individual subjects.

3.1   Structural Connectivity

Boxplot of average structural dissimilarities of subjects with respect to the healthy control population is shown in Fig. 1. We observe that patient groups

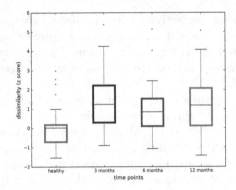

Fig. 1. Z-score of graph dissimilarity of subjects in each group (healthy and patients in 3 time points) with respect to the healthy controls, where the distance is measured over the structural connectome of subjects. Red lines indicate means of distributions. Group differences are significant between patients at all time points and controls. (Color figure online)

Table 1. Correlation between the graph dissimilarity over structural connectomes and cognitive measures for participants with TBI. p-values are shown in parenthesis and are corrected for multiple comparison using false discovery rate (FDR).

	3 Months	6 Months	12 Months
Processing speed	**−0.39(0.027)**	**−0.59(0.003)**	**−0.49(0.020)**
Executive function	−0.17(0.298)	**−0.48(0.013)**	**−0.75(0.000)**
Verbal learning	−0.28(0.116)	**−0.40(0.026)**	−0.40(0.055)
PTA	**0.41(0.027)**	**0.44(0.017)**	**0.56(0.008)**

at 3, 6, and 12 months have higher dissimilarity scores relative to healthy controls, demonstrating significant group differences with effect sizes of 0.91, 0.68, and 0.88, respectively (Student's t-test, $p < 0.01$, effect size calculated using Cohen's d). Distribution of dissimilarities span a larger interval for the patient groups than controls, which can be attributed to heterogeneity of the TBI. We observe a decline in dissimilarity at 6 months which later on increases at 12 months, where the group difference between the 3–6 months and 6–12 months are significant with effect sizes 0.95 and 0.72, respectively ($p < 0.01$). It is also interesting to note that the longitudinal trajectory of processing speed follows that of dissimilarity scores. This pattern resembles the curvilinear recovery trajectory of the PS reported in [26] over the same data set. Overall, these results indicate that the dissimilarity captured by graph matching between the graph representation of structural connectomes distinguishes TBI patients from controls, as well as patient groups at different time points from each other.

We further calculated the correlation between dissimilarity scores of the patients with respect to healthy controls, and their cognitive scores, to investigate whether the dissimilarity score captures cognitive changes in TBI (Table 1).

We observe significant correlations between the dissimilarity scores of patients and their cognitive scores and PTA duration. Specifically, a significant negative correlation was observed for processing speed at all time points, with correlation reaching its peak at 6 months. It is interesting to note the parallelism between the strength of correlation peaking at 6 months and the processing speed recovering at its highest level at 6 months post-injury. In addition, the negative correlation of executive function is significant in 6 and 12 months with an increase in the magnitude of correlation over time. We also observed verbal learning to significantly correlate with dissimilarity at 6 months.

We observe significant negative correlations between PTA and the similarity scores of patients at 3, 6, and 12 months, with a steady increase in the magnitude of the correlation over time. Noting that PTA is a measure of trauma severity which is calculated once for each patient based on the duration of post traumatic amnesia, the increase in correlation over time can be attributed to the structural connectomes at 12 months post-injury revealing the degree of injury severity better due to progressed neurodegeneration. PTA duration is assumed to reflect the overall amount of axonal injury, suggesting that graph matching based dissimilarity score captures the injury level of the patients.

3.2 Functional Connectivity

Boxplots of average functional dissimilarities of subjects with respect to the healthy control population is shown in Fig. 2. We evaluated the functional dissimilarity of subjects using full functional connectome, and positive and negative functional connectomes separately. We observe significant group difference between healthy controls and patients at three time points for full functional connectome with effect sizes 0.99, 0.98, and 1.09 ($p < 0.01$, effect size calculated

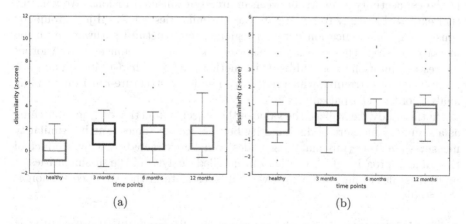

(a)   (b)

Fig. 2. Z-score of graph dissimilarity of subjects in each group (healthy and patients in 3 time points) with respect to the healthy controls. Distance is measured over (a) full and (b) positive functional connectome of subjects. Red lines indicate means of distributions. (Color figure online)

using Cohen's d), respectively (Fig. 2a). A similar pattern is observed in the positive functional connectome where the patients were significantly different than controls at 3 and 12 months with effect sizes 0.51 and 0.65 ($p < 0.05$) (Fig. 2b). We did not observe any significant group difference between the dissimilarity scores of patients and controls over negative functional connectomes. We also calculated the correlation between the dissimilarities using full, positive, and negative functional connectomes and the cognitive scores, and did not observe any significant correlation between them.

4   Conclusions and Future Work

In this paper, we presented a graph matching based approach for calculating dissimilarity between TBI patients and controls in terms of structural and functional connectomes, and demonstrated its utility over a longitudinal TBI dataset. We showed that the proposed dissimilarity metric can distinguish patient groups at three time points from healthy controls as well as differentiating patient groups in different time points from each other using both modalities. We also demonstrated that, the structural dissimilarity of patients correlates well with neuropsychological scores and a clinical measure used to evaluate TBI severity.

We note that, application of the method we presented here can be considered as a proof of concept for the use of graph matching in the calculation of dissimilarities between connectomes to assess neuropathology and plasticity in TBI. We will extend this study to develop a trauma specific dissimilarity measure by focusing on subgraphs with edges that are significantly affected by the disease across the population as suggested in [4].

As the trauma causes disconnections in the structural connectivity of the brain, one might expect to observe larger differences across patients and controls in the connectivity strength between indirectly connected regions. We will further extend this study to investigate the dissimilarities across subject groups by considering communication between regions over structural pathways by using message passing schemes such as shortest path or communicability. Another interesting question to be addressed is whether the brain changes its communication scheme post trauma, which requires an in depth structure-function coupling analysis of the TBI patients.

In this study, we utilized the value of the objective function of graph matching as a summary measure of dissimilarity between connectomes. Further similarity measures, such as (mis)matching nodes, can be extracted as a by-product of the same approach, which can help in detailed analysis of the dissimilarities at multiple scales of network, which we will investigate in the extended version of this study.

Acknowledgements. This work was funded by NIH grants R01HD089390-01A1, 1 R01 NS096606, 5R01NS092398, and 5R01NS065980.

References

1. Johnson, V.E., Stewart, W., Smith, D.H.: Axonal pathology in traumatic brain injury. Exp. Neurol. **246**, 35–43 (2013)
2. Gale, S.D., Johnson, S.C., Bigler, E.D., Blatter, D.D.: Nonspecific white matter degeneration following traumatic brain injury. J. Int. Neuropsychological Soc. **1**(1), 17–28 (1995)
3. Hayes, J.P., Bigler, E.D., Verfaellie, M.: Traumatic brain injury as a disorder of brain connectivity. J. Int. Neuropsychological Soc. **22**(2), 120–137 (2016)
4. Solmaz, B., et al.: Assessing connectivity related injury burden in diffuse traumatic brain injury. Hum. Brain Mapp. **38**(6), 2913–2922 (2017)
5. Bonnelle, V., et al.: Default mode network connectivity predicts sustained attention deficits after traumatic brain injury. J. Neurosci. **31**(38), 13442–13451 (2011)
6. Caeyenberghs, K., et al.: Altered structural networks and executive deficits in traumatic brain injury patients. Brain Struct. Function **219**(1), 193–209 (2014)
7. Bullmore, E.T., Sporns, O., Solla, S.A.: Complex brain networks: graph theoretical analysis of structural and functional systems. Nat. Rev. Neurosci. **10**(3), 186–198 (2009)
8. Sporns, O.: From simple graphs to the connectome: networks in neuroimaging. Neuroimage **62**(2), 881–886 (2012)
9. Livi, L., Rizzi, A.: The graph matching problem. Pattern Anal. Appl. **16**(3), 253–283 (2013)
10. Osmanlıoğlu, Y., Ontañón, S., Hershberg, U., Shokoufandeh, A.: Efficient approximation of labeling problems with applications to immune repertoire analysis. In: 2016 23rd International Conference on Pattern Recognition (ICPR), pp. 2410–2415. IEEE (2016)
11. Ktena, S.I., et al.: Metric learning with spectral graph convolutions on brain connectivity networks. NeuroImage **169**, 431–442 (2018)
12. Raj, A., Mueller, S.G., Young, K., Laxer, K.D., Weiner, M.: Network-level analysis of cortical thickness of the epileptic brain. Neuroimage **52**(4), 1302–1313 (2010)
13. Richiardi, J., Eryilmaz, H., Schwartz, S., Vuilleumier, P., Van De Ville, D.: Decoding brain states from fmri connectivity graphs. Neuroimage **56**(2), 616–626 (2011)
14. Mokhtari, F., Hossein-Zadeh, G.-A.: Decoding brain states using backward edge elimination and graph kernels in fMRI connectivity networks. J. Neurosci. Methods **212**(2), 259–268 (2013)
15. Desikan, R.S., et al.: An automated labeling system for subdividing the human cerebral cortex on MRI scans into gyral based regions of interest. Neuroimage **31**(3), 968–980 (2006)
16. Satterthwaite, T.D., et al.: An improved framework for confound regression and filtering for control of motion artifact in the preprocessing of resting-state functional connectivity data. Neuroimage **64**, 240–256 (2013)
17. Power, J.D., Barnes, K.A., Snyder, A.Z., Schlaggar, B.L., Petersen, S.E.: Spurious but systematic correlations in functional connectivity MRI networks arise from subject motion. Neuroimage **59**(3), 2142–2154 (2012)
18. Wechsler, D., Coalson, D.L., Raiford, S.E.: WAIS-IV: Wechsler Adult Intelligence Scale. Pearson, San Antonio (2008)
19. Benton, A.L., deS. Hamsher, K., Sivan, A.B.: Multilingual Aphasia Examination: Token Test. AJA Associates, Iowa City (1994)
20. Reitan, R.M., Wolfson, D.: The Halstead-Reitan Neuropsychological Test Battery: Theory and Clinical Interpretation, vol. 4. Reitan Neuropsychology (1985)

21. Wechsler, D.: Wechsler Memory Scale Fourth Edition (WMSIV). Pearson, San Antonio (2009)
22. Rey, A.: Memorisation d'une serie de 15 mots en 5 repetitions. L'examen clinique en psychologie (1958)
23. Fornito, A., Zalesky, A., Breakspear, M.: Graph analysis of the human connectome: promise, progress, and pitfalls. Neuroimage 80, 426–444 (2013)
24. Koopmans, T.C., Beckmann, M.: Assignment problems and the location of economic activities. Econometrica: J. Econometric Soc. 25, 53–76 (1957)
25. Kuhn, H.W.: The hungarian method for the assignment problem. Naval Res. Logistics Quart. 2(1–2), 83–97 (1955)
26. Rabinowitz, A.R., Hart, T., Whyte, J., Kim, J.: Neuropsychological recovery trajectories in moderate to severe traumatic brain injury: influence of patient characteristics and diffuse axonal injury. J. Int. Neuropsychological Soc. 24(3), 237–246 (2018)

Multi-scale Convolutional-Stack Aggregation for Robust White Matter Hyperintensities Segmentation

Hongwei Li[1], Jianguo Zhang[3(✉)], Mark Muehlau[2], Jan Kirschke[2], and Bjoern Menze[1]

[1] Technical University of Munich, Munich, Germany
[2] Klinikum rechts der Isar, Munich, Germany
[3] University of Dundee, Dundee, UK
j.n.zhang@dundee.ac.uk

Abstract. Segmentation of both large and small white matter hyperintensities/lesions in brain MR images is a challenging task which has drawn much attention in recent years. We propose a multi-scale aggregation model framework to deal with volume-varied lesions. Firstly, we present a specifically-designed network for small lesion segmentation called *Stack-Net*, in which multiple convolutional layers are 'one-by-one' connected, aiming to preserve rich local spatial information of small lesions before the sub-sampling layer. Secondly, we aggregate multi-scale *Stack-Nets* with different receptive fields to learn multi-scale contextual information of both large and small lesions. Our model is evaluated on recent MICCAI WMH Challenge Dataset and outperforms the state-of-the-art on lesion recall and lesion F1-score under 5-fold cross validation. It claimed the **first place** on the hidden test set after independent evaluation by the challenge organizer. In addition, we further test our pretrained models on a Multiple Sclerosis lesion dataset with 30 subjects under cross-center evaluation. Results show that the aggregation model is effective in learning multi-scale spatial information.

Keywords: White matter hyperintensities · Deep learning

1 Introduction

White matter hyperintensities (WMH) characterized by bilateral, mostly symmetrical lesions are commonly seen on FLAIR magnetic resonance imaging (MRI) of clinically healthy elderly people; furthermore, they have been repeatedly associated with various neurological and geriatric disorders such as mood problems and cognitive decline [1]. Detection of such lesions on MRI has become a crucial criterion for diagnosis and predicting prognosis in early stage of diseases.

Different from brain tumor segmentation [2] in MR images where most of the abnormal regions are large and with spatial continuity, in the task of WMH

© Springer Nature Switzerland AG 2019
A. Crimi et al. (Eds.): BrainLes 2018, LNCS 11383, pp. 199–207, 2019.
https://doi.org/10.1007/978-3-030-11723-8_20

segmentation, both large and small lesions with high discontinuity are commonly found as shown in Fig. 1. Generally, small abnormal region contains relatively less contextual information due to the poor spatial continuity. Furthermore, the feature representation of small lesions tend to be trivial when image features are extracted in a global manner. One solution to tackle this issue is to use an ensemble model or aggregation model [3] to learn different attributes i.e., multi-levels of feature representation from the training data.

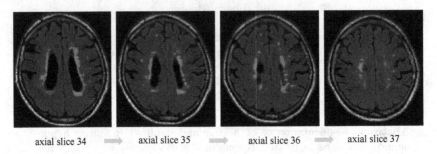

axial slice 34   axial slice 35   axial slice 36   axial slice 37

Fig. 1. From left to right: axial slices from 34 to 37 of one case from the MICCAI WMH Challenge public training set, showing the high discontinuity of white matter hyperintensities. The red pixels indicate the WMH annotated by a neuroradiologist. (Color figure online)

Although there exist various computer-aided diagnostic systems for automatic segmentation of white matter hyperintensities [4,5], the reported results are largely incomparable due to different datasets and evaluation protocols. The MICCAI WMH Segmentation Challenge 2017[1] was the first competition held to compare state-of-the-art algorithms on this task. The winning method [6] of the challenge employed the modified U-Net [7] architecture and ensemble models (*U-Net ensembles* in short). Three U-Net models of same architecture were trained with shuffled data and different weight initializations.

In traditional fully convolutional networks [8], each convolutional layer is followed by a max-pooling operation which causes the loss of spatial information. In the task of WMH segmentation, this sub-sampling operation can be devastating because small-volume hyperintensities with less than 10 voxels are commonly found. Instead of using single convolutional layer before the sub-sampling layer, we hypothesize that a *convolutional stack* with multiple convolutional layer is able to extract rich local information and it would be more effective by propagating the feature maps learned to the high-resolutional *deconvolutional layer* by skip connections similar to the U-Net approach [7].

In this paper, we present a *stacked* architecture of fully convolutional network called *Stack-Net* which aims at preserving the local spatial information of small lesions and propagating them to deconvolutional layers. We further aggregate two *Stack-Nets* with different receptive fields to learn multi-scale spatial information from both large and small abnormal regions. Our method outperforms

---

[1] http://wmh.isi.uu.nl/.

the state-of-the-art in lesion recall by 4% on the MICCAI Challenge Dataset with 60 cases. In addition, we test our pre-trained models on a private Multiple Sclerosis (MS) lesion dataset with 30 subjects. Results further demonstrate the effectiveness of the aggregation idea.

2   Method

2.1   Convolutional Stack and Multi-scale Convolutions

Formulation. Let $f(\cdot)$ represent the nonlinear activation function. The kth output feature map of lth layer Y_{lk} can be computed as: $Y_{lk} = f(W_{lk}^r * x)$ where the input image is denoted by x; the convolutional kernel with fixed size $r \times r$ related to the k feature map is denoted by W_{lk}^r; the multiplication sign refers to the 2D convolutional operator, which is used to calculate the inner product of the filter model at each location of the input image. We now generalize this convolution structure from layer to stack. Let a convolutional stack S contain L convolutional layers, the kth output feature map Y_{Sk} of S can be computed as

$$Y_{Sk} = f(W_{Lk}^r * f(W_{(L-1)k}^r \cdots f(W_{0k}^r * x))\ldots) \tag{1}$$

Obviously, the use of multiple connected convolutional layers to replace single layer would lead to the increase of computational complexity. However, the local spatial information of small lesions could be largely reduced after the first pooling layer (with 2×2 kernel or larger). As a result, we only replaced the first two convolutional layers before sub-sampling layers with convolutional stack as shown in Fig. 2.

We further employed multi-scale convolutional kernels to learn different contextual information from both large and small abnormal regions. In our task, we aggregate the proposed two *Stack-Nets* with different receptive fields. *Stack-Net* with small receptive field i.e., 3×3 kernel, is expected to learn local spatial information of small-volume hyperintensities while *Stack-Net* with large kernel is designed to learn spatial continuity of large abnormal regions. Two models were trained and optimized independently and were aggregated by using a voting strategy during the testing stage. Let P_i be the 3D segmentation probablity masks predicted by one single model M_i. Then the final segmentation probability map of the aggregation of n models is defined as: $P_{aggr} = \frac{1}{n}\sum_i P_i$. The threshold for generating binary mask is set to 0.4.

Architecture. As shown in Fig. 2, we built two *Stack-Nets* with different convolutional kernels, which takes as input the axial slices (2D) of two modalities from the brain MR scans during both training and testing. Different from the winning architecture *U-Net Ensembles* [6] in the MICCAI challenge, we replaced the first two convolutional layers by a convolutional stack, with 3×3 and 5×5 kernel size respectively. Each convolutional stack is followed by a rectified linear unit (ReLU) and a 2×2 max pooling operation with stride 2 for downsampling. The depth of the convolutional stack was set as $L = 5$. In total the network contains 24 convolutional and de-convolutional layers.

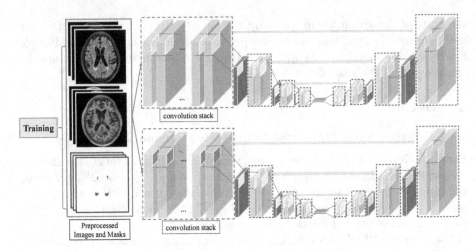

Fig. 2. Overview of the multi-scale convolutional-stack aggregation model. We replaced the traditional single convolutional-layer with convolutional-stack to extract and preserve local information of small lesions. The depth of the convolution-stack was flexible and set to 5 in our experiments. Two convolutional kernels i.e., 3×3 and 5×5 were used in two *Stack-Nets* to learning multi-scale context information. The detailed parameters setting/architecture was presented in Fig. 3.

Training. For the data preprocessing, each slice and the corresponding segmentation mask were cropped or padded to 200×200 to guarantee a uniform input for the model. Then we obtained the brain mask using simple thresholding and mask filling. Gaussian normalization was applied to each subject to rescale the intensities. Dice loss function [9] was employed during the training process. Data augmentation including rotation, shearing and zoom was used during the batch training. The optimal number of epochs was set to 50 by contrasting training loss and validation loss over epochs. The batch size was set to 30 and learning rate was set to 0.0002 throughout all of the experiments.

3 Materials

3.1 Datasets and Experimental Setting

Two clinical datasets: the public MICCAI WMH dataset with 60 cases from 3 centers and a private MS Lesion dataset with 30 cases collected from a hospital in Munich, were employed in our experiments. For each dataset, the FLAIR and T1 modality of each subject were co-registered. Properties of the data were summarised in Table 1. In the experiments reported in Sects. 4.1 and 4.2, five-fold cross-validation setting was used. Specifically, subject IDs were used to split the public training dataset into training and validation sets. In each split, slices from 16 subjects from each center were pooled into training set, and the slices from the remaining 4 subjects from each center for testing. This procedure was

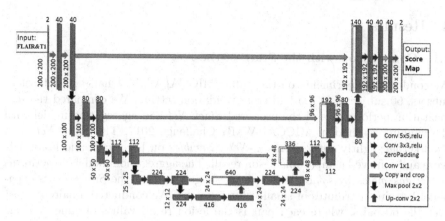

Fig. 3. Detailed parameters setting of the deep networks. The number of stacked layers is 2.

Table 1. Detailed information of *MICCAI WMH challenge* dataset from three centers and private *Multiple Sclerosis* dataset from a hospital in Munich.

Datasets	Lesion type	Subjects	Voxel size (m^3)	Size of FLAIR & T1 Scans
Utrecht	WMH	20	$0.96 \times 0.95 \times 3.00$	$240 \times 240 \times 48$
Singapore	WMH	20	$1.00 \times 1.00 \times 3.00$	$252 \times 232 \times 48$
GE3T	WMH	20	$0.98 \times 0.98 \times 1.20$	$132 \times 256 \times 83$
Munich	Multiple sclerosis	30	$1.00 \times 1.00 \times 0.99$	$240 \times 240 \times 170$

repeated until all of the subjects had been used in testing phase. The Dice score, lesion recall and lesion F1-score of all testing subjects were averaged afterwards.

3.2 Evaluation Metrics

Three evaluation metrics were used to evaluate the segmentation performance of the algorithm in different aspects from MICCAI WMH Challenge. Given a ground-truth segmentation map G and a segmentation map P generated by an algorithm, the evaluation metrics are defined as follows. **Dice score:** $DSC = 2(G \cap P)/(|G| + |P|)$. This metric measures the overlapping volume of G and P. **Recall for individual lesions:** Let N_G be the number of individual lesions delineated in G, and N_P be the number of correctly detected lesions after comparing P and G. Each individual lesion is defined as a 3D connected component. Then the recall for individual lesions is defined as: $Recall = N_P/N_G$. **F1-score for individual lesions:** Let N_P be the number of correctly detected lesions after comparing P and G. N_F be the number of wrongly detected lesions in P. Each individual lesion is defined as a 3D connected component. Then the recall for individual lesions is defined as: $F1 = N_P/(N_P + N_F)$.

4   Results

4.1   Comparison with the State-of-the-Art

We conducted experiments on the public MICCAI WMH Challenge dataset (3 subsets, 60 subjects) in a 5-fold cross validation setting. We compared the segmentation performance of the proposed *Stack-Net* and aggregation model with the winning method in MICCAI WMH Challenge 2017. The *Stack-Net* with 3×3 kernel slightly outperforms *U-Net ensembles* on Dice score and lesion F1-score and achieved comparable lesion recall. The aggregation model outperforms *U-Net ensembles* by 4% on lesion recall, suggesting that the *Stack-Net* is capable of learning attributes of small-volume lesions. We conducted a paired Z-test over the 60 pairs, where each pair is the lesion recall values obtained on one validation scan by the proposed aggregation model and *U-Net ensembles*. Small p-value ($p < 0.01$) indicates that the improvements are statistically significant. Figure 4 shows a segmentation case in which we can see that our aggregation model is more effective in detecting small lesion. Furthermore, the proposed method claimed to be the **first place** (teamname: *sysu_media_2*) on the hidden set after independent evaluation by the challenge organizer. Please see the details in http://wmh.isi.uu.nl/results/.

To better understand how each part of the proposed model worked effectively on the volume-varied lesions, we grouped all the lesions into three types: small, medium and large, by defining the volume range of each type. Three sets $S_{small} = \{s| \ volume(lesion) < 10\}$, $S_{medium} = \{s| \ 10 < volume(lesion) < 20\}$ and $S_{large} = \{s| \ volume(lesion) > 20\}$ were obtained. Then the number of detected lesions of three types was calculated by comparing the predicted segmentation masks and ground-truth segmentation masks on all the test subjects. Figure 5(a) further shows the distribution of detected lesions with small, medium and large volumes respectively. Our aggregation model detected 2008 small lesions while the U-Net ensembles detected 1851, i.e., **8%** improvement over U-Net ensembles. We conducted a paired Z-test over the 60 pairs, where

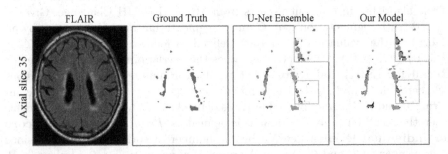

Fig. 4. Segmentation result of a testing case from *Utrecht* by U-Net ensembles and our model respectively. The green area is the overlap between the segmentation result and the ground truth. The red pixels are the false negatives, and the black ones are the false positives. (Color figure online)

each pair is the recall values of small lesion obtained on one validation scan by the proposed aggregation model and *U-Net ensembles*. Small p-value ($p < 0.01$) indicates that the improvements are statistically significant. We also observed that the aggregation model achieved a comparable Dice score which measures the overlapping volumes, demonstrating that it was effective in dealing with both large and small lesions. The *Stack-Net* with 5×5 kernel slightly outperformed U-Net ensembles in the detection of large lesions. It demonstrates that large convolutional kernel is effective in learning contextual information from large abnormality regions with spatial continuity (Table 2).

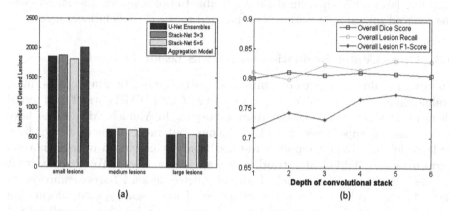

(a)   (b)

Fig. 5. (a) Distribution of small, medium and large lesions detected by U-Net ensembles, each component of our aggregation model; (b) Overall Dice, lesion recall and lesion F1-score achieved by six *Stack-Nets* with different depths.

Table 2. Comparsion with the winning method in MICCAI WHM Challenge 2017. Values in bold indicates results outperforming the state-of-the-art.

Method	Dice score	Lesion recall	Lesion F1-score
U-Net ensembles [6]	80.10%	82.96%	76.41%
Stack-Net with 3×3 kernel (ours)	**80.75%**	82.94%	**77.29%**
Stack-Net with 5×5 kernel (ours)	**80.46%**	80.96%	**76.75%**
Multi-scale aggregation model (ours)	80.09%	**86.96%**	**76.73%**

Table 3. Segmenation performance on the MS lesion dataset. Figures in bold indicate the best performance.

Method	Dice score	Lesion recall	Lesion F1-score
U-Net ensembles [6]	75.95%	93.16%	42.41%
Stack-Net with 3×3 kernel (ours)	75.89%	**94.76%**	42.59%
Stack-Net with 5×5 kernel (ours)	74.71%	93.37%	41.15%
Multi-scale aggregation model (ours)	**76.93%**	93.16%	**49.57%**

4.2 Analysis on the *Stack-Net*

To investigate the effect of the depth in *Stack-Net*, we evaluate six models with 3×3 kernel with depths ranging from 1 to 6 on the MICCAI WMH dataset using 5-fold cross validation. Using the same grouping criteria and calculation strategy as mentioned above, we calculate the averaged Dice, averaged lesion recall and averaged lesion F1-score on all test subjects after 5 splits. As one can observe from Fig. 5(b) that using the *thin* convolutional stack i.e., one or two convolutional layer yields relatively poor segmentation performance on three evaluation metrics. This is because spatial information is reduced drastically after the subsampling layer while the thin stack is not able to preserve rich information and the reduced spatial information is propagated to the deconvolutional layers.

4.3 Cross-Center Evaluation on the MS Lesion Dataset

To further evaluate the idea of multi-scale spatial aggregation in a cross-center-evaluation manner, we trained the models on MICCAI WMH dataset, and tested them on the MS lesion dataset from a hospital in Munich. MS lesions have a very similar appearance with WM lesions, but most of them are medium or large lesions. Table 3 reported resulted from a comparison of segmentation performance of individual network and aggregation model. We observed that the aggregation model achieved significantly better lesion F1-score compared to individual networks, suggesting combination of multi-scale spatial information can help to remove false positives. Interestingly, we found the lesion recall did not improve after aggregating the individual *Stack-Nets*. This is due to the fact that most of the MS lesions are in medium or large size, which made the function of convolutional stack achieve limited improvement over the lesion recall. It further suggested that the aggregation of models with multi-scale receptive field is effective in learning multi-scale spatial information.

5 Conclusions

In this paper, we explored an architecture specifically designed for small lesion segmentation, to learn attributes of small regions. We found the convolutional stack was effective in preserving local information of small lesions and the rich information was propagated to the high-resolution deconvolutional stack. By aggregating multi-scale *Stack-Net* with different receptive fields, our method outperformed the state-of-the-art on MICCAI WMH Challenge dataset. We further showed multi-scale context aggregation model was effective in MS lesion segmentation under a cross-center evaluation.

Acknowledgements. This work was supported in part by NSFC grant (No. 61628212), Royal Society International Exchanges grant (No. 170168), the Macau Science and Technology Development Fund under 112/2014/A3. We gratefully acknowledge the support of NVIDIA Corporation with the donation of the Titan Xp GPU used for this research.

References

1. Debette, S., Markus, H.: Bmj **341**, c3666 (2010)
2. Menze, B.H., et al.: IEEE Trans. Med. Imaging **34**(10), 1993–2024 (2015)
3. Yu, F., Koltun, V.: arXiv preprint arXiv:1511.07122 (2015)
4. Borghesani, P.R., et al.: Neuropsychologia **51**(8), 1435–1444 (2013)
5. Moeskops, P., et al.: NeuroImage: Clin. **17**, 251–262 (2018)
6. Li, H., et al.: arXiv preprint arXiv:1802.05203v1 (2018)
7. Ronneberger, O., Fischer, P., Brox, T.: U-Net: convolutional networks for biomedical image segmentation. In: Navab, N., Hornegger, J., Wells, W.M., Frangi, A.F. (eds.) MICCAI 2015. LNCS, vol. 9351, pp. 234–241. Springer, Cham (2015). https://doi.org/10.1007/978-3-319-24574-4_28
8. Long, J., Shelhamer, E., Darrell, T.: Proceedings of the IEEE Conference on Computer Vision and Pattern Recognition, pp. 3431–3440 (2015)
9. Milletari, F., Navab, N., Ahmadi, S.-A.: 2016 Fourth International Conference on 3D Vision (3DV), pp. 565–571. IEEE (2016)

Holistic Brain Tumor Screening and Classification Based on DenseNet and Recurrent Neural Network

Yufan Zhou[1], Zheshuo Li[1], Hong Zhu[2], Changyou Chen[1], Mingchen Gao[1], Kai Xu[3(✉)], and Jinhui Xu[1(✉)]

[1] Department of Computer Science and Engineering, University at Buffalo, The State University of New York, Buffalo, USA
jinhui@buffalo.edu
[2] School of Medical Information, Xuzhou Medical University, Xuzhou, China
[3] Affiliated Hospital of Xuzhou Medical University, Xuzhou, China
xukai@xzhmu.edu.cn

Abstract. We present a holistic brain tumor screening and classification method for detecting and distinguishing multiple types of brain tumors on MR images. The challenges arise from the significant variations of location, shape, size, and contrast of these tumors. The proposed algorithms start with feature extraction from axial slices using dense convolutional neural networks; the obtained sequential features of multiple frames are then fed into a recurrent neural network for classification. Different from most other brain tumor classification algorithms, our framework is free from manual or automatic region of interests segmentation. The results reported on a public dataset and a population of 422 proprietary MRI scans diagnosed as normal, gliomas, meningiomas and metastatic brain tumors demonstrate the effectiveness and efficiency of our method.

1 Introduction

Brain tumor is one of the most fatal cancers. In the United States, an estimated 700,000 people are living with primary brain and central nervous system tumors. Nearly 80,000 new cases of primary brain tumors are diagnosed yearly, and approximately one-third are malignant [1]. Many different types of brain tumors exist. The most prevalent brain tumor types in adults are gliomas and meningiomas.

Medical imaging plays a central role in diagnosing brain tumors. There are many imaging modalities that can provide information about brain tissue non-invasively, such as Magnetic Resonance Images (MRI), Computed Tomography (CT) and Positron Emission Tomography (PET). MRI has particularly been used frequently in brain tumor detection and identification, due to its high contrast of soft tissue, high spatial resolution and free of radiation. Despite these

---

Y. Zhou, Z. Li and H. Zhu are equally contributed co-first authors.

© Springer Nature Switzerland AG 2019
A. Crimi et al. (Eds.): BrainLes 2018, LNCS 11383, pp. 208–217, 2019.
https://doi.org/10.1007/978-3-030-11723-8_21

facts, brain tumor diagnosis still remains a challenging task. Its detection heavily relies on the experience of radiologists, and diagnosing a large amount of data can be quite time-consuming and sometimes non-reproducible.

Computer-Aided Diagnosis (CAD) can provide tremendous help in brain tumor diagnosis, prognosis and surgery. A typical brain tumor CAD system consists of three main phases, tumor region of interest (ROI) segmentation, feature extraction, and classification (based on the extracted features) [4–6]. Brain tumor segmentation, either manual or automatic, is perhaps the most important and time-consuming phase of such a system. A great deal of effort has been devoted to this problem, e.g., releasing publicly available benchmark datasets and organizing challenges [10]. Many algorithms have been proposed to solve the brain tumor segmentation problem, such as Deep Neural Networks [7] and SVM with Conditional Random Field [3]. Classifications based on SVM and/or ANN are then followed to distinguish different types of brain tumors based on the extracted features from ROIs. An obvious limitation of such frameworks is the need of tracing ROIs, which can cause a few problems. Firstly, since brain tumors can vary dramatically in their shapes, sizes, and locations, tracing ROIs could be quite challenging and often not fully automatic. This may cause significant errors to the segmentation, and be accumulated into the following phases, thus leading to inaccurate classification. Secondly, the tumor-surrounding tissues are suggested to be discriminative between different tumor categories [5]. Thirdly, relying solely on the features of ROIs means complete ignorance of the location information of the tumors, which can affect the classification considerably.

The aforementioned problems motivate us to propose an alternative approach for brain tumor screening and classification, eliminating the segmentation phase completely. Particularly, we propose to use the holistic 3D images directly without detailed annotation at the pixel or slice levels. Our approach models the 3D holistic images as sequences of 2D slices. It first adopts an auto-encoder, based on a deep DenseNet, to extract features of each 2D image. This allows us to avoid using the original noisy and high dimensional data. After features of 2D slices extracted, it is natural to apply a Recurrent Neural Network (RNN), specifically the Long Short Term Memory (LSTM) model to handle the sequential data for the classification. We also apply a purely convolutional model for sequential data, by stacking 2D slices features together to be treated as another image data. This is inspired by a recent work of using purely convolutional auto-encoder for sequence representation learning [12].

Our contributions in this work are three-fold:

- The proposed models only need holistic label of patients other than pixel-wise/slice-wise labeling. Holistic labels are much easier to obtain in clinical routine.
- We have collected a dataset of 422 MRI scans, containing normal control images as well as three types of brain tumors (i.e., meningioma, glioma, and metastasis tumor).[1]

---

[1] The anonymized proprietary dataset will be shared publicly with labels later on.

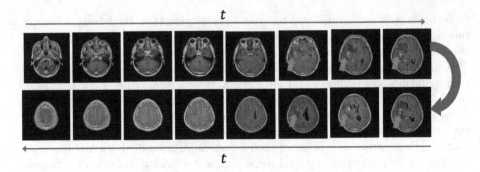

Fig. 1. An MR image sequence of a glioma patient.

– Our deep neural network implements a novel architecture, treating 3D data as sequences of 2D slices, and using RNN or CNN to learn sequence-to-label mapping, with a DenseNet based auto-encoder for feature extraction. Two proposed models **DenseNet-LSTM** and **DenseNet-DenseNet** are demonstrated with two experiments **tumor screening** and **tumor type classification** using both public and proprietary datasets.

2 Preliminaries

2.1 Brain-Tumor Image Representations

Brain tumors are usually diagnosed with MRI or CT images, where patient i is represented by a sequence of 2-D images, denoting as $\mathbf{X}_i = \{x_1^{(i)}, \cdots, x_T^{(i)}\}$ with $x_t^{(i)} \in \mathbb{R}^{\ell_1 \times \ell_2}$ being the t-th frame image. Different from existing label-exhaustive datasets where each 2-D image is associated with a label, in our dataset, each sequence of images $\mathbf{X}_i$ is associated with a single label $y_i \in \{0, 1, \cdots, P\}$, where P is the number of tumor types. As a result, our dataset is represented as $\mathcal{D} \triangleq \{(\mathbf{X}_i, y_i)\}_{i=1}^N$ with N being the total number of image sequences (including patients and normal people). Figure 1 illustrates an example sequence of MRI images from a Glioma patient in our proprietary dataset. Note that there are only a few frames showing the existence of Glioma.

2.2 DenseNet

DenseNet [9] is a recently proposed special type of convolutional neural networks, where the current layer is connected by all its previous layers. The structure has some advantages over existing structures such as alleviating the vanishing-gradient problem, strengthening feature propagation, encouraging feature reuse, and reducing the number of parameters. A deep DenseNet is defined as a set of DenseNets (called dense blocks) connected sequentially, with additional convolutional and pooling operations between consecutive dense blocks. By such a

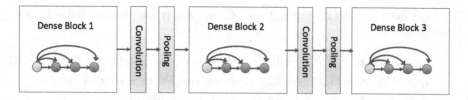

Fig. 2. A deep DenseNet with 3 dense blocks. In each dense block, the input for a particular layer is the concatenation of all outputs from its previous layers; the output is obtained by convolving the input with some kernels to be learned.

construction, we can build a deep neural network flexible enough to represent complicated transformations. An example of the deep DenseNet is illustrated in Fig. 2.

2.3   Recurrent Neural Network (RNN)

RNN is a powerful framework to model sequence-to-sequence data. In our brain tumor application, the input sequence corresponds to features of the MRI images, which are extracted with a DenseNet described above; the output sequence degenerates to a single label, indicating whether the input sequence is diagnosed as tumor or not. Specifi-

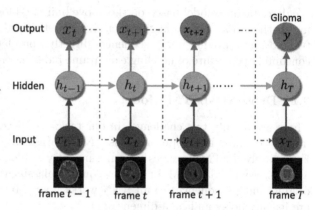

Fig. 3. The RNN structure.

cally, consider an input sequence $\mathbf{X} = \{x_1, \cdots, x_T\}$, where x_t is the input data vector at time t. The corresponding hidden state vector h_t at each time t is recursively calculated by applying a transition function $h_t = \mathcal{H}(h_{t-1}, x_t)$ (specified below). Finally, the output y is calculated by mapping the final state h_T to the label space. Figure 3 illustrates the RNN structure in our setting.

Long Short-Term Memory (LSTM). Vanilla RNN defines $\mathcal{H}$ as a linear transformation followed by an activation function. This simple structure is unable to model long-term dependency from the input, as is the case in our application. Instead, we adopt the more powerful LSTM transition function by introducing a memory cell that is able to preserve the state over long periods [8]. Specifically, each LSTM unit contains a cell c_t at time t, which can be viewed as a memory unit. Reading or writing the cell is controlled through sigmoid gates: input gate

i_t, forget gate f_t, and output gate o_t. Consequently, the hidden units h_t are updated as:

$$i_t = \sigma(\mathbf{W}_i \boldsymbol{x}_t + \mathbf{U}_i \boldsymbol{h}_{t-1} + \boldsymbol{b}_i), \qquad f_t = \sigma(\mathbf{W}_f \boldsymbol{x}_t + \mathbf{U}_f \boldsymbol{h}_{t-1} + \boldsymbol{b}_f),$$
$$o_t = \sigma(\mathbf{W}_o \boldsymbol{x}_t + \mathbf{U}_o \boldsymbol{h}_{t-1} + \boldsymbol{b}_o), \qquad \tilde{c}_t = \tanh(\mathbf{W}_c \boldsymbol{x}_t + \mathbf{U}_c \boldsymbol{h}_{t-1} + \boldsymbol{b}_c),$$
$$c_t = f_t \odot c_{t-1} + i_t \odot \tilde{c}_t, \qquad h_t = o_t \odot \tanh(c_t)$$

where $\sigma(\cdot)$ denotes the logistic sigmoid function, and $\odot$ represents the element-wise matrix multiplication operator. $\mathbf{W}_{\{i,f,o,c\}}$, $\mathbf{U}_{\{i,f,o,c\}}$ and $\boldsymbol{b}_{i,f,o,c}$ are the weights of the LSTM to be learned. Having obtained the hidden unit for the last time step T, we map $\boldsymbol{h}_T$ to y by simply using a linear transformation followed by a softmax-layer, i.e., $p(y = k|\boldsymbol{h}_T) = \mathsf{Softmax}_k(\mathbf{W}_y \boldsymbol{h}_T + \boldsymbol{b}_y)$, where $\mathsf{Softmax}_k(\boldsymbol{a}) \triangleq \frac{\exp(a_k)}{\sum_i \exp(a_i)}$, and $\mathbf{W}_y$ and $\boldsymbol{b}_y$ are the parameters to be learned.

3   Labeling-Free Brain-Tumor Classification

We describe our model based on the above building blocks. Different from existing methods for tumor classification using a standard alone CNN, we propose two models to predict image sequences directly, completely eliminating the time consuming procedure of labeling each frame independently, thus free of labeling.

3.1   DenseNet-LSTM Model

There are mainly two challenges in our task: (*i*) Directly using CNN to tackle image sequences is inappropriate as CNN is originally designed for static data. Fortunately, LSTM provides us a natural way to deal with sequence data. As a result, we adopt LSTM for image-sequence classification. (*ii*) Directly feeding original image sequences to an RNN works poorly because the original images are usually noisy and high-dimensional.

To alleviate this problem, we propose an auto-encoder structure based on the deep *DenseNet* to extract features of the original images. The features from the auto-encoder are then fed to an RNN for classification. Specifically, in an auto-encoder, one trains an encoder and a deconder together, to reconstruct the output the same as input. To train the auto-encoder given brain-tumor images $(\boldsymbol{x}_t^{(i)})_{i,t}$, the objective is to minimize the reconstruction error: $\mathcal{F} = \sum_i \sum_t \left\| \boldsymbol{x}_t^{(i)} - \mathrm{DEC}\left(\mathrm{ENC}(\boldsymbol{x}_t^{(i)})\right) \right\|^2$, where $\| \cdot \|$ is the standard Frobenius norm; $\mathrm{ENC}(\cdot)$ and $\mathrm{DEC}(\cdot)$ denote the encoder and decoder implemented by two deep DenseNets, respectively. After training the auto-encoder, the extracted features for all the images are then used as the input data to train an RNN classifier for holistic brain-tumor classification. We adopt the standard cross-entropy loss function to train the RNN. The whole framework is illustrated in Fig. 4. We denote this model as DenseNet-LSTM.

3.2 DenseNet-DenseNet Model

An alternative way to RNN for sequence classification discovered recently is to replace the RNN with a CNN [12]. We stack the features of a tumor-sequence returned from the auto-encoder as a 2-D tensor, and treat it as input data to a second deep DenseNet for classification. In this way, the inter-frame correlations is translated into column-wise correlations in a single 2-D tensor, which can be effectively modeled by the convolutional operator in a DensetNet. We denote this model as DenseNet-DenseNet.

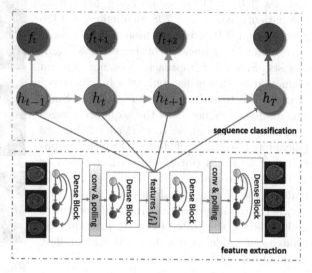

Fig. 4. The proposed DenseNet-LSTM model for labeling-free brain tumor classification.

4 Experiments

We test our proposed framework on two datasets, one public dataset and one proprietary dataset (collected by our collaborators in their hospital). We have two experiments to evaluate the proposed models: Tumor screening and tumor type classification. Tumor screening is for testing the accuracy of our approach on deciding (or screening) whether a 2D sequence image contains a tumor. Tumor type classification is to classify tumors into multiple types.

Our implementation is based on TensorFlow. To alleviate overfitting, we adopt the weight-decay regularization and dropout in the training. The auto-encoder part needs to be trained only once. It takes around 5 h for 10,000 slices from 500 MRI sequences. The second part takes about half an hour for LSTM or one hour for DenseNet. The models were trained on a Nvidia Titan Xp GPU. For all the experiments, we randomly partition the dataset into a training dataset (72%), a test dataset (14%) and a validation dataset (14%). We repeat this process for six times and report the mean and variance of the accuracies. Figure 7 shows some examples of learning curves.

Public Dataset. The public dataset [5] includes 3064 (2D) slices of brain MRI from 233 patients, containing 708 meningiomas, 1426 gliomas, and 930 pituitary tumors. The tumors were manually delineated by experienced radiologists. Since our approach does not rely on segmentation, we utilize only the holistic label of each slice to indicate the tumor type. Since this dataset does not have the sequence images needed by our model, we convert each 2D image (slice) into a sequence of 20 slices by either duplicating it 19 times (for DenseNet-DenseNet) or adding 19 zero matrices (for DenseNet-LSTM). Our purpose of using this dataset is for both validating the robustness of the proposed framework and achieving the state-of-the-art performance, though our model is not designed for handling such 2D datasets.

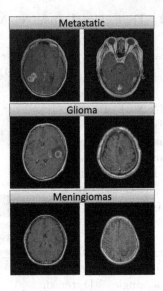

Fig. 5. Examples of the three types of brain tumors.

Proprietary Dataset. We have collected a dataset of 422 MRI scans diagnosed as normal (75), glioma (150), meningiomas (67) and metastatic brain tumors (130). For each patient, T1, T2 and Flair MR images are available. Examples of the three tumor types are depicted in Fig. 5, which shows high variations of tumors in terms of locations, shapes and sizes.

Experimental Setup. In the DenseNet-based auto-encoder, for the encoder, it is a deep DenseNet with 4 dense blocks. In each block, there are 5 convolutional layers with kernel sizes of 3×3 and 1×1. We adopt the same configurations for the decoder. For other parameters of the DenseNet, we adopt the default setting as in [9]. The dimension of the latent space for RNN is set to 128.

Minibatch size is set to 32. We use a validation set to select the learning rates from $\{1e\text{-}1, 1e\text{-}2, 1e\text{-}3, 1e\text{-}4, 1e\text{-}5\}$; the dropout rates for the input-hidden layer and each convolutional layer in the DenseNet from $\{0, 0.05, 0.1, 0.15, 0.2\}$, and the weight-decay rates from $\{1e\text{-}2, 1e\text{-}3, 1e\text{-}4, 1e\text{-}5\}$.

Tumor Screening. The public dataset is not suitable for this task since it only contains images with tumors. We evaluated three models for tumor screening on the proprietary dataset: DenseNet-RNN (with vanilla RNN as a sequence classifier), DenseNet-LSTM and DenseNet-DenseNet. Their accuracies are $87.15\% \pm 3.79\%, 91.09\% \pm 3.62\%, 92.66\% \pm 2.73\%$ respectively. DenseNet-DenseNet presents the best performance for the proprietary dataset.

Tumor Type Classification. For the public dataset, DenseNet-LSTM outperforms all the previous work on this dataset. The baseline methods [5] reports an accuracy of 91.28% for its best model based on a complicated feature engineering and extra data information (from pixel-wise labeling). A recent model based on capsule networks [2] achieves 86.56% accuracy. Furthermore, our models are much more robust and practically useful because they are designed to handle 3D sequence images and is labeling free.

Our proprietary data is significantly more difficult to learn than the public one. Our DenseNet-LSTM is the best among different variations. DenseNet-LSTM is also tested on one versus one tumor type classification, resulting in three groups of experiments. Table 1 summarized the results. Figures 7 and 6 shows the learning curves of our models on proprietary and public dataset, respectively.

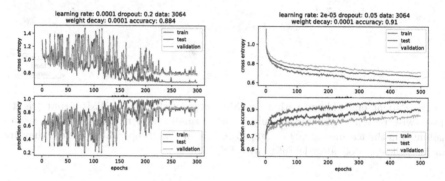

Fig. 6. Learning curves on public dataset. Left: tumor type classication with DenseNet-DenseNet. Right: tumor type classification with DenseNet-LSTM.

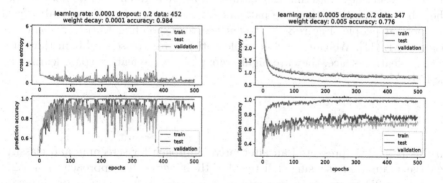

Fig. 7. Learning curves on proprietary dataset. Left: tumor screening with DenseNet-DenseNet. Right: tumor type classification with DenseNet-LSTM.

Table 1. Summary of experimental results on tumor type classification.

Tumor type classification accuracies (three types)					
Models	DenseNet-RNN	DenseNet-LSTM	DenseNet-DenseNet	[5]	[2]
Public	84.61%± 1.87%	**92.13%± 1.59%**	86.68%± 1.54%	91.28%	86.56%
Proprietary	60.00%± 5.70%	**71.10%± 3.82%**	64.95%± 5.16%	-	-

Tumor type classification accuracies (two types) with DenseNet-LSTM		
	Glioma vs Meningiomas	Glioma vs Metastatic	Meningiomas vs Metastatic
Proprietary	80.83%± 6.65%	80.00%± 8.44%	82.50%± 4.18%

Fig. 8. Patient embeddings with DenseNet output (left) and LSTM output (right). Frame-wise patient embeddings (only shows a small number of patients for ease of visibility) in the feature extraction stage (left) are not well-separable; whereas they are almost well-separable after learning with LSTM.

Patient Embeddings with DenseNet and LSTM Features: To illustrate how our proposed framework achieves high discrimination ability, we embed the features from the DenseNet auto-encoder and the LSTM classifier onto a 2-D space, respectively. Note that the features from the auto-encoder do not consider the label information; thus the patients are not expected to be separable from the normal people. Figure 8 illustrates the corresponding feature embeddings using tSNE [11]. We can see that while patients are not separable in the auto-encoder-feature space, they are highly separable in the feature space learned by LSTM.

5   Conclusion

In this paper, we presented an alternative approach for screening and classifying the brain tumors using holistic 3D MR images. Our approach is capable of utilizing 3D sequence images and does not need the pixel-wise or slice-wise labeling. Experiments on public and proprietary datasets indicate that our approach is effective and highly efficient. As future work, we plan to (1) expand our proprietary dataset for more types of brain tumors, and (2) provide model interpretability based on weakly-supervised pathology localization.

Acknowledgement. This work was supported in part by NSF through grants IIS-1422591, CCF-1422324, and CCF-1716400 and by NSFC through grants 81771904 and 61828205. It was also supported in part by start-up funds (for Drs. Mingchen Gao and Changyou Chen) from the Department of Computer Science and Engineering, University at Buffalo, the State University of New York.

References

1. Brian tumor statistics. http://www.abta.org/about-us/news/brain-tumor-statistics/. Accessed 16 Feb 2018
2. Afshar, P., Mohammadi, A., Plataniotis, K.N.: Brain tumor type classification via capsule networks. arXiv preprint arXiv:1802.10200 (2018)
3. Bauer, S., Nolte, L.-P., Reyes, M.: Fully automatic segmentation of brain tumor images using support vector machine classification in combination with hierarchical conditional random field regularization. In: Fichtinger, G., Martel, A., Peters, T. (eds.) MICCAI 2011. LNCS, vol. 6893, pp. 354–361. Springer, Heidelberg (2011). https://doi.org/10.1007/978-3-642-23626-6_44
4. Bauer, S., Wiest, R., Nolte, L.-P., Reyes, M.: A survey of MRI-based medical image analysis for brain tumor studies. Phys. Med. Biol. **58**(13), R97 (2013)
5. Cheng, J., et al.: Enhanced performance of brain tumor classification via tumor region augmentation and partition. PloS One **10**(10), e0140381 (2015)
6. El-Dahshan, E.-S.A., Mohsen, H.M., Revett, K., Salem, A.-B.M.: Computer-aided diagnosis of human brain tumor through MRI: a survey and a new algorithm. Expert Syst. Appl. **41**(11), 5526–5545 (2014)
7. Havaei, M., et al.: Brain tumor segmentation with deep neural networks. Med. Image Anal. **35**, 18–31 (2017)
8. Hochreiter, S., Schmidhuber, J.: Long short-term memory. Neural Comput. **9**(8), 1735–1780 (1997)
9. Huang, G., Liu, Z., van der Maaten, L., Weinberger, K.Q.: Densely connected convolutional networks. In: 2017 IEEE Conference on Computer Vision and Pattern Recognition (CVPR), pp. 2261–2269. IEEE (2017)
10. Menze, B.H., et al.: The multimodal brain tumor image segmentation benchmark (BRATS). IEEE Trans. Med. Imaging **34**(10), 1993–2024 (2015)
11. van der Maaten, L., Hinton, G.: Visualizing high-dimensional data using t-SNE. J. Mach. Learn. Res. **9**(Nov), 2579–2605 (2008)
12. Zhang, Y., Shen, D., Wang, G., Gan, Z., Henao, R., Carin, L.: Deconvolutional paragraph representation learning. In: Advances in Neural Information Processing Systems, pp. 4169–4179 (2017)

3D Texture Feature Learning for Noninvasive Estimation of Gliomas Pathological Subtype

Guoqing Wu[1], Yuanyuan Wang[1,2(✉)], and Jinhua Yu[1,2(✉)]

[1] Department of Electronic Engineering, Fudan University, Shanghai, China
{yywang, jhyu}@fudan.edu.cn
[2] The Key Laboratory of Medical Imaging Computing and Computer Assisted
Intervention of Shanghai, Shanghai, China

Abstract. Pathological subtype saved as an important marker in gliomas has considerable diagnostic and prognostic values. However, previous identification of pathological subtype relies on tumor samples, which is invasive. In this paper, we proposed a 3D texture feature learning method which is based on sparse representation (SR) theory to noninvasively estimate the pathological subtype for gliomas. Firstly, we developed a 3D patch-based SR model to extract 3D tumor texture features form magnetic resonance (MR) images. Then, by considering the physical meaning and characteristics of the extracted features, instead of performing feature selection directly, we further extract some deep features describing the statistical difference of the texture features of different tumors for subtype estimation. 213 subjects are divide into cross validation cohort and independent testing cohort to validate the proposed method. The proposed method achieves encouraging performance, with the accuracy of 91.43% and 88.57% by using T1 contrast-enhanced and T2-Flair MR images, respectively.

Keywords: Gliomas · Pathological subtype · Radiomics ·
Sparse representation

1 Introduction

Glioma is a common and aggressive type of malignant brain tumor. Its pathological subtype plays an important role in the process of therapeutic schedule making and prognosis evaluation [1]. However, the pathological subtype of glioma is currently determined through the histological examination, which is invasive. Hence, an image-based non-invasive method which can analyze global tumor characteristics is undoubtedly required in routine clinical practice [2].

Radiomics has achieved much success in the community of clinical diagnosis and prognosis [3–5] due to its key advantage of non-invasive evaluation on medical images. It first converts medical images into high-throughput quantitative imaging features, then selects few of features with significant statistical correlation with the specific clinical problem and builds the predictive or prognostic models on the selected features. Recently, radiomics has been applied for different cancers such as, head & neck [3],

© Springer Nature Switzerland AG 2019
A. Crimi et al. (Eds.): BrainLes 2018, LNCS 11383, pp. 218–227, 2019.
https://doi.org/10.1007/978-3-030-11723-8_22

colorectal [4], and glioma [5–7]. For glioma, several radiomics models have been proposed to predict survival [5], treatment response [6] and molecular characteristics [7].

Extracting and quantifying high-throughput features play crucial roles in radiomics applications, as good features lead directly to accurate classification and prediction. Gray-level co-occurrence matrix (GLCM)-based and gray-level run-length matrix (GLRLM)-based texture features are widely used for radiomics analysis [3–5]. Recently, our previous work [8] developed a sparse representation (SR) and dictionary learning-based feature extraction model, which has achieved the start-of-the art performance on tumor differentiation and isocitrate dehydrogenase 1 (IDH1) mutation estimation. Instead of directly calculating texture features through GLCM and GLRLM, we first trained texture dictionaries corresponding to each class of images, then used the combined dictionaries to sparsely represent tumor image into quantitative texture features. Because dictionary training captures image characteristics differences more thoroughly than the traditional GLCM like features, SR representation coefficients can reflect intrinsic differences among various image classes, and therefore produce improved performance. In [9], the feature extraction method was applied for overall survival time prediction of gliomas patient, also achieving promising performance.

Although the SR and dictionary learning-based feature extraction has achieved much success in the community of radiomics analysis due to its adaptive learning property, there are still two shortcomings. First, it is a 2D image patch-based texture dictionary learning, hence only the texture features of transverse plane will be learned. However, tumor is a 3D structure. And the texture information of coronal plane and sagittal plane are as same important as that of the transverse plane for tumor' diagnosis and prognosis. Second, it uses feature selection to select a few of features for subsequent classification, which limits to taking full advantage of the features. Because removing a large number of features destroys the inherent correlations among features. And these correlation informations directly reflect the difference between different image classes and are crucial for classification (more details are discussed in Sect. 3).

Hence, in this paper, we proposed two improvements for the shortcomings of the SR-based radiomics in [8], and use the improved radiomics framework to estimate the pathological subtype for gliomas. Specifically, first, we develop 3D texture feature learning method to extract the 3D tumor texture features. Second, by considering the inherent correlations of the texture features, we further extracted some deep statistical features form the texture features. These deep features reflect the difference of different classes of tumor texture more intuitively. Finally, these deep features are directly fed into SR classifier to identify the pathological subtype.

2 Data Acquisition and Preprocessing

2.1 Subjects

A total of 213 subjects with corresponding patient information and high-quality T1-Contrast and T2-Flair MR images were provided by the Neurosurgical Center, Huashan Hospital, Fudan University, Shanghai, China. The patient information is summarized in

Table 1. The T1-Contrast and T2-Flair MR images were acquired on a Magnetom Trio 3T (Siemens AG, Erlangen, Germany) scanner. The voxel resolution and size of the T1 Contrast MR images were 0.488 * 0.488 * 1 mm^3 and 448 * 512 * 176, respectively. The voxel resolution and size of the T2-weighted MR images were 0.468 * 0.468 * 2 mm^3 and 456 * 512 * 66, respectively. The study was approved by the ethics committee of Huashan Hospital. All patients underwent the surgical resection or biopsy for being histological examined. Based on the World Health Organization criteria, the 213 subjects were classified into 3 categories, i.e. Oligodendrocytoma (O), astrocytoma (A) and glioblastoma (GBM). For each class of subjects, we sorted them by time, then divided them into a cross validation cohort and an independent validation cohort at a ratio of approximately 1:1, respectively. The patient characteristics of the two cohorts are reported in Table 1.

Table 1. Characteristics of patients from the Huashan Hospital.

		Cross validation cohort			Independent testing cohort		
	Subtypes	O	A	GBM	O	A	GBM
	Case number	31	41	36	30	40	35
Sex	Male	19	27	21	15	23	21
	Female	12	14	15	15	17	14
Age	0–30	2	7	6	2	6	5
	31–60	26	31	25	28	31	23
	>60	3	3	5	0	3	7
	Mean ± SD	46.64 ± 9.95	40.00 ± 12.11	45.19 ± 14.42	40.33 ± 8.36	42.03 ± 12.65	47.60 ± 16.51

2.2   Tumor Segmentation

We first skull- and scalp-stripped and intensity normalized all of the MRI images in the two cohorts by the SPM12 [10]. Next, a convolutional neural network (CNN)-based tumor segmentation method proposed in our previous work [8] was used to segment the tumor regions. Finally, we performed some post-treatments on the segmentation results, including the removal of regions with less than 50 voxels, finding the largest connected region, and filling the holes in the tumor regions. Since the dataset of this work and the dataset of [8] are form the same imaging machine and hospital, the CNN segmentation models trained in [8] can be directly used to deal with the experimental images of this study. Figure 1 shows the tumor segmentation results on the T1 Contrast (first row) and T2-Flair MR images (second row).

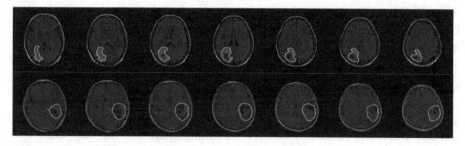

Fig. 1. Illustrations of the tumor segmentation results.

3 The Proposed Method

3.1 3D Texture Dictionary Learning-Based Feature Extraction

Tumor phenotypic information, especially some texture information, has been widely used for tumor classification and prognosis prediction [3–5]. In this paper, based on our previous work [8], sparse representation-based texture feature extraction, we developed a novel 3D texture dictionary learning-based feature extraction method. It first learns 3D texture dictionaries form each of images respectively, then uses the learned dictionaries to sparsely represent the testing images to obtain the corresponding texture features. Figure 2 shows the flowchart of the 3D texture dictionary learning-based feature extraction. It consists of three steps: 3D image patch extraction, 3D texture dictionary learning and sparse representation-based feature extraction.

3D Image Patch Extraction. First, for each tumor, we multiplied the 3D image matrix by the segmentation label matrix to obtain segmented tumor region. Then we extracted 3D image patches form tumor region with the sliding distance of 6 pixels in horizontal direction and 3 pixels in vertical direction. Finally, we pulled each extracted image patch into a column vector and denoted by $\mathbf{Y} \in R^{n \times d}$ the image patch set for a tumor. Where n is the number of voxels in the 3D image patch, and d denotes the number of image patches contained in a tumor.

3D Texture Dictionary Learning. First, we construct three training image patch sets, $\mathbf{Y}_O \in R^{n \times d_o}$, $\mathbf{Y}_A \in R^{n \times d_A}$ and $\mathbf{Y}_{GBM} \in R^{n \times d_{GBM}}$, by separately grouping the image patch sets extracted from the each class of tumor images. Where n is the size of image patch, d_o, d_A and d_{GBM} are the number of image patches in the three sets, respectively. Second, we learn three 3D texture dictionaries from $\mathbf{Y}_O \in R^{n \times d_o}$, $\mathbf{Y}_A \in R^{n \times d_A}$ and $\mathbf{Y}_{GBM} \in R^{n \times d_{GBM}}$, respectively, through the following optimization problem:

$$\arg\min_{\mathbf{D}, \mathbf{\Psi}} \|\mathbf{Y} - \mathbf{D}\mathbf{\Psi}\|_2^2 + \eta \sum_j \|\psi_j\|_0 \tag{1}$$

where $\mathbf{Y}$ represents one of the three training sets, $\mathbf{D}$ is the 3D texture dictionary, $\mathbf{\Psi}$ is the SR coefficients, and ψ_j is j-th column vector in $\mathbf{\Psi}$. Equation (1) can be solved by the K-singular value decomposition algorithm [11]. Once the three 3D dictionaries $\mathbf{D}_O \in R^{n \times k}$, $\mathbf{D}_A \in R^{n \times k}$ and $\mathbf{D}_{GBM} \in R^{n \times k}$ are learned, we put the three learned dictionaries together to construct final feature extraction dictionary $\bar{\mathbf{D}} = [\mathbf{D}_O, \mathbf{D}_A, \mathbf{D}_{GBM}]$. Where k is the number of atoms in the sub-dictionary.

Sparse Representation-Based Feature Extraction. Since the three sub-dictionaries are learned from the corresponding image classes, respectively. Hence, using $\bar{\mathbf{D}}$ to sparsely represent the test image, the statistical distribution of the SR coefficients can naturally reflect the correlation between the testing image and the three classes of images. Suppose $\mathbf{Y} = [\mathbf{y}_1, \cdots \mathbf{y}_i, \cdots \mathbf{y}_m]$ represents the image patch set of a testing image, where $\mathbf{y}_i$ denotes the i-th image patch, m denotes the number of patches. We exploit the following SR-based model to extract the texture feature of $\mathbf{Y}$:

$$\begin{cases} \hat{\Lambda} = \sum_{i=1}^{m} \arg\min_{\alpha_i} \|\mathbf{y}_i - \bar{\mathbf{D}}\alpha_i\|_2^2 + \phi \|\alpha_i\|_0 \\ f = \frac{1}{m} \sum_{i=1}^{m} |\alpha_i| \end{cases}, \quad (2)$$

where $\hat{\Lambda} = [\alpha_1, \cdots \alpha_i, \cdots \alpha_m]$, $\alpha_i \in R^{3k}$ is the SR coefficients corresponding to $\mathbf{y}_i$, ϕ is a regularization parameter. $f \in R^{3k}$ is the extracted texture feature. We use the Orthogonal matching pursuit (OMP) algorithm to solve the SR model in Eq. (2).

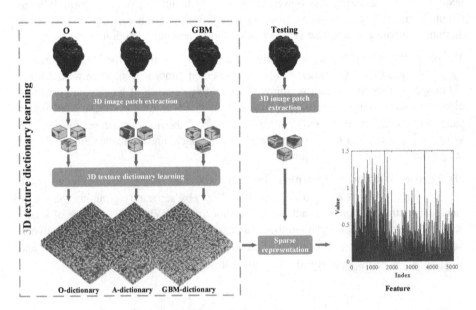

Fig. 2. SR-based feature extraction.

3.2 Deep Feature Extraction

Generally, the extracted high-throughput features contains lots of redundant informa-tion which not only increases the computational complexity, but also has negative effects on classification. Hence, in most conventional radiomics, a feature selection method was often introduced to select the discriminative features for classification. T-test-based *p-value* comparison and sparse representation regression are the two more commonly used feature selection methods in rediomics [3–8]. *p-value* comparison selects a few discriminative features by analyzing the correlation of each feature with the label separately. While sparse representation regression aims to directly select a feature set which is an optimal representation of the label.

Although some discriminative features can be selected by these methods, the inherent correlation between all features are ignored. Especially for SR-based features, these deep correlation informations play critical roles in subsequent classification. Figure 3(a), (b) and (c) show the extracted features of an O, an A and a GBM, respectively. In the three sub-figures, the two red lines divide the features into three parts, which are the SR coefficients of O-dictionary, A-dictionary and GBM-dictionary, respectively. By making overall comparisons on the values of the three parts in the three sub-figures, we can see that the first part is highest in O feature, the second part is highest in A feature and the third part is highest in GBM feature. And this tendency is more obvious in Fig. 4, which shows the sum of each class of subject features. The correlations between the three parts of feature intuitively reflect the class of subject. While it is difficult to mine these correlation informations by using feature selection in conventional radiomics. Hence, we propose to further extract some deep features describing the correlation informations for classification.

We divide a feature into three parts according to the corresponding dictionary and calculate nine statistical features for each part. Specifically, for each part, we calculate the following features: the sum and mean of all features (F_{all}); the sum and number of features with values above t1 (F_{t1}), t2 (F_{t2}) and t3 (F_{t3}), respectively; the number of nonzero features (F_{nz}). These features are summarized in Table 2. Then for each of the 9 types of features, we compare the values of the three parts, and take the location of the part which has the largest value as a location feature. Finally, a total of 36 features including the 27 features in Table 2 and 9 location features are fed into classifier for classification directly. Note that, the t1, t2 and t3 are adaptive. For the features of each sample, we sort them in the descending order, then set the t1, t2 and t3 to the values of the twentieth feature, thirtieth feature and fortieth feature.

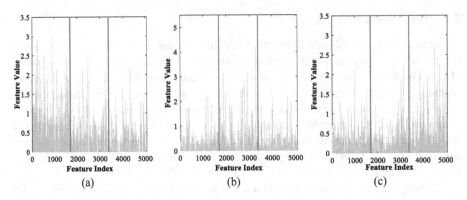

Fig. 3. The extracted features. (a) O feature. (b) A feature. (c) GBM feature.

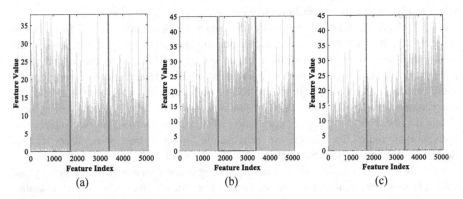

Fig. 4. The sum of the extracted features. (a) O feature. (b) A feature. (c) GBM feature.

Table 2. The 27 deep features.

Target feature	Part1 (P_1)	Part2 (P_2)	Part3 (P_3)
F_{all}	$P_1\_F_{all}\_Sum$	$P_2\_F_{all}\_Sum$	$P_3\_F_{all}\_Sum$
	$P_1\_F_{all}\_Mean$	$P_2\_F_{all}\_Mean$	$P_3\_F_{all}\_Mean$
F_{t1}	$P_1\_F_{t1}\_Sum$	$P_2\_F_{t1}\_Sum$	$P_3\_F_{t1}\_Sum$
	$P_1\_F_{t1}\_No.$	$P_2\_F_{t1}\_No.$	$P_3\_F_{t1}\_No.$
F_{t2}	$P_1\_F_{t2}\_Sum$	$P_2\_F_{t2}\_Sum$	$P_3\_F_{t2}\_Sum$
	$P_1\_F_{t2}\_No.$	$P_2\_F_{t2}\_No.$	$P_3\_F_{t2}\_No.$
F_{t3}	$P_1\_F_{t3}\_Sum$	$P_2\_F_{t3}\_Sum$	$P_3\_F_{t3}\_Sum$
	$P_1\_F_{t3}\_No.$	$P_2\_F_{t3}\_No.$	$P_3\_F_{t3}\_No.$
F_{nz}	$P_1\_F_{nz}\_No.$	$P_2\_F_{nz}\_No.$	$P_3\_F_{nz}\_No.$

3.3 Sparse Representation Classification

Sparse representation classification (SRC) proposed in [12] has achieved much success in pattern classification due to its good properties in handling errors and avoiding over-fitting. Hence, we apply SRC to identify the pathological subtype based on the deep features. Suppose f_t represents the deep features of a testing sample. Suppose $\mathbf{A} = [\mathbf{A}_1, \cdots \mathbf{A}_c, \cdots \mathbf{A}_C]$ represents the feature set of training samples, C is the number of classes, $\mathbf{A}_c$ is the training feature set from class c. The SRC can be described as the following two steps.

1. Sparsely code f_t over $\mathbf{A}$ by solving:

$$\hat{\chi} = \arg\min_{\chi} \|f_t - \mathbf{A}\chi\|_2^2 + \gamma\|\chi\|_1 \tag{3}$$

where $\hat{\chi}$ is the sparse representation coefficient to be calculated. γ is a regularization parameter.

2. Identify the class label of f_t via:

$$ID(f_t) = \arg\min_{c} \|f_t - \mathbf{A}\delta_c(\hat{\chi})\|_2^2 \tag{4}$$

where $\delta_c(\cdot)$ is used to select the coefficients associated with the c-th class.

4 Results

We first extract the 3D texture features of each subject from T1 Contrast and T2-Flair MR images, respectively. Then we calculate the 36 statistical features of these texture features. Finally, the 36 features are fed into classifier to estimate gliomas subtype. Leave-one-out cross validation (LOOCV) is performed on the cross validation set. And then the cross validation set is taken as a training set to directly test the subjects in independent testing set. The size of 3D image patch is set to $13 \times 13 \times 5$. Dictionaries are trained with two-fold redundancy. The sparsity of the OMP algorithm for solving (2) is set to 1/5070. The determination of the sparsity is based on the classification performance of training dataset. We calculate the classification accuracy of the overall testing subjects (ACC) and the classification accuracy of O (O-ACC), A (A-ACC) and GBM (GBM-ACC), respectively, to evaluate the estimation performance. The estimation accuracies of the proposed method are reported in Table 3. For 'T1-Loocv', 'T1-IT', 'T2-Loocv' and 'T2-IT', the parts before and behind dash represent image modality and validation method, respectively. 'IT' represents independent testing. 'T1_T2-IT' represents the independent testing on the combination of T1-contrast and T2-Flair images.

From Table 3, we can see that both T1 contrast-enhanced and T2-Flair modalities have achieved promising estimation results in terms of both LOOCV and independent testing. The T1 contrast-enhanced modality with LOOCV achieved the highest overall accuracy of 93.52%, and the corresponding accuracies of each class reached 93.55%, 97.56% and 88.89%, respectively. For the two modalities, the gaps between the results of two validation methods are within 4%. This demonstrates the stability of the extracted deep features. In addition, the accuracy gaps between the three classes are within 10%. The performance of the combined two modalities outperforms that of each modality. The classification accuracy of GBMs is lower than that of the other two subtypes. This is due to that GBMs are more malignant and have more complicated and varying image structural texture. Hence, it is difficult to differentiate GBMs form the other two subtypes.

Section 3 reports that the texture feature of different tumor classes have some exclusive characteristics on their three parts. And nine types of deep features are designed to quantitatively analyze these characteristics. Figure 5 provides a visualization of the nine types of features on the 105 testing subjects. From top to bottom and left to right, these sub-figures correspond to the sum and mean of F_{all}, the sum and number of Ft1, Ft2 and Ft3, and the number of F_{nz}, respectively. In each sub-figure, X, Y and Z axes represent the features of P1, P2 and P3, respectively. The three colors represent three class of tumors. It can be seen clearly that, the nine types of features of different tumor classes exhibit obvious differences on the three parts.

Table 3. Estimation accuracies of different modalities and validation methods.

Method	ACC	O-ACC	A-ACC	GBM-ACC
T1-Loocv	93.52%	93.55%	97.56%	88.89%
T1-IT	91.43%	93.33%	92.50%	88.57%
T2-Loocv	91.67%	93.55%	95.12%	86.11%
T2-IT	88.57%	90.00%	92.50%	82.68%
T1_T2-IT	92.38%	93.33%	95.00%	88.57%

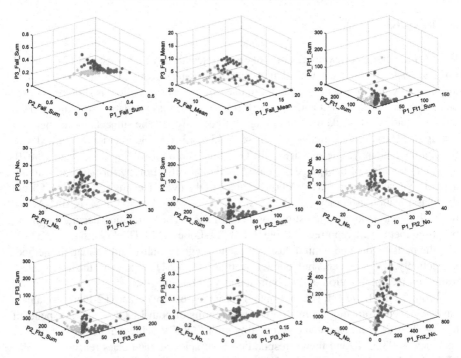

Fig. 5. The visualization of the nine types of features.

5 Conclusions

In this paper, we proposed a SR-based radiomics with 3D texture feature learning to estimate gliomas' pathological subtype. Two improvements has been made to overcome the shortcomings of the previous SR-based radiomics in [8]. On the on hand, instead of training 2D dictionary to extract tumor texture features in transverse plane, we constructed a 3D texture dictionary to extract the 3D texture feature of tumors, which provided more comprehensive description of tumor structure. On the other hand, by considering the inherent correlations of the texture features, we further extracted some deep features form the texture features. These deep features reflect the difference of different classes of tumor texture more intuitively. Both cross validation and independent testing were used to validate the proposed method on 213 subjects. Experimental results shown the encouraging performance of the proposed method.

References

1. Young, R.J., Knopp, E.A.: Brain MRI: tumor evaluation. J. Magn. Reson. Imaging **24**(4), 709–724 (2006)
2. Gao, Y., et al.: Histological grade and type classification of glioma using magnetic resonance imaging. In: CISP-BMEI, pp. 1808–1813 (2017)
3. Huang, Y., et al.: Radiomics signature: a potential biomarker for the prediction of disease-free survival in early-stage (i or ii) non-small cell lung cancer. Radiol. **281**(3), 947–957 (2016)
4. Huang, Y., et al.: Development and validation of a radiomics nomogram for preoperative prediction of lymph node metastasis incolorectal cancer. J. Clin. Oncol. **34**(8), 2157–2164 (2016)
5. Liu, L., Zhang, H., Rekik, I., Chen, X., Wang, Q., Shen, D.: Outcome prediction for patient with high-grade gliomas from brain functional and structural networks. In: Ourselin, S., Joskowicz, L., Sabuncu, M.R., Unal, G., Wells, W. (eds.) MICCAI 2016. LNCS, vol. 9901, pp. 26–34. Springer, Cham (2016). https://doi.org/10.1007/978-3-319-46723-8_4
6. Kickingereder, P., et al.: Large-scale radiomic profiling of recurrent glioblastoma identifies an imaging predictor for stratifying antiangiogenic treatment response. Clin. Cancer Res. **22**(23), 5765–5771 (2016)
7. Kickingereder, P., et al.: Radiogenomics of glioblastoma: machine learning-based classification of molecular characteristics by using multiparametric and multiregional MR imaging features. Radiol. **281**(3), 907–918 (2016)
8. Wu, G., et al.: Sparse representation-based radiomics for the diagnosis of brain tumors. IEEE Trans. Med. Imaging **37**(4), 893–905 (2018)
9. Wu, G., Wang, Y., Yu, J.: Overall survival time prediction for high grade gliomas based on sparse representation framework. In: Crimi, A., Bakas, S., Kuijf, H., Menze, B., Reyes, M. (eds.) BrainLes 2017. LNCS, vol. 10670, pp. 77–87. Springer, Cham (2018). https://doi.org/10.1007/978-3-319-75238-9_7
10. Tzourio-Mazoyer, N., et al.: Automated anatomical labeling of activations in SPM using a macro-scopic anatomical parcellation of the MNI MRI single-subject brain. Neuroimage **15**(1), 273–289 (2002)
11. Elad, M., Aharon, M.: Image denoising via sparse and redundant representations over learned dictionaries. IEEE Trans. Image Process. **15**(12), 3736–3745 (2006)
12. Wright, J., et al.: Robust face recognition via sparse representation. IEEE Trans. Pattern Anal. Mach. Intell. **31**(2), 210–227 (2009)

Pathology Segmentation Using Distributional Differences to Images of Healthy Origin

Simon Andermatt$^{(\boxtimes)}$, Antal Horváth, Simon Pezold, and Philippe Cattin

Department of Biomedical Engineering, University of Basel, Allschwil, Switzerland
`simon.andermatt@unibas.ch`

Abstract. Fully supervised segmentation methods require a large training cohort of already segmented images, providing information at the
pixel level of each image. We present a method to automatically segment
and model pathologies in medical images, trained solely on data labelled
on the image level as either healthy or containing a visual defect. We
base our method on CycleGAN, an image-to-image translation technique,
to translate images between the domains of healthy and pathological
images. We extend the core idea with two key contributions. Implementing the generators as residual generators allows us to explicitly model the
segmentation of the pathology. Realizing the translation from the healthy
to the pathological domain using a variational autoencoder allows us to
specify one representation of the pathology, as this transformation is otherwise not unique. Our model hence not only allows us to create pixelwise
semantic segmentations, it is also able to create inpaintings for the segmentations to render the pathological image healthy. Furthermore, we
can draw new unseen pathology samples from this model based on the
distribution in the data. We show quantitatively, that our method is able
to segment pathologies with a surprising accuracy being only slightly
inferior to a state-of-the-art fully supervised method, although the latter
has per-pixel rather than per-image training information. Moreover, we
show qualitative results of both the segmentations and inpaintings. Our
findings motivate further research into weakly-supervised segmentation
using image level annotations, allowing for faster and cheaper acquisition
of training data without a large sacrifice in segmentation accuracy.

1 Introduction

Supervised segmentation in medical image analysis is an almost solved problem for many applications, where methodological improvements have a marginal
effect on accuracy. Such methods depend on a large annotated training corpus,
where pixelwise labels have to be provided by medical experts. In practice, such

Electronic supplementary material The online version of this chapter (https://
doi.org/10.1007/978-3-030-11723-8_23) contains supplementary material, which is
available to authorized users.

data are expensive to gather. In contrast, weakly labelled data, such as images showing a certain disease are easily obtainable, since they are created on a daily basis in medical practice. We want to take advantage of these data for pathology segmentation, by providing a means to finding the difference between healthy and pathological data distributions. We present a weakly supervised framework capable of pixelwise segmentation as well as generating samples from the pathology distribution, trained on images which are only annotated with a scalar binary label at the image level classifying them as healthy or pathological.

Our idea is inspired by CycleGAN [11], a recently proposed solution for unpaired image to image translation, where the combination of domain-specific generative adversarial networks (GANs) and so-called cycle consistency allow for robust translation. We call our adaptation PathoGAN and count the following contributions: We formulate a model capable of segmentation based on a single label per training sample. We simultaneously train two generative models, able to generate inpaintings at a localized region of interest to transform an image from one domain to the other. We are able to sample healthy anatomy as well as sample possible pathologies for a given anatomical structure. Furthermore, our method enforces domain-specific information to be encoded outside of the image, which omits adversarial "noise" common to CycleGAN [4] to some degree.

We show the performance of our implementation on 2d slices of the training data of the Brain Tumor Segmentation Challenge 2017 [2,9] and compare our segmentation performance to a supervised segmentation technique [1].

CycleGAN has been previously used to segment by transfering to another target modality, where segmentation maps are available (e.g. [10]), or applied to generate training from cheaply generated synthetic labelmaps [6]. Using a Wasserstein GAN, another method directly estimates an additive visual attribution map [3]. To our knowledge, there has not been a method that jointly learns to segment on one medical imaging modality using only image-level labels and generate new data using GANs for both healthy and pathological cases.

2   Methods

2.1   Problem Statement

We assume two image domains $\mathcal{A}$ and $\mathcal{B}$, where the former contains only images of healthy subjects and the latter consists of images showing a specific pathology. We seek to approximate the functions $G_{\mathcal{A}}$ and $G_{\mathcal{B}}$ that perform the mappings $(x_{\mathcal{A}}, \delta_{\mathcal{B}}) \mapsto \hat{y}_{\mathcal{B}}$ and $(x_{\mathcal{B}}, \delta_{\mathcal{A}}) \mapsto \hat{y}_{\mathcal{A}}$, where $x_{\mathcal{A}}, \hat{y}_{\mathcal{A}} \in \mathcal{A}$ and $x_{\mathcal{B}}, \hat{y}_{\mathcal{B}} \in \mathcal{B}$. Vectors $\delta_{\mathcal{B}}$ and $\delta_{\mathcal{A}}$ encode the missing target image information (e.g. the pathology):

$$\hat{y}_{\mathcal{B}} = G_{\mathcal{A}}(x_{\mathcal{A}}, \delta_{\mathcal{B}}), \qquad \hat{y}_{\mathcal{A}} = G_{\mathcal{B}}(x_{\mathcal{B}}, \delta_{\mathcal{A}}). \tag{1}$$

We encourage $G_{\mathcal{A}}, G_{\mathcal{B}}$ to produce results, such that the mappings are realistic (2), cycle-consistent (3), specific (4) and that only the affected part in the image is modified (5):

$$G_A(x_A, \delta_B) \sim \mathcal{B}, \qquad\qquad G_B(x_B, \delta_A) \sim \mathcal{A}, \qquad\qquad (2)$$

$$G_B(G_A(x_A, \delta_B), \delta_A) \approx x_A, \qquad G_A(G_B(x_B, \delta_A), \delta_B) \approx x_B, \qquad (3)$$

$$G_B(x_A, 0) \approx x_A, \qquad\qquad G_A(x_B, 0) \approx x_B, \qquad\qquad (4)$$

$$\arg\min_{\overline{G}_A} |x_A - \overline{G}_A(x_A, \delta_B)| \approx G_A, \qquad \arg\min_{\overline{G}_B} |x_B - \overline{G}_B(x_B, \delta_A)| \approx G_B.$$

$$(5)$$

2.2 Model Topology

To fulfill Eqs. (2–5), we adopt the main setup and objective from CycleGAN: we employ two discriminators, D_A and D_B together with generators G_A and G_B to perform the translation from domain $\mathcal{A}$ to $\mathcal{B}$ and vice versa, formulating two generative adversarial networks (GANs) [7]. In both directions, the respective discriminator is trained to distinguish a real image from the output of the generator, whereas the generator is incentivized to generate samples that fool the discriminator.

In the remaining paper, we will use a short notation for equations which are applicable to both pathways (see Fig. 1) to overcome redundancy due to symmetrical components. We specify placeholder domains $\mathcal{X}$ and $\mathcal{Y}$, where for both pathways, $\mathcal{X}$ denotes the domain the initial image belongs to and $\mathcal{Y}$ stands for the target domain we want to transfer an image into. Domain $\mathcal{X}$ could therefore be either $\mathcal{A}$ or $\mathcal{B}$, and domain $\mathcal{Y}$ either $\mathcal{B}$ or $\mathcal{A}$, respectively. Furthermore, $x_{\mathcal{X}} \in \mathcal{X}$ denotes the image x in the respective initial domain $\mathcal{X}$, $G_{\mathcal{X}}(x_{\mathcal{X}}) = \hat{y}_{\mathcal{Y}}$ its transformation in the target domain $\mathcal{Y}$ and $G_{\mathcal{Y}}(\hat{y}_{\mathcal{Y}}) = \tilde{x}_{\mathcal{X}}$ its reconstruction in $\mathcal{X}$.

Residual Generator. In order to segment pathologies, we seek to only modify a certain part of the image. In contrast to CycleGAN, we model the transformation G from one domain to the other as a residual or inpainting p which is exchanged with part l of the original image. We achieve this by directly estimating $n+1$ feature maps $r_{\mathcal{X}}$ using a residual generator $Z_{\mathcal{X}}$ within $G_{\mathcal{X}}$, where n is the number of image channels used. We obtain labelmap $l_{\mathcal{X}}$ and inpaintings $p_{\mathcal{X}}$, activating $r_{\mathcal{X}}^{(0)}$ with a sigmoid and each $r_{\mathcal{X}}^{(i)}$ with a tanh activation:

$$l_{\mathcal{X}} = \mathrm{S}\left(r_{\mathcal{X}}^{(0)} + \epsilon\right), \qquad p_{\mathcal{X}}^{(i-1)} = \tanh\left(r_{\mathcal{X}}^{(i)}\right), \qquad r_{\mathcal{X}} = Z_{\mathcal{X}}(x_{\mathcal{X}}, \delta_{\mathcal{Y}}), \qquad (6)$$

where $\mathrm{S}(y) = \frac{1}{1+e^{-y}}$ and $i > 0$. With $\epsilon \sim \mathcal{N}(0, I)$, we turn $r_{\mathcal{X}}^{(0)} + \epsilon$ into samples from $\mathcal{N}(r_{\mathcal{X}}^{(0)}, I)$ using the reparameterization trick [8]. This allows reliable calculations of $l_{\mathcal{X}}$ only for large absolute values of $r_{\mathcal{X}}^{(0)}$, forcing $l_{\mathcal{X}}$ to be binary and intensity information to be encoded in the inpaintings. We set ϵ to zero during testing. From $l_{\mathcal{X}}$ and $p_{\mathcal{X}}$ we compute the translated result $\hat{y}_{\mathcal{Y}}$, supposedly in domain $\mathcal{Y}$ now:

$$\hat{y}_{\mathcal{Y}} = l_{\mathcal{X}} \odot p_{\mathcal{X}} + (1 - l_{\mathcal{X}}) \odot x_{\mathcal{X}} = G_{\mathcal{X}}(x_{\mathcal{X}}, \delta_{\mathcal{Y}}).$$

In the following, we detail the computation of $r_{\mathcal{A}}$ and $r_{\mathcal{B}}$ using the two networks $Z_{\mathcal{A}}$ and $Z_{\mathcal{B}}$ for the two possible translation directions. Both translation pathways are visualized in Fig. 1.

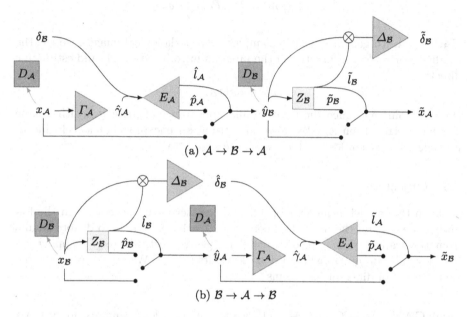

(a) $\mathcal{A} \rightarrow \mathcal{B} \rightarrow \mathcal{A}$

(b) $\mathcal{B} \rightarrow \mathcal{A} \rightarrow \mathcal{B}$

Fig. 1. Proposed architecture: *top to bottom:* directions $\mathcal{A} \rightarrow \mathcal{B} \rightarrow \mathcal{A}$ and $\mathcal{B} \rightarrow \mathcal{A} \rightarrow \mathcal{B}$; $x_{\mathcal{A}}, x_{\mathcal{B}}$ are samples from the two data distributions $\mathcal{A}$ (healthy), $\mathcal{B}$ (pathological) and $\delta_{\mathcal{B}} \sim \mathcal{N}(0, I)$. Red and blue triangles depict decoder and encoder networks. A red square illustrates a simple generator. $\Delta_{\mathcal{B}}$ and $\Gamma_{\mathcal{A}}$ encode features inside and outside the pathological region. $\Delta_{\mathcal{B}}, \Gamma_{\mathcal{A}}$ and $E_{\mathcal{A}}$ form a variational autoencoder, which acts as the residual generator $Z_{\mathcal{A}}$ used in $G_{\mathcal{A}}$. On the other hand, the residual generator $Z_{\mathcal{B}}$ in $G_{\mathcal{B}}$ is implemented as one component, as information about the missing healthy structure is completely inferred from the surroundings without explicitly encoding it. (Color figure online)

$Z_{\mathcal{A}}$:$\mathcal{A} \rightarrow \mathcal{B}$: To map from healthy to pathological data, we estimate $r_{\mathcal{A}}$ (and thus $l_{\mathcal{A}}, p_{\mathcal{A}}$) using a variational autoencoder (VAE) [8]. First, we employ encoders $\Gamma_{\mathcal{A}}$ and $\Delta_{\mathcal{B}}$ to encode anatomical information around and inside the pathological region:

$$\gamma_{\mathcal{A}} = \Gamma_{\mathcal{A}}(x_{\mathcal{A}}), \qquad \delta_{\mathcal{B}} = \Delta_{\mathcal{B}}(l'_{\mathcal{B}} \odot x'_{\mathcal{B}}),$$

where $x_{\mathcal{A}}$ is our healthy image, $l'_{\mathcal{B}}$ and $x'_{\mathcal{B}}$ are the labelmap and pathological image of the previous transformation and $\delta_{\mathcal{B}}, \gamma_{\mathcal{A}} \sim \mathcal{N}(0, I)$. If $l'_{\mathcal{B}}$ and $x'_{\mathcal{B}}$ are not available because $x_{\mathcal{A}}$ is a real healthy image, we simply sample $\delta_{\mathcal{B}}$. Finally, a decoder $E_{\mathcal{A}}$ is applied to $\gamma_{\mathcal{A}}$ and $\delta_{\mathcal{B}}$:

$$r_{\mathcal{A}} = E_{\mathcal{A}}(\gamma_{\mathcal{A}}, \delta_{\mathcal{B}}).$$

The residual generator $Z_{\mathcal{A}}$ is hence composed of encoder $\Gamma_{\mathcal{A}}$ and decoder $E_{\mathcal{A}}$, where $\delta_{\mathcal{B}}$ is either sampled or calculated with the additional encoder $\Delta_{\mathcal{B}}$:

$$Z_{\mathcal{A}}(x_{\mathcal{A}}, \delta_{\mathcal{B}}) = E_{\mathcal{A}}(\Gamma_{\mathcal{B}}(x_{\mathcal{A}}), \delta_{\mathcal{B}}),$$

$Z_{\mathcal{B}}:\mathcal{B} \rightarrow \mathcal{A}$: To generate healthy samples from pathological images, we use the residual generator $Z_{\mathcal{B}}$ directly on the input as introduced in [11] and estimate r directly:

$$r_{\mathcal{B}} = Z_{\mathcal{B}}(x_{\mathcal{B}}).$$

Here, we omit $\delta_{\mathcal{A}}$, since the location and appearance of missing healthy tissue can be inferred from $x_{\mathcal{B}}$. We also omit using an encoding bottleneck due to possible information loss and less accurate segmentation.

2.3   Objective

To train this model, a number of different loss terms are necessary. In the following, we explain the individual components using $\hat{y}$. and $\tilde{x}$. to denote results from the translated and reconstructed images respectively (e.g. mapping $x_{\mathcal{X}}$ into $\mathcal{Y}$ results in $\hat{y}_{y}$, translating it back results in $\tilde{x}_{\mathcal{X}}$). We use λ. to weight the contribution of different loss terms.

CycleGAN. As in CycleGAN [11], we formulate a least squares proxy GAN loss, which we minimize with respect to $G_{\mathcal{X}}$ and maximize with respect to $D_{\mathcal{Y}}$:

$$\mathcal{L}_{\text{GAN}}(D_{\mathcal{Y}}, G_{\mathcal{X}}, x_{\mathcal{X}}, x_{\mathcal{Y}}) = \mathbb{E}[(D_{\mathcal{Y}}(x_{\mathcal{Y}}))^2] + \mathbb{E}[(1 - D_{\mathcal{Y}}(G_{\mathcal{X}}(x_{\mathcal{X}}, \delta_{\mathcal{Y}})))^2]. \quad (7)$$

Likewise, to make mappings reversible, we add the cycle-consistency loss:

$$\mathcal{L}_{\text{CC}}(G_{\mathcal{X}}, G_{\mathcal{Y}}, x_{\mathcal{X}}) = \lambda_{\text{CC}}\|G_{\mathcal{Y}}(G_{\mathcal{X}}(x_{\mathcal{X}}, \delta_{\mathcal{Y}}), \delta_{\mathcal{X}}) - x_{\mathcal{X}}\|_1. \quad (8)$$

$\mathcal{L}_{\text{GAN}}$ and $\mathcal{L}_{\text{CC}}$ encourage the properties defined in Eqs. (2) and (3).

Variational Autoencoder. A variational autoencoder (VAE) is trained by minimizing the KL-divergence of the distribution $q(z|x)$ of encoding z to some assumed distribution and the expected reconstruction error $\log p(x|z)$, where x is the data. In contrast to a classical VAE, we use two distinct encoding vectors $\gamma_{\mathcal{A}}$ and $\delta_{\mathcal{B}}$, encoding the healthy and the pathological part, and produce two separate results, the labelmap $l_{\mathcal{A}}$ and the inpainting $p_{\mathcal{A}}$. We directly calculate the KL-divergence for our two encodings:

$$\mathcal{L}_{\mathrm{KL}}(G_{\mathcal{A}}, G_{\mathcal{B}}, x_{\mathcal{A}}, x_{\mathcal{B}}) = \mathrm{KL}[q(\gamma_{\mathcal{A}}|x_{\mathcal{A}})||\mathcal{N}(0, I)] + \mathrm{KL}[q(\delta_{\mathcal{B}}|x_{\mathcal{B}}, \hat{l}_{\mathcal{B}})||\mathcal{N}(0, I)]. \tag{9}$$

For the expected reconstruction error, we assume that l and p follow approximately a Bernoulli and a Gaussian distribution ($\mathcal{N}(\mu, I)$). We selectively penalize the responsible encoding, by using separate loss functions for the residual region and the rest. Unfortunately, we only ever have access to the ground truth of one of these regions, since we do not use paired data. We solve this, by using the relevant application in the network, where individual ground truths are available, to calculate the approximation of the marginal likelihood lower bound:

$$\mathcal{L}_{\mathrm{VAE}}(G_{\mathcal{X}}, G_{\mathcal{Y}}, x_{\mathcal{X}}, x_{\mathcal{Y}}) =$$
$$- \frac{\lambda_{\mathrm{VAE}}}{N} \sum_{m=1}^{N} (\log p(\tilde{l}_{\mathcal{Y}}|\gamma_{\mathcal{X}}, \delta_{\mathcal{Y}}) + \log p(\hat{p}_{\mathcal{X}}|\gamma_{\mathcal{X}}) + \log p(\tilde{p}_{\mathcal{Y}}|\delta_{\mathcal{Y}})), \tag{10}$$

where $\hat{p}_{\mathcal{X}}$ denotes the inpainting used to translate the original image $x_{\mathcal{X}}$ to domain $\mathcal{Y}$ and N is the total number of pixels. $\tilde{p}_{\mathcal{X}}$ is the inpainting produced when translating an already translated image $\hat{y}_{\mathcal{X}}$ that originated from $\mathcal{Y}$ back to that domain. Similarly, $\hat{l}_{\mathcal{X}}$ and $\tilde{l}_{\mathcal{X}}$ denote the respective labelmaps:

$$\log p(\hat{p}_{\mathcal{X}}|\gamma_{\mathcal{X}}) = \frac{||(1 - \hat{l}_{\mathcal{X}})(\hat{p}_{\mathcal{X}} - x_{\mathcal{X}})||_2}{\omega_1}, \tag{11}$$

$$\log p(\tilde{p}_{\mathcal{X}}|\delta_{\mathcal{Y}}) = \frac{||\tilde{l}_{\mathcal{Y}}(\tilde{p}_{\mathcal{X}} - x_{\mathcal{Y}})||_2}{\omega_2}, \tag{12}$$

where $\omega_1 = \frac{\sum(1-\hat{l}_{\mathcal{X}})+\varepsilon}{N}$ and $\omega_2 = \frac{\sum(\tilde{l}_{\mathcal{Y}})+\varepsilon}{N}$ are considered constant during optimization, with $\varepsilon > 0$. Finally, we use the labelmap produced by the other generator responsible for the opposite transformation $\hat{l}_{\mathcal{Y}}$ as ground truth for $\tilde{l}_{\mathcal{X}}$, where we consider $\hat{l}_{\mathcal{Y}}$ constant in this term:

$$\log p(\tilde{l}_{\mathcal{X}}|\gamma_{\mathcal{X}}, \delta_{\mathcal{Y}}) = \hat{l}_{\mathcal{Y}} \log \tilde{l}_{\mathcal{X}} + (1 - \hat{l}_{\mathcal{Y}}) \log(1 - \tilde{l}_{\mathcal{X}}). \tag{13}$$

To restrict the solution space of our model, we use $\mathcal{L}_{\mathrm{VAE}}$ for both directions.

Identity Loss. We apply an identity loss [11] on labelmap $l_{\mathcal{X},x_{\mathcal{Y}}}$ which results from feeding $G_{\mathcal{X}}$ with the wrong input $x_{\mathcal{Y}}$. In this case $G_{\mathcal{X}}$ should not change anything, since the input is already in the desired domain $\mathcal{Y}$:

$$\mathcal{L}_{\mathrm{Idt}}(G_{\mathcal{X}}, x_{\mathcal{Y}}) = \lambda_{\mathrm{Idt}}||l_{\mathcal{X},x_{\mathcal{Y}}}||_1. \tag{14}$$

Relevancy Loss. By now, we have defined all necessary constraints for a successful translation between image domains. The remaining constraints restrict the location and amount of change, $l_\mathcal{X}$. Fulfilling Eq. (5), we want to entice label map $l_\mathcal{X}$ to be only set at locations of a large difference between inpainting $p_\mathcal{X}$ and image $x_\mathcal{X}$ and penalize large label maps in general:

$$\mathcal{L}_\mathrm{R}(G_\mathcal{X}, x_\mathcal{X}) = \lambda_\mathrm{R} \left[\|| - \log(1 - l_\mathcal{X}^2)\|_1 - \frac{\|l_\mathcal{X}(x_\mathcal{X} - p_\mathcal{X})\|_1}{\|l_\mathcal{X}\|_1} \right]. \tag{15}$$

In order to not reward exaggerated pathology inpaintings, we consider $(x_\mathcal{X} - p_\mathcal{X})$ constant in this expression.

Full PathoGAN Objective. combining all loss terms for direction $\mathcal{X}$ to $\mathcal{Y}$ as $\mathcal{L}_{\mathcal{X} \to \mathcal{Y}}$, we can finally define:

$$\mathcal{L}_{\mathcal{X} \to \mathcal{Y}} = \mathcal{L}_\mathrm{GAN} + \mathcal{L}_\mathrm{CC} + \mathcal{L}_\mathrm{VAE} + \mathcal{L}_\mathrm{Idt} + \mathcal{L}_\mathrm{R}, \tag{16}$$

$$\mathcal{L}_\mathrm{PathoGAN} = \mathcal{L}_{\mathcal{A} \to \mathcal{B}} + \mathcal{L}_{\mathcal{B} \to \mathcal{A}} + \lambda_\mathrm{KL}\mathcal{L}_\mathrm{KL}(x_\mathcal{A}, x_\mathcal{B}). \tag{17}$$

2.4   Network Structure

We describe the networks used in our generators $G_\mathcal{A}$ and $G_\mathcal{B}$ using a shorthand notation which we will define in the following. We denote the number of output feature maps as f, the number of inpaintings including labelmap as r, the smallest image width and height before reshaping to a vector as i and the encoding size as z. The smallest image width and height in our case is 15 and we chose an encoding length of 256. We denote the reshaping operation from a square image to a vector as Q2F and the inverse operation as F2Q. We further denote a convolution operation with stride 1 and a kernel size of k as Ck-f. We use a lower-case c in ck-f if the convolution is followed by an instance norm and an activation with an exponential linear unit (ELU) [5]. A downsampling operation of a stride 2 convolution with kernel size 3, followed with an instance norm and following ELU is coded as df, whereas an upsampling operation using a transposed convolution followed by an instance norm and an ELU activation is denoted as uf. We describe a fully connected layer with l(f) and a residual block as Rf. A residual block results in a sum of its input with a residual computation of a convolution layer of kernel size 3, an instance norm, an ELU and a convolution layer of kernel size 3. Finally, a trailing t or e denotes an additional tanh or ELU activation, respectively.

The generator $G_\mathcal{A}$ consists of the residual generator $Z_\mathcal{A}$, which is composed of two encoder networks $(\Gamma_\mathcal{A}, \Delta_\mathcal{B})$ and a decoder network $(E_\mathcal{A})$. Both $\Gamma_\mathcal{A}$ and $\Delta_\mathcal{B}$ are defined as:

```
c7-64,d128,d256,d512,d1024,C1-15,Q2F,l(z*i)t,l(2*z).
```

The output of these encoders is split in half to produce the mean and log variance vectors of length 256 each.

Similarly, the decoder network $E_\mathcal{A}$ is defined as:

```
l(i*i)e,l(i*i),F2Q,c3-1024,u512,u256,C7-256e,R256,R256,R256,
R256,R256,R256,R256,R256,R256,u128,u64,C7-r.
```

The generator G_B consists of the residual generator network Z_B, which is slightly adjusted compared to the generators used in [11]. Here, we use exponential linear units instead of rectified linear units and double each layers' number of feature maps:

```
c7-64,d128,d256,R256,R256,R256,R256,R256,R256,R256,R256,R256,
    u128,u64,C7-r.
```

We use the same network architecture as in CycleGAN [11] for the discriminator networks.

2.5   Data

We include all training patients of Brats2017 and normalize each brain scan for its non-zero voxels to follow $\mathcal{N}(0, 1/3)$, and clip the resulting intensities to $[-1, 1]$. We select all transverse slices from 60 to 100 in caudocranial direction. In order to create two distinct datasets and relying on the manual segmentations, we label slices without pathology as *healthy*, with more than 20 pixels segmented as *pathological* slices, and discard the rest. For training, we select 1500 unaffected and 6000 pathological slices from a total of 1755 and 9413 respectively[1].

We augment our training data by randomly mirroring the samples, applying a random rotation sampled from $U[-0.1, 0.1]$ and a random scaling $s = 1.1^r$ where r is sampled from $U[-1, 1]$. Furthermore we apply random deformations as described in [1], with a grid spacing of 128 and sample both components of the grid deformation vectors from $\mathcal{N}(0, 5)$.

Table 1. Segmentation Results. *Columns:* Dice, 95th percentile Hausdorff distance (HD95), average Hausdorff distance (AVD) and volumetric Dice per-patient (Dice PP) by stacking all evaluated slices. *Rows:* Scores are shown as mean±std(median) for the weakly-supervised PathoGAN (proposed) and the fully-supervised MDGRU, applied to training (Tr) and testing (Te) data.

	Dice (in %)	HD95 (in pixel)	AVD (in pixel)	Dice PP (in %)
PathoGAN, (Tr)	72.4 ± 24.4(81.0)	40.6 ± 30.7(38.0)	10.3 ± 15.4(4.7)	77.4 ± 14.4(81.2)
PathoGAN, (Te)	72.9 ± 23.8(81.4)	39.4 ± 29.9(37.6)	9.4 ± 13.7(4.6)	77.4 ± 14.4(81.7)
MDGRU, (Tr)	87.8 ± 20.0(94.4)	3.7 ± 9.7(1.0)	1.0 ± 4.7(0.2)	90.8 ± 8.8(93.3)
MDGRU, (Te)	86.3 ± 21.3(93.6)	3.9 ± 9.5(1.0)	1.1 ± 4.9(0.2)	90.6 ± 9.5(93.1)

3   Results

Since the BratS evaluation is volumetric and comparing performance is difficult, we also train a supervised segmentation technique on our data. We chose

---

[1] Thus we would like to stress that the manual segmentations were only used to create the two image domains, but not for the actual training.

MDGRU [1] for this task, a multi-dimensional recurrent neural network, due to code availability and consistent state-of-the-art performance on different datasets.

We trained PathoGAN[2] for 119 epochs using batches of 4 and $\lambda_{\mathrm{KL}} = 0.1, \lambda_{\mathrm{R}} = 0.5, \lambda_{\mathrm{Idt}} = 1, \lambda_{\mathrm{CC}} = 5$ and $\lambda_{\mathrm{VAE}} = 1$. We trained MDGRU[3] as defined in [1], using batches of 4 and 27 500 iterations. Table 1 shows the results on the pathological training and test data. Figure 2 shows an exemplary sample from the test data. On the left, the input data, the generated inpaintings and the translation result are displayed. On the right, the manual segmentation for "whole tumor" and generated labelmaps of the weakly-supervised PathoGAN and fully-supervised MDGRU are presented.

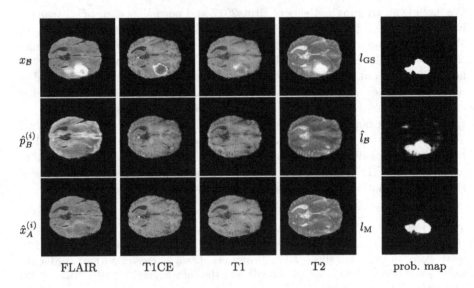

Fig. 2. Qualitative Results of one example from the testing data. *Left columns, top to bottom:* the four available image channels $x_B^{(i)}$, the generated inpaintings $\hat{p}_B^{(i)}$ and the translated images $\hat{y}_A^{(i)}$. *Right column, top to bottom:* The manual segmentation l_{GS}, the probability maps $\hat{l}_B$ from PathoGAN (weakly-supervised, proposed) and l_{M} from MDGRU (fully-supervised) for whole tumor.

4   Discussion

The results in Fig. 2 indicate that our relative weighting of the two inpainting reconstruction losses results in better reconstruction inside the tumor region

---

[2] Our implementation is based on https://github.com/junyanz/pytorch-CycleGAN-and-pix2pix.

[3] We use the implementation of MDGRU at https://github.com/zubata88/mdgru.

than outside. The labelmaps of the supervised method compared to ours in Fig. 2 show great agreement, and both are relatively close to the gold standard. As the 95th-percentile and average Hausdorff measures in Table 1 show, there are some outliers in our proposed method, due to its weakly-supervised nature. Compared to the fully-supervised method, Dice scores for PathoGAN are about 10% smaller for both the per-slice and the per-patient case. Slightly inferior results for the proposed method are not surprising, though, given the drastically reduced information density during its training (one label per image for PathoGAN, one label per pixel for fully-supervised MDGRU). Also, it is important to remember that we segment with the only criterion of being not part of the healthy distribution, which could vary from the subjective measures used to manually segment data. The increase in accuracy and decrease in standard deviation in the per-patient case for both methods is most likely caused by the inferior segmentation performance in slices showing little pathology. The per-patient Dice of the supervised method is in the range of the top methods of BraTS 2017. Although not directly comparable, this suggests that we can use our computed supervised scores as good state-of-the-art reference to compare our results to.

We did only scratch the surface on the possible applications of our proposed formulation. Future work will include unaffected samples that are actually healthy. Furthermore, the model architecture could be drastically simplified using one discriminator for both directions, allowing for larger generator networks as well as using multiple discriminators at different scales to find inpaintings that are not just locally but also globally consistent with the image. A restriction to slices is unfortunate but necessary due to memory requirements. A generalisation of our approach to volumetric data would make it feasible for more real clinical applications.

Conclusion. We presented a new generative pathology segmentation model capable of handling a plethora of tasks: First and foremost, we presented a weakly supervised segmentation method for pathologies in 2D medical images, where it is only known if the image is affected by the pathology and thus no pixel-wise label or classification is provided. Furthermore, we were able to sample from both our healthy as well as our pathology model. We showed qualitatively and quantitatively, that we are able to produce compelling results, motivating further research towards actual clinical applications of PathoGAN.

Acknowledgements. We are grateful to the MIAC corporation for generously funding this work.

References

1. Andermatt, S., Pezold, S., Cattin, P.: Multi-dimensional gated recurrent units for the segmentation of biomedical 3D-data. In: Carneiro, G., et al. (eds.) LABELS/DLMIA -2016. LNCS, vol. 10008, pp. 142–151. Springer, Cham (2016). https://doi.org/10.1007/978-3-319-46976-8_15

2. Bakas, S., et al.: Advancing the cancer genome atlas glioma MRI collections with expert segmentation labels and radiomic features. Nat. Sci. Data **4**, 170117 (2017)
3. Baumgartner, C.F., Koch, L.M., Tezcan, K.C., Ang, J.X., Konukoglu, E.: Visual feature attribution using wasserstein gans. arXiv preprint arXiv:1711.08998 (2017)
4. Chu, C., Zhmoginov, A., Sandler, M.: CycleGAN, a Master of Steganography. arXiv:1712.02950 [cs, stat], December 2017
5. Clevert, D.A., Unterthiner, T., Hochreiter, S.: Fast and accurate deep network learning by exponential linear units (ELUs). arXiv preprint arXiv:1511.07289 (2015)
6. Fu, C., et al.: Fluorescence microscopy image segmentation using convolutional neural network with generative adversarial networks. arXiv preprint arXiv:1801.07198 (2018)
7. Goodfellow, I., et al.: Generative adversarial nets. In: Advances in Neural Information Processing Systems, pp. 2672–2680 (2014)
8. Kingma, D.P., Welling, M.: Stochastic gradient VB and the variational auto-encoder. In: Second International Conference on Learning Representations (2014)
9. Menze, B.H., Jakab, A., Bauer, S., et al.: The multimodal brain tumor image segmentation benchmark (BRATS). IEEE Trans. Med. Imaging **34**(10), 1993–2024 (2015)
10. Chartsias, A., Joyce, T., Dharmakumar, R., Tsaftaris, S.A.: Adversarial image synthesis for unpaired multi-modal cardiac data. In: Tsaftaris, S.A., Gooya, A., Frangi, A.F., Prince, J.L. (eds.) SASHIMI 2017. LNCS, vol. 10557, pp. 3–13. Springer, Cham (2017). https://doi.org/10.1007/978-3-319-68127-6_1
11. Zhu, J.Y., Park, T., Isola, P., Efros, A.A.: Unpaired image-to-image translation using cycle-consistent adversarial networks, pp. 2223–2232 (2017)

Multi-stage Association Analysis
of Glioblastoma Gene Expressions
with Texture and Spatial Patterns

Samar S. M. Elsheikh[1(✉)], Spyridon Bakas[2,3], Nicola J. Mulder[1],
Emile R. Chimusa[4], Christos Davatzikos[2,3], and Alessandro Crimi[5,6]

[1] Computational Biology Group, Department of Integrative Biomedical Sciences,
Institute of Infectious Disease and Molecular Medicine, Faculty of Health Sciences,
University of Cape Town, Cape Town, South Africa
Samar.salah119@gmail.com
[2] Center for Biomedical Image Computing and Analytics (CBICA),
Perelman School of Medicine, University of Pennsylvania, Philadelphia, PA, USA
[3] Department of Radiology, Perelman School of Medicine,
University of Pennsylvania, Philadelphia, PA, USA
[4] Division of Human Genetics, Institute of Infectious Disease
and Molecular Medicine, Faculty of Health Sciences, University of Cape Town,
Cape Town, South Africa
[5] University Hospital of Zürich, Zürich, Switzerland
[6] African Institute for Mathematical Sciences, Biriwa, Ghana

Abstract. Glioblastoma is the most aggressive malignant primary brain
tumor with a poor prognosis. Glioblastoma heterogeneous neuroimaging,
pathologic, and molecular features provide opportunities for subclassifi-
cation, prognostication, and the development of targeted therapies. Magnetic resonance imaging has the capability of quantifying specific phenotypic imaging features of these tumors. Additional insight into disease
mechanism can be gained by exploring genetics foundations. Here, we use
the gene expressions to evaluate the associations with various quantitative imaging phenomic features extracted from magnetic resonance imaging. We highlight a novel correlation by carrying out multi-stage genome-
wide association tests at the gene-level through a non-parametric correlation framework that allows testing multiple hypotheses about the integrated relationship of imaging phenotype-genotype more efficiently and
less expensive computationally. Our result showed several novel genes
previously associated with glioblastoma and other types of cancers, as
the LRRC46 (chromosome 17), EPGN (chromosome 4) and TUBA1C
(chromosome 12), all associated with our radiographic tumor features.

Keywords: Glioblastoma · Gene expression · Brain tumor ·
Radiomics · Radiogenomics

© Springer Nature Switzerland AG 2019
A. Crimi et al. (Eds.): BrainLes 2018, LNCS 11383, pp. 239–250, 2019.
https://doi.org/10.1007/978-3-030-11723-8_24

1   Introduction

Gliomas are the most common type of primary adult brain tumors that arise from glial cells. Gliomas have a very heterogeneous landscape, and they can be classified according to their grade into low-grade glioma, anaplastic glioma, and glioblastoma. The most common and aggressive type of glioma in adults is glioblastoma (GBM), which gives to the affected patient average survival time of only 10 to 18 months. The known molecular classification of GBM into classical, mesenchymal, neural and proneural subtypes is relatively accepted to be related to the expression of EGFR, NF1 and PDGFRA/IDH1 genes [1].

Imaging, specifically magnetic resonance imaging (MRI), can offer data towards promising biomarkers reflecting underlying tumor pathology and biological function. If imaging phenotypes of GBM obtained from routine clinical MRI studies can be associated with specific gene expression signatures, quantitative imaging phenotypes will serve as non-invasive surrogates for cancer genomic events and provide valuable information as to the diagnosis, prognosis, and optimal treatment.

Several radiogenomic studies have been carried out for many diseases [8–16]. For instance for schizophrenia pairs of SNP/Gene and MRI features have been mapped by using PLINK [8], and Parallel-ICA showed promising results [9]. Batmanghelich et al. [10] proposed a Bayesian framework to relate imaging and genetic data to phenotypes exploiting connection among these data modalities simultaneously in Alzheimer. Recently, correlations of connectomic features have been related to genes which are known to be related to Alzheimer progression [11]. In contrast to Alzheimer's disease and schizophrenia, glioma lesions are generally not spread all over the brain, and local features from MRI can be used. An imaging-genomic analysis study [12], performed by using the tumor volume in T2-weighted FLuid-Attenuated Inversion Recovery (T2-FLAIR) images and large-scale genetic and micro-RNA expression probes demonstrated the potential for molecular subtyping and showed that the high median expression of POSTN gene results in a significant decrease in survival, and for that they used ANOVA and Tukey-Kramer test. Other studies [13,14] showed correlations between image feature annotations and expression of genes with glioma molecular subtypes [1]. Specifically, Gutman et al. [13] found a significant association between contrast-enhanced tumor and these molecular subtypes [1], where proneural type expressed by PDGFRA/IDH1 gene showed low levels of contrast enhancement, and the classical type (i.e., primarily described by EGFR amplification) correlates with the increased percentage of contrast enhancement. The study used sher exact statistics.

Recent population-based studies have assessed the anatomical location of GBM in relation to distinct clinically-relevant molecular characteristics, and have identified the spatial distribution of the tumors being descriptive of their molecular status [14, 17–22]. Furthermore, the emerging research direction of radiomics has shown promise that texture analysis of the various tumor sub-regions in radiographic imaging can also be informative of the tumor's molecular characterization [23–25]. Furthermore, using MRI features for GBM lesions, including texture and shape features, Haruka et al. proposed a classification imaging method and found three clusters of GBM patients [35]. In their method, they integrate copy number and gene expression data to estimate the molecular pathway activity and show that the three clusters reveal not only different molecular characteristics but also different survival probabilities.

The purpose of this paper is to identify significant associations between gene expressions, across the whole genome, and quantitative imaging phenomic features extracted from multi-modal MRI brain scans of patients diagnosed with de novo primary GBM. In line with the pre-mentioned studies, here we focus on evaluating the spatial location and texture features of GBM and investigate their associations with gene expressions.

2   Materials and Methods

2.1   Data

For the quantitative association analysis conducted here, we utilized a retrospective cohort of 135 de novo primary GBM patients from the TCGA-GBM collection [6], with available pre-operative multi-modal MRI scans in The Cancer Imaging Archive (TCIA) [7] and corresponding molecular characterization in The Cancer Genome Atlas (TGCA). The multi-modal MRI data we utilized comprise native (T1) and post-contrast T1-weighted (T1Gd), T2-weighted (T2), and T2-FLAIR modalities. The TCGA-GBM subset of 135 patients were identified by Bakas et al. [4] as brain scans without any surgically-imposed cavity, and their co-registered and skull-stripped imaging were provided in the TCIA Analysis Results together with expert manually annotated segmentation labels for the various histologically-distinct tumor sub-regions, i.e. enhancing tumor (ET), non-enhancing tumor (NET), peritumoral edematous/invaded tissue (ED) (Fig. 1) [4,5]. The total sample size of GBM patients reduced to 88 after evaluating patients that had available imaging [6] and corresponding gene expressions. In total, we assessed expression energies for 17815 genes, 11 distinct descriptors of tumor spatial location (Fig. 2), and 517 radiomic/texture features (Fig. 2) for each patient's brain tumor scan [2,4,5].

2.2   Quantitative Imaging Phenomic Features

Radiomic/Texture Features. We extracted an extensive panel of quantitative texture features, volumetrically (in 3D), for each tumor sub-region as provided by the expert annotations, across all available modalities. Specifically, the texture features we evaluated (i) capture global characteristics (i.e., variance, skewness, kurtosis) of each sub-region's intensity distribution on each modality, and (ii) include features based on Gray-Level Co-occurrence Matrix (GLCM) [26] (Fig. 2), Gray-Level Run-Length Matrix (GLRLM) [27–30], Gray-Level Size Zone Matrix (GLSZM) [28–30], and Neighborhood Gray-Tone Difference Matrix (NGTDM) [31].

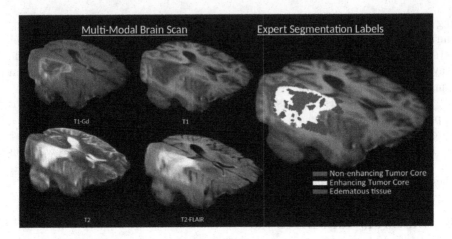

Fig. 1. Example of a multi-modal MRI brain scan and its corresponding expert segmentation labels.

Spatial Distribution Patterns. Beyond texture features, we collected discrete spatial information about the anatomical location of each tumor on each brain scan (Fig. 2). To obtain these spatial distribution patterns we registered all brain tumor scans in a standardized healthy atlas space using an iterative Expectation-Maximization framework [3], while incorporating a biophysical tumor growth model (based on a reaction-diffusion-advection model [32–34]) to account for tumor mass effects in the brain parenchyma. We then retrieved the spatial distribution of each tumor according to the discretized anatomical locations of the (i) specific lobes (i.e., frontal, temporal, parietal, occipital), (ii) insula, (iii) basal ganglia, (iv) fornix, (v) cerebellum, and (vi) brain stem. In addition, we also included as distinct features the distances of (i) the tumor core (defined as the union of ET and NET), and (ii) the ED, from the ventricles.

To produce these quantitative features we have utilized GLISTRboost. Specifically, in the process to produce segmentations of the various tumor sub-regions, the generative part [37] of GLISTRboost, following an Expectation-Maximization framework registers a healthy population probabilistic atlas to glioma patients' brain scans while incorporating a biophysical glioma growth model to account for mass effects. Then, after converting the predicted segmentation in the healthy atlas space, the percentage of the tumor core (i.e., enhancing and non-enhancing tumor) is calculated on each of the brain lobes in this healthy atlas.

2.3 Data Analysis

Initially, we combined the two types of data (imaging - genetics) using the patient ID as a primary column. As a first stage, we used the gene expressions and the spatial distribution patterns to perform a non-parametric test of association. To assess the associations, we computed the Spearman correlation coefficient

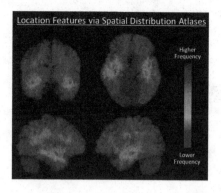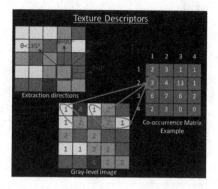

Fig. 2. Illustrative examples of spatial distribution (left) and texture (right) patterns.

(r_s) between the gene expressions, individually, as a with each of the spatial distribution patterns described in Sect. 2.2. We then assessed the significant of the correlation coefficient by calculating the p-values as described below.

For each quantitative feature and each gene, We obtained the p-value associated with Spearman correlation coefficient test statistic. That is, the p-value of the correlation between a single gene expression with a single feature of the tumor's location in the brain. The Spearman correlation coefficient model for a given feature (y) and given gene expression (x) is;

$$r_s = 1 - \frac{6 \sum_{i=1}^{i=N} d_i^2}{N(N^2 - 1)} \tag{1}$$

Where d_i is the difference between the ranks of x_i and y_i, and N is 88; representing the number of GBM patients [38]. r_s can take any real value between $+1$ and -1; $+1$ represents a strong positive association, -1 means a perfect negative association and 0 indicates no association between the ranks of x and y. Our hypothesis of interest is:

H_0: *There is no association between the gene expression and the tumor's feature under study*

vs

H_a: *There is an association between the gene expression and the tumor's feature under study,* alternatively:
H_0: $r_s = 0$ vs H_0: $r_s \neq 0$

To determine the significance of r_s, one can use the t test statistic defined as

$$t_c = r_s \sqrt{\frac{n - 2}{1 - r_s^2}}, \tag{2}$$

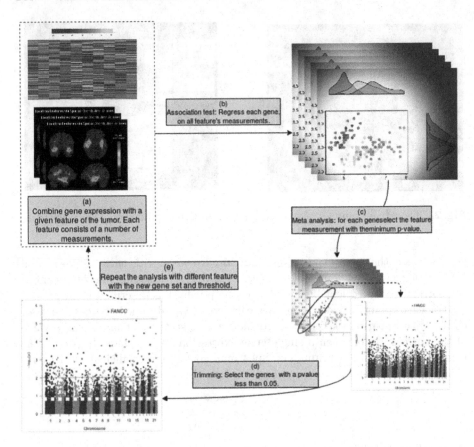

Fig. 3. Schematic representation of the study's analysis workflow.

t_c follows approximately the Student's t distribution with a $N-2$ degrees of freedom under the null hypothesis [38]. At a certain significance level, the calculated value of t_c can be compared to the table value obtained from the Student's t distribution (as described previously). The significance of r_s can also be determined using the p-value, which is simply the integration, or the area under the curve from t_c to infinity.

Briefly, in this first stage, the association test was initially conducted to six features of the tumor location (Sect. 2.2). More specifically, for each gene, we computed six p-values, then considered only the minimum p-value at each gene (see Fig. 3 for the analysis workflow). The latter is referred to as meta-analysis in Fig. 3 (step(c)). All results reported in Sect. 3 use the summary statistics of the meta-analysis. Moreover, out of the all the association results, we excluded all the genes with p-values greater than or equal 0.05. Here we meant to exclude the genes that have very low (and not significant) association with the spatial pattern, which we believe is an important phenotype. This step is referred to as (d) in Fig. 3. In the second stage, we proceeded with all the genes with p-value less than 0.05, excluding the least significant genes, and we carried the same analysis as in the first stage but using the radiomic features (Sect. 2.2. Table 1 shows the thresholds at both 5% and 10% significance level), along with the number of genes used and remained in each stage.

Table 1. Number of genes, 5% and 10% thresholds used at each stage of the analysis.

Feature	No. genes used	5% threshold	10% threshold	Genes after trim
Location	15009	0.000003331 ($3.3e^{-6}$)	0.000006663 ($6.7e^{-6}$)	5401
Texture	5401	0.000009258 ($9.3e^{-6}$)	0.000018515 ($1.9e^{-5}$)	5370

It is worth mentioning that, out of the total number of genes, we were able to annotate 15009 genes and assign them to their defined physical locations in the DNA. We carried on the first stage of the analysis using those genes (Table 1).

3   Results

The incidence of tumors specific for region is summarized in Table 2. The Manhattan plot for the p-values obtained from the meta-analysis is illustrated in Fig. 4. The plot shows two horizontal lines which associate with the thresholds of 5% significance level (top line), and 10% significance level (bottom line), after correcting for multiple comparisons. The x-axis is the physical position of genes in the DNA, and the y-axis is the negative $log10$ of the p-values. Figure 4 also shows the qq-plot of all the genes used in the association analysis. Likewise, each dot corresponds to a p-value of a single gene and $-log10$ of the p-value is used instead. The qq-plot reported with each Manhattan plot, and it compares the observed distribution of p-values (y-axis) to the expected distribution (x-axis), for each gene tested, where the diagonal line is the null distribution.

Table 2. Number and percentage of patients with tumor per brain region

Loc	Vent TC	Vent ED	Frontal	Temporal	Parietal	Basal	Insula	Fornix	Occipital	Cerebellum	Brainstem
No	88	88	63	70	62	55	43	26	35	8	24
%	100.00	100.00	71.59	79.55	70.45	62.50	48.86	29.55	39.77	9.09	27.27

Table 3 shows (only) the highest ten p-values and the corresponding genes of the first stage of the analysis. In this stage, non of the p-values was less than $3.3e^{-6}$ or $6.7e^{-6}$ (see Table 1); therefore, no gene was significantly associated with any of the features. Table 3 reports the gene symbol, its start and end position, the associated p-value and feature, and the chromosome.

We then pruned the genes used in the previous stage to a smaller set, by removing the genes that have p-values less than 0.05. With the 5401 genes

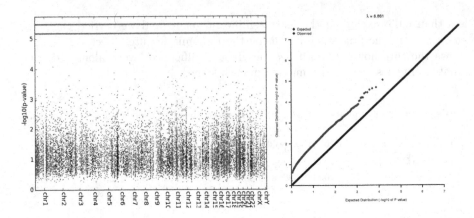

Fig. 4. A Manhattan (left) and qq-plot (right) of the associations between the tumor spatial distribution patterns, and gene expressions. The plot is showing the meta-analysis results.

Table 3. Top 10 genes: non-parametric association between genes and brain tumor location features in glioblastoma ordered according to the absolute value of r_s.

Gene	Start	End	r_s	p-value	Spatial pattern	Chromosome
TCN1	59620272	59634048	0.454	8.814e−06	DIST_Vent_TC	chr11
OR2AE1	99473609	99474680	−0.438	2.010e−05	Basal_G	chr7
KIF13A	17759413	17987854	−0.435	2.271e−05	Basal_G	chr6
NCBP2	196662272	196669468	0.432	2.619e−05	Occipital	chr3
RLN2	5299867	5304969	0.426	3.527e−05	Basal_G	chr9
KCNK9	140613080	140715299	0.426	3.533e−05	Parietal	chr8
B3GALT6	1167628	1170421	−0.423	3.938e−05	Brain_stem	chr1
FOXD3	63788729	63790797	0.414	6.0483e−05	Parietal	chr1
KISS1R	917286	921015	0.414	6.078e−05	Brain_stem	chr19
PLEKHA8	30067019	30170096	−0.413	6.362−05	Insula	chr7

remaining, we took over the second stage and repeated the same analysis with the texture characteristics of the tumor. The Manhattan and qq-plot for the texture features are shown in Fig. 5, and Table 4 shows the top 10 significant genes. Total of significant genes in this stage is 37 (at 5% significance level).

4 Discussion

GBM is a fatal malignant disease that so far is incurable. The identification of genetic risk factors that affect the tumor characteristics improves our understanding of the underlying biological processes for GBM, and contribute to therapeutic discovery. In this study, we proposed a framework that allows quantifying the non-parametric correlations to test associations between gene expressions

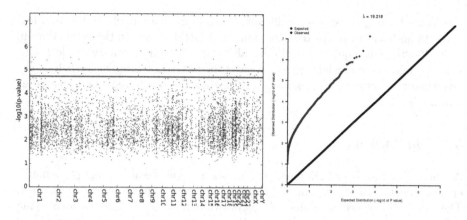

Fig. 5. A Manhattan (left) and qq-plot (right) of the associations between the tumor texture features, and gene expressions. The plot is showing the meta-analysis results.

Table 4. Top 10 significant genes associated with texture features of specific GBM sub-regions from specific modalities ordered according to the absolute value of r_s.

Gene	Start	End	r_s	p-value	Feature	Sub-region (MRI)	Chromosome
LRRC46	45908992	45915079	0.537	7.102e−08	GLCM Variance	ED (T2)	chr17
USP38	144106069	144144983	−0.511	3.648e−07	GLSZM SZLGE	ED (T1Gd)	chr4
EPGN	75174189	75181024	0.501	6.542e−07	GLSZM LGZE	ED (T1Gd)	chr4
TUBA1C	49582518	49667114	0.4999	7.096e−07	GLRLM RLV	NET (T1)	chr12
ZNF284	44576296	44593766	−0.498	7.907e−07	GLRLM LGRE	NET (T1Gd)	chr19
IPO8	30781921	30848920	−0.490	1.243e−06	GLRLM GLV	ET (T2)	chr12
MMP7	102391238	102401484	0.490	1.260e−06	GLCM Auto Corr	ET (T1Gd)	chr11
TLL2	98124362	98273675	0.489	1.342e−06	GLSZM LGZE	NET (T1Gd)	chr10
TRIM55	67039130	67087720	0.488	1.408e−06	GLSZM LGZE	ED (T1Gd)	chr8
UBAP1	34179002	34252521	−0.486	1.582e−06	GLSZM SZLGE	ET (T2)	chr9

and different quantitative imaging phenomic characteristics of GBM. Our result has shown a high genetic enrichment through the Manhattan and qq-plots, especially for the texture features (Fig. 5).

Our results highlighted several genes that significantly associated with the tumor texture features, including *LRRC46, USP38, EPGN, TUBA1C, ZNF284, IPO8, MMP7, TLL2, TRIM55* and *UBAP1*, as the top ten significant genes (Table 4). However, there are, in total, 37 genes are significantly associated with the texture features (Fig. 5). EPGN expression associates significantly ($r_s = 0.501$, p-value = 6.542e−07) with GLSZM LGZE in the T1Gd modality (Table 4). EPGN previously reported to be one of the top ten upregulated genes after EBLN1 silencing in oligodendroglia cells [39]. Moreover, the emergence of EPGN was marked in another study by Duhem-Tonnelle et al. in EGF ligands expression profile, between glioblastoma cell lines and biopsies [40]. Located at chromosome 4, USP38 ($r_s = -0.511$, p-value = 3.648e−07) [41]. Moreover, as it is illustrated in the Manhattan plot of the spatial features of the tumor (Fig. 4

and Table 3), no gene shows significant association with any of the location features. In addition to the latter, the number of GBM lesions in the cerebellum in clinical settings are quite rare [36], as also shown in our summary Table 2. Our study can give some insight into this rare type of GBM lesion. Nevertheless, the investigation excluding the patients having those lesions have to be repeated as a future work.

5   Conclusion

As the understanding of gliomagenesis grows, several medical imaging biomarkers and genetic variations can be identified, and new hypotheses can be formed. The hereby proposed genome-wide association framework aims at identifying differentially expressed genes that significantly correlate with various aspects of GBM. The identification of such genes may contribute to the development of targeted therapies that focus on the resistance mechanisms of individual patients.

Through the systematic testing of associations and shrinking of the number of genes at every stage, this pipeline facilitates the evaluation of various hypotheses and reduces the computational complexity. In future work, we plan to extend the study by integrating more quantitative imaging phenomic tumor characteristics, inclusive of morphological, intensity, and volumetric descriptors, as well as parameters derived by biophysical tumor growth modeling.

Acknowledgement. Research reported in this publication was partly supported by the National Institutes of Health (NIH) under award numbers NIH/NINDS:R01NS042645, NIH/NCI:U24CA189523, NIH/NCATS:UL1TR001878, the ITMAT of the University of Pennsylvania as well as by the Swedish International Development Cooperation Agency (SIDA) through the Organization for Women in Science for the Developing World (OWSD). Computations were performed using facilities provided by the University of Cape Town's ICTS High Performance Computing team: hpc.uct.ac.za.

References

1. Verhaak, R.G.W., et al.: Integrated genomic analysis identifies clinically relevant subtypes of glioblastoma characterized by abnormalities in PDGFRA, IDH1, EGFR, and NF1. Cancer Cell **17**(1), 98–110 (2010)
2. Davatzikos, C., et al.: Cancer imaging phenomics toolkit: quantitative imaging analytics for precision diagnostics and predictive modeling of clinical outcome. J. Med. Imaging **5**(1), 011018 (2018)
3. Bakas, S., et al.: GLISTRboost: combining multimodal MRI segmentation, registration, and biophysical tumor growth modeling with gradient boosting machines for glioma segmentation. In: Crimi, A., Menze, B., Maier, O., Reyes, M., Handels, H. (eds.) BrainLes 2015. LNCS, vol. 9556, pp. 144–155. Springer, Cham (2016). https://doi.org/10.1007/978-3-319-30858-6_13
4. Bakas, S., et al.: Advancing the cancer genome atlas glioma MRI collections with expert segmentation labels and radiomic features. Nat. Sci. Data **4**, 170117 (2017)

5. Bakas, S., et al.: Segmentation labels and radiomic features for the pre-operative scans of the TCGA-GBM collection. The Cancer Imaging Archive (2017). https://doi.org/10.7937/K9/TCIA.2017.KLXWJJ1Q

6. Scarpace, L., et al.: Radiology data from the cancer genome atlas glioblastoma multiforme [TCGA-GBM] collection. Cancer Imaging Arch. **11**, 4 (2016)

7. Clark, K., et al.: The Cancer Imaging Archive (TCIA): maintaining and operating a public information repository. J. Digit. Imaging **26**(6), 1045–1057 (2013)

8. Stein, J.L., et al.: Voxelwise genome-wide association study (vGWAS). Neuroimage **53**(3), 1160–1174 (2010)

9. Liu, J., et al.: Combining fMRI and SNP data to investigate connections between brain function and genetics using parallel ICA. Hum. Brain Mapp. **30**(1), 241–255 (2009)

10. Batmanghelich, N.K., Dalca, A.V., Sabuncu, M.R., Golland, P.: Joint modeling of imaging and genetics. In: Gee, J.C., Joshi, S., Pohl, K.M., Wells, W.M., Zöllei, L. (eds.) IPMI 2013. LNCS, vol. 7917, pp. 766–777. Springer, Heidelberg (2013). https://doi.org/10.1007/978-3-642-38868-2_64

11. Elsheikh, S., et al.: Relating connectivity changes in brain networks to genetic information in Alzheimer patients. In: 2018 IEEE 15th International Symposium on Biomedical Imaging (ISBI 2018). IEEE (2018)

12. Zinn, P.O., et al.: Radiogenomic mapping of edema/cellular invasion MRI-phenotypes in glioblastoma multiforme. PLoS ONE **6**(10), e25451 (2011)

13. Gutman, D.A., et al.: MR imaging predictors of molecular profile and survival: multi-institutional study of the TCGA glioblastoma data set. Radiology **267**(2), 560–569 (2013)

14. Macyszyn, L., et al.: Imaging patterns predict patient survival and molecular subtype in glioblastoma via machine learning techniques. Neuro-oncology **18**(3), 417–425 (2015)

15. Binder, Z., et al.: Epidermal growth factor receptor extracellular domain mutations in glioblastoma present opportunities for clinical imaging and therapeutic development. Cancer Cell **34**, 163–177 (2018)

16. Bakas, S., et al.: In vivo detection of EGFRvIII in glioblastoma via perfusion magnetic resonance imaging signature consistent with deep peritumoral in ltration: the φ-index. Clin. Cancer Res. **23**, 4724–4734 (2017)

17. Cancer Genome Atlas Research Network: Comprehensive, integrative genomic analysis of diffuse lower-grade gliomas. N. Engl. J. Med. **372**(26), 2481–2498 (2015)

18. Ellingson, B.M., et al.: Probabilistic radiographic atlas of glioblastoma phenotypes. Am. J. Neuroradiol. **34**(3), 533–540 (2012)

19. Ellingson, B.M.: Radiogenomics and imaging phenotypes in glioblastoma: novel observations and correlation with molecular characteristics. Curr. Neurol. Neurosci. Rep. **15**(1), 506 (2015)

20. Steed, T.C., et al.: Differential localization of glioblastoma subtype: implications on glioblastoma pathogenesis. Oncotarget **7**(18), 24899 (2016)

21. Bilello, M., et al.: Population-based MRI atlases of spatial distribution are specific to patient and tumor characteristics in glioblastoma. NeuroImage: Clin. **12**, 34–40 (2016)

22. Akbari, H., et al.: In vivo evaluation of EGFRvIII mutation in primary glioblastoma patients via complex multiparametric MRI signature. Neuro-Oncology **20**(8), 1068–1079 (2018)

23. Aerts, H.J.W.L.: The potential of radiomic-based phenotyping in precision medicine: a review. JAMA Oncol. **2**(12), 1636–1642 (2016)

24. Lambin, P., et al.: Radiomics: extracting more information from medical images using advanced feature analysis. Eur. J. Cancer **48**(4), 441–446 (2012)
25. Aerts, H.J.W.L., et al.: Decoding tumour phenotype by noninvasive imaging using a quantitative radiomics approach. Nat. Commun. **5**, 4006 (2014)
26. Haralick, R.M., et al.: Textural features for image classification. IEEE Trans. Syst. Man Cybern. **3**, 610–621 (1973)
27. Galloway, M.M.: Texture analysis using grey level run lengths. Comput. Graph. Image Process. **4**, 172–179 (1975)
28. Chu, A., et al.: Use of gray value distribution of run lengths for texture analysis. Pattern Recogn. Lett. **11**, 415–419 (1990)
29. Dasarathy, B.V., Holder, E.B.: Image characterizations based on joint gray level-run length distributions. Pattern Recogn. Lett. **12**, 497–502 (1991)
30. Tang, X.: Texture information in run-length matrices. IEEE Trans. Image Process. **7**, 1602–1609 (1998)
31. Amadasun, M., King, R.: Textural features corresponding to textural properties. IEEE Trans. Syst. Man Cybern. **19**, 1264–1274 (1989)
32. Hogea, C., et al.: An image-driven parameter estimation problem for a reaction-diffusion glioma growth model with mass effects. J. Math. Biol. **56**, 793–825 (2008)
33. Hogea, C., et al.: A robust framework for soft tissue simulations with application to modeling brain tumor mass effect in 3D MR images. Phys. Med. Biol. **52**, 6893–6908 (2007)
34. Hogea, C., Davatzikos, C., Biros, G.: Modeling glioma growth and mass effect in 3D MR images of the brain. In: Ayache, N., Ourselin, S., Maeder, A. (eds.) MICCAI 2007. LNCS, vol. 4791, pp. 642–650. Springer, Heidelberg (2007). https://doi.org/10.1007/978-3-540-75757-3_78
35. Itakura, H., et al.: Magnetic resonance image features identify glioblastoma phenotypic subtypes with distinct molecular pathway activities. Sci. Transl. Med. **7**(303), 303ra138 (2015)
36. Drabycz, S., et al.: An analysis of image texture, tumor location, and MGMT promoter methylation in glioblastoma using magnetic resonance imaging. Neuroimage **49**(2), 1398–1405 (2010)
37. Gooya, A., et al.: GLISTR: glioma image segmentation and registration. IEEE Trans. Med. Imaging **31**(10), 1941–1954 (2012)
38. Kendall, M.G.: The advanced theory of statistics. In: The Advanced Theory of Statistics, 2nd edn (1946)
39. He, P., et al.: Knock-down of endogenous bornavirus-like nucleoprotein 1 inhibits cell growth and induces apoptosis in human oligodendroglia cells. Int. J. Mol. Sci. **17**(4), 435 (2016)
40. Duhem-Tonnelle, V., et al.: Differential distribution of erbB receptors in human glioblastoma multiforme: expression of erbB3 in CD133-positive putative cancer stem cells. J. Neuropathol. Exp. Neurol. **69**(6), 606–622 (2010)
41. Carminati, P.O., et al.: Alterations in gene expression profiles correlated with cis-platin cytotoxicity in the glioma U343 cell line. Genet. Mol. Biol. **33**(1), 159–168 (2010)

Ischemic Stroke Lesion Image Segmentation

Stroke Lesion Segmentation with 2D Novel CNN Pipeline and Novel Loss Function

Pengbo Liu$^{(\boxtimes)}$

Beijing University of Technology, Beijing, China
pengbo18555@bjut.edu.cn

Abstract. Recently, CT perfusion (CTP) has been used to triage ischemic stroke patients in the early stage, because of its speed, availability, and lack of contraindications. But CTP data alone, even with the generated perfusion maps is not enough to describe the precise location of infarct core or penumbra. Considering the good performance demonstrated on Diffusion Weighted Imaging (DWI), We propose a CTP data analysis technique using Generative Adversarial Networks (GAN) [2] to generate DWI, and segment the regions of ischemic stroke lesion on top of the generated DWI based on convolutional neutral network (CNN) and a novel loss function. Specifically, our CNN structure consists of a generator, a discriminator and a segmentator. The generator synthesizes DWI from CT, which generates a high quality representation for subsequent segmentator. Meanwhile the discriminator competes with the generator to identify whether its input DWI is real or generated. And we propose a novel segmentation loss function that contains a weighted cross-entropy loss and generalized dice loss [1] to balance the positive and negative loss in the training phase. And the weighted cross entropy loss can highlight the area of stroke lesion to enhance the structural contrast. We also introduce other techniques in our network, like GN [13], Maxout [14] in the proposed network. Data augmentation is also used in the training phase. In our experiments, an average dice coefficient of 60.65% is achieved with four-fold cross-validation. From the results of our experiments, this novel network combined with the proposed loss function achieved a better performance in CTP data analysis than traditional methods using CTP perfusion parameters.

Keywords: Generation · Segmentation · Cross-entropy loss · Generalized dice loss

1 Introduction

Stroke is the second leading cause of death worldwide, accounting for 6.24 million deaths globally in 2015 [4]. And more than 80% of stroke cases are ischemic. However, only very few of stroke patients receive recombinant tissue plasminogen

© Springer Nature Switzerland AG 2019
A. Crimi et al. (Eds.): BrainLes 2018, LNCS 11383, pp. 253–262, 2019.
https://doi.org/10.1007/978-3-030-11723-8_25

activator (rtPA) therapy despite proven effectiveness in reducing stroke disability. Defining location and extent of irreversibly damaged brain tissue is a critical part of the decision-making process in acute stroke. Magnetic resonance images (MRI) using diffusion and perfusion imaging can be used to distinguish between infarcted core and the penumbra. Because of the advantages in speed, availability, and contraindications, CT perfusion have been used to triage with acute stroke instead of MRI, which may shorten the scanning time for the stroke patients [3]. Automatic methods, including many commercial software have been developed to measure the perfusion maps of stroke patients, but such methods may not be good enough to solve the heterogeneities of the stroke patients. Therefore, there is a great need for advanced data analysis techniques that could help to diagnosis stroke accurately and precisely with repetitiveness, and support decision-making for treatments. In the literature, more often a optimal threshold of the CTP parameter is used, like $rCBF < 30\%$ within $delaytime > 3\,\mathrm{s}$ is the threshold of the region of core [3]. But this pre-defined threshold suffers from several drawbacks. First, progression of stroke is patient specific and a population threshold may not work in some cases. Second, thresholding means a fixed value for all patients across all scanners in all hospitals, without considerations of site differences.

In the past decade, deep learning technology, especially Convolutional Neural Network (CNN) has achieved huge successes in various computer vision tasks, like classification [7], detection [8] and segmentation [9]. And the power of CNN is demonstrated in medical imaging more and more. Ciresan et al. [10] firstly introduced CNNs to medical image segmentation by predicting a pixel's label based on the raw pixel values in a square window centered on it. But this method is quite slow because the network must run separately for every pixel within every single image and there is a lot of redundancy due to overlapping windows actually. Later on, Ronneberger et al. proposed U-Net [11], which consists of a contracting path to capture context and a symmetric expanding path that enables precise localization and can be trained end-to-end from very few images built upon the famous Fully Convolutional Network (FCN) framework [9]. Specifically, Nielsen [6] introduce the deep CNN in Acute Ischemic Stroke segmentation task with a simple encoder-decoder structure to predict the final infarct.

In this paper, different from Anne Nielsen, we proposed a novel structure of network combined with the GAN approach to solve this task. Our proposed network structure contains a generator, a discriminator and a segmentator. The structure of our work flow is shown in Fig. 1. Generator is used to synthesize the DWI images from the CT data. Discrimintator is used to perform the classification of the generated DWI image and the true DWI image. Segmentator is finally used to segment the lesion of brain on the generated DWI image.

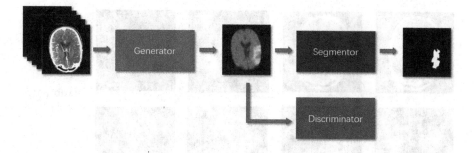

Fig. 1. Structure of our ischemic stroke lesion segmentation work flow. Generator is used to synthesize the DWI images from the CT data. Discrimintator is used to perform the classification of the generated DWI image and the true DWI image. Segmentator is finally used to segment the lesion of brain on the generated DWI image.

2   Method

2.1   Dataset and Data Preprocessing

Our framework is trained and tested using ISLES 2018 Segmentation Challenge dataset, which include imaging data from acute stroke patients in two centers who presented within 8 h of stroke onset and underwent an MRI DWI within 3 h after CTP. The training data set consists of 63 patients, and the testing set includes 40 patients. Because some patients have two slabs to cover the stroke lesion (These are non-, or partially-overlapping brain regions). In the end, we have 94 and 62 cases in training phase and testing phase separately. Each case in training have eight modalities or parametric maps, including CT, CT_4DPWI, CT_MTT, CT_Tmax, CT_CBF, CT_CBV, MR_DWI, and ground truth, OT. All modalities are shown in Fig. 2. ISLES 2018 dataset have the same size, (256 * 256), and the same spacing, (1 * 1 mm), in x and y dimension, but with a very different slice number in z dimension. Most cases of the dataset have only two slices, which is hard to exploit the information from the 3D data via 3D-CNN method. Thus we separate the 3D images into 2D slices along z axis, which equivalently augment the data size, as more training data are available in a 2D manner with a 2D CNN network. We concatenate the channels of CT_MTT, CT_Tmax, CT_CBV, CT_CBF, and the maximal value of CT_4DPWI along time dimension, which we call PWI_max, and normalize them in the training process.

Additionally, We also perform on-the-fly data augmentation when feeding training samples. Augmentation operations include scaling, flipping, rotation, and translation are applied during the training.

2.2   Construction of Network

The whole workflow of our approach is shown in Fig. 1. The generator and segmentator are both designed based on U-Net [11] as illustrated in Fig. 4. It is a fully convolutional network, which consists of a encoder structure (left side) and

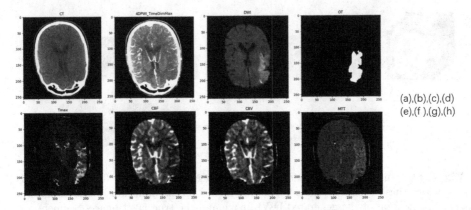

(a),(b),(c),(d)
(e),(f),(g),(h)

Fig. 2. An example of ISLES 2018 training data: (a) CT, (b) max value of 4DPWI, (c) DWI, (d) OT, (e) Tmax, (f) CBF, (g) CBV, (h) MTT.

a decoder structure (right side). The encoder part and decoder part consist of repeated layers of two 3×3 convolutions, with either MaxPooling in encoder part or UpConv layer in decoder part. With the limited GPU memory, a degradation problem is encounted in batch normalization due to the small batch size. Group normalization [13] is thus introduced in our network to solve this problem. GN's performance is relatively independent of the batch sizes used, and its accuracy is stable in a wide range of batch sizes. Inspired by the idea of maxout from Goodfellow et al. [14], maxout operation is performed in the final prediction. Normally, there are two output channels predicted for binary classification tasks, followed by softmax operation. Instead, We predict four layers, including three layers as background prediction for false positive reduction and one layer for foreground prediction. Then the max of three background channels are combined along channel dimension, to construct the final background channel for each pixel. We show a schematic drawing in Fig. 3. Through this maxout operation, many false positives could be reduced.

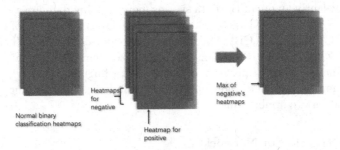

Fig. 3. Maxout for segmentation.

The discriminator in our work flow is in its simplest form, because we want to save the GPU memory used as much as we can. It only consists of five convolutional layers followed by group normalization [13], with Relu and a average pooling layer in the end. We use kernels with size 5 * 5 and stride 2 * 2, with no padding.

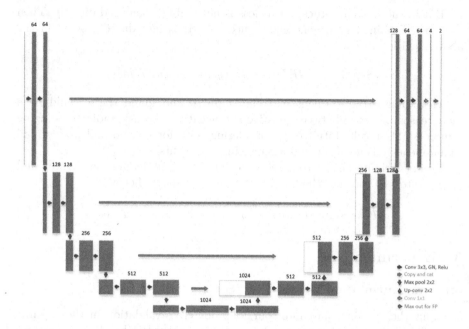

Fig. 4. U-Net with GN and Maxout, the backbone Structure of our generator and segmentator (Maxout is only used in segmentator).

2.3   Loss Function

Our loss function have three parts, which come from the generator, discriminator and segmentor, respectively. For the generator network, we measure how effectiveness the network transfer CT modalities to DWI using traditional mean square error (MSE) loss compared with the ground truth DWI. For the discrimination network, borrowed from LSGAN [15], we calculate the gap between prediction and ground truth via MSE Loss as well. During the training of the discriminator, DLoss is calculated with Eq. (2). In training of the generator and segmentator, DLoss is in the form of Eq. (3).

$$GLoss = MSELoss(Pred, Target) \qquad (1)$$

$$DLoss = MSELoss(FakeDWI, 0) + MSELoss(TrueDWI, 1) \qquad (2)$$

$$DLoss = MSELoss(FakeDWI, 1) \qquad (3)$$

In the segmentation task, people usually use either cross entropy (CE) loss or dice loss alone. In Kaggle Carvana Image Masking Challenge, one challenger come up with a novel loss function result in a good performance with the form:

$$SegLoss = BCELoss - log(diceloss) \tag{4}$$

But in the training process, this loss is not stable when used directly. Then CE Loss combined with generalized dice loss [1] is introduced instead in the above equation:

$$SegLoss = CELoss - log(generalizeddiceloss) \tag{5}$$

We evaluate the gradient in training phase and found there would be a more reasonable ratio between positive and negative regions' gradients, bringing a more stable result. Intuitively, considering both foreground and background simultaneously is good for performance improvements.

Proper weights need to be set for each loss while training generator and segmentator. In the end, the loss function used is shown below.

$$Loss = \omega_g * GLoss + \omega_d * DLoss + \omega_s * SegLoss \tag{6}$$

3   Experiment

3.1   Implementation Details

We split the training data into four folds for cross-validation in the training phase. The input samples' size is (B, 5, 256, 256) where B is batch-size. First, we train the discriminator using generated DWI, which is synthesized via generator from the input CT data and the perfusion parameters, and the true DWI with pre-prepared labels. The discriminator's parameters were optimized using RMS optimizer with $DLoss$ (Eq. (2)). Similar as in the discriminator, we also use RMS optimizer to optimize the generator and segmetator's parameters simultaneously, with $Loss$ (Eq. (6)). The initial learning rate and γ is 0.0001 and 0.9, respectively. We train these two branches alternately 700 epoch in total. Learning rate will be halved in epoch 100, 300, 500. The ratio of weights, $GLoss:DLoss:SLoss$, is *0.002:0.5:1*. In GN layers, we fix the number of channels per group as 16. Because the GN layers can have different channel numbers, the group number can change across layers in this setting. The batch size is 6 in our exerments.

3.2   Results

We implemented our framework using PyTorch with cuDNN, and ran all experiments on a GPU server with 256 GB of memory, and a Nvidia GTX 1080Ti 11G GPUs. We analysed the segmentation loss function in the early stage. So the experiments result below are all based on the loss functions mentioned in Sect. 2.3.

Table 1. Result in different experiment stage.

Stage	First	Second	Third	Fourth
Dice	0.8473	0.5463	0.5829	0.6065

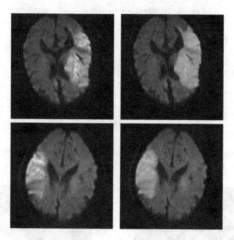

Fig. 5. Visualization of validation results of model trained on MR_DWI directly. Image in the left is the raw DWI image. DWI image overlayed by ground truth and prediction of model is in the right (green is ground truth, cyan is prediction). (Color figure online)

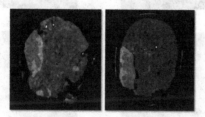

Fig. 6. Visualization of validation results of model trained on MR_MTT, MR_Tmax, MR_CBV, MR_CBF directly. MTT image overlayed by ground truth and prediction of model is used to show. (green is ground truth, yellow is prediction). (Color figure online)

The inspiration of our novel work flow is as the following. At the very beginning, first stage (Table 1), we don't know there is no MR_DWI modality in testing phase. So MR_DWI is used as input to train the algorithm with a raw U-Net only. And the result is pretty good (Fig. 5) with dice coefficient equal to 0.8473. Then at the second stage (Table 1), we tried training a U-Net with MTT, Tmax, CBV and CBF, getting a much worse result of dice coefficient of 0.5463. The big gap in performance of segmentation between CTP image and DWI image drives me to transfer CTP images to a DWI image with a GAN approach. LSGAN [15] is a variation of traditional GAN [2] with a more stable performance and

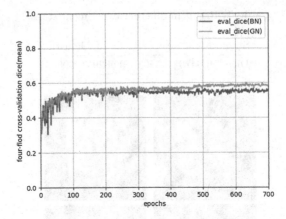

Fig. 7. Four-flod cross-validation dice (mean). Comparation between BN and GN.

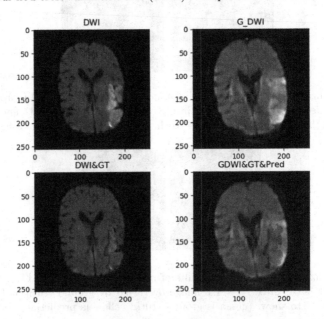

Fig. 8. Visualization of final result of model trained on our whole pipeline. (Top): Image in the left is the raw DWI image. Image in the right is our generalized DWI modality. (Bottom): Image overlayed by ground truth and prediction of model (green is ground truth, cyan is prediction). (Color figure online)

quicker convergence by changing log loss to L2 loss. In the third stage, after adding LSGAN in our work flow, the final dice coefficient score had a great improvement, from 0.5456 to 0.5811. Meanwhile, we concatenate a new channel, max of CT_4DPWI along time dimension, into the input. So, this work flow is determined to be our pipeline. In the fourth stage (Table 1), we mainly did some network structure modifications, like self attention, Maxout, GN, etc, to stable the training process and reduce the FP regions. The average dice score of four-fold cross-validation is 0.6065 in the end. The visualization of our final result is shown in Fig. 8. And the average dice score of the four fold validation sets is shown in Fig. 7. Batch normalization using BN or GN is also compared. The curve of network with GN is more stable than the curve of network with BN (Fig. 6).

4 Conclusion

In this paper, we propose a deep learning based method combined with adversarial networks to locate the region of ischemic stroke lesion automatically. And we proposed a novel pipeline for stroke lesion segmentation through DWI modality generation from CT perfusion data, which have a potentially promising prospective. This novel pipeline have a improvement over direct segmentation with a large margin. And a novel loss function is proposed as well, with maxout operation used in segmentation to achieve the state of art in ISLES challenge 2018.

References

1. Sudre, C.H., Li, W., Vercauteren, T., et al.: Generalised dice overlap as a deep learning loss function for highly unbalanced segmentations (2017)
2. Goodfellow, I., et al.: Generative adversarial nets. In: Advances in Neural Information Processing Systems (2014)
3. Yu, Y., Han, Q., Ding, X., et al.: Defining core and penumbra in ischemic stroke: a voxel- and volume-based analysis of whole brain CT perfusion. Sci. Rep. **6**, 20932 (2015)
4. The top 10 causes of death (2017). http://www.who.int/mediacentre/factsheets/fs310/en/. Accessed 30 June 2017
5. Mouridsen, K., Hansen, M.B., Østergaard, L., Jespersen, S.N.: Reliable estimation of capillary transit time distributions using DSC-MRI. J. Cereb. Blood Flow Metab. **34**, 1511–1521 (2014). https://doi.org/10.1038/jcbfm.2014.111
6. Nielsen, A., Hansen, M.B., Tietze, A., et al.: Prediction of tissue outcome and assessment of treatment effect in acute ischemic stroke using deep learning. Stroke **49**, 1349–1401 (2018)
7. Krizhevsky, A., Sutskever, I., Hinton, G.E.: ImageNet classification with deep convolutional neural networks. In: International Conference on Neural Information Processing Systems, pp. 1097–1105. Curran Associates Inc. (2012)
8. Ren, S., He, K., Girshick, R., et al.: Faster R-CNN: towards real-time object detection with region proposal networks. IEEE Trans. Pattern Anal. Mach. Intell. **39**(6), 1137–1149 (2017)
9. Long, J., Shelhamer, E., Darrell, T.: Fully convolutional networks for semantic segmentation. IEEE Trans. Pattern Anal. Mach. Intell. **PP**(99), 1 (2014)
10. Ciresan, D., Giusti, A., Gambardella, L.M., Schmidhuber, J.: Deep neural networks segment neuronal membranes in electron microscopy images. In: Advances in Neural Information Processing Systems, pp. 2843–2851 (2012)

11. Ronneberger, O., Fischer, P., Brox, T.: U-Net: convolutional networks for biomedical image segmentation. In: Navab, N., Hornegger, J., Wells, W.M., Frangi, A.F. (eds.) MICCAI 2015. LNCS, vol. 9351, pp. 234–241. Springer, Cham (2015). https://doi.org/10.1007/978-3-319-24574-4_28

12. Glorot, X., Bordes, A., Bengio, Y.: Deep sparse rectifier neural networks. In: International Conference on Artificial Intelligence and Statistics, pp. 315–323 (2011)

13. Wu, Y., He, K.: Group normalization (2018)

14. Goodfellow, I.J., Warde-Farley, D., Mirza, M., et al.: Maxout networks. arXiv preprint arXiv:1302.4389 (2013)

15. Mao, X., Li, Q., Xie, H., et al.: Least squares generative adversarial networks (2016)

Contra-Lateral Information CNN for Core Lesion Segmentation Based on Native CTP in Acute Stroke

Jeroen Bertels[(✉)], David Robben, Dirk Vandermeulen, and Paul Suetens

Medical Image Computing (ESAT/PSI), KU Leuven, Leuven, Belgium
jeroen.bertels@kuleuven.be

Abstract. Stroke is an important neuro-vascular disease, for which distinguishing necrotic from salvageable brain tissue is a useful, albeit challenging task. In light of the Ischemic Stroke Lesion Segmentation challenge (ISLES) of 2018 we propose a deep learning-based method to automatically segment necrotic brain tissue at the time of acute imaging based on CT perfusion (CTP) imaging. The proposed convolutional neural network (CNN) makes a voxelwise segmentation of the core lesion. In order to predict the tissue status in one voxel it processes CTP information from the surrounding spatial context from both this voxel and from a corresponding voxel at the contra-lateral side of the brain. The contra-lateral CTP information is obtained by registering the reflection w.r.t. a sagittal plane through the geometric center. Preprocessed training data was augmented during training and a five-fold cross-validation was used to experiment for the optimal hyperparameters. We used weighted binary cross-entropy and re-calibrated the probabilities upon prediction. The final segmentations were obtained by thresholding the probabilities at 0.50 from the model that performed best w.r.t. the Dice score during training. The proposed method achieves an average validation Dice score of 0.45. Our method slightly underperformed on the ISLES 2018 challenge test dataset with the average Dice score dropping to 0.38.

Keywords: Stroke · CTP · Core · CNN

1 Introduction

Stroke is the main cause of neurological disability in older adults [2]. Up to four out of five acute stroke cases is ischemic. In these cases, the ischemia is the result

J. Bertels is part of NEXIS, a project that has received funding from the European Union's Horizon 2020 Research and Innovations Programme (Grant Agreement #780026).
D. Robben is supported by an innovation mandate of Flanders Innovation & Entrepreneurship (VLAIO).

A. Crimi et al. (Eds.): BrainLes 2018, LNCS 11383, pp. 263–270, 2019.
https://doi.org/10.1007/978-3-030-11723-8_26

of a sudden occlusion of a cerebral artery. If a large artery is occluded, there is a possibility that mechanical thrombectomy (i.e. the intra-arterial removal of the cloth) can improve patient outcome [4]. Important biomarkers for the correct selection of patients are the volumes of necrotic and salvageable tissue, respectively the core and penumbra. In that light, this year's Ischemic Stroke Lesion Segmentation (ISLES) challenge [1] asked for methods that can automatically segment core tissue based on CT (perfusion) imaging.

With respect to the detection of core tissue, a standard CT is unable to capture early necrotic changes and will therefor result in underestimation [5]. A more sensitive approach is to acquire a perfusion scan and extract certain perfusion parameters of the parenchymal tissue via deconvolution analysis [3].

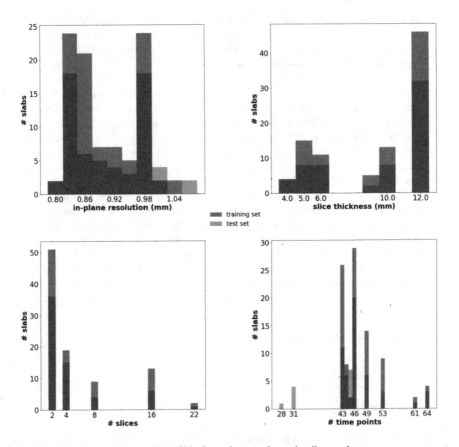

Fig. 1. Both the training data (*blue*) and test data (*red*) are heterogeneous w.r.t. certain image properties. The y-axis shows the number of data samples (i.e. slabs). *Top left*: The in-plane resolution ranges from 0.785 to 1.086 mm. *Top right*: The slice thickness ranges from 4.0 to 12.0 mm. *Bottom left*: The number of slices ranges from 2 to 22. *Bottom right*: The number of time points (at the initial resolution of 1 image/1 s) ranges from 28 to 64. (Color figure online)

As such, core tissue is characterized with an increase in Tmax and MTT, and a decrease in CBF and CBV. During last year's ISLES challenge it was shown that convolutional neural networks (CNNs) can be used to predict the final infarction based on these parameter maps (and additional patient and treatment information). In parallel, it has been found that CNNs can be used to estimate the perfusion parameters directly, hence bypassing the mathematically ill-posed deconvolution problem [7]. Nonetheless, these CNN-based methods still require an arterial input function (AIF) to do the deconvolution.

This work makes a contribution on two levels. First, we are the first to strictly limit the CNN to use only CT perfusion (CTP) data as input in the prediction of core tissue. This holds both during training and testing. Second, we explore the alternative use of contra-lateral information instead of the explicit manual or (semi-)automatic selection of the AIF. We identify the ISLES 2018 challenge as the perfect setup to compare the performance directly with other state-of-the-art methods. In the next section we briefly introduce the dataset and some initial preprocessing. In Sect. 3, we highlight the extraction of contra-lateral information and the experimental setup, including the architecture of our CNN. In Sect. 4, we show the validation results on the ISLES 2018 training set as well as the results on the test set.

2 Data

The ISLES 2018 dataset consists of acute CT and CTP images, and the derived Tmax, CBF and CBV perfusion maps, from 125 patients. The ground truth cores were delineated on the MR DWI images, which were acquired soon thereafter. The participants only have access to the training dataset. For some patients there are two slabs to cover the lesion. We consider each slab independently, resulting in 94 slabs present in the training dataset and 62 in the test dataset.

The in-plane resolutions are isotropic and we notice resolutions ranging from 0.796 to 1.039 mm in the training set and from 0.785 to 1.086 mm in the test set (Fig. 1). We consider these spatial resolutions similar enough and avoid resampling the images. The spatial resolution in the axial direction ranges from 4 to 12 mm and the number of slices from 2 to 22. We therefore opt to work in 2D. We further notice the discrepancy in the available number of time points, ranging from 43 to 64 and from 28 to 64, both at a temporal resolution of 1 image/1 s, for training and test sets respectively. We first resample the signal along the time axis to a temporal resolution of 1 image/2 s by using a smoothing kernel of [1/4, 2/4, 1/4] with a stride of 2. We then pad the signal by repeating the final value until we have 32 time points (i.e. 64 s). The resulting volumes have lower temporal noise and identical shape.

3 Method

As we hypothesised in Sect. 1, we will investigate whether CTP information only, but complemented with contra-lateral information, can be used to predict the

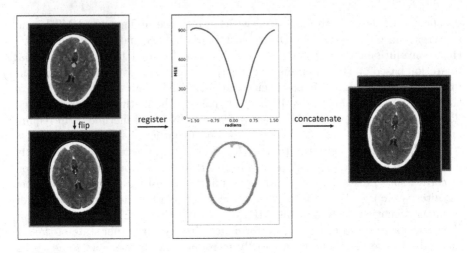

Fig. 2. In order to complement each voxel with contra-lateral information, we perform the following operations. *Left*: We flip the entire original volume (*blue*) w.r.t. a sagittal axis through the geometrical center (*yellow*) to obtain the contra-lateral volume (*red*). *Middle*: We rotate the clipped (in [0, 100] HU) contra-lateral volume for the lowest mean squared error (MSE). *Right*: We apply this angle (here 0.28 radians for case 4 of the training dataset) to the contra-lateral, unclipped volume and concatenate with the original to obtain 64 features for each voxel. (Color figure online)

core lesion. We will use DeepVoxNet [6] as a framework for doing segment-based training and extensive parallel data augmentation on the CPU. First, we explain how to construct the input for our CNN. Then, the architecture of our CNN will be discussed. Finally, we detail some further training methodologies.

3.1 Contra-Lateral Information

Because an ischemic stroke typically occurs uni-laterally, we want to enrich the information in one voxel (i.e. currently a time series of 32 points depicting the passage of contrast) with the information present in a corresponding voxel at the contra-lateral side of the brain (Fig. 2). This way each voxel contains information from at least one healthy voxel, thus how the perfusion or contrast passage could look like in similar but healthy parenchyma. For this purpose, we average each CTP over time and clip values in the range of [0, 100] HU. We then flip the image laterally (i.e. w.r.t. a sagittal plane) across its geometrical center and rotate the flipped image back to minimize the mean squared error (MSE). This rotation is applied to the unclipped, flipped volume and the result is concatenated to the original CTP volume. Each voxel now has 64 features. Before we let the network crunch this data, we make sure to suppress the influence of extreme outliers (e.g. streak artifacts) via clipping in the range of [0, 150] HU. We furthermore normalize the data to zero mean and unit variance.

3.2 Data Augmentation

To artificially increase the number of training samples and regularize the network as such we augment input images on the level of the image and on the level of the segments, which are extracted from the image. At the image level, the image is rotated with 0.5 probability according to $\mathcal{N}(0, 10°)$ and Gaussian noise $\mathcal{N}(0, 0.03)$ is added to the original and flipped part independently. At the segment level, the segments can be flipped laterally with 0.5 probability, the intensities can be shifted $\mathcal{N}(0, 0.01)$ and scaled $\mathcal{N}(1, 0.01)$, and the time series can be shifted forward or backward in time $\mathcal{U}(-6, 6)$. We also perform contrast scaling of the segments, where we scale the intensity differences of a certain time point w.r.t. the first time point according to a Lognormal distribution with zero mean and a standard deviation of 0.3 [5].

3.3 Class Balancing

Based on the images obtained before normalization we construct a binary head mask from where the intensities lie in the range [0, 150] HU. This head mask is used to mask the output of the network both during training and testing. Network input segment centers are sampled uniformly from within this region. In order to further balance the class observations we use a weighted binary cross-entropy with weights calculated to equalize the prior class probabilities across the training data. Upon prediction we re-calibrate the probabilities.

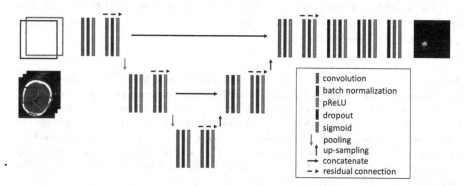

Fig. 3. The CNN takes as input a 97 × 97 2D segment with 64 features, sampled from the concatenated original and contra-lateral volumes (see Fig. 2) and outputs lesion probabilities for the corresponding 29 × 29 segment. Our U-Net-like architecture consists of two pooling (*green* arrows) and two up-sampling layers (*red* arrows). Together with all the 3 × 3 convolution layers (*blue* layers; each of which has 64 filters) involved, this results in a receptive field of 69 × 69 voxels. The final three convolution layers are fully-convolutional with 128 filters. (Color figure online)

3.4 CNN Architecture

We use a U-Net-like [8] architecture with two pooling and two up-sampling layers, which respectively downsample or upsample by a factor of three (Fig. 3). Before each pooling or up-sampling layer (and after the final one), we perform two consecutive convolutions with 64 filters of size 3×3. Two fully-convolutional layers precede the final layer, each with 128 filters. The parametric ReLU (pReLU) is used as the non-linear activation function. We use batch normalization and local skip connections to improve learning. The CNN is characterized with valid padding, an in-plane receptive field size of 69×69 voxels and a total number of 470,657 trainable parameters. Both during training and testing we use input segments with a fixed size of 97×97 with 64 features to predict an output segment of size 29×29.

3.5 Experiments

We train the CNN using five-fold cross validation on the ISLES 2018 training set. The predictions of each model will be evaluated and averaged during test time. We use the ADAM optimizer for 4000 epochs with an initial learning rate of 10^{-4} and decrease the learning rate with a factor of 10 after each 1000 epochs. Both decay rate parameters for ADAM were fixed at 0.9. Here, one epoch iterates through 1280 segments, extracted from 16 subjects and processed in batches of 64 segments. We use L1 and L2 weight regularization with weights of 10^{-4} and 10^{-2}, respectively. Every 20 epochs we run the model on the (left-out) validation set. The model that optimizes the validation cross-entropy and the model that optimizes the validation Dice score, both at a threshold of 0.50, were stored, $\mathrm{CE}^{0.50}$ and $\mathrm{D}^{0.50}$ respectively.

Table 1. The results of our models w.r.t. different metrics. *Rows*: The evaluated metrics: Dice score, Hausdorff distance (in voxels), average distance (in voxels), precision, recall, average volume difference (in voxels; except for the value for $\mathrm{D}^{0.50}$ @ test, which is in ml). *Columns*: The models with the best binary cross-entropy measures on the validation set during training thresholded at 0.50 ($\mathrm{CE}^{0.50}$) or at 0.11 (the optimal Dice threshold; $\mathrm{CE}^{0.11}$). The models with the best Dice scores on the validation set during training thresholded at 0.50 ($\mathrm{D}^{0.50}$) or at 0.30 (the optimal Dice threshold; $\mathrm{D}^{0.30}$). The results of model $\mathrm{D}^{0.50}$ on the test set ($\mathrm{D}^{0.50}$ @ test).

	$\mathrm{CE}^{0.50}$	$\mathrm{CE}^{0.11}$	$\mathrm{D}^{0.50}$	$\mathrm{D}^{0.30}$	$\mathrm{D}^{0.50}$ @ test
Dice score	0.32 ± 0.25	$\mathbf{0.50 \pm 0.24}$	0.46 ± 0.25	0.49 ± 0.24	0.38 ± 0.30
Hausdorff dst. (vxls.)	$\mathbf{32.2 \pm 21.6}$	44.1 ± 30.1	35.8 ± 26.8	44.9 ± 29.1	/
Average dst. (vxls.)	9.0 ± 9.1	7.6 ± 8.4	$\mathbf{7.2 \pm 11.6}$	9.0 ± 11.8	/
Precision	$\mathbf{0.55 \pm 0.35}$	0.43 ± 0.24	0.51 ± 0.29	0.44 ± 0.25	0.47 ± 0.35
Recall	0.24 ± 0.23	$\mathbf{0.58 \pm 0.30}$	0.40 ± 0.26	0.54 ± 0.29	0.44 ± 0.34
Avg. vol. dff. (vxls.)	-1546 ± 2858	-1546 ± 2858	$\mathbf{296 \pm 2830}$	$\mathbf{296 \pm 2830}$	17.2 ± 16.9

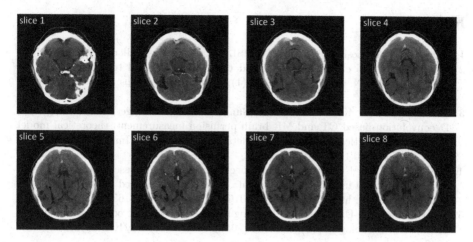

Fig. 4. An example output of model $D^{0.50}$ for slab number 5 from the validation data. In each of the eight slices we visualize the expert delineations in *blue* and the thresholded segmentations of our model in *red*. This is a rather good segmentation with a Dice score of 0.68. (Color figure online)

4 Results and Conclusion

We will compare the results of the models that were found optimal during training w.r.t. binary cross-entropy ($CE^{0.50}$) and Dice score ($D^{0.50}$), and of their optimal threshold derivatives w.r.t. the Dice score, $CE^{0.11}$ and $D^{0.30}$ respectively, by varying the segmentation threshold. In Table 1 these results are listed, as well as the result of model $D^{0.50}$ on the test set ($D^{0.50}$ @ test). Without post Dice score threshold optimization we notice the better results for the $D^{0.50}$ model compared to the $CE^{0.50}$ model. Especially the Dice score and the average volume difference stand out. Optimizing the thresholds for each of those models w.r.t. the Dice score could further improve the performance on this metric. Especially the $CE^{0.50}$ model benefits from this, explained by the discrepancy between its precision and recall. Although other methods performed better w.r.t. Dice score, we opted for model $D^{0.50}$ for participating the challenge because of lower surface distances and the lower average volume difference. In Fig. 4 one example segmentation with a Dice score of 0.68 on the validation dataset for our submitted model is depicted.

Although results are promising, further research is needed. For example, the registration of the contra-lateral information is far from ideal. The immediate concatenation could hinder the network to learn the correct use of this contra-lateral information.

References

1. Ischemic Stroke Lesion Segmentation (ISLES). www.isles-challenge.org
2. Berkhemer, O.A., et al.: A randomized trial of intraarterial treatment for acute ischemic stroke. N. Engl. J. Med. **372**(1), 11–20 (2015). https://doi.org/10.1056/NEJMoa1411587
3. Fieselmann, A., Kowarschik, M., Ganguly, A., Hornegger, J., Fahrig, R.: Deconvolution-based CT and MR brain perfusion measurement: theoretical model revisited and practical implementation details. Int. J. Biomed. Imaging **2011** (2011). https://doi.org/10.1155/2011/467563
4. Goyal, M., et al.: Endovascular thrombectomy after large-vessel ischaemic stroke: a meta-analysis of individual patient data from five randomised trials. Lancet **387**, 1723–1731 (2016). https://doi.org/10.1016/S0140-6736(16)00163-X
5. von Kummer, R., Dzialowski, I.: Imaging of cerebral ischemic edema and neuronal death. Neuroradiology **59**(6), 545–553 (2017). https://doi.org/10.1007/s00234-017-1847-6
6. Robben, D., Bertels, J., Willems, S.: DeepVoxNet. Technical report, KU Leuven: KUL/ESAT/PSI/1801 (2018)
7. Robben, D., Suetens, P.: Perfusion parameter estimation using neural networks and data augmentation. MICCAI SWITCH (2018). https://arxiv.org/abs/1810.04898
8. Ronneberger, O., Fischer, P., Brox, T.: U-Net: convolutional networks for biomedical image segmentation. In: Navab, N., Hornegger, J., Wells, W.M., Frangi, A.F. (eds.) MICCAI 2015. LNCS, vol. 9351, pp. 234–241. Springer, Cham (2015). https://doi.org/10.1007/978-3-319-24574-4_28

Dense Multi-path U-Net for Ischemic Stroke Lesion Segmentation in Multiple Image Modalities

Jose Dolz[✉], Ismail Ben Ayed, and Christian Desrosiers

Laboratory of Imaging, Vision and Artificial Intelligence,
Ecole de techologie supérieure, Montreal, Canada
`jose.dolz@etsmtl.ca`

Delineating infarcted tissue in ischemic stroke lesions is crucial to determine the extend of damage and optimal treatment for this life-threatening condition. However, this problem remains challenging due to high variability of ischemic strokes' location and shape. Recently, fully-convolutional neural networks (CNN), in particular those based on U-Net [27], have led to improved performances for this task [7]. In this work, we propose a novel architecture that improves standard U-Net based methods in three important ways. First, instead of combining the available image modalities at the input, each of them is processed in a different path to better exploit their unique information. Moreover, the network is densely-connected (i.e., each layer is connected to all following layers), both within each path and across different paths, similar to HyperDenseNet [11]. This gives our model the freedom to learn the scale at which modalities should be processed and combined. Finally, inspired by the Inception architecture [32], we improve standard U-Net modules by extending inception modules with two convolutional blocks with dilated convolutions of different scale. This helps handling the variability in lesion sizes. We split the 93 stroke datasets into training and validation sets containing 83 and 9 examples respectively. Our network was trained on a NVidia TITAN XP GPU with 16 GBs RAM, using ADAM as optimizer and a learning rate of 1×10^{-5} during 200 epochs. Training took around 5 h and segmentation of a whole volume took between 0.2 and 2 s, as average. The performance on the test set obtained by our method is compared to several baselines, to demonstrate the effectiveness of our architecture, and to a state-of-art architecture that employs factorized dilated convolutions, i.e., ERFNet [26].

1 Introduction

Stroke is one the leading causes of global death, with an estimate of 6 million cases each year [19,28]. It is also a major cause of long-term disability, resulting in reduced motor control, sensory or emotional disturbances, difficulty understanding language, and memory deficit. Cerebral ischemia, which comes from the blockage of blood vessels in the brain, represents approximately 80% of all stroke cases [13,31]. Brain imaging methods based on Computed Tomography

© Springer Nature Switzerland AG 2019
A. Crimi et al. (Eds.): BrainLes 2018, LNCS 11383, pp. 271–282, 2019.
https://doi.org/10.1007/978-3-030-11723-8_27

(CT) and Magnetic Resonance Imaging (MRI) are typically employed to evaluate stroke patients [34]. Early-stage ischemic strokes appear as a hypodense regions in CT, making them hard to locate with this modality. MRI sequences, such as T1 weighted, T2 weighted, fluid-attenuated inversion recovery (FLAIR), and diffusion-weighted imaging (DWI), provide a clearer image of brain tissues than CT, and are preferred modalities to assess the location and evolution of ischemic stroke lesions [2,3,18].

The precise delineation of stroke lesions is critical to determine the extend of tissue damage and its impact on cognitive function. However, manual segmentation of lesions in multi-modal MRI data is time-consuming as well as prone to inter and intra-observer variability. Developing methods for the automatic segmentation can thus contribute to having more efficient and reliable tools to quantify stroke lesions over time [24]. Over the years, various semi-automated and automated techniques have been proposed for segmenting lesions [21,25]. Recently, deep convolutional neural networks (CNNs) have shown high performance for this task, outperforming standard segmentation approaches on benchmark datasets [4,14,17,20,35].

Multi-modal image segmentation based on CNNs is typically addressed with an *early fusion* strategy, where multiple modalities are merged from the original input space of low-level features [10,16,17,22,33,39]. This strategy assumes a simple relationship (e.g., linear) between different modalities, which may not correspond to reality [30]. For instance, the method in [39] learns complementary information from T1, T2 and FA images, however the relationship between these images may be more complex due to the different image acquisition processes. To better account for this complexity, Nie et al. [23] proposed a *late fusion* approach, where each modality is processed by an independent CNN whose outputs were fused in deep layers. The authors showed this strategy to outperform early fusion on the task of infant brain segmentation.

More recently, Aygün et al. explored different ways of combining multiple modalities [1]. In this work, all modalities are considered as separate inputs to different CNNs, which are later fused at an 'early', 'middle' or 'late' point. Although it was found that 'late' fusion provides better performance, as in [23], this method relies on a single-layer fusion to model the relation between all modalities. Nevertheless, as demonstrated in several works [30], relations between different modalities may be highly complex and they cannot easily be modeled by a single layer. To account for the non-linearity in multi-modal data modeling, we recently proposed a CNN that incorporates dense connections not only between the pairs of layers within the same path, but also between those across different paths [9,11]. This architecture, known as *HyperDenseNet*, obtained very competitive performance in the context of infant and adult brain tissue segmentation with multiple MRI data.

Despite the remarkable performance of existing methods, the combination of multi-modal data at various levels of abstraction has not been fully exploited for the segmentation of ischemic stroke lesions. In this paper, we adopt the strategy presented in [9,11] and propose a multi-path architecture, where each modality is employed as input of one stream and dense connectivity is used between layers in the same and different paths. Furthermore, we also extend the standard convolutional module of InceptionNet [32] by including two additional dilated convolutional blocks, which may help to learn larger context. Experiments on 103 ischemic stroke lesion multi-modal scans from the Ischemic Stroke Lesion Segmentation (ISLES) Challenge 2018 shows our model to outperform architectures based on early and late fusion, as well as state-of-art segmentation networks.

2   Methodology

The proposed models build upon the UNet architecture [27], which has shown outstanding performance in various medical segmentation tasks [8,12,29]. This network consists of a contracting and expanding path, the former collapsing an image down into a set of high level features and the latter using these features to construct a pixel-wise segmentation mask. Using skip connections, outputs from early layers are concatenated to the input of subsequent layers with the objective of transferring information that may be lost in the encoding path.

2.1   Proposed Multi-modal UNet

Disentangling Input Data. Figure 1 depicts our proposed network for ischemic stroke lesion segmentation in multiple image modalities. Unlike most UNet-like architectures, the encoding path is split into N streams, which serve as input to each image modality. The main objective of processing each modality in separated streams is to disentangle information that otherwise would be fused from an early stage, with the drawbacks introduced before, i.e., limitation to capture complex relationships between modalities.

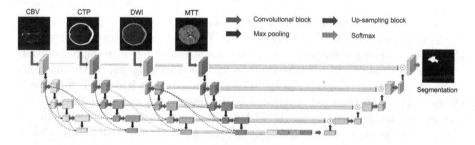

Fig. 1. Proposed architecture multi-path dense UNet. Dotted lines represent some of the dense connectivity patterns adopted in this extended version of UNet.

Hyper-Dense Connectivity. Inspired by the recent success of densely and hyper-densely connected networks in medical image segmentation works [5,9,11, 37], we propose to extend UNet to accommodate hyper-dense connections within

Modalitiy 1   Modalitiy 2

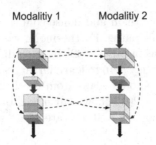

Fig. 2. Detailed version of a section of the proposed dense connectivity in multi-modal scenarios. For simplicity two image modalities are considered in this example.

the same and between multiple paths. In addition to better modeling relationships between different modalities, employing dense connections also brings the three following benefits [15]. First, direct connections between all layers help improving the flow of information and gradients through the entire network, alleviating the problem of vanishing gradient. Second, short paths to all the feature maps in the architecture introduce an implicit deep supervision. And third, dense connections have a regularizing effect, which reduces the risk of over-fitting on tasks with smaller training sets.

In standard CNNs, the output of the lth layer, denoted as x_l, is typically obtained from the output of the previous layer x_{l-1} by a mapping H_l:

$$x_l = H_l(x_{l-1}). \tag{1}$$

where H_l commonly integrates a convolution layers followed by a non-linear activation. In a densely-connected network, all feature outputs are concatenated in a feed-forward manner,

$$x_l = H_l([x_{l-1}, x_{l-2}, \ldots, x_0]), \tag{2}$$

where $[\ldots]$ denotes a concatenation operation.

As in HyperDenseNet [9,11], the outputs from layers in different streams are also linked. This connectivity yields a much more powerful feature representation than early or late fusion strategies in a multi-modal context, as the network learns the complex relationships between the modalities within and in-between all the levels of abstractions. Considering the case of only two modalities, let x_l^1 and x_l^2 denote the outputs of the lth layer in streams 1 and 2, respectively. In general, the output of the lth layer in a stream s can then be defined as follows:

$$x_l^s = H_l^s([x_{l-1}^1, x_{l-1}^2, x_{l-2}^1, x_{l-2}^2, \ldots, x_0^1, x_0^2]). \tag{3}$$

Inspired by the recent findings in [6,38,40], where shuffling and interleaving feature map elements in a CNN improved the efficiency and performance, while serving as a strong regularizer, we concatenate feature maps in a different order for each branch and layer:

$$x_l^s = H_l^s(\pi_l^s([x_{l-1}^1, x_{l-1}^2, x_{l-2}^1, x_{l-2}^2, \ldots, x_0^1, x_0^2])), \tag{4}$$

with π_l^s being a function that permutes the feature maps given as input. Thus, in the case of two image modalities, we have:

$$x_l^1 = H_l^1\left([x_{l-1}^1, x_{l-1}^2, x_{l-2}^1, x_{l-2}^2, \ldots, x_0^1, x_0^2]\right)$$
$$x_l^2 = H_l^2\left([x_{l-1}^2, x_{l-1}^1, x_{l-2}^2, x_{l-2}^1, \ldots, x_0^2, x_0^1]\right)$$

A detailed example of hyper-dense connectivity for the case of two image modalities is depicted in Fig. 2.

2.2 Extended Inception Module

Salient regions in a given image can have extremely large variation in size. For example, in ischemic stroke lesion segmentation, the area occupied by a lesion highly varies from one image to another. Therefore, choosing the appropriate kernel size is not trivial. While a smaller kernel is better for local information, a larger kernel is preferred to capture information that is distributed globally. InceptionNet [32] exploits this principle by including convolutions with multiple kernel sizes which operate on the same level. Furthermore, in versions 2 and 3, convolutions of the shape n×n are factorized to a combination of 1×n and n×1 convolutions, which have demonstrated to be more efficient. For example, a 3×3 convolution is equivalent to a 1×3 followed by a 3×1 convolution, which was found to be 33% cheaper.

We also extended the convolutional module of InceptionNet to facilitate the learning of multiple context. Particularly, we included two additional convolutional blocks, with different dilation rates, which help the module to learn from multiple receptive fields and to increase the context with respect to the original inception module. Since dilated convolutions were shown to be better alternative to max-pooling when capturing global context [36], we removed the latter operation in the proposed module. Our extended inception modules are depicted in Fig. 3.

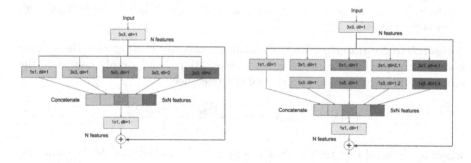

Fig. 3. Proposed extended inception modules. The module on the left employs standard convolutions while the module on the right adopts the idea of asymmetric convolutions [32].

3   Materials

3.1   Dataset

The training dataset, composed of 103 ischemic stroke lesion multi-modal scans, was provided by the ISLES organizers. We split the 94 stroke datasets into training and validation sets containing 83 and 11 examples, respectively. Each scan contains: Diffusion maps (DWI) and Perfusion maps (CBF, MTT, CBV, Tmax and CTP source data). In addition, the manual ground truth segmentation is provided only for the training samples. Detailed information about the dataset can be found in the ISLES website[1].

3.2   Evaluation Metrics

Dice Similarity Coefficient (DSC). We first evaluate performance using Dice similarity coefficient (DSC), which compares volumes based on their overlap. Let V_{ref} and V_{auto} be, respectively, the reference and automatic segmentations of a given tissue class and for a given subject, the DSC for this subject is defined as:

$$DSC(V_{ref}, V_{auto}) = \frac{2 \mid V_{ref} \cap V_{auto} \mid}{\mid V_{ref} \mid + \mid V_{auto} \mid} \tag{5}$$

Modified Hausdorff Distance (MHD). The second metric measures the accuracy of the segmentation boundary. Let P_{ref} and P_{auto} denote the sets of voxels within the reference and automatic segmentation boundary, respectively. MHD is given by

$$MHD(P_{ref}, P_{auto}) = \max \left\{ \max_{q \in P_{ref}} d(q, P_{auto}), \max_{q \in P_{auto}} d(q, P_{ref}) \right\}, \tag{6}$$

where $d(q, P)$ is the point-to-set distance defined by: $d(q, P) = \min_{p \in P} \|q - p\|$, with $\|.\|$ denoting the Euclidean distance. In the MHD, the 95th percentile is used for the estimation of the maximum distance value. Low MHD values indicate high boundary similarity.

Volumetric Similarity (VS). Volumetric similarity (VS) ignores the overlap between the predicted and reference segmentations, and simply compares the size of the predicted volume to that of the reference:

$$VS(V_{ref}, V_{auto}) = 1 - \frac{\mid |V_{ref}| - |V_{auto}| \mid}{|V_{ref}| + |V_{auto}|}. \tag{7}$$

where a VS equal to 1 reflects that the predicted segmentation is the same size as the reference volume.

---

[1] http://www.isles-challenge.org.

3.3 Implementation Details

Baselines. To demonstrate the effectiveness of hyper-dense connectivity in deep neural networks we compare the proposed architecture to the same network with early and late fusion strategies. For early fusion, all the MRI image modalities are merged into a single input, which is processed through a unique path, as many current works. On the other hand, each image modality is treated as an independent signal and processed by separate branches in the later fusion strategy, where features are fused at a higher level. The details of the late fusion architecture are depicted in Table 1. In both cases, i.e., early and late fusion, the left module depicted in Fig. 3 is employed. Furthermore, feature maps from the skip connections are summed before being fed into the convolutional modules of the decoding path, instead of concatenating them, as in standard UNet.

Proposed Network. The proposed network is similar to the architecture with the late fusion strategy. Nevertheless, as introduced in Sect. 2.1, feature maps from previous layers and different paths are concatenated and fed into the subsequent layers. The details of the resulted architecture are reported in Table 1, most-right columns. The first version of the proposed network employs the same convolutional module than the two baselines. The second version, however, adopts asymmetric convolutions instead (Fig. 3).

Table 1. Layers disposal of the architecture with late fusion and the proposed hyper dense connected UNet.

		Late fusion		HyperDense connectivity	
	Name	Feat maps (input)	Feat maps (output)	Feat maps (input)	Feat maps (output)
Encoding path (each modality)	Conv layer 1	$1 \times 256 \times 256$	$32 \times 256 \times 256$	$1 \times 256 \times 256$	$32 \times 256 \times 256$
	Max-pooling 1	$32 \times 256 \times 256$	$32 \times 128 \times 128$	$32 \times 256 \times 256$	$32 \times 128 \times 128$
	Layer 2	$32 \times 128 \times 128$	$64 \times 128 \times 128$	$128 \times 128 \times 128$	$64 \times 128 \times 128$
	Max-pooling 2	$64 \times 128 \times 128$	$64 \times 64 \times 64$	$64 \times 128 \times 128$	$64 \times 64 \times 64$
	Layer 3	$64 \times 64 \times 64$	$128 \times 64 \times 64$	$384 \times 64 \times 64$	$128 \times 64 \times 64$
	Max-pooling 3	$128 \times 64 \times 64$	$128 \times 32 \times 32$	$128 \times 64 \times 64$	$128 \times 32 \times 32$
	Layer 4	$128 \times 32 \times 32$	$256 \times 32 \times 32$	$896 \times 32 \times 32$	$256 \times 32 \times 32$
	Max-pooling 4	$256 \times 32 \times 32$	$256 \times 16 \times 16$	$256 \times 32 \times 32$	$256 \times 16 \times 16$
	Bridge	$1024 \times 16 \times 16$	$512 \times 16 \times 16$	$1920 \times 16 \times 16$	$512 \times 16 \times 16$
Decoding path	Up-sample 1	$512 \times 16 \times 16$	$256 \times 32 \times 32$	$512 \times 16 \times 16$	$256 \times 32 \times 32$
	Layer 5	$256 \times 32 \times 32$	$256 \times 32 \times 32$	$256 \times 32 \times 32$	$256 \times 32 \times 32$
	Up-sample 2	$256 \times 32 \times 32$	$128 \times 64 \times 64$	$256 \times 32 \times 32$	$128 \times 64 \times 64$
	Layer 6	$128 \times 64 \times 64$	$128 \times 64 \times 64$	$128 \times 64 \times 64$	$128 \times 64 \times 64$
	Up-sample 3	$128 \times 64 \times 64$	$64 \times 128 \times 128$	$128 \times 64 \times 64$	$64 \times 128 \times 128$
	Layer 7	$64 \times 128 \times 128$	$64 \times 128 \times 128$	$64 \times 128 \times 128$	$64 \times 128 \times 128$
	Up-sample 4	$64 \times 128 \times 128$	$32 \times 256 \times 256$	$64 \times 128 \times 128$	$32 \times 256 \times 256$
	Layer 8	$32 \times 256 \times 256$	$32 \times 256 \times 256$	$32 \times 256 \times 256$	$32 \times 256 \times 256$
	Softmax layer	$32 \times 256 \times 256$	$2 \times 256 \times 256$	$32 \times 256 \times 256$	$2 \times 256 \times 256$

Training. Network parameters were optimized via Adam with β_1 and β_2 equal to 0.9 and 0.99, respectively and training is run during 200 epochs. Learning rate is initially set to 1×10^{-4} and reduced after 100 epochs. Batch size was equal to 4. For a fair comparison, the same hyper-parameters were employed across all the architectures. The proposed architectures were implemented in pytorch. Experiments were performed on a NVidia TITAN XP GPU with 16 GBs RAM. While training took around 5 h, inference on a single 2D image was done in 0.1 s, as average. No data augmentation was employed. Images were normalized between 0 and 1 and no other pre- or post-processing steps were used. As input to the architectures we employed the following four image modalities in all the cases: CBV, CTP, DWI and MTT.

4   Results

Table 2 reports the results obtained by the different networks that we investigated in terms of mean DSC, MHD and VS values and their standard deviation. First, we compare the different multi-modal fusion strategies with the baseline UNet employed in this work. We can observe that fusing learning features in a higher level provides better results in all the metrics than early fusion strategies. Additionally, if hyper-dense connections are adopted in the late fusion architecture, i.e., interconnecting convolutional layers from the different image modalities, the segmentation performance is significantly improved, particularly in terms of DSC and VS. Specifically, while the proposed network outperforms the late fusion architecture by nearly 5% in both DSC and VS, the mean MHD is decreased by almost 1 mm, obtaining a mean MHD of 18.88 mm. On the other hand, replacing the standard convolutions of the proposed module (Fig. 3, *left*) by asymmetric convolutions (Fig. 3, *right*), brings another boost on performance on the proposed hyper-dense UNet. In this case, the mean DSC and MHD are the best ones among all the architectures, with mean values of 0.635% and 18.64 mm, respectively.

Table 2. Mean DSC,MHD and VS values, with their corresponding standard deviation, obtained by the evaluated methods on the independent validation group.

	Validation		
Architecture	DSC (%)	MHD (mm)	VS (%)
Early fusion	0.497 ± 0.263	21.30 ± 13.25	0.654 ± 0.265
Late fusion	0.571 ± 0.221	19.72 ± 12.29	0.718 ± 0.235
Proposed	0.622 ± 0.233	18.88 ± 14.87	0.764 ± 0.247
Proposed (asymmetric conv)	$\mathbf{0.635} \pm 0.186$	$\mathbf{18.64} \pm 14.26$	0.796 ± 0.162
ERFNet [26]	0.540 ± 0.258	21.73 ± 11.46	$\mathbf{0.823} \pm 0.119$

Then, we also compare the results obtained by the proposed network to another state-of-the-art network that includes factorized convolution modules, i.e., ERFNet. Even though its performance outperforms the baseline with early fusion, results are far from those obtained by the proposed network, except for the volume similarity, where both ERFNet and the proposed network with asymmetric convolutions obtain similar performances.

Qualitative evaluation of the proposed architecture is assessed in Fig. 4, where ground truth and automatic CNN contours are visualized on MTT images. We can first observe that, by employing strategies where learned features are merged at higher levels, unlike *early fusion*, the region of the ischemic stroke lesion is generally better covered. Furthermore, by giving freedom to the architecture to learn the level of abstraction at which the different modalities should be combined segmentation results are visually improved, which is in line with the results reported in Table 2.

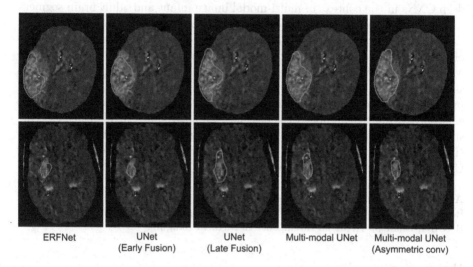

| ERFNet | UNet
(Early Fusion) | UNet
(Late Fusion) | Multi-modal UNet | Multi-modal UNet
(Asymmetric conv) |

Fig. 4. Visual results for two subjects on the validation set. While the area in yellow represents the ground truth, purple contours depict the automatic contours for each of the architectures. (Color figure online)

5 Discussion

In this work we extended the well-known UNet to leverage information in multi-modal data. Particularly, following recent work in multi-modal learning for medical image segmentation [9,11], we processed each image modality in different streams on the encoding path and densely connected all the convolutional layers from all the streams. Thus, each convolutional layer received as input the features maps from all the previous layers within the same stream, i.e., same modality, but also the learned features from previous layers in every different stream. In this way, the network has the freedom to learn any pattern at any level of abstraction of the network, which seems to improve its representation learning power.

Results obtained in this work demonstrate that better strategies to model multi-modal information can bring a boost on performance when compared to more naive fusion strategies. These results are in line with recent studies in multi-modal image segmentation on the medical field [9,11,23]. For instance, in [23], a *late fusion* strategy was proposed to combine high-level features to better capture the complex relationships between different modalities. They used an independent convolutional network for each modality, and fused the outputs of the different networks in higher-level layers, showing better performance than early fusion in the context infant brain segmentation. More recently, we demonstrated that hyper dense connectivity can strength the representation power of deep CNNs in the context of multi-modal image infant and adult brain segmentation, surpassing the performance of several features fusion strategies [9,11].

One of the limitations of this work is that volumes were treated as a stack of 2D slides, where each slide was processed independently. Thus, 3D context was discarded, which might have improved the segmentation performance, as shown by recent works that employ 3D convolutions. One of the reasons for privileging 2D convolutions is that some of the volumes on the ISLES dataset contained a limited number of slides, i.e., 2 and 4 slides in many cases. One strategy to explore in the future could be to employ Long-Short Term Memory (LSTM) networks to propagate the spatial information extracted from the 2D CNN through the third dimension.

Acknowledgments. This work is supported by the National Science and Engineering Research Council of Canada (NSERC), discovery grant program, and by the ETS Research Chair on Artificial Intelligence in Medical Imaging.

References

1. Aygün, M., Şahin, Y.H., Ünal, G.: Multi modal convolutional neural networks for brain tumor segmentation. arXiv preprint arXiv:1809.06191 (2018)
2. Barber, P., et al.: Imaging of the brain in acute ischaemic stroke: comparison of computed tomography and magnetic resonance diffusion-weighted imaging. J. Neurol. Neurosurg. Psychiatry **76**(11), 1528–1533 (2005)
3. Chalela, J.A., et al.: Magnetic resonance imaging and computed tomography in emergency assessment of patients with suspected acute stroke: a prospective comparison. Lancet **369**(9558), 293–298 (2007)
4. Chen, L., Bentley, P., Rueckert, D.: Fully automatic acute ischemic lesion segmentation in DWI using convolutional neural networks. NeuroImage: Clin. **15**, 633–643 (2017)
5. Chen, L., Wu, Y., DSouza, A.M., Abidin, A.Z., Wismüller, A., Xu, C.: MRI tumor segmentation with densely connected 3D CNN. In: Medical Imaging 2018: Image Processing. International Society for Optics and Photonics (2018)
6. Chen, Y., Wang, H., Long, Y.: Regularization of convolutional neural networks using shufflenode. In: 2017 IEEE International Conference on Multimedia and Expo (ICME), pp. 355–360. IEEE (2017)

7. Choi, Y., Kwon, Y., Lee, H., Kim, B.J., Paik, M.C., Won, J.H.: Ensemble of deep convolutional neural networks for prognosis of ischemic stroke. In: Crimi, A., Menze, B., Maier, O., Reyes, M., Winzeck, S., Handels, H. (eds.) BrainLes 2016. LNCS, vol. 10154, pp. 231–243. Springer, Cham (2016). https://doi.org/10.1007/978-3-319-55524-9_22

8. Çiçek, Ö., Abdulkadir, A., Lienkamp, S.S., Brox, T., Ronneberger, O.: 3D U-Net: learning dense volumetric segmentation from sparse annotation. In: Ourselin, S., Joskowicz, L., Sabuncu, M.R., Unal, G., Wells, W. (eds.) MICCAI 2016. LNCS, vol. 9901, pp. 424–432. Springer, Cham (2016). https://doi.org/10.1007/978-3-319-46723-8_49

9. Dolz, J., Ben Ayed, I., Yuan, J., Desrosiers, C.: Isointense infant brain segmentation with a hyper-dense connected convolutional neural network. In: 2018 IEEE 15th International Symposium on Biomedical Imaging (ISBI 2018), pp. 616–620. IEEE (2018)

10. Dolz, J., Desrosiers, C., Wang, L., Yuan, J., Shen, D., Ayed, I.B.: Deep CNN ensembles and suggestive annotations for infant brain MRI segmentation. arXiv preprint arXiv:1712.05319 (2017)

11. Dolz, J., Gopinath, K., Yuan, J., Lombaert, H., Desrosiers, C., Ayed, I.B.: HyperDense-Net: a hyper-densely connected CNN for multi-modal image segmentation. arXiv preprint arXiv:1804.02967 (2018)

12. Dong, H., Yang, G., Liu, F., Mo, Y., Guo, Y.: Automatic brain tumor detection and segmentation using U-Net based fully convolutional networks. In: Valdés Hernández, M., González-Castro, V. (eds.) MIUA 2017. CCIS, vol. 723, pp. 506–517. Springer, Cham (2017). https://doi.org/10.1007/978-3-319-60964-5_44

13. Feigin, V.L., Lawes, C.M., Bennett, D.A., Anderson, C.S.: Stroke epidemiology: a review of population-based studies of incidence, prevalence, and case-fatality in the late 20th century. Lancet Neurol. $2(1)$, 43–53 (2003)

14. Guerrero, R., et al.: White matter hyperintensity and stroke lesion segmentation and differentiation using convolutional neural networks. NeuroImage: Clin. 17, 918–934 (2018)

15. Huang, G., Liu, Z., Van Der Maaten, L., Weinberger, K.Q.: Densely connected convolutional networks. In: CVPR, vol. 1, p. 3 (2017)

16. Kamnitsas, K., Chen, L., Ledig, C., Rueckert, D., Glocker, B.: Multi-scale 3D convolutional neural networks for lesion segmentation in brain MRI. Ischemic Stroke Lesion Segment. 13, 46 (2015)

17. Kamnitsas, K., et al.: Efficient multi-scale 3D CNN with fully connected CRF for accurate brain lesion segmentation. Med. Image Anal. 36, 61–78 (2017)

18. Lansberg, M.G., Albers, G.W., Beaulieu, C., Marks, M.P.: Comparison of diffusion-weighted MRI and CT in acute stroke. Neurology $54(8)$, 1557–1561 (2000)

19. Lopez, A.D., Mathers, C.D., Ezzati, M., Jamison, D.T., Murray, C.J.: Global and regional burden of disease and risk factors, 2001: systematic analysis of population health data. Lancet $367(9524)$, 1747–1757 (2006)

20. Maier, O., et al.: ISLES 2015-a public evaluation benchmark for ischemic stroke lesion segmentation from multispectral MRI. Med. Image Anal. 35, 250–269 (2017)

21. Maier, O., Schröder, C., Forkert, N.D., Martinetz, T., Handels, H.: Classifiers for ischemic stroke lesion segmentation: a comparison study. PloS One $10(12)$, e0145118 (2015)

22. Moeskops, P., Viergever, M.A., Mendrik, A.M., de Vries, L.S., Benders, M.J., Išgum, I.: Automatic segmentation of MR brain images with a convolutional neural network. IEEE Trans. Med. Imaging $35(5)$, 1252–1261 (2016)

23. Nie, D., Wang, L., Gao, Y., Sken, D.: Fully convolutional networks for multi-modality isointense infant brain image segmentation. In: 2016 13th International Symposium on Biomedical Imaging (ISBI), pp. 1342–1345. IEEE (2016)

24. Praveen, G., Agrawal, A., Sundaram, P., Sardesai, S.: Ischemic stroke lesion segmentation using stacked sparse autoencoder. Comput. Biol. Med. **99**, 38–52 (2018)

25. Rekik, I., Allassonnière, S., Carpenter, T.K., Wardlaw, J.M.: Medical image analysis methods in MR/CT-imaged acute-subacute ischemic stroke lesion: segmentation, prediction and insights into dynamic evolution simulation models. A critical appraisal. NeuroImage: Clin. **1**(1), 164–178 (2012)

26. Romera, E., Alvarez, J.M., Bergasa, L.M., Arroyo, R.: ERFNet: efficient residual factorized ConvNet for real-time semantic segmentation. IEEE Trans. Intell. Transp. Syst. **19**(1), 263–272 (2018)

27. Ronneberger, O., Fischer, P., Brox, T.: U-Net: convolutional networks for biomedical image segmentation. In: Navab, N., Hornegger, J., Wells, W.M., Frangi, A.F. (eds.) MICCAI 2015. LNCS, vol. 9351, pp. 234–241. Springer, Cham (2015). https://doi.org/10.1007/978-3-319-24574-4_28

28. Seshadri, S., Wolf, P.A.: Lifetime risk of stroke and dementia: current concepts, and estimates from the framingham study. Lancet Neurol. **6**(12), 1106–1114 (2007)

29. Sirinukunwattana, K., et al.: Gland segmentation in colon histology images: the glas challenge contest. Med. Image Anal. **35**, 489–502 (2017)

30. Srivastava, N., Salakhutdinov, R.: Multimodal learning with deep Boltzmann machines. J. Mach. Learn. Res. **15**, 2949–2980 (2014)

31. Sudlow, C., Warlow, C.: Comparable studies of the incidence of stroke and its pathological types: results from an international collaboration. Stroke **28**(3), 491–499 (1997)

32. Szegedy, C., Vanhoucke, V., Ioffe, S., Shlens, J., Wojna, Z.: Rethinking the inception architecture for computer vision. In: CVPR, pp. 2818–2826 (2016)

33. Valverde, S., et al.: Improving automated multiple sclerosis lesion segmentation with a cascaded 3D convolutional neural network approach. NeuroImage **155**, 159–168 (2017)

34. Van der Worp, H.B., van Gijn, J.: Acute ischemic stroke. N. Engl. J. Med. **357**(6), 572–579 (2007)

35. Winzeck, S., et al.: ISLES 2016 and 2017-benchmarking ischemic stroke lesion outcome prediction based on multispectral MRI. Front. Neurol. **9** (2018)

36. Yu, F., Koltun, V.: Multi-scale context aggregation by dilated convolutions. arXiv preprint arXiv:1511.07122 (2015)

37. Yu, L., et al.: Automatic 3D cardiovascular MR segmentation with densely-connected volumetric ConvNets. In: Descoteaux, M., Maier-Hein, L., Franz, A., Jannin, P., Collins, D.L., Duchesne, S. (eds.) MICCAI 2017. LNCS, vol. 10434, pp. 287–295. Springer, Cham (2017). https://doi.org/10.1007/978-3-319-66185-8_33

38. Zhang, T., Qi, G.-J., Xiao, B., Wang, J.: Interleaved group convolutions. In: CVPR, pp. 4373–4382 (2017)

39. Zhang, W., et al.: Deep convolutional neural networks for multi-modality isointense infant brain image segmentation. NeuroImage **108**, 214–224 (2015)

40. Zhang, X., Zhou, X., Lin, M., Sun, J.: ShuffleNet: an extremely efficient convolutional neural network for mobile devices. arXiv preprint arXiv:1707.01083 (2017)

Multi-scale Deep Convolutional Neural Network for Stroke Lesions Segmentation on CT Images

Liangliang Liu[1], Shuai Yang[2], Li Meng[2], Min Li[1], and Jianxin Wang[1(✉)]

[1] School of Information Science and Engineering, Central South University,
Changsha 410083, People's Republic of China
liuliang_double@csu.edu.cn, {limin,jxwang}@mail.csu.edu.cn
[2] Department of Radiology, Xiangya Hospital, Central South University,
Changsha 410008, People's Republic of China

Abstract. Ischemic stroke is the top cerebral vascular disease leading to disability and death worldwide. Accurate and automatic segmentation of lesions of stroke can assist diagnosis and treatment planning. However, manual segmentation is a time-consuming and subjective for neurologists. In this study, we propose a novel deep convolutional neural network, which is developed for the segmentation of stroke lesions from CT perfusion images. The main structure of network bases on U-shape. We embed the dense blocks into U-shape network, which can alleviate the over-fitting problem. In order to acquire more receptive fields, we use multi-kernel to divide the network into two paths, and use the dropout regularization method to achieve effective feature mapping. In addition, we use multi-scale features to obtain more spatial features, which will help improve segmentation performance. In the post-processing stage of soft segmentation, we use image median filtering to eliminate the specific noises and make the segmentation edge smoother. We evaluate our method in Ischemic Stroke Lesion Segmentations Challenge (ISLES) 2018. The results of our approach on the testing data places hight ranking.

Keywords: Stroke · CT perfusion images · Dropout · Multi-scale · U-shape network

1 Introduction

The stroke, one of the leading causes of death and disability worldwide, is triggered by an obstruction in the cerebrovascular system preventing the blood to reach the brain regions supplied by the blocked blood vessel directly. Ischemic stroke is the commonest subtype of stroke, which is a disease with sudden onset and high mortality. It prevents blood flow in small vessels. When the blood flow interruption is too long, cell will undergo necrosis and irreversibly injured infarct core is formed [7]. Defining location and extend of the infarct core is a critical

A. Crimi et al. (Eds.): BrainLes 2018, LNCS 11383, pp. 283–291, 2019.
https://doi.org/10.1007/978-3-030-11723-8_28

part of the decision making process in acute stroke. In clinical diagnosis, CT image is a speed, availability, and lack of contraindications manner to triage stroke patients. If we can locate the location and size of the lesion quickly, it is the key to save some viable brain tissue [24]. In traditional medical diagnosis, the lesion tissue is accomplished by manual segmentation on medical images. However, manual delineation of stroke lesions a time-consuming and very tedious task [8]. Automatic and accurate quantification of stroke lesions is an important metric for planning treatment strategies, monitoring disease progression.

Over the past decades, Unsupervised methods and shallow machine learning methods are traditional methods of image analysis, such as: multi-modal generative based mixture-model [1], image cross-saliency approach [3], spatial decision forests approach [5] and multi-atlas segmentation method [19], and so on, those methods had been successful. However, there are also some limitations in these methods. For example, some of those methods are designed specifically require and heavily dependent on handcrafted lesions segmentation [11,12] or improve the accuracy of segmentation depend on multi-atlas label [23].

Recent years, deep convolution neural networks (DCNNs) are one of the most competitive approach used for medical image semantic segmentation. The DCNN models are capable of learning features from raw images and extracting context information. The feature sets filtered by DCNN often outperform pre-defined and hand-crafted feature sets. For example, Ronneberger et al. proposed a novel U-net model based on DCNN architecture [25]. U-net combined the down-sampling layers and up-sampling layers with skip connections, this architecture can reuse the context information of the down-sampling layers and greatly improve the performance of the segmentation. Long et al. proposed a novel framework to automatically segment stroke lesions. This framework consists of two deep convolutional neural networks, and it achieved state-of-the-art performance on an acute ischemic stroke dataset [21]. Zhang et al. used a custom DCNN to automatic segmentation acute ischemic stroke from DWI modality, in the network, they used dense connectivity to relieve the problems of deep network, and the network outperforms other state-of-the-art methods by a large margin [28]. Li et al. developed an automatic intervertebral discs (IVDs) segmentation method based on fully convolution networks [20], they used multi-scale and feature dropout learning technology to segment region of interest (ROI) from multi-modality MRI images, this method achieved the 1st place in the MICCAI challenge in 2016. Others methods based on DCNN which are applied in medical images of various diseases, such as: stroke image segmentation [22], brain tumor image segmentation [17], WMH segmentation [9], and optic disc segmentation [4], and so on. Most these methods are based on magnetic resonance imaging (MRI). Especially, the segmentation methods of stroke lesions is seldom used in CT images.

In this paper, we propose a novel multi-scale features deep convolution neural network (MS-DCNN) for stroke lesions segmentation on CT images. The whole neural network consists of a series of convolution layers, dense blocks [13], transition blocks and upsampling blocks. We use the dropout regularization method to alleviate neural network from over-fitting. We use random rotated and distortion to increase the number of training samples. He network with the main contributions as follows:

1. We propose an end-to-end deep convolution neural network base on two symmetrical U-shape networks [25], and embedded dense blocks into the U-shape [13]. This strategy can improve the information on the sampling and improve the feature reuse.

2. We use the dropout regularization method in dense blocks and transition blocks. It's a simple method to prevent neural network from over-fitting and improve the neural network efficiency. Proper use of dropout can help improve the accuracy of segmentation.

3. We employ dual parallel kernel pathways in our framework to process input CT images. This design can help extract the image features fully, and finally combine the two pathways before output, it helps to improve the performance of the segmentation [27]. We evaluate our method on ISLES 2018 challenge.

2 Material and Method

2.1 Data

Ischemic Stroke Lesion Segmentations Challenge (ISLES) 2018 offers a platform for participants to compare their methods directly and fair. ISLES 2018 challenge offers 103 stroke patients, which is based on acute CT perfusion data. Each patient has 5 CT sequences (CBV, CBF, MTT, TMAX, CTP). Imaging data from acute stroke patients in two centers who presented within 8 h of stroke onset and underwent an Magnetic Resonance Imaging (MRI) DWI within 3 h after CTP were included. The challenge's training data set consists of 63 patients, some patient cases have two slabs to cover the stroke lesion, finally, we got 94 samples in the training dataset. The testing dataset consists of 40 patients. Some patient cases have two slabs to cover the stroke lesion. We got 62 testing samples. In this challenges, the training data set and the ground truth are opened to all participants. The testing data set only open the CT images which is to be predicted, without the ground truth is distributed on the challenge web pages. Participants should submitted their final segmentation results to the organizers, who scored the segmentation results.

2.2 MS-DCNN

A traditional image-processing CNN is composed of one input layer, many convolution layers and one output layer. Features are transmitted by single line between layers, which leads to inadequate extraction features. We propose the MK-DCNN framework is based on the U-net architecture [25] and we embed dense structure as a block into the U-shape framework [13], both two methods include jump layer which can help to improve feature reuse.

Figure 1 illustrates the pipeline of our proposed segmentation network. Our network is based on two symmetric U-shape structures, and we use dense block to implement the down-sampling operation in contracting path of U-shape, and after completes up-sampling operation in expansive path, we concatenate two symmetric networks and output the predicted result. We use multi-scale features strategy to enhance the feature extraction sufficiently. In the first layer, we use dual parallel kernels in two symmetric pathway to extract different features. To handle the problem of over-fitting of DCNNs, we not only use dense block to resist over-fitting, but also use dropout regularization method to alleviate over-fitting and improve the efficiency of neural network.

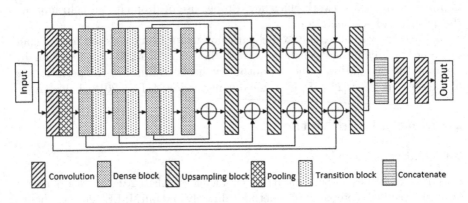

Fig. 1. Architecture of the MS-DCNN.

As shown in Fig. 1. Our network consists of 3 separate convolution layers, 1 pooling layer, 4 dense blocks, 3 transition blocks and 4 up-sampling blocks. We extend the deep of DenseNet-121 to 123 layers in dense blocks. Each dense block contains several micro-dense units, each dense unit is composed of a batch normalization (BN) [16] layer, a rectified linear units (ReLU) [6] layer and a convolution (Conv) layer, the concatenation operation is required before result output. A n-layer dense block consists of a dense unit or several continuous dense units. Figures 2 and 3 illustrate the basic implementation of a dense unit and a n-layer dense block, respectively. In dense block, each dense unit is regarded as one layer, all layers inside the block are directly connected. The transition block consists of a BN layer, a ReLU layer and an average pooling layer [18]. We embed the dropout regularization method into the both dense block and transition block. The up-sampling block consists of a concatenate layer, a BN layer, a ReLU layer and a Conv layer, we use bilinear interpolation technology to realize image zooming. Then, we concatenate the two un-sampling results which come from different paths. Finally, after two convolution layers, we use the sigmoid function to complete segmentation task and output the final lesion information.

In our network, we only use 4 CT modality sequences (CBV, CBF, TMAX and MTT). According to the clinical prior knowledge, 4 modalities play different roles in stroke diagnosis, we divide the 4 modalities into two groups

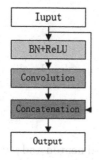

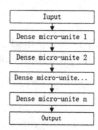

Fig. 2. A dense unit.

Fig. 3. Architecture of the n-layer dense block.

(TMAX+CBF+CBV and MTT). First, We set different dropout rates for the two groups in our network. Then, we concatenate the 2 unsampling results which come from 2 pathways of MS-DCNN. Finally, after two convolution operations, we use the sigmoid function to complete segmentation task and out the final lesion information.

2.3 Dropout Regularization Method for Effective Learning

The regularization is a popular method to prevent over-fitting and filter important features. It is a very important and effective technology to reduce generalization error in machine learning. Regularization can automatically weak the unimportant feature variables. Dropout is one of a general and concise regularization methods which performs well in many tasks [10,26]. In our study, we use dropout to reduce redundant features produced by multi-scale method and to alleviate the problem of duplicate feature acquisition from the same area of the image. We use dropout regularization method in dense block and transition block. The application of dropout on a generic i-th neuron in the n layer is shown below:

$$Q_i = x_i a(\sum_{k=1}^{d_i} w_k x_k + b_k)(0 \leq i \leq h), \tag{1}$$

where Q_i is the retained probability of the i-th neuron, x_i is the i-th neuron, $a()$ is an activation function, $k \in [1, i]$ is unit number, w_k and b_k are the k-th unit weight and bias. d denotes dimensional, x_{di} denotes x_i is a Bernoulli variables with d dimensional. $\sum_{k-1}^{d_i} w_k x_k$ is the sum of the product of all neurons weights w_k and x_k before i-th neuron.

In our network, we need to dropout a set of neurons of a layer. Let the j-th layer has n neurons, in a cycle, the neural network can be regarded as the integration n times of Bernoulli's experiments, and the probability of each neuron being retained is q and the dropout probability is p. Thus, the number of neurons retained in layer j-th is as follows:

$$Y = \sum_{i=1}^{d_j} x_i, \tag{2}$$

where x_i is a retained neuron (a Bernoulli random variable). In the n experiments, the probability of retaining k neurons was:

$$f(k; n, p) = \binom{n}{k} p^k q^{(n-k)}, \tag{3}$$

where $q = 1-p$, q represents the probability of a retained neuron and p represents the probability of a neuron turn off, $p^k q^{(n-k)}$ is the probability of obtaining k neurons successful sequence in the n test and $(n - k)$ failures, while $\binom{n}{k}$ is the binomial coefficient used to calculate the number of possible successful sequences.

In our lesion segmentation network, we use fixed dropout ratio to handle the feature filtering in each training iteration. The dropout ratio of group TMAX+CBF+CBV is set to 0.01, the dropout ratio of MTT is set to 0.5.

2.4   Loss Function

In image segmentation tasks, Dice coefficient (DC) is one of the classic indexes for evaluating the segmentation effect, and it can also be used as a loss function to measure the gap between the result of the segmentation and the ground truth. In binary image segmentation, we use the continuous softmax function outputs to replace the predicted binary labels, we Combine DC with cross entropy function, a pseudo DC loss function proposed in this paper is defined as:

$$L = 1 - \frac{1}{C} \sum_{c=1}^{C} \left(\frac{2 \sum_{n=1}^{N} (p(x_n)^c q(x_n)^c)}{\sum_{n=1}^{N} q(x_n)^c + \sum_{n=1}^{N} p(x_n)^c} \right), \tag{4}$$

where C is the class number, $c \in C$ is the pixel class, N is the pixel number, x_n is the n-th pixel. $p(x_n)^c$ is a binary value (label) of pixel x_n belongs class c, and $q(x_n)^c$ represents the probability of pixel x_n predicted by softmax function belongs class c. In order to measure the loss contribution of each class, aggregating DC from different classes C as an average. In the traditional single type lesion segmentation task, C is usually set to 1.

3   Experiments and Result

3.1   Experiments

We apply MS-DCNN in the ISLES 2018 challenge. The network architecture has shown in Fig. 1, i.e. a dual-pathway DCNN. For ISLES challenge, all CT sequences are resized to 160×160. We use images slices flipped and randomly rotated methods to augment the training images. In training process, the hyperparameter kept constant: batch size is set to 4, epoch is set to 70, and learning rate is set to 0.001. In our experiment, when the dropout ratio was set as 0.01, the segmentation results are close to optimal on training dataset. In testing process, network inherits the weight of the training model and realizes the automatic lesions segmentation. After testing, we use the affine transform method to restore the size of all prediction images to the original size. A post-processing step to refine the networks output, we use image median filtering algorithm [14] to alleviate noises and preserve the edge details of images. Finally, we synthesize the 2D slice images into 3D images.

3.2 Results

In this challenge, online evaluation is provided with the Dice coefficient (DC) [2], Hausdorff distance (HD) [15], Precision, Recall and AVD as quality metrics. We won't able to see the Ground Truth of the testing dataset. After uploading the segmentation results for the testing dataset, results of each participating team and their ranking be revealed on the challenge websites in a frozen table. We have obtained the scores presented in Table 1.

Table 1. The results of our network on ISLES 2018 challenge. Values correspond to the mean (and standard deviation)

Team	DC [0,1]	HD [mm]	Precision [0,1]	Recall [0,1]	AVD [mm]
songt1	0.51 (0.31)	19354856.39 (39507890.41)	0.55 (0.36)	0.55 (0.34)	10.24 (9.94)
pengl1	0.49 (0.31)	19354856.66 (39507890.28)	0.56 (0.37)	0.53 (0.33)	10.08 (10.58)
cheny11	0.47 (0.31)	19384856.55 (39507890.34)	0.56 (0.37)	0.49 (0.33)	11.14 (12.74)
pengl1	0.47 (0.32)	16129055.93 (36779842.04)	0.53 (0.36)	0.47 (0.32)	10.37 (14.42)
Ours	0.44 (0.30)	19354857.71 (39507889.77)	0.54 (0.36)	0.44 (0.32)	10.99 (12.70)

Among the 38 submissions on ISLES 2018, our submission have a superior performance, and ranks fifth. This task is simply too complex and variable for our algorithms to solve. In our training process, our model performs well in segmentation of large lesions. However, smaller and less pronounced lesions are the challenges for our model. As Table 1 shown, compared with DC, Precision and Recall, the values of Hausdorff distance is too hight, this may be due to the fact that some lesions are not detected, or there are many outlier points in the our segmentation result. Further work to improve the segment result will consist in optimizing, the particularity of CT image segmentation and incorporating other post-processing to improve the Hausdorff distance.

4   Conclusion

In this paper, we proposed the MS-DCNN is an automatic medical image segment network, it surpasses mostly state-of-the-art on ISLES 2018 challenge. Our network inherits previous work and integrates dense blocks. The architecture of U-shape is used to improve the feature locate accurately and semantics capture. The dense block is used to reuse previous features and alleviate over-fitting. In addition, two different dropout rate pathways are used to reduce the number of features between layers and retain important features. Different CT modal sequences play different roles in diagnosis. We will assign different dropout rates to each CT sequence to improve the performance of the current model. At present, our model does not provide precise segmentation for physicians and clinical researchers in this challenge, but it can be used as a support tool.

References

1. Cardoso, M.J., Sudre, C.H., Modat, M., Ourselin, S.: Template-based multimodal joint generative model of brain data. In: Ourselin, S., Alexander, D.C., Westin, C.-F., Cardoso, M.J. (eds.) IPMI 2015. LNCS, vol. 9123, pp. 17–29. Springer, Cham (2015). https://doi.org/10.1007/978-3-319-19992-4_2
2. Dice, L.R.: Measures of the amount of ecologic association between species. Ecology **26**(3), 297–302 (1945)
3. Erihov, M., Alpert, S., Kisilev, P., Hashoul, S.: A cross saliency approach to asymmetry-based tumor detection. In: Navab, N., Hornegger, J., Wells, W.M., Frangi, A.F. (eds.) MICCAI 2015. LNCS, vol. 9351, pp. 636–643. Springer, Cham (2015). https://doi.org/10.1007/978-3-319-24574-4_76
4. Fakhry, A., Zeng, T., Ji, S.: Residual deconvolutional networks for brain electron microscopy image segmentation. IEEE Trans. Med. Imaging **36**(2), 447–456 (2017)
5. Geremia, E., Menze, B.H., Clatz, O., Konukoglu, E., Criminisi, A., Ayache, N.: Spatial decision forests for MS lesion segmentation in multi-channel MR images. Neuroimage **57**(2), 378–390 (2011)
6. Glorot, X., Bordes, A., Bengio, Y.: Deep sparse rectifier neural networks. In: International Conference on Artificial Intelligence and Statistics, pp. 315–323 (2011)
7. Gonzalez, R.G., Hirsch, J.A., Koroshetz, W.J., Lev, M.H., Schaefer, P.: Acute ischemic stroke: imaging and intervention. J. Neuroradiol. **33**(3), 193 (2006)
8. Grimaud, J., et al.: Quantification of MRI lesion load in multiple sclerosis: a comparison of three computer-assisted techniques. Magn. Reson. Imaging **14**(5), 495–505 (1996)
9. Guerrero, R., et al.: White matter hyperintensity and stroke lesion segmentation and differentiation using convolutional neural networks. NeuroImage: Clin. **17**(C), 918–934 (2017)
10. Hinton, G.E., Srivastava, N., Krizhevsky, A., Sutskever, I., Salakhutdinov, R.R.: Improving neural networks by preventing co-adaptation of feature detectors. Comput. Sci. **3**(4), 212–223 (2012)
11. Hoover, A., Goldbaum, M.: Locating the optic nerve in a retinal image using the fuzzy convergence of the blood vessels. IEEE Trans. Med. Imaging **22**(8), 951–958 (2003)
12. Hoover, A.D., Kouznetsova, V., Goldbaum, M.: Locating blood vessels in retinal images by piecewise threshold probing of a matched filter response. IEEE Trans. Med. Imaging **19**(3), 203–210 (2000)
13. Huang, G., Liu, Z., Maaten, L.V.D., Weinberger, K.Q.: Densely connected convolutional networks. In: IEEE Conference on Computer Vision and Pattern Recognition, pp. 2261–2269 (2017)
14. Huang, T., Yang, G., Tang, G.: A fast two-dimensional median filtering algorithm. IEEE Trans. Acoust. Speech Signal Process. **27**(1), 13–18 (1979)

15. Huttenlocher, D.P., Klanderman, G.A., Rucklidge, W.A.: Comparing images using the Hausdorff distance. IEEE Trans. Pattern Anal. Mach. Intell. **15**(9), 850–863 (1993)
16. Ioffe, S., Szegedy, C.: Batch normalization: accelerating deep network training by reducing internal covariate shift. arXiv:1502.03167v3
17. Kamnitsas, K., et al.: Efficient multi-scale 3D CNN with fully connected CRF for accurate brain lesion segmentation. Med. Image Anal. **36**, 61 (2016)
18. Lcun, Y., Bottou, L., Bengio, Y., Haffner, P.: Gradient-based learning applied to document recognition. Proc. IEEE **86**(11), 2278–2324 (1998)
19. Ledig, C., et al.: Robust whole-brain segmentation: application to traumatic brain injury. Med. Image Anal. **21**(1), 40 (2015)
20. Li, X., et al.: 3D multi-scale FCN with random modality voxel dropout learning for intervertebral disc localization and segmentation from multi-modality MR images. Med. Image Anal. **45**, 41–54 (2018)
21. Long, J., Shelhamer, E., Darrell, T.: Fully convolutional networks for semantic segmentation. In: IEEE Conference on Computer Vision and Pattern Recognition, pp. 3431–3440 (2015)
22. Maier, O., et al.: ISLES 2015 - a public evaluation benchmark for ischemic stroke lesion segmentation from multispectral MRI. Med. Image Anal. **35**, 250–269 (2017)
23. Rao, A., Ledig, C., Newcombe, V., Menon, D., Rueckert, D.: Contusion segmentation from subjects with traumatic brain injury: a random forest framework. In: IEEE International Symposium on Biomedical Imaging, pp. 333–336 (2014)
24. Rekik, I., Allassonnire, S., Carpenter, T.K., Wardlaw, J.M.: Medical image analysis methods in MR/CT-imaged acute-subacute ischemic stroke lesion: segmentation, prediction and insights into dynamic evolution simulation models. A critical appraisal. NeuroImage: Clin. **1**(1), 164–178 (2012)
25. Ronneberger, O., Fischer, P., Brox, T.: U-Net: convolutional networks for biomedical image segmentation. In: Navab, N., Hornegger, J., Wells, W.M., Frangi, A.F. (eds.) MICCAI 2015. LNCS, vol. 9351, pp. 234–241. Springer, Cham (2015). https://doi.org/10.1007/978-3-319-24574-4_28
26. Srivastava, N., Hinton, G., Krizhevsky, A., Sutskever, I., Salakhutdinov, R.: Dropout: a simple way to prevent neural networks from overfitting. J. Mach. Learn. Res. **15**(1), 1929–1958 (2014)
27. Szegedy, C., et al.: Going deeper with convolutions. In: IEEE Conference on Computer Vision and Pattern Recognition, pp. 1–9 (2015)
28. Zhang, R., et al.: Automatic segmentation of acute ischemic stroke from DWI using 3D fully convolutional denseNets. IEEE Trans. Med. Imaging, 1 (2018)

Ischemic Stroke Lesion Segmentation Using Adversarial Learning

Mobarakol Islam[1,2], N. Rajiv Vaidyanathan[2,3], V. Jeya Maria Jose[2,4],
and Hongliang Ren[2(✉)]

[1] NUS Graduate School for Integrative Sciences and Engineering (NGS),
National University of Singapore, Singapore, Singapore
mobarakol@u.nus.edu

[2] Department of Biomedical Engineering, National University of Singapore,
Singapore, Singapore
rajiv.vaidyanathan4@gmail.com, jeyamariajose7@gmail.com, ren@nus.edu.sg

[3] Department of Mechanical Engineering, NIT, Tiruchirappalli, India

[4] Department of Instrumentation and Control Engineering, NIT,
Tiruchirappalli, India

Abstract. Ischemic stroke occurs through a blockage of clogged blood
vessels supplying blood to the brain. Segmentation of the stroke lesion is
vital to improve diagnosis, outcome assessment and treatment planning.
In this work, we propose a segmentation model with adversarial learning
for ischemic lesion segmentation. We adopt U-Net with skip connection
and dropout as segmentation baseline network and a fully connected network (FCN) as discriminator network. Discriminator network consists of
5 convolution layers followed by leaky-ReLU and an upsampling layer to
rescale the output to the size of the input map. Training a segmentation network along with an adversarial network can detect and correct
higher order inconsistencies between the segmentation maps produced
by ground-truth and the Segmentor. We exploit three modalities (CT,
DPWI, CBF) of acute computed tomography (CT) perfusion data provided in ISLES 2018 (Ischemic Stroke Lesion Segmentation) for ischemic
lesion segmentation. Our model has achieved dice accuracy of 42.10%
with the cross-validation of training and 39% with the testing data.

1 Introduction

A stroke occurs due to an interruption in blood flow to the brain. The most
common form of stroke is ischemic stroke [13] which occurs due to a reduction
of blood flow, a condition known as ischemia when the brain arteries become
narrow or get blocked. It is a medical emergency and undergoes in different
disease stages ($acute : 0 - -24h$, $sub - acute : 24h - 2w$ and $chronic : > 2w$)
with time [4]. At present, Ischemic stroke is assessed by manually delineating the
lesion from computed tomography (CT) or magnetic resonance imaging (MRI)
including other factors like age, blood pressure, speech, and headache. However,
manual delineation is a time-consuming and tedious task which prone to human

© Springer Nature Switzerland AG 2019
A. Crimi et al. (Eds.): BrainLes 2018, LNCS 11383, pp. 292–300, 2019.
https://doi.org/10.1007/978-3-030-11723-8_29

error and inter-rater variability. Given the severity of the stroke, it is necessary to detect it as quickly as possible. With the recent advancement in the field of deep learning, convolutional neural networks can perform real-time predictions way faster than humans which have improved the process of detection of diseases.

There are few works on developing machine learning and deep learning approaches to detect, localize and segment the stroke lesion. Core and penumbra regions have been characterized using machine learning techniques in sub-acute and acute stage [9]. ISLES 2016 and 2017 [14] presents various deep learning models for stroke lesion segmentation from multi-modal MRI data. A comparison study among different machine learning and convolutional neural network (CNN) has done for stroke segmentation [10]. Islam et al. [7] exploits class-balanced PixelNet [1] to segment multi-modal neurological images (MRI) including ischemic stroke in an efficient way. However, these models have not obtained excellent accuracy by considering the visibility of the ischemic stroke in MRI or CT.

In this paper, we propose an architecture for the ischemic stroke lesion segmentation using adversarial learning. The concept of adversarial learning was first introduced in generative adversarial networks (GANs) [5]. It was initially used widely for image synthesis tasks. From then on, the concept of adversarial networks was exploited for usage in a lot of tasks, especially in segmentation. Luc et al. [8] introduce adversarial learning for the semantic segmentation. Consecutively, semi-supervised adversarial learning has been utilized in segmentation which outperforms previous approach [6]. In this work, we have proposed a variant of the adversarial network for segmentation problems by inspiring [6,8]. We have also discussed its performance for the segmentation of ischemic stroke lesion.

2   Proposed Model

In this section, we discuss about the in detail architecture of the proposed model. The model consists of two deep networks working simultaneously, a concept adopted from the generative adversarial network [5] and adversarial learning for segmentation [6,8]. We simultaneously train two models: a generative model G that generates the synthesized model, and a discriminative model D that estimates the probability that a sample came from the original data rather than G. The training procedure for G is to maximize the probability of D making a mistake. In our proposed model, the generator is replaced with a segmentor so that it can take in input CT data and produce segmentation label maps as its output. The discriminator is not changed.

2.1   Segmentor

U-Net [3,12] has been used as the segmentor. The network architecture has been illustrated in Fig. 1. It is of an architecture wherein the convolution layers in the left side contract the input image and the right side expands the feature maps produced by the last convolution layer in the left side. In the left side, the

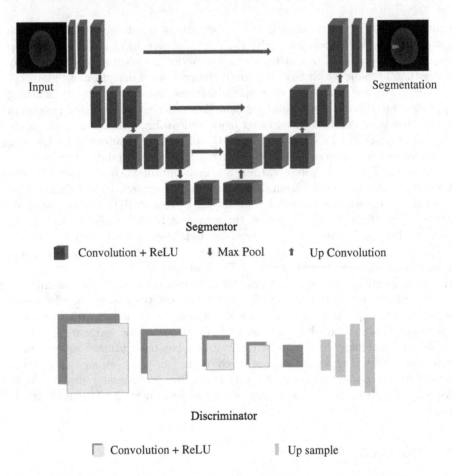

Fig. 1. UNet for adversarial training to segment ischemic stroke from CT images.

convolutional layers are of 3×3 kernel size with stride 1 followed by a rectified linear unit (ReLU) [11]. A 2×2 max pooling layer is used for downsampling with a stride 2. We double the number of feature channels at the same downsampling stage. In the right side of the network, upsampling layers are illustrated. They follow a 2×2 up-convolution which reduces the feature maps by half. A kernel size of 1×1 has been utilized at the final layer to produce feature maps of desire number classes. At the final layer, a 1×1 convolution is used to map each 64 component feature vector to the desired number of classes. In total the network has 23 convolutional layers.

2.2 Discriminator

Fully Convolutional Network [6] is used as the Discriminator. It consists of four layers of 2D Convolution followed by ReLu activation. The number of channels increases in each convolution. In the final convolution layer, 512 channels are convolved into 2 confidence maps which consist information about the probability of real or fake data. These confidence maps are then up-sampled to the original size where each upsampling layer has a scale factor of 2.

3 Experiments

3.1 Dataset

We use ISLES 2018 (Ischemic Stroke Lesion Segmentation) CT dataset to perform all the experiments. ISLES 2018 consists 6 modalities of CT perfusion images. Such as CT, DPWI, CBF, CBV, MTT, and Tmax. Each modality has 3D CT scan of 256×256 axial dimension and a variable slice of 2 to 18 (approximately). There are 94 and 62 cases for training and testing set respectively. The ground-truth for 94 training cases is given in the training phase. The annotation contains 2 classes where 1 and 0 denote stroke lesion and the healthy tissue respectively.

3.2 Training

We exploit only 3 modalities namely CT,DPWI and CBF. We split the training data into train and valid data in an 80/20 ratio to find the best epoch for segmentation. For testing, another 62 cases are used. The 3D CT image is first converted into stacks of 2D CT images. These images are given as input to the segmentor which gives probability maps of segmented image as its output. The synthesized segmented probability maps from generator is fed to the discriminator as one of the inputs along with the ground truth. The discriminator is simultaneously trained on the ground truth which is labelled as 1 corresponding to real. 0 corresponds to fake data which is the initial data produced by the segmentor. So, when the probability maps from generator are fed to the discriminator, it gives confidence maps from which the cross-entropy loss is found with respect to two classes. With this adversarial loss along with the segmentor's loss, we seek to train the segmentation network to fool the discriminator by maximizing the probability of the segmentation prediction [6]. The total loss function is denoted as

$$\chi = \chi_{seg} + \lambda_{adv}\chi_{adv}$$

where χ is the total loss that is used to backpropagate through both the segmentor and the discriminator. χ_{seg} is the loss that is got in the segmentor by comparing the output probability map directly with the ground truth. χ_{adv} is the loss that is got in the discriminator which is actually the cross-entropy loss calculated between the two classes. λ_{adv} is the factor introduced to increase or decrease the amount by which adversarial loss will affect the total loss. λ_{adv} for ischemic stroke lesion data set is taken as 0.1. The total loss is backpropagated in both the segmentor and discriminator.

3.3   Testing

While testing, the discriminator network is removed and the segmentation labels are got by inputting the test image to the segmentor network. 2D slices have been fed to the model to predict lesion. Finally, all the predicted slices have converted to 3D CT to measure the performances.

4   Results

The cross-validation dice accuracy of different segmentation architectures has been compared with our proposed model in Table 1. Table 2 gives details about the performance of our model on the test data set. Figures 2 and 3 show the predictions of the proposed model for the validation and test data set respectively.

Table 1. Cross validation performance of different models with training dataset

Models	Ours	PixelNet [7]	U-Net [12]	Deeplab v2 [2]	ICNet [15]	PSPNet [16]
Dice	**0.421**	0.409	0.419	0.373	0.387	0.319

Table 2. Performance of our model with testing dataset

Dice	Hausdorff	Avg distance	Precision	Recall	AVD
0.39	17741954.64	17741938.19	0.55	0.36	10.90

5   Discussion

From the above results, it is evident that even though the network gives a good segmentation prediction, there is still room for improvement. With the same architecture, the network would perform better with a more consistent data set. This data set is inconsistent and has certain defects as some data had just two 2D slices and some images were of poor contrast or shaken. The data set is still less in number as it is difficult to find the best-converged network with such fewer cross-validation data. Even the performance gap in different folds of the cross-validation is very high.

From Table 1, it can be noted that U-Net, when trained in an adversarial way, gave a better dice accuracy when compared to normal U-Net. Training a segmentation network along with an adversarial network can detect and correct higher order inconsistencies between the segmentation maps produced by ground-truth and the Segmentor. This is due to the extra addition to the loss

Input (CT)   Ground-Truth   Prediction

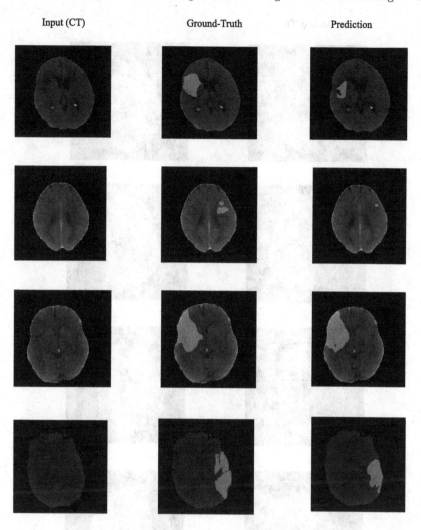

Fig. 2. Qualitative results of cross validation on training data

function that is given due to adversarial learning. Adversarial loss improves the performance of both the segmentor and discriminator. As the training is similar to a min-max game and as the segmentor does not only have its loss but also the extra loss from the discriminator, adversarial training of any segmentation architecture gives a better result when compared to its normal training.

Input (CT)   Prediction

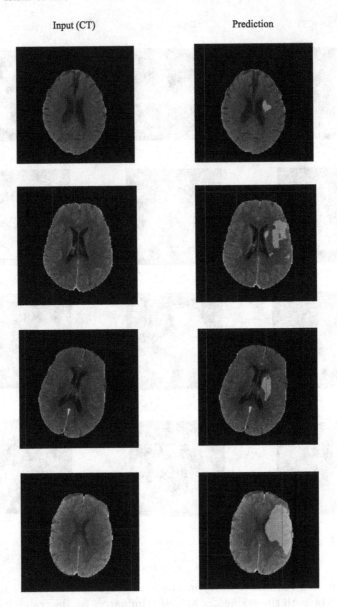

Fig. 3. Qualitative results of on testing data

6   Conclusion

In this paper, we have proposed an automatic ischemic stroke lesion segmentation using adversarial learning scheme. We have demonstrated than adversarial learning can improve the performance of the segmentation architecture like U-Net [12]. Our model has achieved better performance comparing to other state of the art models. However, the overall performance of the model is not excellent because of poor visibility of the lesion area in the CT imaging. Moreover, 3D deep learning models cannot be suitable for this dataset where dept slices are inconsistency and very small. As a future work, we can reslice or resample the CT into a higher dimension and try to exploit 3D models directly on 3D CT images.

Acknowledgement. This work is supported by the Singapore Academic Research Fund under Grant R-397-000-227-112, NUSRI China Jiangsu Provincial Grant BK20150386 and BE2016077 and NMRC Bedside & Bench under grant R-397-000-245-511 awarded to Dr. Hongliang Ren.

References

1. Bansal, A., Chen, X., Russell, B., Gupta, A., Ramanan, D.: PixelNet: representation of the pixels, by the pixels, and for the pixels. arXiv preprint arXiv:1702.06506 (2017)
2. Chen, L.-C., Papandreou, G., Kokkinos, I., Murphy, K., Yuille, A.L.: DeepLab: semantic image segmentation with deep convolutional nets, atrous convolution, and fully connected CRFs. IEEE Trans. Pattern Anal. Mach. Intell. **40**(4), 834–848 (2018)
3. Drozdzal, M., Vorontsov, E., Chartrand, G., Kadoury, S., Pal, C.: The importance of skip connections in biomedical image segmentation. In: Carneiro, G., et al. (eds.) LABELS/DLMIA -2016. LNCS, vol. 10008, pp. 179–187. Springer, Cham (2016). https://doi.org/10.1007/978-3-319-46976-8_19
4. González, R.G., Hirsch, J.A., Koroshetz, W.J., Lev, M.H., Schaefer, P.W.: Acute Ischemic Stroke. Springer, Heidelberg (2011)
5. Goodfellow, I., et al.: Generative adversarial nets. In: Advances in Neural Information Processing Systems, pp. 2672–2680 (2014)
6. Hung, W.-C., Tsai, Y.-H., Liou, Y.-T., Lin, Y.-Y., Yang, M.-H.: Adversarial learning for semi-supervised semantic segmentation. arXiv preprint arXiv:1802.07934 (2018)
7. Islam, M., Ren, H.: Class balanced pixelnet for neurological image segmentation. In: Proceedings of the 2018 6th International Conference on Bioinformatics and Computational Biology, pp. 83–87. ACM (2018)
8. Luc, P., Couprie, C., Chintala, S., Verbeek, J.: Semantic segmentation using adversarial networks. arXiv preprint arXiv:1611.08408 (2016)
9. Maier, O., et al.: ISLES 2015-a public evaluation benchmark for ischemic stroke lesion segmentation from multispectral MRI. Med. Image Anal. **35**, 250–269 (2017)
10. Maier, O., Schröder, C., Forkert, N.D., Martinetz, T., Handels, H.: Classifiers for ischemic stroke lesion segmentation: a comparison study. PloS One **10**(12), e0145118 (2015)
11. Nair, V., Hinton, G.E.: Rectified linear units improve restricted boltzmann machines. In: Proceedings of the 27th International Conference on Machine Learning (ICML-10), pp. 807–814 (2010)
12. Ronneberger, O., Fischer, P., Brox, T.: U-Net: convolutional networks for biomedical image segmentation. In: Navab, N., Hornegger, J., Wells, W.M., Frangi, A.F. (eds.) MICCAI 2015. LNCS, vol. 9351, pp. 234–241. Springer, Cham (2015). https://doi.org/10.1007/978-3-319-24574-4_28

13. World Health Organization (WHO), et al.: Cause-specific mortality estimates 2000–2012 (2014)
14. Winzeck, S., et al.: ISLES 2016 and 2017-benchmarking ischemic stroke lesion outcome prediction based on multispectral MRI. Front. Neurol. **9** (2018)
15. Zhao, H., Qi, X., Shen, X., Shi, J., Jia, J.: ICNet for real-time semantic segmentation on high-resolution images. arXiv preprint arXiv:1704.08545 (2017)
16. Zhao, H., Shi, J., Qi, X., Wang, X., Jia, J.: Pyramid scene parsing network. In: IEEE Conference on Computer Vision and Pattern Recognition (CVPR), pp. 2881–2890 (2017)

V-Net and U-Net for Ischemic Stroke Lesion Segmentation in a Small Dataset of Perfusion Data

Gustavo Retuci Pinheiro[1]([⊠]), Raphael Voltoline[1], Mariana Bento[2], and Leticia Rittner[1]

[1] School of Electrical and Computing Engineering (FEEC),
University of Campinas (UNICAMP), Campinas, SP, Brazil
`gustavorp@dca.fee.unicamp.br`
[2] Calgary Image Processing and Analysis Centre (CIPAC),
Department of Radiology and Clinical Neuroscience,
University of Calgary, Calgary, Canada

Abstract. Ischemic stroke is the result of an obstruction within a brain blood vessel, blocking the fresh blood flow, resulting in a tissue lesion. Early prediction of the ischemic stroke lesion region is important because it can help to choose the most suitable treatment. However, that is not trivial since current medical data, such as CT and MRI, have no explicit information about the future extension of the permanent lesion. A step towards efficiently using these data to predict the lesions is the use of Deep Convolutional Neural Networks as they are able to extract "hidden" information from the data when a reasonable labeled dataset is available and the deep networks are used properly. In order to try to extract this information, we have tested two different deep network architectures that are the state of the art in segmentation problems: V-net and U-net. In both networks, we tried different configurations, such as depth variations, pixel interpolations, MRI image combinations, among others. Experiments showed the following: normalizing the voxels sizes results in better training and predictions; deeper U-Net performs slightly better than the shallower U-Net, however it requires much more computation for only a small gain in accuracy; the inclusion of CT modality improved slightly the results; the use of only perfusion maps brought much better results than the use of raw perfusion data; smaller lesions are harder to detect properly.

Keywords: Ischemic stroke · Lesion segmentation · Deep learning · U-Net · V-Net · Perfusion

1 Introduction

Ischemic stroke occurs when there is an obstruction of a brain blood vessel, usually of small caliber, that irrigates the brain. This obstruction is named

Supported by CNPq and CAPES.

microangiopathy and it causes a decrease or cessation of blood circulation, fast degeneration of brain tissue and tissue lesion. These brain lesions observed in ischemic stroke work as biomarkers of the disease, aiding in the diagnosis and treatment.

The most suitable imaging modality to detect and analyze these brain lesions is MRI because it presents an excellent contrast in soft issues, allowing the detection of subtle abnormalities in early stages of the disease [1]. Early prediction of the ischemic stroke lesion region is relevant to the early patient diagnosis and selection of the most suitable treatment strategy [2].

We developed an automatic approach for ischemic stroke lesion by using two deep net architectures that are the state of the art regarding medical image segmentation: V-Net [3] and U-Net [4].

This paper is organized as follows: The dataset is presented and explained at Sect. 2. The methods, including the architectures and parameters description, are described in Sect. 3. The experimental setting and results are presented in Sect. 4 and discussed in Sect. 5. Our conclusions are presented in Sect. 6.

2   Dataset

Two datasets were used in the development of this project, both from ISLES challenge: ISLES2017 and ISLES2018 [5,6].

In the first dataset (ISLES2017), the training set comprises data and ischemic lesion segmentation masks of 43 patients, data from another 32 patients with no ground truth was also available as testing set in the challenge. This dataset is composed of Apparent Diffusion Coefficient (ADC), Perfusion Weighted Images (PWI), and Perfusion maps, including Cerebral Blood Volume (CBV), Cerebral Blood Flow (CBF), Mean Transit Time (MTT), Time to Peak Concentration of the contrast agent (TTP), and the time need at which the residue function reaches its maximum value (Tmax).

The second dataset (ISLES2018) contains 63 patients (split in 94 cases/volumes) to train and 40 patients (split in 62 cases/volumes) to the test. Different from the ISLES2017 data, this data has no ADC and PWI, and it has the addition of CT Perfusion (CTP) data. Another difference is that the ISLES2018 dataset has kept only slices that contain lesions, thus, some subjects could have more than one slab to cover the lesion.

All the data were acquired during the ischemic stroke acute stage (within 8 h of the stroke). The ground-truth was manually drawn on $T2$ or FLAIR, when the stroke lesion had stabilized, for ISLES2017 and on DWI for ISLES2018.

Both datasets are provided in NIfTI format and already pre-processed with skull stripping, anonymization and co-registration for each subject individually.

3   Methods

The initial step in our proposed method is to create patches with 64×64 pixels. All the patches must contain lesions, at least partially (Fig. 1) in order to the dataset not be unbalanced.

Fig. 1. Example of 64×64 pixels resolution patches; ground truth of the ischemic stroke lesion highlighted over Perfusion MTT map. (Color figure online)

After patches are ready, two deep networks were applied: V-Net and U-Net. V-Net (Fig. 2(a)) is a fully Convolutional Neural Networks (CNN) for volumetric medical image segmentation, therefore, it is originally a $3D$ deep net architecture. U-Net (Fig. 2(b)) is a fully CNN that was developed for biomedical image segmentation. In its original configuration, it works only in $2D$ images, requiring an independent prediction for every slice of the volume.

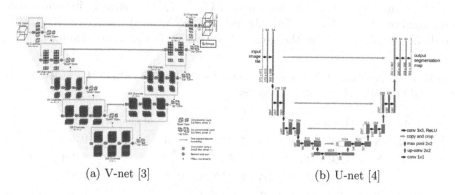

(a) V-net [3]   (b) U-net [4]

Fig. 2. Used deep neural network architectures for our proposed segmentation approach

Initially these CNNs were trained using a leave-out-out approach [7], in which the training dataset was randomly splitted into training and validation sets in a 80–20 ratio to avoid overfitting issues. After this initial experiment, we applied a different approach: k-fold cross validation, using k equal to 4. This change was made to have all the data in the training group, thus, increasing the accuracy in the test dataset. As only the prediction done in the test data was required to be submitted to the challenge platform (ISLES2018), the k-fold approach was done only in this data.

Another important step in the training was data augmentation. In addition to the patching, we have done flipping in the training patches in 50% of the cases. This flip means that half of the times a patch and its respective mask enter in the train batch, they are horizontally mirrored, thus, inputting a "new" valid data in the training.

In both training methods, the used parameters were: optimizer RMSprop [8]; learning rate 0.0005, momentum 0.9; up to 300 epochs.

4   Experiments and Results

As shown in Sect. 2, the datasets have a variety of available image modalities, including CT and MRI. We have tested different combinations to achieve the best result. Each image modality or measure was a channel in the image for both CNNs.

In the ISLES2017 dataset, we have separately tested the networks with Perfusion Weighted Images (PWI) and Perfusion Maps. Since PWI is a $4D$ image that has up to 40 volumes, this dimension is taken as channels of the image in the CNNs. In the case of Perfusion Maps, the channels are each different measure (CBV, CBF, MTT, TTP, TMAX) plus ADC map.

In the ISLES2018 dataset, we had a similar approach, however the channels were CBF, MTT, CBV, TMAX, and CTP. We also tested the CNNs without the CTP.

The results show that the inclusion of CTP (Table 1) slightly improves the performance of the CNN. Differently, in the CNN trained using only PWI, the prediction were worse than the same CNN with Perfusion maps (Table 2).

Table 1. Comparison of CNNs performance for ISLES2018 dataset: Perfusion Maps only and Perfusion Maps plus CT Perfusion.

CNN	Data	n° of channels	DICE
32U-NET	CBF, MTT, CBV, TMAX	4	0.324
17U-NET	CBF, MTT, CBV, TMAX	4	0.341
32U-NET	CBF, MTT, CBV, TMAX, CTP	5	0.351
17U-NET	CBF, MTT, CBV, TMAX, CTP	5	0.333

4.1   Architectures Variations

V-net and U-net were originally too deep to be used to analyze our patches with ischemic stroke lesions, since the pooling layers reduce the image size, completely eliminating it before reaching the middle layer. To overcome this issue, we have trimmed these CNNs by removing a few layers.

In this scenario, we used 2 different depth for each CNN: 32U-Net and 17U-Net; 10V-Net and 6V-Net, where the number refers to the amount of convolutional layer in the CNN. We also have add another dimension to the U-Net in order to have it in $3D$ form, allowing the direct comparison with V-Net.

The experiments with the CNN architectures (Table 2) showed that the $3D$ U-Net always outperforms the V-Net. We also can see that the $2D$ U-Net performs better than its $3D$ version and that the depth variation in the U-Net have only a small effect in the prediction accuracy.

Table 2. Comparison of architecture and data type combinations (ISLES2017 dataset): average DICE value for V-Net and U-Net on raw Perfusion images (PWI) and Perfusion Maps.

CNN	Data	DICE	CNN	Data	DICE
32U-NET 3D	PWI	0.359	**10V-NET 3D**	PWI	0.336
17U-NET 3D	PWI	0.357	**6V-NET 3D**	PWI	0.203
32U-NET 3D	P Maps	**0.526**	**10V-NET 3D**	P Maps	0.479
17U-NET 3D	P Maps	**0.518**	**6V-NET 3D**	P Maps	0.510
32U-NET 2D	PWI	0.414	-	-	-
17U-NET 2D	PWI	0.429	-	-	-
32U-NET 2D	P Maps	**0.609**	-	-	-
17U-NET 2D	P Maps	**0.560**	-	-	-

4.2 Voxel Interpolation

The dataset is not consistent regarding the voxel size, thus, the CNNs have to deal with different voxels resolution. One of our findings based on previous experiments is that the prediction results on testing images with the same voxel size as the majority of the train data is better than results achieved in testing images with different size. Our approach to minimize this effect was to normalize the size of the voxels among the whole dataset. By doing this, we have added another parameter to tune, but were able to improve the results.

We used a trilinear interpolation of the voxels to the sizes of $0.5 \times 0.5 \times 6\,mm$, $1.0 \times 1.0 \times 6\,mm$, $2.0 \times 2.0 \times 6\,mm$, and $2.5 \times 2.5 \times 6\,mm$. We have not tested variation of the Z axis because the datasets were more consistent in this dimension and about 6 mm in height.

The results (Table 3) showed that the voxel size of $2.5 \times 2.5 \times 6$ mm presented the best results. It is also shown, by the standard deviation, that normalizing the voxel size reduces the variability in the prediction quality.

4.3 Computational Environment

The experiments were performed using Python 3.6 and PyTorch on Jupyter Notebook. They were locally run on a machine with Intel I7 3.3 GHz processor, 8 GB RAM, and Nvidia GeForce GTX TITAN with 6 GB GDDR5.

Table 3. Comparison of different voxel size interpolations on the 17U-Net on ISLES2018 dataset: voxel interpolation size, average DICE, and standard deviation for the whole dataset.

Voxel size (mm)	CNN	$\overline{\text{DICE}}$	σ
Original size (no interpolation)	**17U-NET 2D**	0.341	0.272
$\mathbf{0.5 \times 0.5 \times 6}$	**17U-NET 2D**	0.369	0.254
$\mathbf{1.0 \times 1.0 \times 6}$	**17U-NET 2D**	0.438	0.237
$\mathbf{2.0 \times 2.0 \times 6}$	**17U-NET 2D**	0.448	0.254
$\mathbf{2.5 \times 2.5 \times 6}$	**17U-NET 2D**	**0.524**	0.247

4.4 Training Time

Different data combination and changes in the CNN architecture have a considerable influence in the training time (Table 4). For example, the 32U-Net $2D$ consumes more than twice the time for each epoch when compared to the 17U-Net $2D$. Moreover, the use of PWI instead Perfusion Maps increases the time by more than 5 times. The same CNN spends more than 10 times in the $3D$ version than in its $2D$ version.

Table 4. Epoch duration for each CNN architecture

CNN	Data	Time	CNN	Data	Time
32U-NET 3D	PWI	205.8	**10V-NET 3D**	PWI	265.7
17U-NET 3D	PWI	203.6	**6V-NET 3D**	PWI	238.8
32U-NET 3D	P Maps	42.3	**10V-NET 3D**	P Maps	83.1
17U-NET 3D	P Maps	34.2	**6V-NET 3D**	P Maps	55.5
32U-NET 2D	PWI	17.5	-	-	-
17U-NET 2D	PWI	13.4	-	-	-
32U-NET 2D	P Maps	7.0	-	-	-
17U-NET 2D	P Maps	3.3	-	-	-

4.5 Prediction

Since the networks are fully convolutional, the slices or volumes can simply be processed at once, independently of the size of the patch and the images used during training.

At this point we have defined the training method, the best data arrangement, the deep net architecture, the voxel size, and the prediction method, therefore, we are able to train the net to make the prediction on the test dataset.

We have defined the number of epochs to avoid overfitting by using the leave-one-out experimental approach. Then, we changed to k-fold training in order to increase the amount of data in the training while ensuring a non overfit, improving the generalization capability of the model.

With the k-fold model trained, we did the prediction on the training dataset to analyze the results qualitatively (Fig. 3). The prediction showed that for medium to larger lesion the CNNs are performing very well, with a slight tendency of overestimation. On the other hand, the predictor makes mistakes when segmenting small lesions, either in position or in extension.

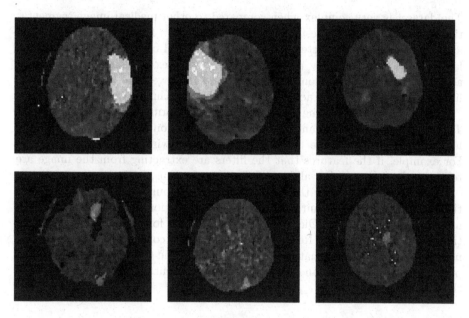

Fig. 3. Example of Ischemic Stroke Predictions, in axial slices, on validation group: 3 best predictions (top) and 3 worst prediction (bottom); correct prediction (yellow), false negative (green), and false positive (red), all over Perfusion MTT map. (Color figure online)

4.6   Challenge Results

The ISLES2018 challenge had 24 participating teams that were ranked by averaging the segmentation rank for every subject. With the results achieve by our team (Table 5), we were in the 8th global position.

Table 5. Results of the segmentation done on test dataset (ISLES2018) by the selected model: 17U-Net 2D with 2.5 mm interpolation using k-fold training method.

Dice		Precision		Recall		AVD	
avg	σ	avg	σ	avg	σ	avg	σ
0.43	0.32	0.50	0.38	0.49	0.34	14.69	16.83

5 Discussion

The dataset is very complex when compared to other medical image segmentation problems, given the best achieved Dice was 0.51[1]. Besides, with the limited amount of data, we restricted our model in terms of complexity and depth. A larger dataset would allow us to train more complex or deeper models. And although data augmentation was applied and improved the results, there is a limitation on what can be achieved by using such techniques.

Another relevant finding is related with the normalization of voxels size. This step plays an important role in our segmentation solution. This probably has to deal with the scale that the convolutional windows analyze the data. For example, if the features that the filters are extracting from the image are textures, they may not be valid in a different scale, thus, confusing the predictor.

When discussing about the models architecture, our experiments had shown that $3D$ architectures requires much more computational power than $2D$ with no significant gain in the Dice coefficient, so at least for our segmentation solution is not recommended. Regarding the amount of convolutional layers in the networks, it was verified that shallower versions of the CNNs are comparable in performance and present an expressive gain in computational efficiency against the deeper version.

6 Conclusion

In this paper, we have proposed the investigation of the V-Net and the U-Net in the context of ischemic stroke lesion segmentation. We have concluded that U-Net on MRI Perfusion maps plus CT Perfusion and voxel normalization ($2.5 \times 2.5 \times 6$ mm) is the best combination to estimate the extension of the stroke lesion. However, the use of CT Perfusion must be further investigate in order to determine its role in the results.

The use of raw Perfusion Weighted Images led to poor results. As this data is too complex, there is a need for a much larger dataset in order to the CNN to be able to extract the necessary features. When we compute the Perfusion Maps from PWI, we need a simpler CNN because we are, analogously, already extracting and feeding the net with relevant features by classical methods.

---

[1] http://www.isles-challenge.org/.

Additionally, voxel size standardization is crucial to improve the performance of the predictor. Furthermore, downsampling the images has improved the performance of the trained model. If this step is not done, the CNN would also have to deal with scale.

Finally, U-Net always outperform V-Net in this particular problem. Even in the case where the U-Net is in the $3D$ for directly comparison, V-Net performs worse.

References

1. Vernooij, M.W., et al.: Incidental findings on brain MRI in the general population. N. Engl. J. Med. **357**(18), 1821–1828 (2007). https://doi.org/10.1056/NEJMoa070972
2. Ginsberg, M.D.: Neuroprotection for ischemic stroke: past, present and future. Neuropharmacology **55**, 363–389 (2008). PubMed
3. Milletari, F., Navab, N., Ahmadi, S.A.: V-Net: fully convolutional neural networks for volumetric medical image segmentation. In: Proceedings of 2016 4th International Conference on 3D Vision, 3DV, pp. 565–571 (2016)
4. Ronneberger, O., Fischer, P., Brox, T.: U-Net: convolutional networks for biomedical image segmentation. In: Navab, N., Hornegger, J., Wells, W.M., Frangi, A.F. (eds.) MICCAI 2015. LNCS, vol. 9351, pp. 234–241. Springer, Cham (2015). https://doi.org/10.1007/978-3-319-24574-4_28
5. Maier, O., et al.: ISLES 2015 - a public evaluation benchmark for ischemic stroke lesion segmentation from multispectral MRI. Med. Image Anal. **35**, 250–269 (2017). https://doi.org/10.1016/j.media.2016.07.009
6. Kistler, M., et al.: The virtual skeleton database: an open access repository for biomedical research and collaboration. JMIR **15**(11), e245 (2013). https://doi.org/10.2196/jmir.2930
7. Arlot, S., Celisse, A.: A survey of cross-validation procedures for model selection. ArXiv e-prints (2009)
8. Hinton, G., Srivastava, N., Swersky K.: Neural networks for machine learning. http://www.cs.toronto.edu/~tijmen/csc321/slides/lecture_slides_lec6.pdf. Accessed 10 Oct 2018

Integrated Extractor, Generator and Segmentor for Ischemic Stroke Lesion Segmentation

Tao Song$^{(\boxtimes)}$ and Ning Huang

SenseTime Inc., Shanghai, China
{songtao,huangning}@sensetime.com

Abstract. The challenge of Ischemic Stroke Lesion Segmentation 2018 asks for methods that allow the segmentation of stroke lesion based on acute CT perfusion data, and provided a data set of 103 stroke patients and matching expert segmentations. In this paper, a novel deep learning framework with extractor, generator and segmentor for ischemic stroke lesion segmentation has been proposed. Firstly, the extractor is to extract the feature map from processed perfusion weighted imaging (PWI). Secondly, the output of extractor, cerebral blood volume (CBV), cerebral blood flow (CBF), mean transit time (MTT) and time of peak of the residue function (Tmax), etc. as the input of the generator to generated the Diffusion weighted imaging (DWI) modality. Finally, the segmentor is to precisely segment the ischemic stroke lesion using the generated data. In order to overcome the over-fitting, the data augmentation (e.g. random rotations, random crop and radial distortion) is used in training phase. Therefore, generalized dice combined with cross entropy were used as loss function to handle unbalanced data. All networks are trained end-to-end from scratch using the 2018 Ischemic Stroke Lesion Challenge dataset which contains training set of 63 patients and testing set of 40 patients. Our method achieves state-of-the-art segmentation accuracy in the testing set.

Keywords: Extractor · Generator · Segmentor · Generalized dice

1 Introduction

CT perfusion (CTP) [1] is an important diagnostic method in ischemic stroke. It enables differentiation of salvageable ischemic brain tissue (penumbra) from irrevocably damaged infarcted brain (infarct core). This is useful when assessing a patient for treatment (clot retrieval or thrombolysis). Compared with CTP, magnetic resonance images (MRI) is more sensitive to the early parenchymal changes of infarction. But its clinical application has been limited due to difficulties in timely access of MRI in many hospitals. For ischemic stroke patients, rapid imaging is especially important in the clinical treatment workflow.

The quantitative perfusion parameters of CTP (included CBV, CBF, MTT, Tmax, etc.) are usually used to identify the ischemic penumbra and the infarct core. The infarct core is defined as an area with prolonged MTT or Tmax, with markedly decreased CBF and CBV. The ischemic penumbra, which in most cases surrounds the

A. Crimi et al. (Eds.): BrainLes 2018, LNCS 11383, pp. 310–318, 2019.
https://doi.org/10.1007/978-3-030-11723-8_31

infarct core, also has prolonged MTT or Tmax (typically >6 s), but in contrast has only moderately decreased CBF and, importantly, near normal or even increased CBV [16]. Although the parameters of CTP can provide abundantly information in treatment of ischemic stroke, the accuracy is affected by factors such as the placement of arterial input function (AIF), and the deconvolution method used. In this work, we are also trying the extract the perfusion information directly from the CTP data, instead of indirectly from the perfusion parameters extracted.

In recent years, deep convolutional neural networks (CNNs) [2], have achieved great successes in image classification, segmentation and detection tasks and rapidly become the most popular technique in the medical imaging analysis. Image segmentation plays an important role in medical imaging, so the CNNs were firstly applied on medical image segmentation using patch-wise pixel classification. Later on, the global and local information are considered in the fully convolutional network (FCN) [3], which have encoder, decoder and show state-of-the-art performance in segmentation. The U-Net [4] is developed based on the FCN framework and uses the skip-connections to combine the low-level feature maps with higher-level feature maps, which has achieved better result in breast cancer segmentation in pathology. Cross-entropy (CE) was commonly used as loss function in segmentation networks of medical imaging, where background pixels are in majority which could cause serious class-imbalance problem, therefore the Dice loss function was proposed in [5] to alleviate this problem.

In this paper, we proposed an integrated network of ischemic stroke lesion segmentation for CTP data, which consists of an extractor, a generator and a segmentor. The FCN-like framework, with encoder and decoder, is used in extractor, generator and segmentor, respectively. This network is trained and tested on 2018 Ischemic Stroke Lesion Challenge [11], and achieved the first place of this challenge. Figure 1 shows the overall pipeline of our method, which contains three networks: (1) the extractor is to learn the representative image or the most important information from CT perfusion images; (2) the generator is to generate the DWI data using the output of the extractor

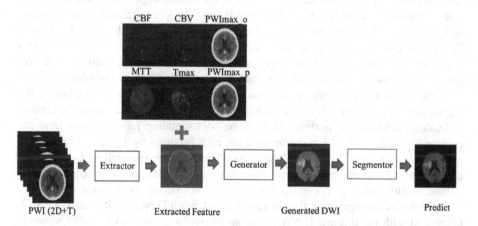

Fig. 1. The overall pipeline of our integrated segmentation framework. Here, *PWImax_o* and *PWImax_p* represents the max value in the time dimension of original PWI and preprocessed PWI, respectively.

and the perfusion parameters, which provides a better input for the segmentor; (3) the segmentor is to precisely segment the ischemic stroke lesion using the generated data. All networks are trained end-to-end, and the infarct core will be automatically predicted in the inference phase.

2 Method

2.1 Dataset and Preprocessing

The integrated segmentation framework is trained and tested using the 2018 Ischemic Stroke Lesion Challenge dataset. Imaging data from acute stroke patients in two centers who underwent CTP within 8 h of stroke onset, and were further scanned with MRI with DWI sequence within 3 h after CTP were included. Infarcted brain tissue is hyperintense on the DWI images. The training set consist of 63 patients, which contains, plain CT, DWI and CTP as well as perfusion maps of CBF, CBV, MTT and Tmax. Ground-truth segmentations were also provided. 40 patients with perfusion maps (CBF, CBV, MTT, Tmax), CT and CTP are provided in the testing phase, with no DWI or ground truth. The perfusion maps are calculated from original PWI using deconvolution method. The preprocessing of CTP data contains three stages: (1) firstly, the pixel values of a certain CTP frame are summed at any given time point, which forms a single time intensity curve for the whole CTP data of any case; (2) secondly, this time intensity curve is smoothed using a Gaussian smoothing filter, with a kernel size of 5. (3) finally, the 11 frames of CTP are selected, which are sampled between the onset of contrast injection and the end of the first pass. For normalization of the input data, the Batch Normal layer with no parameters is chosen.

2.2 Network Architecture

Our integrated segmentation framework consists of extractor, generator and segmentor, with all networks adapted from U-Net, which is a fully convolutional neural network and uses skip connections to combine low-level feature maps with higher-level ones. The U-Net consists of four blocks in the downsampling stage, each block has two 3×3 convolutions, each followed by a rectified linear unit (ReLU) and a 2×2 max pooling operation with a stride of 2. At each downsampling step, the number of feature channels doubles. Every step in the expansive path consists of an upsampling of the feature map followed by a 2×2 convolution ("deconvolution") that halves the number of feature channels, a concatenation with the correspondingly feature map from the contracting path, and two 3×3 convolutions, each followed by a ReLU. The network of extractor is a small U-Net that halves number of feature channels of U-Net in the downsampling and upsampling stage. The network of generator is an original U-Net without modification, except in the first convolution to adapt to the input data.

With regard to segmentor, the network is an attention U-Net as illustrated in Fig. 2. Compared with the original U-Net, it has an inserted network block, called squeeze-and-excitation networks (SE Block) [6], and with a switchable normalization (SN) [7] layer. The SE Block adaptively recalibrates channel-wise feature responses by explicitly modelling interdependencies between channels using the attention mechanism. And the SN layer is to learn the weights of batch-wise, channel-wise and layer-wise normalizers for normalization, it is robust to a wide range of batch sizes, maintaining high performance even when the batch size is small.

In the training phase, the same two small networks are used to help train the generator following the generator, each network consists of five 3×3 convolutions with stride 2, each followed a ReLU, and two adaptive average pooling layers in the end of network with size 7×1 and 1×7, respectively.

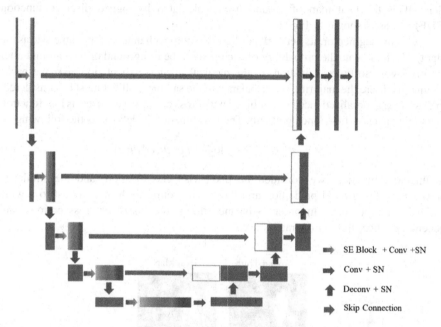

SE Block + Conv +SN

Conv + SN

Deconv + SN

Skip Connection

Fig. 2. The architecture of our segmentation network (Attention U-Net) adapted from 2D U-Net.

2.3 Loss Function

The output of extractor is a single channel feature map with the same size of input, which is calculated before sigmoid activation in the final layer. Through the sigmoid activation, this single channel feature map indicated the confidence probability of each pixel to be foreground. The L1 loss function is used for training in the extractor to regress the confidence probability and be expressed as

$$L_e = \lambda_1 * \|p - y\|_1 \tag{1}$$

where p and y represent the confidence probability of predicted and ground-truth respectively.

The loss function of generator has two parts: one part is the L2 loss to constrain the distance between the generated DWI and real DWI; the other part is to calculate the L2 distance of feature maps extracted from the generated DWI and the real DWI. Thus the whole loss function of generator can be written as

$$L_g = \lambda_2 * W * \left\| DWI_g - DWI_r \right\|_2 + \lambda_3 * \left(\left\| F_A - F_B \right\|_2 \right) \tag{2}$$

where the DWI_g and DWI_r is the generated DWI and real DWI, respectively. F_A is high-level feature map, extracted from generated DWI using model A, and F_B is the high-level feature map, extracted from real DWI using model B, the model A and B have the same network to extract high-level information, similar as perceptual loss. Here, W is the heat map of ground truth, calculated by signed distance function (SDF) [9], as is shown in Fig. 3.

As for the segmentor, its network predicts two output channels of the same size as the input, which indicate the probability of each pixel to be foreground or background after the pixel-wise softmax activation. In the medical image segmentation task, the background pixels are the majority, so the balance of the sample gradient must be considered. In this work, generalized dice [8] combined with cross entropy were used as loss function to handle the class-imbalance problem. The loss function is defined as the following

$$L_s = \lambda_4 * \{ W * CE - \log(generalized\,dice) \} \tag{3}$$

In this loss function, we consider the pixel-wise classification, and the similarity between the foreground prediction and the given ground truth, it is robust to a wide variance of input data. In order to balance the gradient size of cross entropy and generalized dice, the log operate is used.

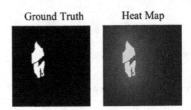

Ground Truth   Heat Map

Fig. 3. The heat map of ground truth.

2.4   Training and Testing

As mentioned above, the extractor, generator and segmentor are the main components of our integrated segmentation framework, also the same two small networks (model A and B) need to be trained in the training phase. The five networks are trained end-to-end from scratch using the 2018 Ischemic Stroke Lesion Challenge dataset which contains training set with 63 patients and testing set with 40 patients. The training set is divided to four subsets to validate the trained models using the cross-validation method. Firstly, the extractor with input size 256 * 256 * 11 is used to extract the feature map using the regression of confidence probability. Then a concatenation with the extractor's feature map and other perfusion maps with a size of 256 * 256 * 7 as input of generator. In the end, the segmentor is used to predict the foreground region (probability) using generated DWI with a size of 256 * 256 * 1. In training, the weights of all networks are initialized using Xavier initialization [12] and updated using RMSprop [14] optimizer with a batch size of 5 samples. The strategy of warm-up [13] and step-by-step learning rate decay are used, and the learning rate is initialized at 0.002 and reduced by factor 0.2 after 180, 300 epochs, and the λ_1, λ_2, λ_3 and λ_4 we set at 1.0, 0.002, 1.2 and 1.0, respectively.

In the testing phase, we preprocess a testing case to get the 11 slices of PWI as the input of extractor, and concatenate the extractor's feature map, CBF, CBV, MTT, Tmax, *PWImax_o* and *PWImax_p* to feed into the network of generator, then the Segmentor to predict the probability of infarct core using the generated DWI. The final segmentation region is an ensemble result of the four cross-validation models by computing their mean probability, and post-processing method of connected-component analysis is used to ensure the continuity of predicted area in space.

3   Result

The integrated segmentation framework is implemented by PyTorch [10] with cuDNN, and all experiments are performed on a workstation with 32 GB of memory, Intel Core i7 6700 k @ 4.0 GHz, and four Nvidia GTX 1080Ti 11G GPUs. In the training stage of 2018 ischemic stroke lesion challenge, the training datasets were divided into four subsets to validate the trained models by cross-validation, so all analysis of the trained model is performed on the cross-validation dataset. The detailed statistics are listed in Table 1.

Perfusion maps without DWI is provided in the testing phase, so we firstly attempt to feed the perfusion maps (CBV, CBF, MTT, Tmax) into the network of U-Net and use cross entropy after softmax activation for training, which caused a lot of false positives with a Dice score of only 0.53. In order to reduce the gradient of background to balance the gradient in the training phase, a novel loss function was designed using a combination of cross entropy and generalized dice, it improved the result with a Dice score of 0.55 and made the training more stable. Meanwhile, we designed a novel network (Attention U-Net), adapted form U-Net, which used an attention block (SE block) to get better performance with a Dice score of 0.56. An initial experiment using DWI as input achieves a dice score of 0.83, which inspired us to propose a two stage segmentation framework, which contains a generator with U-Net and a segmentor with Attention U-Net. In this framework, the *PWImax_p*, CBF, CBV, MTT and Tmax are used as the input of the generator to generate a pseudo-DWI, then the generated DWI is used to predict the region of infarct core. It increased the dice score by two percentage points on the original result. The heat map, calculated by sign distance function, is used as loss function for the generator and the segmentor, which makes the network focus more on the region of infarct core, so a Dice score of 0.59 was achieved. With the network becoming larger and larger, smaller batch number is needed in the training phase, so we replace Batch Norm (BN) [15] with Switch Norm (SN) in the networks. The SN is to learn the probabilities of different normalizers for normalization, it is robust to a wide range of batch sizes, maintaining high performance even when the batch size is small. Therefore, a Dice score of 0.60 is obtained by the SN.

Next, we try to extract the useful information from original CTP data to generate a more realistic DWI, firstly we extract 6 slices from PWI in time dimension, then a small U-Net was used to extract the representative images to feed into the generator's network. Compared with previous result, using the extractor makes the result increased by one percentage point. Later on, the two same small networks (model A and model B) are used following the generator, it makes the gradients descent faster and the training more stable. To calculate a more reasonable mean and standard deviation, the batch normal layer without learned parameters, designed before first convolution layer, is used to normalize the input data. Finally, the Dice score 0.62 is obtained in the validation set.

As shown as Fig. 4, the predictions are compared with ground truths, and the generated DWI is compared with the original DWI, respectively. It demonstrates the effectiveness of our integrated segmentation framework.

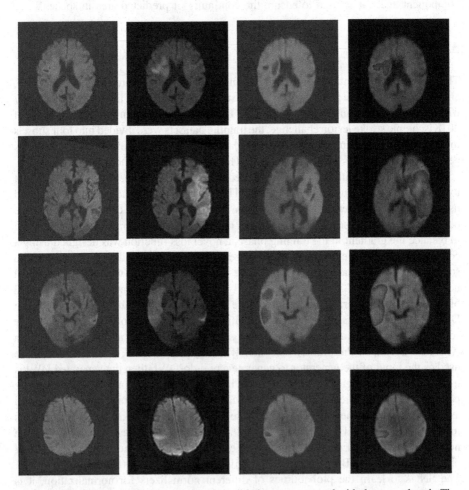

Fig. 4. Segmentation results of 4 cases in the validation set compared with the ground truth. The ground truths and predictions are given in green and red, respectively. From left to right, the original DWI, ground truth superimposed on original DWI, the generated DWI and the predictions superimposed on the generated DWI. (Color figure online)

Table 1. Ablation study of segmentation results using different methods. Here, AU-Net is the attention U-Net, and GD is generalized dice. And the E and G is the extractor and generator, respectively.

Stage	Method	Dice
One	U-Net + CE	0.53
One	U-Net + CE + GD	0.55
One	AU-Net + CE + GD	0.56
Two	AU-Net + CE + GD + G	0.58
Two	AU-Net + CE + GD + G + SDF	0.59
Two	AU-Net + CE + GD + G + SDF + SN	0.60
Three	AU-Net + CE + + GD + G + SDF + SN + E	0.61
Three	AU-Net + CE + GD + G + SDF + SN + E + BN	0.62

4 Conclusion

This paper detailed an integrated segmentation framework, which consists of extractor, generator and segmentor, for ischemic stroke lesion segmentation. First, the extractor is to extract the representative image from the original CTP data. Secondly, the generator is to generate the DWI from extractor's output and perfusion maps. Finally, the segmentor is to precisely segment the infarct core using the generated data. The network achieved a dice coefficient of 0.62 in cross validation stage and won the first place in the 2018 ischemic stroke lesion challenge in the test stage.

References

1. Eastwood, J.D., Lev, M.H., Azhari, T.: CT perfusion scanning with deconvolution analysis: pilot study in patients with acute middle cerebral artery stroke. Radiology **222**(1), 227–236 (2002)
2. Krizhevsky, A., Sutskever, I., Hinton, G.E.: ImageNet classification with deep convolutional neural networks. In: International Conference on Neural Information Processing Systems, pp. 1097–1105 (2012)
3. Long, J., Shelhamer, E., Darrell, T.: Fully convolutional networks for semantic segmentation. In: Proceedings of the IEEE Conference on Computer Vision and Pattern Recognition, pp. 3431–3440 (2015)
4. Ronneberger, O., Fischer, P., Brox, T.: U-Net: convolutional networks for biomedical image segmentation. In: Navab, N., Hornegger, J., Wells, W.M., Frangi, A.F. (eds.) MICCAI 2015. LNCS, vol. 9351, pp. 234–241. Springer, Cham (2015). https://doi.org/10.1007/978-3-319-24574-4_28
5. Zhang, J., Shen, X., Zhuo, T., et al.: Brain tumor segmentation based on refined fully convolutional neural networks with a hierarchical dice loss. arXiv preprint arXiv:1712.09093 (2018)
6. Luo, P., Ren, J., Peng, Z., Zhang, R., Li, J.: Differentiable learning-to-normalize via switchable normalization. arXiv preprint arXiv:1806.10779 (2018)
7. Hu, J., Shen, L., Sun, G.: Squeeze-and-excitation networks. arXiv preprint arXiv:1709.01507 (2017)

8. Sudre, C.H., Li, W., Vercauteren, T., Ourselin, S., Jorge Cardoso, M.: Generalised dice overlap as a deep learning loss function for highly unbalanced segmentations. In: Cardoso, M.J., et al. (eds.) DLMIA/ML-CDS - 2017. LNCS, vol. 10553, pp. 240–248. Springer, Cham (2017). https://doi.org/10.1007/978-3-319-67558-9_28

9. https://en.wikipedia.org/wiki/Signed_distance_function

10. Pytorch. http://pytorch.org/

11. http://www.isles-challenge.org/

12. Jia, Y., et al.: Caffe: convolutional architecture for fast feature embedding. arXiv preprint arXiv:1408.5093 (2014)

13. He, K., Zhang, X., Ren, S., Sun, J.: Deep residual learning for image recognition. arXiv preprint arXiv:1512.03385 (2015)

14. Tieleman, T., Hinton, G.: Lecture 6.5-rmsprop: divide the gradient by a running average of its recent magnitude. COURSERA: Neural Netw. Mach. Learn. 4(2), 26–31 (2012)

15. Ioffe, S., Szegedy, C.: Batch normalization: accelerating deep network training by reducing internal covariate shift. arXiv preprint arXiv:1502.03167 (2015)

16. https://radiopaedia.org/articles/ct-perfusion-in-ischaemic-stroke

ISLES Challenge: U-Shaped Convolution Neural Network with Dilated Convolution for 3D Stroke Lesion Segmentation

Alzbeta Tureckova[1(✉)] and Antonio J. Rodríguez-Sánchez[2]

[1] Faculty of Applied Informatics, Tomas Bata University in Zlin,
Nam T.G. Masaryka 5555, 760 01 Zlin, Czech Republic
tureckova@utb.cz
[2] Intelligent and Interactive Systems, Institute of Computer Science,
University of Innsbruck, Innsbruck, Austria

Abstract. In this paper, we propose the algorithm for stroke lesion segmentation based on a deep convolutional neural network (CNN). The model is based on U-shaped CNN, which has been applied successfully to other medical image segmentation tasks. The network architecture was derived from the model presented in Isensee et al. [1] and is capable of processing whole 3D images. The model incorporates the convolution layers through upsampled filters – also known as dilated convolution. This change enlarges filter's field of the view and allows the net to integrate larger context into the computation. We add the dilated convolution into different parts of network architecture and study the impact on the overall model performance. The best model which uses the dilated convolution in the input of the net outperforms the original architecture in nearly all used evaluation metrics. The code and trained models can be found on the GitHub website: http://github.com/tureckova/ISLES2018/.

Keywords: Medical image segmentation ·
Deep convolutional neural networks · U-Net · Dilated convolution

1 Introduction

According to the World Health Organization (WHO) stroke is the world's second biggest killer after the ischemic heart disease [2]. The stroke is caused by low blood flow which results in the death of brain cells. The precise localisation of the affected tissue is an essential step in the diagnostic procedure. Magnetic Resonance Imaging (MRI) using diffusion and perfusion is perceived as the best current technique for distinguishing between infarcted tissue (core) and hypoperfused lesion tissue (penumbra). On the other hand, stroke patients are recently often first imaged by Computer Tomography (CT) due to its speed, availability, and lack of contraindications. Although it is usually challenging to distinguish

© Springer Nature Switzerland AG 2019
A. Crimi et al. (Eds.): BrainLes 2018, LNCS 11383, pp. 319–327, 2019.
https://doi.org/10.1007/978-3-030-11723-8_32

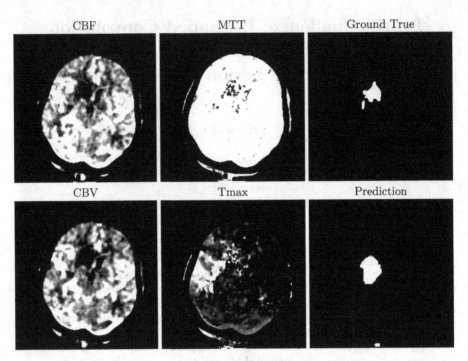

Fig. 1. Data example. The left and the middle columns show different input modalities and the right column shows ground true and the prediction.

soft tissues in CT images, it was proposed that CT perfusion could be used to identify irreversibly damaged, infarcted tissue (core) [3]. Automatic segmentation can be very useful in this context due to its speed, objectivity and ability to take multiple modalities as an input. This advantages may result in a more accurate description of the stroke lesion.

The Ischemic Stroke Lesion Challenge [3,4] aims at encouraging the development of state of the art methods for stroke segmentation by providing a large manually annotated dataset of stroke patients. Unlike last years, this year the ISLES 2018 challenge provides acute stroke CT perfusion imaging scans and manually outlined core lesions on MRI DWI scans acquired soon after that. The dataset contains images from 103 patients, but some patient cases have two slabs to cover the stroke lesion. Therefore there are 156 cases each containing CT perfusion (CTP) source data and four different perfusion maps obtained from this raw data for clinical interpretation: cerebral blood volume (CBV), cerebral blood flow (CBF), time to peak of the residue function (Tmax) and mean transit time (MTT). All these modalities plus a plain CT image serve as input to the segmentation algorithm. An example data of perfusion maps can be seen in Fig. 1.

Other recent approaches have been presented to solve the stroke segmentation problem. Usinskas and Gleizniene [5] introduced a method based on 2D

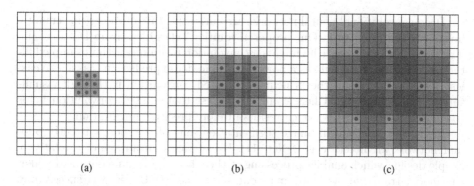

Fig. 2. Visualisation of dilated convolutions, adopted from [10]. Systematic dilation supports an exponential expansion of the receptive field without loss of resolution or coverage. (a) F1 is produced from F0 by a 1-dilated convolution; each element in F1 has a receptive field of 3×3. (b) F2 is produced from F1 by a 2-dilated convolution; each element in F2 has a receptive field of 7×7. (c) F3 is produced from F2 by a 4-dilated convolution; each element in F3 has a receptive field of 15×15. The number of parameters associated with each layer is identical. The receptive field grows exponentially while the number of parameters grows linearly.

CT images. It applies statistical measures like the mean, the standard deviation, the co-occurrence matrix and a histogram to detect the centre of the stroke region. From this centre, it traces 64 rays which along the way checks the similarity of pixels and find the stroke contour if the similarity ratio is bigger than some threshold. Rajini and Bhavani [6] implemented a segmentation technique based on k-means clustering followed by decision tree classifiers which distinguish between healthy and ischemic brain. Both papers work with plain CT images, not with CT perfusion.

Reviews regarding stroke segmentation can be found in Maier et al. [7] and Winzeck et al. [8]. Both papers nicely document the continuous transition from classical image processing to machine learning and finally deep learning. The overall best segmentation results presented in [7] were achieved by the tuned Extra Tree Forests closely followed by Neural Networks. The ISLES challenge dataset was considerably enlarged in the years 2016 and 2017 which encouraged the deployment of data-driven approaches; all participating teams of ISLES 2017 suggested a deep learning technique, with top-ranked methods featuring CNN architectures [8]. The presented results also indicate that the ensembles of different systems together tend to perform better, although the results are not statistically strong enough [8]. Despite the great efforts and accomplishments present at ISLES, automatic segmentation of stroke lesions remain challenging tasks, and none of the methods is capable of achieving results comparable to a human expert. Both summary papers deal with the MRI perfusion sequences, and therefore the outcomes cannot be directly compared to the results of ISLES challenge 2018.

Deep convolutional neural networks are currently state of the art in semantic image segmentation task. Out proposed solution takes ideas from DeepLab [9]. The DeepLab architecture highlights a convolution technique with upsampled filters, or *dilated convolution*, as a powerful tool in dense prediction tasks. The easiest way to understand the principle of dilated convolution is through visualisation; Consider Fig. 2. Chen et al. [9] suggest that dilated convolution helps to enlarge the filters view which results in more precise segmentation.

In the field of medical image segmentation, one of the most currently popular network architecture is U-Net [11]. U-Net takes advantage of a relatively simple design, which enables processing of the whole 3D volume data and makes the model trainable even on small datasets. The 3D U-Net was successfully applied for different segmentation tasks in medicine, i.e. segmentation of brain tumour [12]; lungs and their parts [13]; or even multiclass segmentation of chest radiographs [14].

This paper proposes the ischemic stroke segmentation from the CT perfusion image based on a deep fully convolutional neural network. Inspired by DeepLab architecture [9], we investigated the impact of the dilated convolution incorporation into U-net architecture. Several models with a different ratio of the dilated convolution used in architecture were constructed, and their performance was tested on ISLES 2018 challenge dataset. The experimental works are implemented using Python and Keras library. The code and trained models can be found on the GitHub website: http://github.com/tureckova/ISLES2018/.

2   Methods

This chapter describes the whole process of the automatic segmentation. The subsections that follow provide the details on data pre-processing, model architecture, training pipeline and evaluation metrics.

2.1   Data Pre-processing

Medical images are known for the presence of a low-frequency intensity nonuniformity named as bias, inhomogeneity, illumination nonuniformity, or gain field [15]. To overcome this potential cause of problems in image analysis, first, the N4ITK [15] bias correction was applied to the training and the testing data. Moreover, the medical data usually came from different research institutes using different scanners; are non-standardized; and are acquired in slightly different conditions, i.e. with different slice spacing, different acquiring protocols etc. Therefore, the normalisation is a critical step in automatic segmentation. This statement is particularly true for neural networks where we need to ensure that the value ranges match between patients as well as between the modalities. To this purpose mean and standard derivation (std) were counted for each modality, and then each modality was normalised by subtraction of mean and division by std. This way the values are normalised to be between zero and one. Finally, all 3D images were re-sized to dimensions $(128, 128, 32)$ (for axes x, y, z).

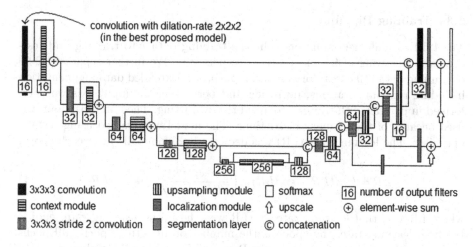

Fig. 3. A network architecture as proposed by Isensee et al. [1]. Dilated convolutions were used in different parts of the net, but the best performing model uses dilated convolution only in the first network layer.

2.2 Network Architecture

The model architecture is a fully convolutional neural network and is mainly based on the architecture presented in Isensee et al. [1] which follows on U-Net design [11]. It consists of a contracting part (encoder), which encodes increasingly abstracted representations of input, followed by the localisation part (decoder), which decodes these representations and combines them with the shallower features to precisely localise the structures of interest.

In the encoder part of the network, the spatial resolution is four times decreased by a convolution with a stride value of two, resulting in the lower spatial resolution sixteen times smaller than the input. The decoder part reversely mimics the encoder part using upsampling followed by a convolution. On each level, the output of the encoder part is concatenated to the output of the upsampling layer of the decoder part enabling the feature vectors of different resolutions to contribute to the final output. The activations are computed by context modules in the encoder part and by the localisation modules in the decoder each following the layer which is responsible for the change of the spatial resolution. Finally, the segmented output of each level is recombined with the upsampled segmentation output of the lower levels via element-wise summation. For more detail about the architecture, please refer to Isensee et al. [1]. The model architecture is shown in Fig. 3. We replaced some convolutions by the convolution using dilated filters. This change enlarges the filter's field of the view and allows the model to integrate larger context information into the computation. Throughout the network, we use leaky ReLu activations for all feature maps computing convolutions. Additionally, we replace the traditional batch normalisation with instance normalisation [16], to compensate the stochasticity induced by small batch sizes used due to memory restrictions.

2.3 Training Pipeline

The ISLES challenge organisers split the training data into training and testing part consisting of 94 and 62 cases respectively. We further separated 20% of training data (19 cases) for validation purposes. Extended data augmentation including random rotations, mirroring and permuting the input axis was performed in parallel with training to avoid the overfitting. The batch size four was used because of the memory restrictions. The model was trained using ADAM optimiser. The learning rate (LR) was decreased polynomially after each epoch:

$$LR = (initLR - endLR) * \left(\frac{1 - epoch}{allEpochs} \right)^{0.9} + endLR \qquad (1)$$

where LR is actual learning rate, initLR was the initial learning rate, endLR the final learning rate, the epoch is the number of an actual epoch, allEpochs is the overall number of epochs. InitLR value of 5e−3, endLR value of 5e−10 were empirically found as a compromise between parameters recommended by Isensee [1] and Chen et al. [9]. The training was stopped after 300 epochs before the vast over-fitting starts to occur.

2.4 Evaluation Metrics

Different evaluation metrics were used to compare the performance of the presented models. Each of the metric is briefly explained below.

Dice coefficient [17] equals twice the number of elements common to both sets divided by the sum of the number of elements in each set. If we denote $|X|$ and $|Y|$ the cardinalities of the two sets, the dice coefficient can be written as:

$$Dice = \frac{2|X \cap Y|}{|X| + |Y|} \qquad (2)$$

Hausdorff Distance [18] measures how far two subsets of a metric space are from each other.

Precision and Recall [19] in the context of segmentation compare the results of the classifier under test with trusted external judgments using the terms *true positives (TP)*, *true negatives (TN)*, *false positives (FP)*, and *false negatives (FN)*. The terms positive and negative refer to the classifier's prediction, and the terms true and false refer to whether that prediction corresponds to the external judgment. Precision and recall are then defined as:

$$precision = \frac{TP}{TP + FP} \qquad (3)$$

$$recall = \frac{TP}{TP + FN} \qquad (4)$$

Average Distance [20] is the mean distance between the segmented object and ground true segmentation.

Absolute Volume Difference [20] counts the absolute number of voxels segmented differently by model than is stated in ground true segmentation.

3   Results

Several models with different dilated convolutions were designed to explore its impact on the performance of the deep convolution neural network. All the models were trained in the same conditions, as described in the previous Sect. 2.3, using the same training and validation data split. The cross-validation technique was used for evaluation to minimise the influence of the train-test sample splitting. Therefore every model was trained five times, every time using different testing samples. The overall performance of the model was obtained as the average from partial results of each model. We use test augmentation in the form of mirroring and average the softmax outputs. The threshold value 0.5 was used for conversion of the label map to output mask.

Average cross-validation results are reported in Table 1. The example input data and qualitative segmentation result of our best performing model can be seen in Fig. 1. We compare the results of these model variants:

1. Original Isensee [1] (in Table 1 marked as "Isensee")
2. Dilated filters in first convolution of input (in Table 1 marked as "dil. in.")
3. Dilated filters in first convolution of input, in context module and in localization module (in Table 1 marked as "dil. in.&cont.&local.")
4. Dilated filters in all convolutions except for the ones responsible for spatial resolution change (in Table 1 marked as "dil. all")

Table 2 shows the test results of the best performing model counted by the ISLES 2018 challenge evaluation server.

Table 1. Cross validation results for modules with the different incorporation of dilated convolution, for detail description of models, please see the list above.

EXP	Dice coefficient	Hausdorff distance	Precision	Recall	Average distance	Absolute volume difference
Isensee [1]	0.5084	**34.2000**	0.4783	0.6507	5.5614	18.5286
dil. in.	**0.5144**	34.7591	0.4737	**0.7065**	**4.5142**	**17.8350**
dil. in.&cont.&local.	0.5140	37.1488	**0.4823**	0.6921	5.3290	21.3382
dil. all	0.5061	37.4597	0.4759	0.6612	5.5793	20.2391

Table 2. Test results of the best proposed model. The values were computed by the ISLES 2018 challenge evaluation server.

EXP	Dice coefficient	Hausdorff distance	Precision	Recall	Average distance	Absolute volume difference
input	0.37	19354868.84	0.44	0.44	19354843.18	24.95

4   Conclusion

This paper investigates the influence of dilated convolutions on the performance of U-shaped convolution neural network aimed at ischemic lesion segmentation. The results section compares the model presented by Isensee et al. [1] and three derived models with dilated convolution used in different parts of the net. The incorporation of dilated convolutions aims to enlarge the filters field of view enabling them to use a broader context in the computation. The best performing model uses dilated convolution only in the input layer. This result suggests that the perception of greater context is most important in the first overall view of the segmented image. On the other hand, the usage of dilated convolution deeper in the network architecture does not have a significant impact on the overall model performance.

Our best model outperforms the original architecture in nearly all used evaluation techniques except Hausdorff Distance in which the results are comparable. Notably better is the proposed model's Recall value. The original architecture [1] failed in four cases entirely, resulting in zero Dice coefficient. In contrast, our model was able to find in all cases at least some intersection of a segmented area, i.e. obtain nonzero Dice coefficient. We hypothesise that the reason for this is that the model can spot the lesion but struggles in localising exact segmentation borders. A possible solution for making segmentation more precise might be to extend the context or the localisation module. These ideas will be considered for future work.

Acknowledgments. This work was supported by the Internal Grant Agency of Tomas Bata University under the Project no. IGA/CebiaTech/2018/003 and further by resources of A. I. Lab (https://ailab.fai.utb.cz/) and IIS group at the University of Innsbruck (https://iis.uibk.ac.at/).

References

1. Isensee, F., Kickingereder, P., Wick, W., Bendszus, M., Maier-Hein, K.H.: Brain tumor segmentation and radiomics survival prediction: contribution to the BRATS 2017 challenge. In: Crimi, A., Bakas, S., Kuijf, H., Menze, B., Reyes, M. (eds.) BrainLes 2017. LNCS, vol. 10670, pp. 287–297. Springer, Cham (2018). https://doi.org/10.1007/978-3-319-75238-9_25
2. Global Health Estimates 2016: Deaths by cause, sex, by country and by region, 2000–2016. World Health Organization, Geneva (2018)

3. Maier, O., et al.: ISLES 2015: a public evaluation benchmark for ischemic stroke lesion segmentation from multispectral MRI. Med. Image Anal. (2016). https://doi.org/10.1016/j.media.2016.07.009. ISSN 1361-8415

4. Kistler, M., Bonaretti, S., Pfahrer, M., Niklaus, R., Büchler, P.: The virtual skeleton database: an open access repository for biomedical research and collaboration. JMIR (2013). https://doi.org/10.2196/jmir.2930

5. Usinskas, A., Gleizniene, R.: Ischemic stroke region recognition based on ray tracing. In: Proceedings of International Baltic Electronics Conference (2006). https://doi.org/10.1109/BEC.2006.311103

6. Rajini, N.H., Bhavani, R.: Computer aided detection of ischemic stroke using segmentation and texture features. Measurement 46, 1865–1874 (2013)

7. Maier, O., Schröder, C., Forkert, N.D., Martinetz, T., Handels, H.: Classifiers for ischemic stroke lesion segmentation: a comparison study. PLoS One 10(12), e0145118 (2015). https://doi.org/10.1371/journal.pone.0145118

8. Winzeck, S., et al.: ISLES 2016 and 2017-benchmarking ischemic stroke lesion outcome prediction based on multispectral MRI. Front. Neurol. 9 (2018). https://doi.org/10.3389/fneur.2018.00679. ISSN 1664-2295

9. Chen, L.-Ch., Papandreou, G., Kokkinos, I., Murphy, K., Yuille, A.L.: DeepLab: semantic image segmentation with deep convolutional nets, atrous convolution, and fully connected CRFs. In: CoRR (2016)

10. Yu, F.; Koltun, V.: Multi-scale context aggregation by dilated convolutions. In: CoRR (2015)

11. Ronneberger, O., Fischer, P., Brox, T.: U-Net: convolutional networks for biomedical image segmentation. In: Navab, N., Hornegger, J., Wells, W.M., Frangi, A.F. (eds.) MICCAI 2015. LNCS, vol. 9351, pp. 234–241. Springer, Cham (2015). https://doi.org/10.1007/978-3-319-24574-4_28

12. Amorim, P.H.A., et al.: 3D U-Nets for brain tumor segmentation in MICCAI 2017 BraTS challenge. In: 2017 International MICCAI BraTS Challenge (2017)

13. Gordienko, Y., et al.: Deep learning with lung segmentation and bone shadow exclusion techniques for chest x-ray analysis of lung cancer. In: CoRR (2017)

14. Novikov, A.A., et al.: Fully convolutional architectures for multi-class segmentation in chest radiographs. In: CoRR (2017)

15. Tustison, N.J., et al.: N4ITK: improved N3 bias correction. IEEE Trans. Med. Imaging 29(6), 1310–1320 (2010). https://doi.org/10.1109/TMI.2010.2046908

16. Ulyanov, D., Vedaldi, A., Lempitsky, V.S.: Instance normalization: the missing ingredient for fast stylization. In: CoRR (2016)

17. Dice, L.R.: Measures of the amount of ecologic association between species. Ecology. 26(3), 297–302 (1945). https://doi.org/10.2307/1932409

18. Rockafellar, R.T., Wets, R.J.-B.: Variational Analysis, p. 117. Springer, Heidelberg (2005). https://doi.org/10.1007/978-3-642-02431-3. ISBN 3-540-62772-3

19. Olson, D.L., Delen, D.: Advanced Data Mining Techniques, 1st edn, p. 138. Springer, Heidelberg (2008). https://doi.org/10.1007/978-3-540-76917-0. ISBN 3-540-76916-1

20. Taha, A.A., Hanbury, A.: Metrics for evaluating 3D medical image segmentation: analysis, selection, and tool. BMC Med. Imaging 15, 29 (2015). https://doi.org/10.1186/s12880-015-0068-x

Fully Automatic Segmentation for Ischemic Stroke Using CT Perfusion Maps

Vikas Kumar Anand, Mahendra Khened, Varghese Alex,
and Ganapathy Krishnamurthi[✉]

Indian Institute of Technology Madras, Chennai, India
gankrish@iitm.ac.in

Abstract. We propose an algorithm for automatic segmentation of
ischemic lesion using CT perfusion maps. Our method is based on
encoder-decoder fully convolutional neural network approach. The pre-
processing step involves skull stripping and standardization of perfusion
maps and extraction of slices with lesions as the training data. These CT
perfusion maps are used to train the proposed network for automatic segmentation of stroke lesions. The network is trained by minimizing the
weighted combination of cross entropy and dice losses. Our algorithm
achieves 0.43, 0.53 and 0.45 Dice, precision, and recall respectively on
challenge test data set.

Keywords: Deep learning · CNN · CT perfusion · Ischemic stroke

1 Introduction and Related Work

Ischemic stroke is one of the most frequent cerebrovascular disease and vital cause
of death and disability in the world. Defining location and extent of irreversibly
damaged brain tissue is a critical part of the decision-making process in acute
stroke. When compared to normal tissues, the infarcted tissues in the volume
differ on the basis of diffusion & perfusion, hence making diffusion and perfusion
MR imaging the gold standard for ischemic lesions. Recently, due to high speed,
availability and lack of contradiction, Computed Tomography (CT) has been
used to evaluate the order of treatment for ischemic stroke. Though the contrast
of ischemic stroke lesions are poor in brain CT images, CT perfusion imaging
could be exploited to identify & locate the ischemic stroke lesions. Qualitative
images derived from CT perfusion imaging comprises of Cerebral Blood Flow
(CBF), Cerebral Blood Volume (CBV), time to peak of the residue function
(Tmax) and Mean Transit Time (MTT). CBV is defined as the volume of blood
in a given amount of brain tissue. CBF is defined as CBV per unit time. The
ratio of CBV and CBF is expressed as MTT.

The work presented in the paper is an attempt to develop an algorithm for
fully automatic segmentation for ischemic stroke using CT perfusion maps. Several perfusion maps (CBV, CBF, MTT, Tmax) were used as inputs to the algorithm whereas ground truth used was generated from MRI Diffusion Weighted

© Springer Nature Switzerland AG 2019
A. Crimi et al. (Eds.): BrainLes 2018, LNCS 11383, pp. 328–334, 2019.
https://doi.org/10.1007/978-3-030-11723-8_33

Imaging (DWI) of ischemic stroke lesion. On DWI maps, infarcted brain tissues are portrayed as hyper-intense regions.

Convolutional neural networks (CNNs) [7] have been applied to wide variety of pattern recognition tasks, most common ones are image classification [2,6,12] and semantic segmentation using fully convolutional networks (FCN) [8]. CNNs have also been applied to medical image segmentation and classification [1,9]. In this paper, we propose a CNN based architecture for segmentation of the stroke lesion from CT perfusion maps and CT images of the brain. Our network's connectivity pattern was inspired from DenseNets [3].

2   Materials and Methods

2.1   Data Pre-processing Pipeline

Skull Stripping and Brain Mask Generation. Skull stripping is an important pre-processing step in brain image analysis. In this process, non brain tissues like skull, eyes, scalp, dura were removed. This process helped in enhancing the segmentation accuracy and lowering the execution time required by segmentation algorithms. Brain extraction tool (BET) [13] was utilized for removal of skull and another non brain tissues. This was followed by Hounsfield windowing (0–100) of brain tissues so as to improve the contrast.

A brain mask was generated from skull stripped CT images of brain by means of simple thresholding. The brain mask was used to remove non-brain regions from CBF, CBV, MTT and Tmax maps. Different steps in the data pre-processing pipeline are illustrated in Fig. 1.

Data Normalization. Volumes were normalized to have zero mean and unit variance using Eq. 1.

$$X_{norm} = \frac{X - \mu}{\sigma} \tag{1}$$

where X is the data, μ and σ are global mean and global standard deviation associated with X. These were calculated from entire training dataset.

2.2   Proposed Network

We utilize an encoder-decoder architecture for the task of segmenting ischmeic lesion from the CT perfusion maps. The encoder in the network comprise of dense connectivity pattern & was inspired from the DenseNet-121 architecture. The decoder was composed of the bi-linear up-sampling module and convolutional layers. Features learnt in the down-sampling path were concatenated with the features learnt in up-sampling path using long skip connection. The architecture of the network is illustrated in Fig. 2.

The input to the network comprised of 5 channels for feeding CT image and corresponding 4 different CT perfusion maps. The first convolutional layer composed of 64 different 7×7 kernels. The resultant feature maps were then passed

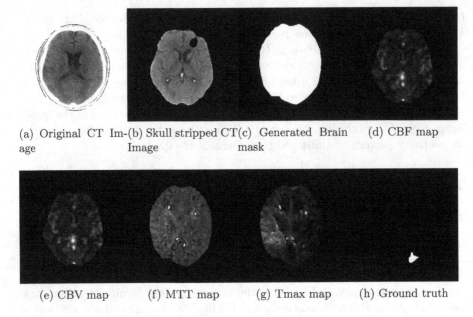

(a) Original CT Im- (b) Skull stripped CT (c) Generated Brain   (d) CBF map
age   Image   mask

(e) CBV map   (f) MTT map   (g) Tmax map   (h) Ground truth

Fig. 1. Data pre-processing steps involved (a) CT image after Hounsfield windowing, (b) Skull stripped image, (c) Mask generated using skull stripped image, (d) Cerebral blood flow information, (e) Cerebral blood volume information, (f) MTT, (g) Tmax and (h) Ground-truth image

to a batch normalization [4] and a non linearity layer (ReLU) [11]. This was followed by a max pooling layer with the kernel size and stride set to 3×3 and 2 respectively. Max-pooling aided in reducing the spatial dimension of the generated features. The features were passed through a series of dense block and

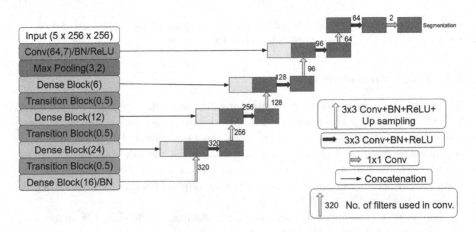

Fig. 2. Proposed network

transition layers. A dense block was composed of convolutional layers, wherein each convolutional layer received inputs from all the preceding layers in the network, Fig. 3. In a dense block, batch normalization and a non-linearity layer preceded a convolutional layer in the network, while a 2-D dropout layer ($p = 0.5$) succeeded the convolutional layer, Fig. 4. In a dense block, the number of convolutional layers was an hyper-parameter. In this work, the network composed of 4 dense blocks and had 6, 12, 24, and 16 convolutional layers respectively. Transition Down block was composed of $[1 \times 1]$ convolutional layer followed by dropout with $p = 0.5$ and $[2 \times 2]$ max pooling layer and aided in reducing the spatial dimension of the features learnt by the network Fig. 5.

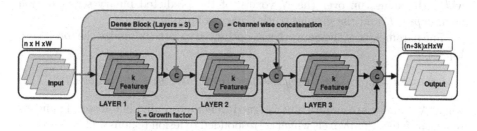

Fig. 3. Dense block with 3 layers

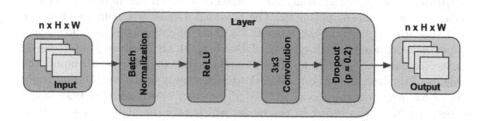

Fig. 4. Layer of dense block

2.3 Loss Function

Lesions are represented by a minuscule proportion of voxels in a medical volume by thereby leading to class imbalance. This issue was circumvented by training the network to minimize a hybrid loss function. The hybrid cost function comprised of weighted cross entropy & dice loss [10].

The dice co-efficient is an overlap metric used for assessing the quality of segmentation maps. The dice coefficient between two binary volumes can be written as:

$$DICE = \frac{2\sum_i^N p_i g_i}{\sum_i^N p_i^2 + \sum_i^N g_i^2} \tag{2}$$

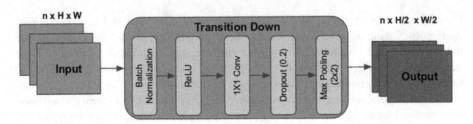

Fig. 5. Transition block

where the sums run over the N voxels, of the predicted binary segmentation volume $p_i \in P$ and the ground truth binary volume $g_i \in G$.

The parameters of the network was optimized so as to minimize the $total\_loss$, Eq. (3).

$$total\_loss = \lambda(cross\_entropy\_loss) + \gamma(dice\_loss\_bg) + \delta(dice\_loss\_fg) \quad (3)$$

where λ, δ and γ are empirically assigned weights to individual losses, fg and bg represent foreground voxels which corresponded to lesion regions and background voxels which corresponded to non-lesion regions respectively. In this work we set $\gamma = 0.25$, $\delta = 0.25$ and $\lambda = 0.50$.

Training. The proposed model was trained on a batch size of 6 for 90 epochs using ADAM [5] as the optimizer. The provided dataset was split into training and validation subsets in the ratio of 7:3. The weights of the encoding layers of our network were initialized using DenseNet-121 architecture pre-trained on ImageNet dataset. Since DenseNet-121 architecture took 3 channel input, the weights of the first layer of our network were randomly initialized. During training of our network, the model was saved on every epoch and the best model selection criteria was based on the model which gave the highest dice score on the validation set.

3   Experimental Setup and Result

3.1   Data Sets and Evaluation Criteria

Imaging data from acute stroke patients in two centers who presented within 8 h of stroke onset and underwent an MRI DWI within 3 h after CT perfusion were included. To assess cerebral perfusion, a contrast agent (CA) was administered to the patient and its temporal change was captured in dynamic scans acquired 1–2 s apart. There were 63 patients in the training data set. Each patient data comprised of brain CT image, CBV map, CBF map, MTT map, Tmax map with ground truth that was generated on DWI map. To access our segmentation algorithm's performance, the organizers employed Dice, precision, recall, Hausdroff distance, Average distance and Absolute volume difference.

3.2   Experimental Results

See Fig. 6 and Table 1.

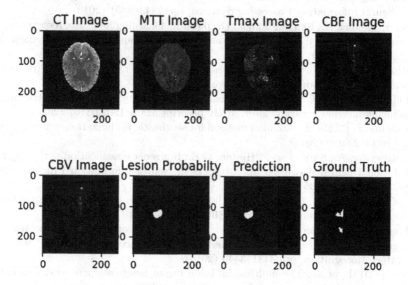

Fig. 6. The figure illustrates our model's prediction on held-out test set.

Table 1. The table provides the list of evaluation metrics used for evaluating our segmentation on the challenge test dataset (n = 62). HD - Hausdorff Distance, AD - Average Distance, AVD - Absolute Volume Difference. The values provided are mean (standard deviation).

Dice	Precision	Recall
0. 43 (0.30)	0.53 (0.38)	0.45 (0.33)
HD	AD	AVD
16129052.53 (36779843.53)	16791858.37 (36853432.68)	13.13 (14.82)

3.3   Conclusion

We developed an automatic segmentation algorithm for ischemic stroke lesions using CT perfusion maps. CT imaging has several advantage such as less scan time, low cost and ease of availability over MRI. The analysis of CT perfusion maps have shown to be useful for early treatment planning on the onset of stroke.

References

1. Ciresan, D., Giusti, A., Gambardella, L.M., Schmidhuber, J.: Deep neural networks segment neuronal membranes in electron microscopy images. In: Advances in Neural Information Processing Systems, pp. 2843–2851 (2012)
2. He, K., Zhang, X., Ren, S., Sun, J.: Deep residual learning for image recognition. In: Proceedings of the IEEE Conference on Computer Vision and Pattern Recognition, pp. 770–778 (2016)
3. Huang, G., Liu, Z., Van Der Maaten, L., Weinberger, K.Q.: Densely connected convolutional networks. In: CVPR, vol. 1, p. 3 (2017)
4. Ioffe, S., Szegedy, C.: Batch normalization: accelerating deep network training by reducing internal covariate shift. arXiv preprint arXiv:1502.03167 (2015)
5. Kingma, D.P., Ba, J.: Adam: a method for stochastic optimization. arXiv preprint arXiv:1412.6980 (2014)
6. Krizhevsky, A., Sutskever, I., Hinton, G.E.: Imagenet classification with deep convolutional neural networks. In: Advances in Neural Information Processing Systems, pp. 1097–1105 (2012)
7. LeCun, Y., Bottou, L., Bengio, Y., Haffner, P.: Gradient-based learning applied to document recognition. Proc. IEEE **86**(11), 2278–2324 (1998)
8. Long, J., Shelhamer, E., Darrell, T.: Fully convolutional networks for semantic segmentation. In: Proceedings of the IEEE Conference on Computer Vision and Pattern Recognition, pp. 3431–3440 (2015)
9. Menze, B.H., et al.: The multimodal brain tumor image segmentation benchmark (BRATS). IEEE Trans. Med. Imaging **34**(10), 1993 (2015)
10. Milletari, F., Navab, N., Ahmadi, S.A.: V-NET: fully convolutional neural networks for volumetric medical image segmentation. In: 2016 Fourth International Conference on 3D Vision (3DV), pp. 565–571. IEEE (2016)
11. Nair, V., Hinton, G.E.: Rectified linear units improve restricted Boltzmann machines. In: Proceedings of the 27th International Conference on Machine Learning (ICML 2010), pp. 807–814 (2010)
12. Simonyan, K., Zisserman, A.: Very deep convolutional networks for large-scale image recognition. arXiv preprint arXiv:1409.1556 (2014)
13. Smith, S.M.: Fast robust automated brain extraction. Hum. Brain Mapp. **17**(3), 143–155 (2002)

Combining Good Old Random Forest and DeepLabv3+ for ISLES 2018 CT-Based Stroke Segmentation

Lasse Böhme[1], Frederic Madesta[1,2], Thilo Sentker[1,2], and René Werner[1,2(✉)]

[1] Department of Computational Neuroscience,
University Medical Center Hamburg-Eppendorf, Hamburg, Germany
lasse.boehme@stud.uke.uni-hamburg.de, r.werner@uke.de
[2] DAISYlabs, Forschungszentrum Medizintechnik Hamburg (fmthh),
Hamburg, Germany

Abstract. Recent years' segmentation challenges on Ischemic Stroke Lesion Segmentation (ISLES) attracted great interest in the medical image computing domain, reflected in >80 citations of the 2017 summary article of the initial ISLES 2015 challenge [1]. While 2015–2017 ISLES challenges focussed on MRI images, the 2018 challenge takes into account clinical relevance of (perfusion) CT to triage stroke patients. Thus, from a methodological point of view, it is now to be analyzed whether and to what extent the 2015–2017 methods can be adapted to automated core lesion segmentation using acute stroke CT perfusion imaging.

We strive to deliver a baseline for ISLES 2018 by using two well established machine learning-based segmentation approaches already applied for the initial ISLES 2015 challenge: random forest (RF) with classical hand-crafted image features (i.e. the most frequently used type of algorithm in ISLES 2015) and encoder-decoder-style convolutional neuronal networks (CNNs). In detail, for CNN-based segmentation, we employ the DeepLabv3+ architecture. The performance of the individual as well as a combination of the segmentation approaches is evaluated based on the ISLES 2018 training data set, and respective results are presented. Aiming at an ISLES 2018-specific performance baseline, we do neither make use of additional data other than the provided challenge data nor perform extensive data augmentation. The results highlight the potential to improve stroke lesion segmentation accuracy by combining RF and CNN information.

Keywords: ISLES challenge · Stroke segmentation · Random forest · Convolutional neural networks (CNN)

1 Introduction

Ischemic stroke is one of the most prevalent causes for death and disability worldwide [2]. It is caused by an occlusion of the cerebral arteries, which results

A. Crimi et al. (Eds.): BrainLes 2018, LNCS 11383, pp. 335–342, 2019.
https://doi.org/10.1007/978-3-030-11723-8_34

in hypoxia and finally in death of the affected brain tissue. Brain imaging plays a crucial part for diagnosis of ischemic stroke and respective therapeutic decisions. Detection and evaluation of stroke lesions require, however, a significant amount of time from the radiologist. To overcome or at least alleviate this problem, reliable methods to automatically segment the lesions are strongly desired.

The Ischemic Stroke Lesion Segmentation (ISLES) [3] strives to provide open benchmark data to objectively evaluate and compare respective algorithms. The first ISLES challenges and evaluation data sets focussed on magnetic resonance imaging (MRI) data [1,4]. Compared to computed tomography (CT), MRI comes with the advantages of better visibility of stroke lesions and the absence of radiation; yet, CT has higher availability in most hospitals [5]. Taking related clinical relevance of (perfusion) CT to triage stroke patients into account, the 2018 ISLES challenge therefore now focuses on methods to segment acute ischemic stroke lesions in native CT images and contrast medium-enhanced perfusion maps. Thus, it is now to be answered whether and to what extent the 2015–2017 methods can be adapted to core lesion segmentation using such data. Therefore, a training dataset including radiologists' ground-truth segmentations and a testing dataset without known ground truth were provided by the challenge organizers. The participants were asked to upload their stroke segmentation results for the testing data set.

In this article, we describe our participation in the ISLES 2018 challenge. The final choice of the applied method(s) was based on the original MRI-based 2015 ISLES challenge [1]. At that time, the majority of participants applied (either as single solution or as part of a pipeline) random forest for segmentation [6–15]. In addition, first convolutional neural network (CNN)-based approaches were also applied [16,17], and it seems to be the case that (at least at the moment) encoder-decoder-style CNNs [18] manifest as quasi-standard for medical image segmentation. We therefore decided to implement both approaches and to report on their performance on the ISLES 2018 training data set – either applied individually or in combination. Being interested in the pure algorithm performance, we did neither make use of additional data other than the provided challenge training data nor performed extensive data augmentation. Algorithmic details are described in depth in succeeding sections.

2   Materials and Methods

2.1   Image Data Description

The ISLES 2018 training data contained 94 individual data sets (subsequently called 'cases') from 63 patients, i.e. image data from patients with separated lesions were split into different cases. The individual cases consisted of only slices around the stroke lesions, with the number of slices per case ranging from 2 to 16. For each case, native CT image data as well as contrast medium enhanced 4D CT image data were provided. From the latter, cerebral blood flow (CBF), cerebral blood volume (CBV), mean transit time (MTT), and time-to-maximum flow (T_{max}) perfusion maps were calculated and also provided.

Ground truth data were binary masks manually drawn by radiologists on the basis of co-registered diffusion weighted imaging (DWI)-MRI data.

The DWI-MRI data were not available for the ISLES 2018 challenge testing data; we therefore did not consider them for training purposes and the experiments described in the present paper. As dealing with the 4D perfusion image series is also challenging from a computational perspective (e.g. GPU memory handling), we did not make use of the respective explicit temporal information; we only computed a corresponding maximum intensity projection (MIP) along the temporal axis as additional image data.

2.2 Image Preprocessing

Image preprocessing consisted of two parts: intensity normalization and fast brain segmentation. Image intensity normalization was applied independently for each case and image sequence by rescaling the original image intensity dynamics to the range [0; 1]. Fast brain segmentation was based on a 3D-connected component (CC) analysis of the CBV images. The largest non-zero CC was considered an initial brain segmentation, which was further refined for the individual axial slices by filtering out small 2D-CCs based on their size and position (CC centroid should be rather posterior than anterior; the latter corresponded, e.g., to close-to-eyes CCs in inferior slices).

2.3 Random Forest (RF) Implementation

For random forest (RF) implementation, we used the Scikit-learn Python library [19], Gini impurity as impurity measure, and 150 trees. For RF feature extraction, we applied the Python medical image processing library MedPy [20]. Features were computed for each case and voxel inside the case-specific brain mask (see Sect. 2.2); class imbalance due to small lesions (compared to healthy brain tissue) was not accounted for.

As voxel features, we used the image intensity as well as Gaussian-weighted local mean intensity values (standard deviations $\sigma = 3, 7, 13$). These intensity features were computed for each of the rescaled image sequences (i.e. native CT, CBF, CBV, MTT, and T_{max}), resulting in 20 features. Despite the fact that the MedPy implementation to derive hemispheric difference values may lead to partially erroneous values due to the images not being normalized/registered to symmetric atlas space, we also computed voxel-wise hemispheric difference values for the different image sequences, yielding additional five features. Thirdly, we extracted a 10-bin local histogram (5×5 neighborhood) from the T_{max} and used the ten frequency values as additional features. Fourth and finally, we computed the distance of the voxel to the brain mask boundary and the distance of the voxel to the image center. Thus, in total, we extracted 37 features for each voxel.

For classification of test cases, i.e. application of the trained RF, case-specific RF probability maps were generated based on the voxel-wise probability of belonging to the lesion class. The probability map was rescaled to the range

[0; 1] in such a way that the 99th percentile (and values above) of the probability map values corresponded to a value of 1. For lesion segmentation, the resulting map was thresholded by a value of 0.5. Given the resulting binary map, we excluded all connected areas with size smaller than 70 voxels to avoid false positives related to noise or image artifacts. Finally, a hole-filling algorithm was applied for each axial slice.

2.4   CNN Architecture and Workflow

Different CNN architectures were proposed in recent years for semantic image segmentation via deep learning. In this study, we decided to employ the currently popular DeepLabv3+ network that is designed to learn multi-scale contextual image features and controlling signal decimation [21]. For this purpose, it utilizes three key concepts:

Feature extraction. Feature extraction is conducted by a ResNet, pre-trained on the ImageNet data set, which employs residual learning to allow for efficient training of deeper neural networks [22].

Atrous convolutions. Efficient computation of feature responses by adding larger context information without increasing the number of parameters, i.e. sparse convolution kernels, is achieved using atrous convolutional layers.

Atrous spatial pyramid pooling. In order to combine the multi-scale information yielded by different atrous convolution layers, the atrous spatial pyramid pooling is applied.

For the problem defined by the ISLES 2018 challenge, the original DeepLabv3+ structure had to be slightly modified. To convert the network output into a binary segmentation, a convolutional filter was randomly chosen from the 21 pre-trained filters regarding the last convolutional layer. Further, we had to change the number of input color channels from 3 to 6 as we concatenated the given native CT, CBF, CBV, MTT, T_{max} and the computed MIP image data, yielding a tensor for every case and slice of shape $n_x \times n_y \times 6$. To allow for a robust and efficient training, 256 patches of shape $64 \times 64 \times 6$ of each (n_x, n_y)-slice were randomly selected and used as input data set. Based on this data, the model was re-trained in 5 epochs.

2.5   Combining RF and CNN Results

To analyze potential segmentation improvement by combining RF and CNN outputs, we averaged the RF probability map after rescaling by the 99th percentile and the CNN probability map. Thresholding and post-processing of the binary maps remained unchanged compared to previous explanations.

2.6   Experiments

To evaluate the implemented stroke segmentation approaches, we performed a 5-fold cross validation based on the underlying patient collective of the ISLES

Table 1. RF-only, CNN-only and results (mean ± standard deviation over cases) after combining respective probability maps for the individual folds and corresponding average data. HD and ASSD values refers multiples of the in-plane pixel side length.

Approach	Fold	Dice	HD	ASSD
RF-only	Fold 1	0.42 ± 0.25	50.59 ± 29.89	7.15 ± 5.33
	Fold 2	0.43 ± 0.26	34.98 ± 21.07	6.18 ± 6.52
	Fold 3	0.31 ± 0.24	56.25 ± 23.82	11.18 ± 8.57
	Fold 4	0.52 ± 0.23	41.47 ± 30.69	6.04 ± 6.73
	Fold 5	0.47 ± 0.24	55.53 ± 32.57	7.12 ± 7.23
	Avg	**0.42 ± 0.26**	**47.33 ± 28.94**	**7.52 ± 7.29**
CNN-only	Fold 1	0.51 ± 0.16	50.52 ± 27.77	6.58 ± 5.51
	Fold 2	0.41 ± 0.26	36.42 ± 24.93	5.70 ± 6.09
	Fold 3	0.31 ± 0.29	37.03 ± 23.99	6.23 ± 8.82
	Fold 4	0.49 ± 0.29	47.20 ± 29.68	8.07 ± 10.22
	Fold 5	0.48 ± 0.25	44.57 ± 33.40	6.62 ± 5.85
	Avg	**0.44 ± 0.27**	**43.30 ± 28.82**	**6.60 ± 7.46**
Combination	Fold 1	0.51 ± 0.21	44.63 ± 28.02	5.73 ± 5.45
	Fold 2	0.49 ± 0.27	31.45 ± 19.47	6.02 ± 7.82
	Fold 3	0.39 ± 0.26	48.21 ± 25.53	10.09 ± 10.34
	Fold 4	0.58 ± 0.22	39.76 ± 31.23	5.21 ± 6.05
	Fold 5	0.54 ± 0.24	42.83 ± 36.36	6.05 ± 6.93
	Avg	**0.50 ± 0.26**	**41.12 ± 28.85**	**6.69 ± 7.86**

training dataset. Splitting the 63 patients instead of the 94 cases ensured that the same patient was not part of both the training and testing data, and thereby prevented a related bias and overestimation of the segmentation performance.

The same folds were used for RF-only training and evaluation, CNN-only training and evaluation, and evaluation of the combination of RF and CNN output. Following previous ISLES contributions, we focused our evaluation on three metrics: Dice coefficient, Hausdorff distance (HD), and average symmetric surface distance (ASSD).

3   Results

The results for RF-only, CNN-only, and combined RF and CNN stroke lesion segmentation are summarized in Table 1. RF-only segmentation resulted in a mean Dice coefficient of 0.42 ± 0.26 (averaged over all testing cases and folds). The average CNN-only Dice coefficient was slightly higher (0.44 ± 0.27); however, this was consistently observed for all folds. In contrast, the Dice coefficient after combining RF and CNN outputs was for all folds at least as high as the higher single-approach value, indicating the potential of improving segmentation

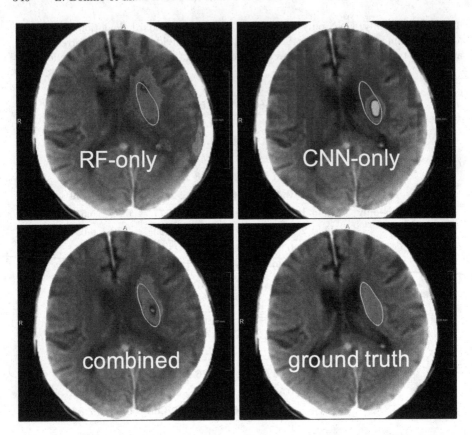

Fig. 1. Illustration of complementary information by RF-only and CNN-only stroke lesion segmentation, and the potential by combining respective outputs. All images are shown for case 40 and the fifth fold. Top row, left: probability map (blue: low probability; ref: max. values) and ground truth segmentation contours for RF-only (Dice 0.44); top row, right: same information for CNN-only (Dice 0.31). Bottom row, left: combined probability map (Dice 0.66); bottom row, right: ground truth segmentation. (Color figure online)

quality by combining the two approaches. Similar observations hold true for the Hausdorff distance values. ASSD was, in turn, on average lower for CNN-only than for RF-only and the combined approach. An illustration of the complementary information of the individual approaches can be found in Fig. 1.

4   Discussion and Conclusion

We applied and evaluated a classical random forest classifier employing hand-crafted image features commonly used in the context of stroke lesion segmentation [1], a state-of-the-art CNN architecture for image segmentation, and a combination of the two algorithms. Given the still relatively small amount of training data especially for deep learning purposes, the underlying hypothesis was that the two methodically different approaches extract complimentary information and that, as a consequence thereof, the performance of the joint approach leads to an improved segmentation performance. The presented results (especially in terms of the Dice coefficient) support the hypothesis. Further improvement of segmentation performance could, for instance, be reached by using additional image data during training, exploiting data augmentation strategies and extensive use of additional methods (i.e. consideration of larger ensembles); nevertheless, the presented baseline performance is considered promising.

Acknowledgements. Supported by Forschungszentrum Medizintechnik Hamburg, grant 02fmthh2017. We further thank NVIDIA for donating the applied Titan Xp GPU.

References

1. Maier, O., Menze, B.H., von der Gablentz, J., Hani, L., Heinrich, M.P., et al.: ISLES 2015 - a public evaluation benchmark for ischemic stroke lesion segmentation from multispectral MRI. Med. Image Anal. **35**, 250–269 (2017)
2. Thrift, A.G., Thayabaranathan, T., Howard, G., Howard, V.J., Rothwell, P.M., et al.: Global stroke statistics. Int. J. Stroke **12**(121), 13–32 (2017)
3. ISLES challenge(s). http://www.isles-challenge.org
4. Winzeck, S., Hakim, A., McKinley, R., Pinto, J., Alves, V., et al.: ISLES 2016 and 2017-benchmarking ischemic stroke lesion outcome prediction based on multispectral MRI. Front. Neurol. **9**, 679 (2018)
5. Tatlisumak, T.: Is CT or MRI the method of choice for imaging patients with acute stroke? Why should men divide if fate has united? Stroke **33**, 2144–2145 (2002)
6. Wang, C.-W., Lee, J.-H.: Stroke lesion segmentation of 3D brain MRI using multiple random forests and 3D registration. In: Proceedings of MICCAI-ISLES 2015, pp. 35–38 (2015)
7. Chen, L., Bentley, P., Rueckert, D.: A novel framework for sub-acute stroke lesion segmentation based on random forest. In: Proceedings of MICCAI-ISLES 2015, pp. 9–12 (2015)
8. Maier, O., Wilms, M., Handels, H.: Random forests with selected features for stroke lesion segmentation. In: Proceedings of MICCAI-ISLES 2015, pp. 17–22 (2015)
9. Reza, S.M.S., Pei, L., Iftekharuddin, K.M.: Ischemic stroke lesion segmentation using local gradient and texture features. In: Proceedings of MICCAI-ISLES 2015, pp. 23–26 (2015)
10. Robben, D., Christiaens, D., Rangarajan, J.R., Gelderblom, J., Joris, P., et al.: ISLES challenge 2015: a voxel-wise, cascaded classification approach to stroke lesion segmentation. In: Proceedings of MICCAI-ISLES 2015, pp. 27–30 (2015)
11. Halme, H.-L., Korvenoja, A., Salli, E.: ISLES (SISS) challenge 2015: segmentation of stroke lesions using spatial normalization, random forest classification and contextual clustering. In: Proceedings of MICCAI-ISLES 2015, pp. 31–34 (2015)
12. Goetz, M., Weber, C., Maier-Hein, K.: Input data adaptive learning (IDAL) for sub-acute ischemic stroke lesion segmentation. In: Proceedings of MICCAI-ISLES 2015, pp. 39–42 (2015)

13. Mahmood, Q., Basit, A.: Automatic ischemic stroke lesion segmentation in multi-spectral MRI images using random forests classifier. In: Proceedings of MICCAI-ISLES 2015, pp. 43–46 (2015)

14. Jesson, A., Arbel, T.: Hierarchical segmentation of normal and lesional structures combining an ensemble of probabilistic local classifiers and regional random forest classification. In: Proceedings of MICCAI-ISLES 2015, pp. 57–62 (2015)

15. Muschelli, J.: Prediction of ischemic lesions using local image properties and random forests. In: Proceedings of MICCAI-ISLES 2015, pp. 63–65 (2015)

16. Kamnitsas, K., Chen, L., Ledig, C., Rueckert, D., Glocker, B.: Multi-scale 3D convolutional neural networks for lesion segmentation in brain MRI. In: Proceedings of MICCAI-ISLES 2015, pp. 13–16 (2015)

17. Dutil, F., Havaei, M., Pal, C., Larochelle, H., Jodoin, P.-M.: A convolutional neural network approach to brain lesion segmentation. In: Proceedings of MICCAI-ISLES 2015, pp. 51–56 (2015)

18. Ronneberger, O., Fischer, P., Brox, T.: U-Net: convolutional networks for biomedical image segmentation. In: Navab, N., Hornegger, J., Wells, W.M., Frangi, A.F. (eds.) MICCAI 2015. LNCS, vol. 9351, pp. 234–241. Springer, Cham (2015). https://doi.org/10.1007/978-3-319-24574-4_28

19. Pedregosa, F., Varoquaux, G., Gramfort, A., Michel, V., Thirion, B., et al.: Scikit-learn: machine learning in python. J. Mach. Learn. Res. **12**, 2825–2830 (2011)

20. Maier, O.: MedPy – medical image processing in Python. https://pypi.org/project/MedPy

21. Chen, L.-C., Zhu, Y., Papandreou, G., Schroff, F., Adam, H.: Encoder-decoder with atrous separable convolution for semantic image segmentation. In: Ferrari, V., Hebert, M., Sminchisescu, C., Weiss, Y. (eds.) ECCV 2018. LNCS, vol. 11211, pp. 833–851. Springer, Cham (2018). https://doi.org/10.1007/978-3-030-01234-2_49

22. He, K., Zhang, X., Ren, S., Sun, J.: Deep residual learning for image recognition. In: Proceedings of the IEEE Conference on Computer Vision and Pattern Recognition, pp. 770–778 (2016)

Volumetric Adversarial Training for Ischemic Stroke Lesion Segmentation

Hao-Yu Yang[1,2(✉)]

[1] Cura Cloud Cooperation, Seattle, WA 98104, USA
haoyuy@curacloudcorp.com
[2] Yale University, New Haven, CT 06511, USA

Abstract. Ischemic stroke is one of the most common and yet deadly cerebrovascular diseases. Identifying lesion area is an essential step for stroke management and outcome assessment. Currently, manual delineation is the gold standard for clinical diagnosis. However, inter-annotator variances and labor-intensive nature of manual labeling can lead to observer bias or potential disagreement of between annotators. While incorporating a computer-aided diagnosis system may alleviate these issues, other challenges such as highly varying shapes and difficult boundaries in the lesion area make the designing of such system non-trivial. To address these issues, we propose a novel adversarial training paradigm for segmenting ischemic stroke lesion. The training procedure involves the main segmentation network and an auxiliary critique network. The segmentation network is a 3D residual U-net that produces a segmentation mask in each training iteration while critique network enforces high-level constraints on the segmentation network to produce predictions that mimic the ground truth distribution. We applied the proposed model on the 2018 ISLES stroke lesion segmentation challenge dataset and achieved competitive results on the training dataset.

Keywords: 3D convolution neural networks · Adversarial training · Ischemic Stroke Lesion Segmentation

1 Introduction

Stroke is one of the leading cause of death in developed countries. The disease is caused by either blockage (ischemic stroke) or rupture of a blood vessel (hemorrhagic stroke). Among the two types of stroke, ischemic stroke takes up roughly 80% [1]. The prevailing imaging modalities for diagnosing brain strokes are magnetic resonance imaging (MRI) and computed tomography (CT). Different MRI sequences such as T1 weighted, T2 weighted, Diffusion Weighted Imaging (DWI) and Fluid Attenuated Inversion Recovery (FLAIR) are utilized for specialized applications. DWIs are especially suitable for ischemic strokes since it is highly sensitive to lesion changes [8].

Segmentation of brain area affected by ischemic stroke lesion plays a crucial role in treatment assessment and prognosis. Producing accurate predictions is

© Springer Nature Switzerland AG 2019
A. Crimi et al. (Eds.): BrainLes 2018, LNCS 11383, pp. 343–351, 2019.
https://doi.org/10.1007/978-3-030-11723-8_35

challenging due to the variability in the shapes and sizes of the targets. Recent studies [2] have shown that perfusion Computed Tomography (CT) shows potential improvement in speed, availability and lack of contraindications compared to MRI. Computer-aided diagnosis (CAD) system using perfusion CT may help clinicians with faster and more accurate diagnosis. In previous works, models such as random forests, support vector machines and autoencoders [9] have been employed to segment ischemic stroke lesion and have shown successful results.

Computer vision tasks such as image recognition, detection, and segmentation have had significant advances in the past few years due to the rise of deep learning, specifically in Convolution Neural Networks (CNN). Medical applications of deep learning have also seen profound successes. As neural networks tend to get deeper as we harness more computational power, the problem of vanishing gradients problem ensues. Vanishing gradients occur when gradients become too small to change the weights of the neuron in back-propagation-trained neural networks. The residual learning networks [4] (ResNet) solves this problem by introducing stacked identity mappings in the form of residual blocks. These residual connections allow the neural network to collapse into a few layers during initialization and gradually expand in the feature space as training takes place. Recently, generative adversarial networks (GAN) [3] have been utilized extensively throughout image generation tasks. Recent studies [6,10] have shown that GANs can also be used in a critique framework for semantic segmentation tasks. The benefits of using such networks include comparing the higher level of inconsistencies between ground truth and predictions and enforcing spatial continuity. In this framework, generating pixel-wise segmentation masks are modeled as a generative procedure and the discriminator of the model attempts to distinguish between real and fake segmentation masks.

In this paper, we've developed a neural-network with adversarial training to segment irreversibly damaged brain area caused by ischemic stroke. The proposed model is trained and validated on the Ischemic Stroke Lesion Segmentation(ISLES) challenge dataset [7]. The ISLES challenge aims at providing a unified platform and high-quality data for training and evaluating models for automatic stroke lesion segmentation. In order to model the variability in the true distribution and improve prediction accuracy, adversarial training. For preprocessing, each modality is normalized and stacked as multi-channel inputs. The overall loss function consists of three terms: negative dice coefficient and binary cross-entropy between the ground truth mask and prediction plus the discriminator loss between real and generated segmentation masks. Our method produced promising results and achieved an average DICE coefficient of 0.87 on the ISLES training dataset.

2   Method

The detailed model architecture and training procedure of the proposed methods are described in this section. First, we address the necessary steps for preprocessing the data. Then we introduce the architecture of the segmentation network.

Finally, we illustrate two adversarial paradigms proposed for training the segmentation model.

2.1   Data

We performed training and validation on the 2018 ISLES challenge dataset. The training dataset contained a total of 63 patients each with 5 different perfusion maps: cerebral blood flow (CBF), Mean transit time (MTT), cerebral blood volume (CBV), time to peak of residue function (TMAX) and computed tomography angiography (CTP). An example of the training data can be found in Fig. 1. The training data also included gold standard diffusion-weighted imaging (DWI) maps that are not available in testing data. The ground truth segmentation masks were derived from the DWI. The data provided are in Neuroimaging Informatics Technology Initiative (NIfTI) format. We used Insight Segmentation and Registration Toolkit (ITK) [12] for data inspection and visualization.

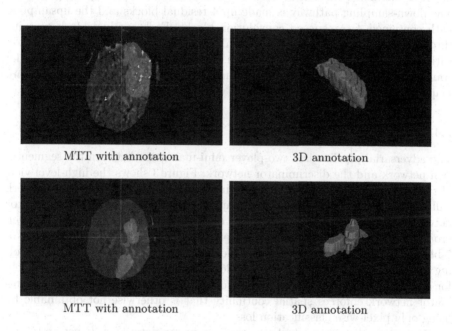

MTT with annotation   3D annotation

MTT with annotation   3D annotation

Fig. 1. Example of training data and corresponding 3-D annotation

2.2   Preprocess

Preprocessing is necessary due to the significant cross-modality variance. There are also substantial deviation in the spatial resolution as dimension of the z-axis ranges from 2 to 16 for different subjects. First, we conducted bicubic spline interpolation [5] to resize each volume to the same dimension. During training and testing, each modality is then normalized respectively by subtracting the mean intensity and divide by the standard deviation as shown in the following equation:

$$x'_m(i,j,k) = \frac{x_m(i,j,k) - \mu_m}{\sigma_m} \tag{1}$$

Where m denotes the modality, μ_m denotes mean intensity and σ_m denotes the standard deviation. i, j, k is the coordinates of the pixel to be normalized. Finally, the normalized whole volume are stacked as multi-channel inputs for the segmentation network.

2.3   3D Residual U-Net

The backbone of the segmentation network is a 3D U-net with residual connection [11]. Network structure and details of the residual block of the can be found in Fig. 2 The U-net consists of both down-sampling and up-sampling pathways. The down-sampling pathway is made up 4 residual blocks and the upsampling path contained 4 transposed convolution blocks. Each residual block contains three $3 \times 3 \times 3$ convolution layer, batch normalization and activation function with leaky reciftied liner unit in between. The up-sampling pathway contained 4 transpose convolution operation and concatenation with corresponding feature maps from the down-sampling pathway.

2.4   Adversarial Training

The adversarial pipeline is a two-player mini-max game between the segmentation network and the discriminator network. Figure 3 shows the high-level view of the training procedure. In each training iteration, the segmentation network will generate a pixel-wise probability map which is then fed to the discriminator network as inputs. The objective of the discriminator is to distinguish between ground truth segmentation mask and predicted mask. The discriminator is a 7-block network containing 3 residual blocks similar to the 3D U-net. Maxpooling was conducted after every residual block. The discriminator network is solely for auxiliary purposes and therefore removed during testing phase. The discriminator network enforces spatial continuity that is otherwise not obtainable by using only pixel-wise classification loss.

We denote the ground truth mask as y, image data as x, U-net as U and discriminator as D, the solution to the mini-max game can, therefore, be written as:

$$\min_U \max_D E_{y \sim p(y)}[\log D(y)] + E_{x \sim p(x)}[\log(1 - D(U(x)))] \tag{2}$$

There are different ways that the adversarial training can be carried out. We proposed two training paradigm for the adversarial pipeline, namely:

- Integrated loss
- Second back propagation

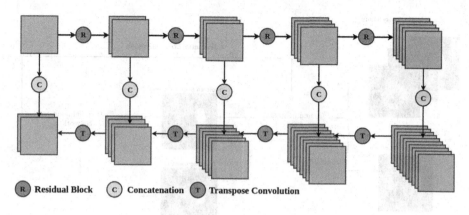

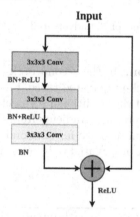

Fig. 2. Top figure shows the architecture of the 3-D residual U-net. Bottom figure shows a single residual block

Integrated Loss. The integrated loss paradigm adds the adversarial loss to the traditional segmentation losses and forms a integrated loss term. Back-propagation are carried out based on the gradients of the integrated loss term. The discriminator network are back-propagated by the errors of not recognizing true label and misclassifying synthetic label as true. The integrated loss function for the segmentation network contained a total of three terms: binary cross entropy loss, negative dice score and adversarial loss as seen in the following equation:

$$\mathcal{L}_{total} = \alpha \mathcal{L}_{adver} + \beta \mathcal{L}_{BCE} + \gamma \mathcal{L}_{dice} \tag{3}$$

Where the α, β, γ are coefficients for each loss terms. We initialized all three coefficients as 1. The coefficients are adjusted by weight decay mechanism which we describe in implementation details section. Detailed algorithm can be found in Algorithm 1.

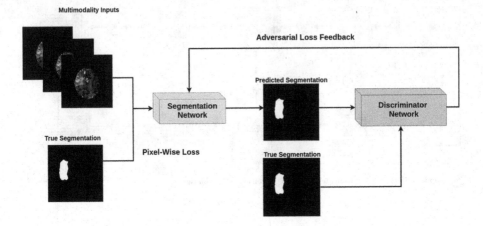

Fig. 3. Overview of the adversarial training paradigm

Second Back-Propagation. In the second back-propagation pardigram, the segmentation network is back-propagated twice. First, the weights are adjusted according to the gradients of the traditional segmentation loss. At the adversarial training phases of each iteration, gradients from the adversarial loss are then passed onto the segmentation network for a second back-propagation. The discrimination network is back-propagated only once. Detailed training algorithm can be found in Algorithm 2.

Algorithm 1. Training procedure with integrated loss function

---

Require: Initialize parameters of U-net θ_U, Discriminator θ_D
Data: Sample images $x_1 \ldots x_n$ and corresponding masks $y_1 \ldots y_n$ from data
Initialize parameters of U-net θ_U, Discriminator θ_D
while *While θ not converged* **do**
    for *Each x_i and y_i in mini-batch* **do**
        Compute mask $\hat{y_i}$ by $U(\hat{y_i}|x_i)$
        Compute $\mathcal{L}_{Dice}(\hat{y_i}, y_i)$, $\mathcal{L}_{BCE}(\hat{y_i}, y_i)$
        Compute $\mathcal{L}_{adver}$ by $\log(1 - D(\hat{y_i}))$
        Compute $\mathcal{L}_{total}$ by $\mathcal{L}_{adver} + \lambda\mathcal{L}_{BCE} + \mathcal{L}_{dice}$
        $\theta_D \leftarrow \nabla[\log D(y_i) + \log(1 - D(\hat{y_i}))]$
        $\theta_U \leftarrow \nabla\mathcal{L}_{total}$
    end
end

Algorithm 2. Training procedure with second back-propagation

Require: Initialize parameters of U-net θ_U, Discriminator θ_D
Data: Sample images $x_1 \ldots x_n$ and corresponding masks $y_1 \ldots y_n$ from data
while *While θ not converged* **do**
    for *Each x_i and y_i in mini-batch* **do**
        Compute mask $\hat{y}_i$ by $U(\hat{y}_i|x_i)$
        Compute $\mathcal{L}_{Dice}(\hat{y}_i, y_i)$
        $\theta_U \leftarrow \nabla \mathcal{L}_{Dice}$
    end
    for *Each x_i and y_i in mini-batch* **do**
        Compute mask $\hat{y}_i$ by $U(\hat{y}_i|x_i)$
        $\theta_D \leftarrow \nabla[\log D(y_i) + \log(1 - D(\hat{y}_i)]$
        $\theta_U \leftarrow \nabla \log(1 - D(\hat{y}_i))$
    end
end

2.5 Implementation Detail

The proposed model was established with python under the pytorch deep learning framework. The learning rate was set differently for the segmentation network and the critique network to avoid collapsing in early epochs, which is a common phenomenon in GANs. Learning rates were initialized at 0.001 for the segmentation model and 0.0005 for the discriminator network. Learning rate decay will take place if there were no improvements of the loss function 5 consecutive epochs. Each learning rate decay reduces the learning rate to 80% of the previous iteration. Early termination will take place if no improvements were seen for 20 consecutive epochs. The mini-batch size was set at 8. GPU training was conducted on 4 NVIDIA Tesla V100. The total training time for the entire pipeline that included segmentation network and discrimination network was approximately 24 h.

3 Results

In this section, we present quantitative results of the proposed model and qualitative comparison of the adversarial training effects. Several matrices including the mean of Dice score, the standard deviation of Dice score, mean of Hausdorff distance and standard deviation of Hausdorff distance were used for model evaluation. Figure 4 is a visualization of the adversarial training effects. As shown in the figure, models with adversarial training are able to capture subtle differences between ground truth and predictions. Table 1 shows that by incorporating adversarial training, dice score increased and Hausdorff distance reduced.

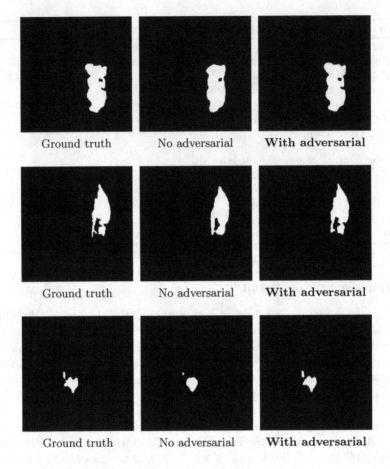

Fig. 4. Effects of adversarial training

Table 1. Dice and Hausdorff distance comparison between three training paradigm

	DICE mean	DICE std	Hausdorff mean	Hausdorff std
Without adversarial	0.78	0.19	27.45	**31.23**
With second back-propagation	0.85	**0.16**	20.78	33.81
With integrated adversarial loss	**0.87**	0.21	**18.89**	37.32

4   Discussion

In this paper, we've presented an automatic ischemic stroke lesion segmentation model using multiple CT perfusion maps with varying dimensions as inputs. We proposed two adversarial training paradigm, namely integrated loss function and second back-propagation. We've demonstrated that by incorporating a discriminator network in the training procedure, the segmentation model is able to mimic subtle inconsistencies between ground truth and prediction that cannot be corrected using only pixel-wise loss functions such as binary cross entropy and dice score. Quantitatively, employing adversarial training increases dice score and reduces Hausdorff distance.

References

1. Feigin, V.L., Lawes, C.M., Bennett, D.A., Anderson, C.S.: Stroke epidemiology: a review of population-based studies of incidence, prevalence, and case-fatality in the late 20th century. Lancet Neurol. **2**(1), 43–53 (2003). https://doi.org/10. 1016/S1474-4422(03)00266-7. http://www.sciencedirect.com/science/article/pii/ S1474442203002667

2. Gillebert, C.R., Humphreys, G.W., Mantini, D.: Automated delineation of stroke lesions using brain ct images. NeuroImage: Clin. **4**, 540–548 (2014). https:// doi.org/10.1016/j.nicl.2014.03.009. http://www.sciencedirect.com/science/article/ article/pii/S2213158214000394

3. Goodfellow, I.J., et al.: Generative Adversarial Networks. arXiv e-prints, June 2014

4. He, K., Zhang, X., Ren, S., Sun, J.: Deep residual learning for image recognition. In: 2016 IEEE Conference on Computer Vision and Pattern Recognition (CVPR), pp. 770–778, June 2016. https://doi.org/10.1109/CVPR.2016.90

5. Keys, R.: Cubic convolution interpolation for digital image processing. IEEE Trans. Acoust. Speech Signal Process. **29**(6), 1153–1160 (1981). https://doi.org/10.1109/ TASSP.1981.1163711

6. Luc, P., Couprie, C., Chintala, S., Verbeek, J.: Semantic segmentation using adversarial networks. CoRR abs/1611.08408 (2016). http://arxiv.org/abs/1611.08408

7. Maier, O., et al.: ISLES 2015 - a public evaluation benchmark for ischemic stroke lesion segmentation from multispectral MRI. Med. Image Anal. **35**, 250–269 (2017). https://doi.org/10.1016/j.media.2016.07.009. http://www.sciencedirect.com/sci ence/article/pii/S1361841516301268

8. Moseley, M.E., et al.: Early detection of regional cerebral ischemia in cats: comparison of diffusion- and T2-weighted MRI and spectroscopy. Magn. Reson. Med. **14**(2), 330–346 (1990)

9. Praveen, G., Agrawal, A., Sundaram, P., Sardesai, S.: Ischemic stroke lesion segmentation using stacked sparse autoencoder. Comput. Biol. Med. **99**, 38–52 (2018). https://doi.org/10.1016/j.compbiomed.2018.05.027. http://www.sciencedirect. com/science/article/pii/S0010482518301409

10. Quan, T.M., Hildebrand, D.G.C., Jeong, W.: FusionNet: a deep fully residual convolutional neural network for image segmentation in connectomics. CoRR abs/1612.05360 (2016). http://arxiv.org/abs/1612.05360

11. Ronneberger, O., Fischer, P., Brox, T.: U-net: convolutional networks for biomedical image segmentation. CoRR abs/1505.04597 (2015). http://arxiv.org/abs/1505. 04597

12. Yushkevich, P.A., et al.: User-guided 3D active contour segmentation of anatomical structures: significantly improved efficiency and reliability. Neuroimage **31**(3), 1116–1128 (2006)

Ischemic Stroke Lesion Segmentation in CT Perfusion Scans Using Pyramid Pooling and Focal Loss

S. Mazdak Abulnaga[1,2]([✉]) and Jonathan Rubin[2]

[1] Computer Science and Artificial Intelligence Lab, MIT, Cambridge, MA, USA
abulnaga@mit.edu
[2] Philips Research North America, Cambridge, MA, USA
jonathan.rubin@philips.com

Abstract. We present a fully convolutional neural network for segmenting ischemic stroke lesions in CT perfusion images for the ISLES 2018
challenge. Treatment of stroke is time sensitive and current standards
for lesion identification require manual segmentation, a time consuming and challenging process. Automatic segmentation methods present
the possibility of accurately identifying lesions and improving treatment
planning. Our model is based on the PSPNet, a network architecture
that makes use of pyramid pooling to provide global and local contextual information. To learn the varying shapes of the lesions, we train our
network using focal loss, a loss function designed for the network to focus
on learning the more difficult samples. We compare our model to networks trained using the U-Net and V-Net architectures. Our approach
demonstrates effective performance in lesion segmentation and ranked
among the top performers at the challenge conclusion.

1 Introduction

We present a model for segmenting stroke lesions in CT perfusion (CTP) data
for the 2018 ischemic stroke lesion segmentation (ISLES) challenge. Ischemic
stroke is caused by an obstruction of blood supply to the brain. Treatment of
stoke is time sensitive, requiring tissue reperfusion within less than 4–6 h of stroke
onset [17]. Current standards for evaluating stroke requires manual segmentation
in MRI or CT images [1,9,17], a challenging and time consuming task, due to the
changing appearance of lesions over time and their presence in various locations
in the brain [17,25]. There is a growing need for automatic segmentation methods
to accurately identify lesions and to help plan effective treatment.

The 2018 ISLES challenge is the first to use CTP data. Currently, MR with
diffusion-weighted imaging (DWI) is considered the most accurate and earliest
at detecting acute stroke [1,9]. CTP however is advantageous in cost, speed,
and availability in acute care units [9]. Furthermore, CTP is emerging as an
effective means to detect the infarct (irreversible) core with high sensitivity and
specificity [1,9]. Detection relies on quantitative evaluation metrics derived from

© Springer Nature Switzerland AG 2019
A. Crimi et al. (Eds.): BrainLes 2018, LNCS 11383, pp. 352–363, 2019.
https://doi.org/10.1007/978-3-030-11723-8_36

the CTP data. For example, a drop in cerebral blood flow (CBF) is a result of an occlusion of blood supply [3]. In this work, we use the CT image, CBF, cerebral blood volume (CBV), time to peak (TTP) and mean transit time (MTT) of the contrast agent injection as signals to identify the infarct core (Fig. 1).

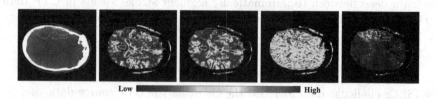

Fig. 1. Example CTP data from one subject in the study, with stroke lesion segmentation overlaid in orange. From left to right: CT, CBV, CBF, MTT, Tmax images. Each image had different units of measure, so we demonstrate for visualization only. (Color figure online)

2   Related Work

Deep learning approaches that utilize fully convolutional neural network (CNN) architectures [18, 22] have become the de facto standard for semantic segmentation tasks in 2D and 3D medical imaging [15]. The ISLES challenge was established to fairly compare approaches in stroke lesion segmentation and characterization [17, 25], resulting in the development of effective CNN models for this task [4, 10, 13]. All previous challenges focused on multispectral MRI data [17, 25]. Many early approaches focused on analyzing patches, in part due to memory issues. A top performer of the 2015 challenge developed a 3D patch-based CNN architecture that used two parallel pathways, allowing the network to process patches at different scales [13, 17]. More recent efforts in the 2016 and 2017 challenges investigated extensions to dual pathway 3D networks [21] and ensembles of multi-scale networks [6]. In the 2017 challenge, the authors of [6] investigated an additional network based on *pyramid scene parsing* [27]. The models we present in this work also make use of pyramid pooling as in [27], though we focus on CTP data. There have been few works exploring the automatic segmentation of stroke lesions using CT data [20], and to the best of our knowledge, none have made use of deep neural networks. Many previous methods have relied on histogram-based classifiers [20], or make use of statistical comparisons for lesion detection [9].

Many other medical image segmentation tasks have benefited from using deep learning-based classifiers, for example in pancreas segmentation using CT [19], prostate segmentation using MRI [18], and multi-organ segmentation in whole-body CT [2]. Finally, it is also worth mentioning that public datasets and challenges, such as The Pascal Visual Object Classes Challenge (VOC) [8], have resulted in significant improvements in natural image semantic segmentation using CNNs. Advances in natural images have also informed medical image segmentation tasks. In this work, we utilize a modified 2D fully convolutional architecture that was pre-trained on natural images from the PASCAL VOC dataset. We build upon models that have demonstrated effectiveness in both medical and natural image segmentation tasks. Furthermore, we are one of the first to develop a deep network to automatically segment stroke lesions in CTP data, and demonstrate strong performance compared to other challenge participants.

3   Dataset

The ISLES challenge data included the CT scan, the CTP source data, and the CBF, CBV, MTT, and Tmax derived perfusion maps, though we did not use the CTP source data. Images were acquired within 8 h of stroke onset. An MRI DWI was then acquired within 3 h after the CTP scan. The infarct core lesions were manually drawn using the corresponding MRI DWI scans. CTP scans were acquired as slabs covering sparse areas (5 mm axial spacing) with stroke lesion in the brain. As a result, the scans had varying depth in the axial dimension, ranging from 2 to 22 slices. Each slice was a 256×256 image. Furthermore, some patients had two non-overlapping or partially-overlapping slabs covering regions within the brain. The training set contained 63 subjects and 94 scans, and the test set contained 40 subjects with 62 scans.

4   Methods

Fully convolutional neural network architectures were trained to predict ischemic stroke lesion masks. We constructed both 2D and 3D CNN models, but found stronger performance in 2D per-slice models given the variable and limited number of axial slices in the scans. The input to the network was a multi-channel 2D image created by stacking a CT slice together with its four corresponding perfusion map slices (Tmax, CBF, CBV, MTT). Cross entropy and focal loss [14] were evaluated as loss functions. We developed models based on the pyramid scene parsing network (PSPNet), [27], the U-Net (2D and 3D) [7,22], and the 3D V-Net [18] architectures. Our final model is based on the PSPNet with focal loss. The PSPNet employs *pyramid pooling* (explained below) within a fully convolutional neural network.

4.1   Data Augmentation

Data augmentation was used to artificially increase the size of our limited training set. We augmented the images sagitally and coronally to reflect likely variations in appearance of the brain and stroke lesions. The augmentation would randomly rotate the images by $[-10°, 10°]$, translate by $[-10\%, 10\%]$ of the image size, flip, and scale by a factor of $[0.9, 1.1]$. The sampling was done uniformly and the order of these operations was chosen randomly.

4.2   Pyramid Scene Parsing Network

The pyramid scene parsing architecture was chosen as it achieves state-of-the-art performance on segmentation tasks in natural images. In particular, it achieved first place at the 2016 ImageNet scene parsing challenge [23].

PSPNet combines a ResNet-based [11] fully convolutional neural network architecture [16] together with dilated convolutions [5, 26]. Further, the PSPNet introduces a pyramid pooling module that performs region-based context aggregation. The pyramid pooling module is designed to capture global information about an input image from different regions of a network's receptive field and at various scales. To do so, pooling kernels of varying sizes and strides are applied to a network's final feature map layer. We adopt the same dimensions of the four level pyramid pooling module as described in [27].

Pyramid Pooling Module. Figure 2 illustrates the module graphically. Consider the last convolutional layer of a network, L_{final} that consists of n_{out} feature maps, $F_{final} \in \mathbb{R}^{n_{out} \times w \times h}$. At the coarsest level, global average pooling (represented by circular arrows in Fig. 2) is applied to F_{final} resulting in $n_{out} \times 1 \times 1$ feature maps. Further average pooling operations are also applied that result in $n_{out} \times 2 \times 2$, $n_{out} \times 3 \times 3$ and $n_{out} \times 6 \times 6$ feature map sizes. The final features of the pyramid pooling module are derived by applying a 1×1 convolution to each of the resultant feature maps (to ensure equal weighting for each pooling kernel) and upsampling (using bilinear interpolation) to match the dimensions of the final layer feature maps, $F_{psp} \in \mathbb{R}^{n_{psp} \times w \times h}$. The original final layer feature maps are then concatenated to those derived from the pyramid pooling module $(F_{final} \oplus F_{psp})$ to give a collection of feature maps that capture both local and global context information at varying sub-regions of the input image.

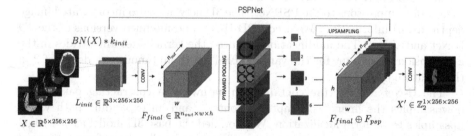

Fig. 2. Architecture of a fully convolutional neural network with a pyramid pooling module for segmenting ischemic stroke lesions. Circular arrows represent average pooling operations. Input to the model is a stack of 256×256 multi-modal CT perfusion maps. The model outputs a $\mathbb{Z}_2 = \{0, 1\}$, single channel 256×256 prediction mask.

4.3   Transfer Learning

We used a pre-trained PSPNet that was trained on natural images from the Pascal VOC dataset [8]. As the original network architecture accepted 3 input channels for processing RGB images, the network was modified to include an additional *initializer* layer that could accept multi-modal CT perfusion slices. Given a collection of stacked CT perfusion maps, $X \in \mathbb{R}^{5 \times 256 \times 256}$ the initializer layer, L_{init}, first applies batch normalization [12] to standardize channel features to a common mean and variance within the batch. Following this, a 1×1 convolutional kernel, $k_{init} \in \mathbb{R}^{5 \times 3 \times 1 \times 1}$ is learned to reduce the channel dimension from 5 to 3. These steps are summarized in Eq. (1), where $BN(\cdot)$ refers to batch normalization and $*$ refers to the convolution operation.

$$L_{init} = BN(X) * k_{init}. \tag{1}$$

The resulting feature maps in layer $L_{init} \in \mathbb{R}^{3 \times 256 \times 256}$, are ready to be processed using the pre-trained PSPNet weights. Correspondingly, the final layer of the network was modified to replace the 21 class prediction channels, used in Pascal VOC, with binary output channels to predict the presence or absence of ischemic stroke lesions.

Initial fine-tuning took place for all new layers introduced into the network architecture, where pre-trained weights were frozen and the weights in the newly introduced layers were updated with a learning rate set to 10^{-2}. Following this, all weights in the network were unfrozen and the network continued training with a learning rate of 10^{-4}.

4.4   U-Net and 3D Networks

We also developed classification models based on the U-Net [7,22] and V-Net [18] architectures, both commonly used in medical image segmentation. These served as natural comparisons to the PSPNet. The 3D networks also incorporated image depth (axial) information. We used a 2D U-Net and modified versions of the 3D U-Net and V-Net. The modifications reduced the kernel sizes and strides in the z dimension to account for the varying axial depth and minimum depth of 2 in the input images.

The U-Net architecture contains two pathways, a contracting path which downsamples the image and captures context, and an expanding path which upsamples to perform localization. We modified the base 3D model as described in [7]. In the contracting layer, we reduced the max pooling layer from a size and stride of 2^3 to $(2 \times 2 \times 1)$, down sampling the image in the x, y dimensions by a factor of 2 but leaving the z dimension unchanged. Similarly, in the expanding pathway, we modified the up sampling operation (a transposed convolution) to have a kernel size and stride of $(2 \times 2 \times 1)$. For the 2D network, we use the base model as described in [22]. For both models, we modify the input layer for the 5-channel images.

The V-Net model contains a similar contraction and expansion pathway. We modified input to the the base architecture [18] to have a kernel size of 3^3 with unit-padding as the original 5^3 kernel is too large for our images. Similarly, we modified the convolution and de-convolution (transposed convolution) layers to have kernels with size and stride $(2 \times 2 \times 1)$ instead of 2^3.

4.5   Loss Function

We trained the networks using the cross entropy or focal loss [14] functions. Given the true image label for pixel i, $y_i \in \{0, 1\}$, and a predicted class membership probability $p_i \in [0, 1]$, the cross entropy loss is formulated as

$$CE(p, y) = -y \log(p) - (1 - y) \log(1 - p), \tag{2}$$

and the total loss $\mathcal{L}_{CE}$ is summed over all N pixels,

$$\mathcal{L}_{CE} = \frac{1}{N} \sum_{i=1}^{N} CE(p_i, y_i). \tag{3}$$

Since our labels are imbalanced, we used a weighted version of the cross entropy loss,

$$WCE(p, y) = -wy \log(p) - (1 - w)(1 - y) \log(1 - p), \tag{4}$$

where w is the empirical measure of lesions in the training dataset, $w \in [0, 1]$.

The focal loss [14] was introduced as an extension to the cross entropy loss, designed to focus the training on hard to classify examples, by down-weighting easily classified examples, i.e. those with high class membership probability. It is formulated as

$$FL(p, y) = -y(1 - p)^\gamma \log(p) - (1 - y)p^\gamma \log(1 - p), \tag{5}$$

where γ is the focusing parameter that governs the down-weighting of the easily classified examples. Note that for $\gamma = 0$, the focal loss is the same as the cross entropy loss. With increasing values of γ, the loss function is smaller for larger values of p. Additionally, the function approaches 0 for smaller values of p, allowing the network to focus on the less-confidently classified examples.

4.6   Implementation Details

RMSProp [24] was used as the optimization routine. The dice coefficient on the validation set was monitored for improvement after every training epoch. If no improvement was observed for 20 epochs, the learning rate was reduced by a factor of 10. A patience flag was set at 50 epochs and if no improvement in the validation dice metric was observed after 50 epochs, early stopping was invoked. For the U-Net and V-Net, the networks were trained from scratch for 200 epochs. The batch size was set to 8 for the 2D networks, and set to 1, using a full image, for the 3D. All models were trained using the PyTorch library on a single Nvidia Titan Xp GPU.

5   Experiments and Results

We conducted experiments to determine the optimal network architecture and loss function. The dataset was split into 5 folds. Per-subject folds were created, ensuring no overlap of subjects between folds, i.e. subjects with multiple scans existed only within the same fold. We created 5 separate models per architecture, each validated on a distinct fold. We trained different networks using focal loss with $\gamma = 1$, and weighted and unweighted cross entropy loss. We evaluated the networks using the Dice Similarity Coefficient (DSC). Given a predicted image label X and the ground truth image label Y, the DSC is defined as

$$DSC(X,Y) = 2\frac{|X \cap Y|}{|X| + |Y|}, \tag{6}$$

where $|X|$ denotes the cardinality of binary image X.

5.1   5-Fold Cross Validation Results

The model parameters for each network were selected based on the best validation dice score. We performed a 5-fold cross-validation to determine model performance. The results are shown on Fig. 3. The pre-trained PSPNet with focal loss demonstrated the strongest 5-fold cross-validation results ($DSC = 0.54$). The 2D U-Net and PSPNet trained from scratch had a similar performance, so the pretraining using our additional input layer improved performance by approximately 7 dice points. Additionally, the focal loss improved the pre-trained network substantially. Table 1 shows the per fold DSC results for the pre-trained PSPNet, trained using focal loss and cross entropy. From the table, it can be seen that some validation folds are more challenging than others, leading to varied DSC scores. Overall, usage of focal loss led to an improved overall DSC (0.54 ± 0.09) compared to cross entropy (0.49 ± 0.11). We hypothesize this is due to the fact that the pretraining helps classify the obvious stroke lesion examples, but the focal loss forces the network to learn the more difficult samples.

Table 1. DSC results per cross validation fold for the PSPNet (pre-trained). Focal loss and cross entropy loss are compared.

Fold	Focal loss	Cross entropy loss
1	0.64	0.64
2	0.42	0.37
3	0.48	0.50
4	0.55	0.54
5	0.58	0.41
Total	**0.54 ± 0.09**	**0.49 ± 0.11**

The 3D networks performed poorly. We observed that their increased number of parameters resulted in more overfitting. Additionally, they were unable to take full advantage of the third image dimension, due to the large number of scans with only 2 axial slices. For the two best models, the 2D U-Net and pre-trained PSPNet, we observe the focal loss improved model performance. We demonstrate in Fig. 4 that the focal loss predicted more fine details in the lesions that were missed by cross entropy in the pre-trained PSPNet. The cross entropy loss network often over-predicted larger lesions than the focal loss network, and the focal loss network was able to more closely predict the fine appearance features of the lesions, and predict areas that cross entropy completely missed.

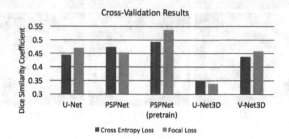

Fig. 3. 5-fold cross validation results on each network architecture, using cross entropy loss (blue) or focal loss (red). The pre-trained PSPNet with focal loss demonstrates the strongest results, with a dice score of 0.54. (Color figure online)

5.2 ISLES 2018 Challenge Results

The challenge evaluated the test data set using the DSC, the Hausdorff Distance (HD), the Average Symmetric Surface Distance (ASSD), precision, recall, and the absolute volume difference (AVD) [17]. The DSC is defined in (6). The HD measures the maximum distance between the two surfaces X_s and Y_s,

$$HD(X_s, Y_s) = \max \left\{ \max_{x \in X_s} \min_{y \in Y_s} d(x, y), \max_{y \in Y_s} \min_{x \in X_s} d(y_s, x_s) \right\}, \quad (7)$$

where $d(\cdot, \cdot)$ is the euclidean distance measure. The ASSD is defined in terms of the average surface distance (ASD),

$$ASD(X_s, Y_s) = \frac{\sum_{x \in X_s} \min_{y \in Y_s} d(x, y)}{|X_s|}, \quad (8)$$

and $ASSD = \frac{1}{2} (ASD(X_s, Y_s) + ASD(Y_s, X_s))$.

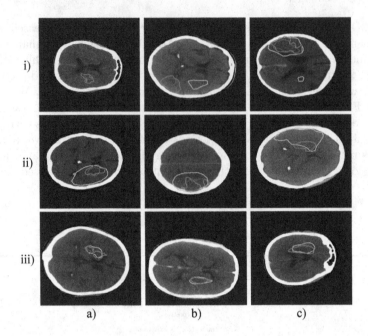

Fig. 4. Example predicted segmentations on 9 subjects using the pre-trained PSPNet. The ground truth is shown in orange, the network trained with focal loss in red, and the network trained with cross entropy in cyan. In row (i), the focal loss network is able to identify difficult lesions and better match the shape of the lesions than the cross entropy network. In the second row, we observe 3 cases where the cross entropy network over-predicts the lesions. Finally, in the the third row, the focal loss network demonstrates closer shape matching than the cross entropy counterpart. (Color figure online)

Our final submission to the ISLES challenge was an ensemble of ten models that included all five PSPNet (pre-trained) models trained with focal loss, combined with a further five PSPNet (pre-trained) models trained using cross entropy loss. The ensemble achieved a final 5-fold cross validation score of $DSC = 0.57$ on the training data leaderboard and $DSC = 0.44$ on the testing data leaderboard. The full results of our final model evaluated on the ISLES test set is shown in Table 2. The ISLES challenge uses a weighted ranking based on the DSC and Hausdorff Distance to rank submissions. We compare our approach to the performance of the top ranking submission for each metric. Out of a total of 38 submissions to the challenge leaderboard, our approach ranked 6th on DSC, HD and ASSD metrics. Our approach also achieved the second best score on the AVD metric.

Table 2. Results of the proposed model compared to the top scores from the ISLES leaderboard, accessed October 2018. Arrows in the header indicate whether lower or higher values are better. *Values normalized by 1,0000,000.

	DSC ↑	Hausdorff distance ↓	ASSD ↓	Precision ↑	Recall ↑	AVD ↓
Ours	0.44	1.62*	1.62*	0.59	0.43	10.18
Best	0.51	0.97*	0.97*	0.62	0.58	10.08
Place	$6^{th}/38$	$6^{th}/38$	$6^{th}/38$	$3^{rd}/38$	$18^{th}/38$	$2^{nd}/38$

6 Conclusion

In this work, we developed fully convolutional neural network models for segmenting ischemic stroke lesions using CTP data. Our model made use of the focal loss function, which demonstrated the ability to identify more fine features in the lesions by focusing on hard to classify examples. We compared models used commonly in medical image segmentation, namely the U-Net and V-Net, with the PSPNet, which was developed for natural image segmentation.

In future work, we plan to further investigate the role of generative adversarial networks (GANs) in medical image semantic segmentation. In particular GANs can potentially be used to create additional synthetic data for model training. Furthermore, inclusion of a generative loss component within the training procedure could also be investigated. Alternative future work, will also focus on bridging the gap between 2D and 3D CNN models. In general, 3D models did not perform well on the ISLES 2018 dataset, given the limited number of axial slices. Utilizing pre-trained 2D models to better initialize 3D models may hold some potential for improving 3D model performance.

References

1. Biesbroek, J., et al.: Diagnostic accuracy of CT perfusion imaging for detecting acute ischemic stroke: a systematic review and meta-analysis. Cerebrovasc. Dis. **35**(6), 493–501 (2013)
2. Brosch, T., Saalbach, A.: Foveal fully convolutional nets for multi-organ segmentation. In: Medical Imaging 2018, Image Processing, vol. 10574, p. 105740U. International Society for Optics and Photonics (2018)
3. Campbell, B.C., et al.: Cerebral blood flow is the optimal CT perfusion parameter for assessing infarct core. Stroke **42**(12), 3435–3440 (2011)
4. Chen, L., Bentley, P., Rueckert, D.: Fully automatic acute ischemic lesion segmentation in DWI using convolutional neural networks. NeuroImage: Clin. **15**, 633–643 (2017)
5. Chen, L.C., Papandreou, G., Kokkinos, I., Murphy, K., Yuille, A.L.: Semantic image segmentation with deep convolutional nets and fully connected CRFs. arXiv preprint arXiv:1412.7062 (2014)

6. Choi, Y., Kwon, Y., Lee, H., Kim, B.J., Paik, M.C., Won, J.H.: Ensemble of deep convolutional neural networks for prognosis of ischemic stroke. In: Crimi, A., Menze, B., Maier, O., Reyes, M., Winzeck, S., Handels, H. (eds.) BrainLes 2016. LNCS, pp. 231–243. Springer, Cham (2016). https://doi.org/10.1007/978-3-319-55524-9_22

7. Çiçek, Ö., Abdulkadir, A., Lienkamp, S.S., Brox, T., Ronneberger, O.: 3D U-Net: learning dense volumetric segmentation from sparse annotation. In: Ourselin, S., Joskowicz, L., Sabuncu, M.R., Unal, G., Wells, W. (eds.) MICCAI 2016. LNCS, vol. 9901, pp. 424–432. Springer, Cham (2016). https://doi.org/10.1007/978-3-319-46723-8_49

8. Everingham, M., Van Gool, L., Williams, C.K., Winn, J., Zisserman, A.: The pascal visual object classes (VOC) challenge. Int. J. Comput. Vis. **88**(2), 303–338 (2010)

9. Gillebert, C.R., Humphreys, G.W., Mantini, D.: Automated delineation of stroke lesions using brain CT images. NeuroImage: Clin. **4**, 540–548 (2014)

10. Guerrero, R., et al.: White matter hyperintensity and stroke lesion segmentation and differentiation using convolutional neural networks. NeuroImage: Clin. **17**, 918–934 (2018)

11. He, K., Zhang, X., Ren, S., Sun, J.: Deep residual learning for image recognition. In: Proceedings of the IEEE Conference on Computer Vision and Pattern Recognition, pp. 770–778 (2016)

12. Ioffe, S., Szegedy, C.: Batch normalization: accelerating deep network training by reducing internal covariate shift. arXiv preprint arXiv:1502.03167 (2015)

13. Kamnitsas, K., et al.: Efficient multi-scale 3D CNN with fully connected CRF for accurate brain lesion segmentation. Med. Image Anal. **36**, 61–78 (2017)

14. Lin, T.Y., Goyal, P., Girshick, R., He, K., Dollár, P.: Focal loss for dense object detection. IEEE Trans. Pattern Anal. Mach. Intell. (2018)

15. Litjens, G., et al.: A survey on deep learning in medical image analysis. Med. Image Anal. **42**, 60–88 (2017)

16. Long, J., Shelhamer, E., Darrell, T.: Fully convolutional networks for semantic segmentation. In: Proceedings of the IEEE Conference on Computer Vision and Pattern Recognition, pp. 3431–3440 (2015)

17. Maier, O., et al.: ISLES 2015-A public evaluation benchmark for ischemic stroke lesion segmentation from multispectral MRI. Med. Image Anal. **35**, 250–269 (2017)

18. Milletari, F., Navab, N., Ahmadi, S.A.: V-net: fully convolutional neural networks for volumetric medical image segmentation. In: 2016 Fourth International Conference on 3D Vision (3DV), pp. 565–571. IEEE (2016)

19. Oktay, O., et al.: Attention U-Net: learning where to look for the pancreas. arXiv preprint arXiv:1804.03999 (2018)

20. Rekik, I., Allassonniére, S., Carpenter, T.K., Wardlaw, J.M.: Medical image analysis methods in MR/CT-imaged acute-subacute ischemic stroke lesion: segmentation, prediction and insights into dynamic evolution simulation models. A critical appraisal. NeuroImage: Clin. **1**(1), 164–178 (2012)

21. Robben, D., Suetens, P.: Dual-scale fully convolutional neural network for final infarct prediction (2017)

22. Ronneberger, O., Fischer, P., Brox, T.: U-Net: convolutional networks for biomedical image segmentation. In: Navab, N., Hornegger, J., Wells, W.M., Frangi, A.F. (eds.) MICCAI 2015. LNCS, vol. 9351, pp. 234–241. Springer, Cham (2015). https://doi.org/10.1007/978-3-319-24574-4_28

23. Russakovsky, O., et al.: Imagenet large scale visual recognition challenge. Int. J. Comput. Vis. **115**(3), 211–252 (2015)

24. Tieleman, T., Hinton, G.: Lecture 6.5–RMSProp: divide the gradient by a running average of its recent magnitude. COURSERA Neural Netw. Mach. Learn. **4**, 26–31 (2012)
25. Winzeck, S., et al.: Isles 2016 and 2017-benchmarking ischemic stroke lesion outcome prediction based on multispectral MRI. Front. Neurol. **9** (2018)
26. Yu, F., Koltun, V.: Multi-scale context aggregation by dilated convolutions. arXiv preprint arXiv:1511.07122 (2015)
27. Zhao, H., Shi, J., Qi, X., Wang, X., Jia, J.: Pyramid scene parsing network. In: IEEE Conference on Computer Vision and Pattern Recognition (CVPR), pp. 2881–2890 (2017)

Grand Challenge on MR Brain Segmentation

MixNet: Multi-modality Mix Network for Brain Segmentation

Long Chen⬤ and Dorit Merhof[✉]⬤

Institute of Imaging and Computer Vision, RWTH Aachen University,
Aachen, Germany
{long.chen,dorit.merhof}@lfb.rwth-aachen.de
https://www.lfb.rwth-aachen.de/

Abstract. Automated brain structure segmentation is important to many clinical quantitative analysis and diagnoses. In this work, we introduce MixNet, a 2D semantic-wise deep convolutional neural network to segment brain structure in multi-modality MRI images. The network is composed of our modified deep residual learning units. In the unit, we replace the traditional convolution layer with the dilated convolutional layer, which avoids the use of pooling layers and deconvolutional layers, reducing the number of network parameters. Final predictions are made by aggregating information from multiple scales and modalities. A pyramid pooling module is used to capture spatial information of the anatomical structures at the output end. In addition, we test three architectures (MixNetv1, MixNetv2 and MixNetv3) which fuse the modalities differently to see the effect on the results. Our network achieves the state-of-the-art performance. MixNetv2 was submitted to the MRBrainS challenge at MICCAI 2018 and won the 3rd place in the 3-label task. On the MRBrainS2018 dataset, which includes subjects with a variety of pathologies, the overall DSC (Dice Coefficient) of 84.7% (gray matter), 87.3% (white matter) and 83.4% (cerebrospinal fluid) were obtained with only 7 subjects as training data.

Keywords: Brain segmentation · CNN · Multi-modality

1 Introduction

Accurate automated segmentation of brain structures, e.g., white matter (WM), gray matter (GM), and the cerebrospinal fluid (CSF) forms the basis for high-throughput quantitative analyses and associated diagnoses. while computed tomography (CT) and positron emission tomography (PET) is also used for brain structure analysis, magnetic resonance imaging (MRI) is the most popular choice [1]. We will only talk about MRI in this work.

As the deep learning approaches are becoming mature, they gradually outperforms previous methods [2–5]. Based on the network architecture, these deep learning approaches can be roughly divided into two categories: the patch-wise [6–8] and semantic-wise [9] architecture. The patch-wise approach takes

© Springer Nature Switzerland AG 2019
A. Crimi et al. (Eds.): BrainLes 2018, LNCS 11383, pp. 367–377, 2019.
https://doi.org/10.1007/978-3-030-11723-8_37

a local patch around a pixel as input. Most of the current works use this strategy, because of its efficiency of using the training dataset. Compared to the semantic-wise approach, the patch-wise approach can extract large number of patches from the MRI subjects for training. But unlike unstructured segmentation, brain structures preserve same relative positions in all subjects and patch-wise approaches ignores that information. Some works like [8] make up for this by augmenting the network input with coordinates of voxels, but semantic-wise methods still have advantages in nature.

In addition to the overall architecture, we can also use input dimensions to distinguish between different methods. The 3D networks leverage the spatial information more efficiently than 2D networks by sharing kernels across three dimensions. The cost is longer runtime and limited network size. As discussed in Sect. 2.3, the 2D network can observe the 3D MRI volume from different directions, that is, more 2D slices as training data. This strategy does not only provides more training images but also plays the role of a ensemble model. By fusing the results obtained from 2D slices along different orientations the segmentation should be more robust and spatially consistent as well.

We propose a 2D semantic-wise CNN to handle the brain structure segmentation problem in Sect. 2. Three structures are tested to see the effect of different ways of mixing multiple modalities. We call them MixNetv1, MixNetv2 and MixNetv3 in Sect. 2.2. The experiments are performed with the MICCAI challenge MRBrainS2018 dataset. The dataset contains annotated multi-sequence (T1-weighted, T1-weighted inversion recovery and T2-FLAIR) scans of 30 subjects. Seven of them are distributed as training data, while the rest subjects are kept unreleased for test. For a limited training dataset, the transfer learning [10] usually boosts the overall segmentation results. But this is achieved by using extra data implicitly. Our experiment works with only 7 subjects of the MRBrainS2018 training dataset.

The code developed for this work and trained models will be available online: https://github.com/looooongChen/MRBrainS-Brain-Segmentation.

2 Method

In Sect. 2.1 we introduce the residual dilated convolution unit. Except the initial convolution layer and the output module, MixNet is composed of residual dilated convolution units connected in series or parallel. Section 2.2 discusses different ways of using multi-modalities. Section 2.3 describes the method of acquiring more 2D training slices from the 3D MRI volume.

2.1 Basic Units of the Nets

As shown in Figs. 4, 5 and 6, the networks are composed of three types of basic units: the InitUnit (Fig. 1), the DilateResUnit (Fig. 2) and the OutputUnit (Fig. 3). In this section, we will described them in detail.

Initial Unit (InitUnit). The InitUnit consists of a single 5×5 convolutional layer and an optional pooling layer. Depending on the input channels, the convolution kernels can be of different sizes. In Fig. 4, three modalities are stacked together, while mixNetv2 (Fig. 5) and mixNetv3 (Fig. 6) have three input streams. Thus, the kernel sizes are $5 \times 5 \times 3$ and $5 \times 5 \times 1$, respectively. In addition, the pooling layer aims to reduce memory usage when necessary. If the pooling layer in the InitUnit is used, the upscaling layer in the OutputUnit should also be activated. In this work, we use a 2×2 pooling with stride 2.

Residual Dilated Convolution Unit (DilateResUnit). The training difficulty varies in different network architectures. For example, the degradation phenomenon arises in practice for a deeper plain CNN, although it includes the solution space of a shallower one. [11] conjecture that the deep plain CNN may have exponentially slow convergence rates and provides empirical evidence showing that a network composed of residual units is easier to optimize. The proceeding work [12] argues that the training procedure benefits from a "direct" path for information propagation, not only within a residual unit but through the whole network. Inspired by the successful works [11,12], we construct a deep residual learning network (DilateResUnit) with 'clear' paths through the layers and multiple modality streams for information propagation.

Fig. 1. InitUnit

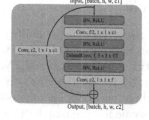

Fig. 2. DilateResUnit

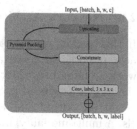

Fig. 3. OutputUnit

As shown in Fig. 2, the shortcut lets the input feature map pass through the unit directly and only the differences between inputs and outputs are learned. When such units are connected to form a network, these short paths will also be interlinked throughout the network. Compared to the residual unit in [12], the second convolutional layer is replaced by a dilated convolutional layer. Alternating convolutional layers and polling layers are a CNN common structure. The Pooling layer increases the receptive field efficiently while keeping the computational workload reasonable. However, the pooling layer loses localization information which is critical for segmentation tasks. Deconvolutions [13] and dilated convolutions (also known as atrous convolution) [14] are possible solutions. Different from the deconvolution where extra layers are involved to recover lost resolution, the dilated convolution keeps the resolution unchanged through the forward propagation. Extra layers mean more parameters. Assuming a network with less parameters is easier to train, we adopt the dilated convolution in this work.

A DilateResUnit is determined by four parameters: $c1, c2, f$ and d. The number of filters and the filter size of the dilated convolutional layer is $f/2$ and f, while d is the dilation factor. The first and last 1×1 convolutional layers are determined by the channels of the input and output feature map. When the inputs and outputs are of different sizes, a 1×1 convolutional layer will be inserted on the shortcut. Since we use the same f through the network, only the units before and after a concatenation in Fig. 5 (except the final concatenation) have such shortcut convolutions.

Output Unit (OutputUnit). As discussed in Sect. 1, anatomical structures preserve certain relative positions. Thus, the OutputUnit augments the input feature map with a global prior first, and then outputs results through a 3×3 convolutional layer. The global prior is captured by a pyramid pooling module [15]. The pyramid pooling module separates the input feature map into subregions and forms representation by average pooling. Then, bilinear interpolation is performed to get the same size as the original feature map. In this work, we use a four-level pyramid with 2×2, 4×4, 6×6, 12×12 bins respectively.

Finally, the upscaling is performed to recover the original resolution, only when the pooling layer in the InitUnit is used. If the network can fit into the memory, pooling and upscaling are not necessary.

2.2 Network Architecture

In this section, we discuss three styles of using multiple modalities: stacked channels, periodic summarization and parallel streams. Correspondingly, three network architectures (MixNetv1, MixNetv2 and MixNetv3) are constructed with the units introduced in Sect. 2.1 to test the effect on the results.

At the output end, all three networks aggregate features form different levels. A multi-modality, mutli-scale feature map is then passed to the OutputUnit. which augments the feature map with a global prior and makes the final prediction. Detailed network parameters are listed in Table 1.

To train the network, we compute the cross-entropy loss of each pixel in an image and accumulate them as the training loss. In this work, all pixels are treated equally, ingoring the label imbalance. The training process can run streadily in this way, but labels of a relatively small number may not receive enough attention. Weighing pixels of different labels is an approach worth trying.

Stacked Channels (MixNetv1). A straightforward way to fuse multiple modalities is to stack them as different channels. Thus, the input of MixNetv1 is a batch of 3-channel images. The forward propagation path is composed of serially connected DilateResUnits. Since the output of a DilateResUnit has a similar resolution with the input, we set the the filter number of all units to the same. In this way, the feature map size and the corresponding computation are balanced throughout different layers.

Periodic Summarization (MixNetv2). MixNetv2 is a network architecture between MixNetv1 and MixNetv3. MixNetv1 fuses the multiple modalities at the very beginning, while MixNetv3 keeps different modality streams independent until the final output. In MixNetv2, periodic summarization of multi-modality information is performed. As shown in Fig. 5, Level 1, Level 3 and Level 5 play such a role. The summarization is then fed back to each modality stream.

Parallel Streams (MixNetv3). Three modality streams propagate forward independently in MixNetv3. Features from three streams are only collected when the OutputUnit makes the final prediction. Actually, the solution space of MixNetv3 is contained in MixNetv1. Each neuron in MixNetv1 has connection to all three modalities (indirect connections considered). If we force each neuron to connect to only one modality by setting some network parameters to 0, MixNetv1 can be equivalent to MixNetv3. However, MixNetv3 performs better than MixNetv1 based on our experiments. Experiment results are demonstrated in Sect. 3.

Table 1. Parameters of three MixNet versions. The input channel $c1$, filter number f, dilation factor d and output channel $c2$ of the DilateResUnit are listed with respect to the network level. As described in Sect. 2.1, the DilatedResUnit is fully determined by these four parameters.

MixNetv1	Level 1	Level 2	Level 3	Level 4	Level 5
Input	$120 \times 120 \times 72$	$120 \times 120 \times 72$	$120 \times 120 \times 72$	$120 \times 120 \times 72$	$120 \times 120 \times 72$
Filters	72	72	72	72	72
Dilation	2	1	4	1	8
Output	$120 \times 120 \times 72$	$120 \times 120 \times 72$	$120 \times 120 \times 72$	$120 \times 120 \times 72$	$120 \times 120 \times 72$
MixNetv2	Level 1	Level 2	Level 3	Level 4	Level 5
Input	$120 \times 120 \times 72$	$120 \times 120 \times 48$	$120 \times 120 \times 72$	$120 \times 120 \times 48$	$120 \times 120 \times 72$
Filters	24	24	24	24	24
Dilation	2	1	4	1	8
Output	$120 \times 120 \times 24$	$120 \times 120 \times 24$	$120 \times 120 \times 24$	$120 \times 120 \times 24$	$120 \times 120 \times 24$
MixNetv3	Level 1	Level 2	Level 3	Level 4	Level 5
Input	$120 \times 120 \times 24$	$120 \times 120 \times 24$	$120 \times 120 \times 24$	$120 \times 120 \times 24$	$120 \times 120 \times 24$
Filters	24	24	24	24	24
Dilation	2	1	4	1	8
Output	$120 \times 120 \times 24$	$120 \times 120 \times 24$	$120 \times 120 \times 24$	$120 \times 120 \times 24$	$120 \times 120 \times 24$

2.3 View MRI Volume from Different Directions

For a 2D CNN, the 3D MRI volume can be observed from any direction. The most commonly used are the three anatomical planes: the sagittal plane, the

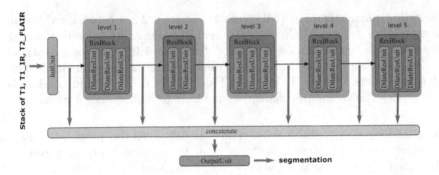

Fig. 4. MixNetv1: multiple modalities are stacked at the very beginning.

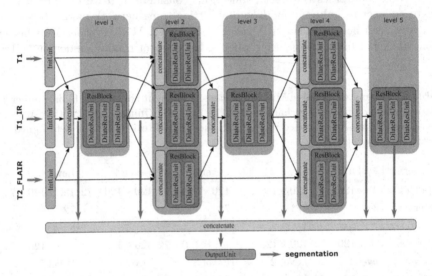

Fig. 5. MixNetv2: summarization of multi-modality information is performed periodically, then the summarization is fed back to each modality stream.

coronal plane and the transverse plane. By viewing the MRI volume from different directions, multiple batches of 2D slices can be acquired for training. For example, a $120 \times 120 \times 120$ volume will generate 360 images of the three anatomical planes. In fact, more directions can be included.

On one hand, changing the observation direction provides more training images. On the other hand, fusing predictions is actually an ensemble model, which improves the algorithm robustness and benefit the spatial consistency.

The annotation resolution of the MRBrainS2018 dataset is anisotropic in three directions. Therefore, this strategy cannot be fully utilized. We train three networks on the sagittal, coronal, transverse plane and fuse the predictions. Further tests can be performed by training a single classifier with images acquired along different orientations.

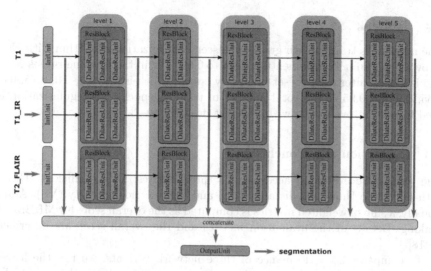

Fig. 6. MixNetv3: modality streams are kept separate until the OutputUnit aggregates information from each stream.

3   Results

The experiments are performed with the MICCAI challenge MRBrainS2018 dataset. The challenge releases 7 MRI scans (including T1-weighted, T1-weighted inversion recovery and T2-FLAIR) as the training data. Another 23 scans are kept secret for test. We test the three networks using leave-one-out cross validation strategy with the training dataset. MixNetv2 is submitted to the challenge and an evaluation of MixNetv2 on the test dataset is performed by the challenge organizers.

3.1   Preprocessing and Data Augmentation

Bias field correction [16] and image registration are performed by the challenge organizer. In addition to this, we linearly scale each modality image of each scan to have zero mean and unit variance.

To train the very deep network, the data is heavily augmented with elastic deformation [17], scaling, rotation and translation. As for the sagittal and coronal plane, the resolution in horizontal and vertical directions are four times different. Thus, we only apply flipping and translation.

It is worth mention that excessive elastic deformation and scaling may lead to an unstable training. We use scaling factors of 0.9, 0.95, 1.05 and 1.1, elastic deformation factor $\alpha = 10$ and $\sigma = 4$ [17] in this work. Rotation is performed around the image center with 8 degrees: 0°, 45°, 90°, 135°, 180°, 225°, 270° and 315°. The random translation is limited to 0.15% of the image size. We use all augmentation methods separately, that is, no images are generated from augmented images.

3.2 Training

The network is trained with gradient descent optimization algorithm with Nesterov momentum. The momentum is set to 0.99. The initial learning rate is 2e−4 and is halved after each preset boundary epoch, which is 0.2, 0.4, 0.6, 0.75, 0.8, 0.85, 0.9 and 0.95 of the total number of training epochs. L2 regularization is used to prevent overfitting with a weight decay of 1e−3.

3.3 Evaluation and Conclusion

The results are evaluated according to three metrics: Dice coefficient (Dice), 95th-percentile Hausdorff distance (HS) and Volumetric similarity (VS). Additionally, a sum of weighted metrics is computed as the overall score for MRBrainS ranking. Details of the evaluation metrics and the overall score are described in [18].

To compare the performance of three network variants, we run the leave-one-out cross validation as a 3-label segmentation problem (GM, WM and CSF) on the MRBrainS2018 training dataset. As shown in Table 2, MixNetv3 gives the best results. The cross validation results of MixNetv1 and MixNetv2 are quite close. But MixNetv2 has a lower validation loss (see Fig. 7). As discussed in Sect. 2.2, MixNetv1 contains the solution space of MixNetv3. However, the results of MixNetv1 is worse. We conjecture that the architecture of parallel modality streams can learn complementary features more easily.

By MixNetv2_multi, three classifiers are trained on the sagittal plane, the coronal plane and the transverse plane, respectively. Results are obtained by fusing predictions of three MixNetv2 classifiers with the corresponding weights 1:1:4. The weights are empirically chosen based on the fact that the transverse plane resolution is 4 times higher. Although the classifiers on the sagittal plane and the coronal plane performs much worse, the fused results still improves.

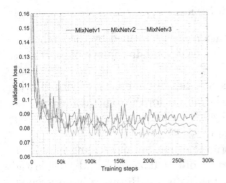

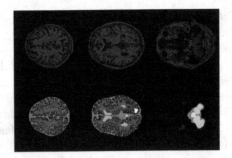

Fig. 7. Validation loss during training (subject 1 as the validation data).

Fig. 8. Qualitative segmentation results of 8 brain structures.

MixNetv2_multi was also trained with the full training dataset as a 3-label and 8-label task. Figure 8 shows the qualitative results of 8-label predictions by MixNetv2_multi. Trained models were submitted to the challenge. Figures 9 and 10 show the evaluation performed by the challenge organizer on the test dataset. We notice a performance drop between the validation results and the evaluation results (about 0.02). That is reasonable, because the relatively small training dataset may not cover all the cases very well.

In the 8-label segmentation task, our network has difficulties in classifying WHM and basal ganglia. One possible reason is the imbalance of labels in the training data. We do not use any methods of balancing the labels during training, that is, labels with a small number in the training data are easy to ignore. The

Table 2. Cross validation results of MixNetv1, MixNetv2 and MixNetv3, performed on the MRBrainS2018 training dataset. The network is trained as a 3-label segmentation task (WM, GM and CSF).

	GM			WM			CSF		
	Dice	HD	VS	Dice	HD	VS	Dice	HD	VS
MixNetv1	.8524	.9583	.9728	.9000	1.9167	.9759	.8599	1.9167	.9508
MixNetv2	.8500	.9583	.9772	.8966	1.9167	.9626	.8609	1.9167	.9506
MixNetv2_multi	.8511	.9583	.9762	.9001	1.3553	.9689	.8624	1.9167	.9447
MixNetv3	.8557	0.9583	.9789	.9049	1.3552	.9743	.8609	1.9167	.9578

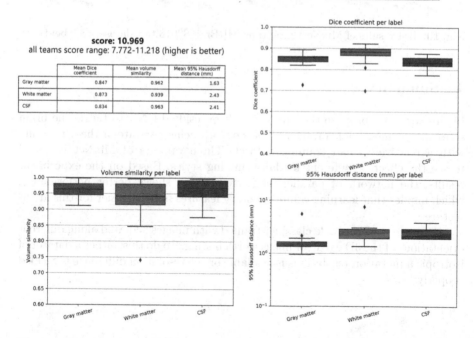

Fig. 9. Test results of MixNetv2_multi on MRBrainS2018 test dataset (3-label task).

8-label methods taking part in the MRBrainS2018 challenge differ mainly in the performance of segmenting WHM and basal ganglia. This problem deserves further study.

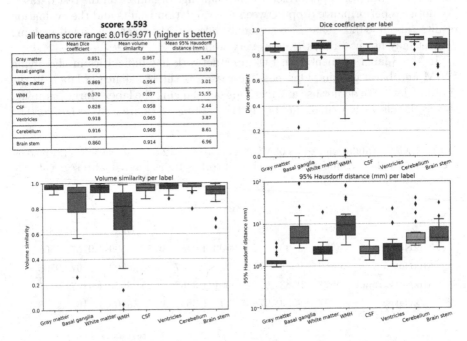

score: 9.593

all teams score range: 8.016-9.971 (higher is better)

	Mean Dice coefficient	Mean volume similarity	Mean 95% Hausdorff distance (mm)
Gray matter	0.851	0.967	1.47
Basal ganglia	0.728	0.846	13.90
White matter	0.869	0.954	3.01
WMH	0.570	0.697	15.55
CSF	0.828	0.958	2.44
Ventricles	0.918	0.965	3.87
Cerebellum	0.916	0.968	8.61
Brain stem	0.860	0.914	6.96

Fig. 10. Test results of MixNetv2_multi on MRBrainS2018 test dataset (8-label task).

4   Summary

In this work, we propose the MixNet, a deep residual CNN to tackle the brain structure segmentation problem. The network achieves state-of-the-art results with a relatively small training dataset. Three variants of MixNet is tested to see the effect of different modality mixing styles. Based on the experiment results, the network of parallel modality streams shows better performance, which implies that learning complementary features may be easier for this architecture.

As future work, a single classifier trained with images acquired along different orientations of the 3D MRI volume is worth testing. To do this, either a dataset of isotropic annotation resolutions is available or the resolution difference is tackled properly.

References

1. Akkus, Z., Galimzianova, A., Hoogi, A., Rubin, D.L., Erickson, B.J.: Deep learning for brain MRI segmentation: state of the art and future directions. J. Digit. Imaging **30**(4), 449–459 (2017)
2. Dale, A.M., Fischl, B., Sereno, M.I.: Cortical surface-based analysis: I. Segmentation and surface reconstruction. NeuroImage **9**(2), 179–194 (1999)
3. Fischl, B., Sereno, M.I., Dale, A.M.: Cortical surface-based analysis: II: inflation, flattening, and a surface-based coordinate system. NeuroImage **9**(2), 195–207 (1999)
4. Jenkinson, M., Beckmann, C.F., Behrens, T.E.J., Woolrich, M.W., Smith, S.M.: FSL. NeuroImage **62**(2), 782–790 (2012)
5. Ashburner, J., Friston, K.J.: Unified segmentation. NeuroImage **26**(3), 839–851 (2005)
6. De Brébisson, A., Montana, G.: Deep neural networks for anatomical brain segmentation. In: 2015 IEEE Conference on Computer Vision and Pattern Recognition Workshops (CVPRW), pp. 20–28 (2015)
7. Zhang, W., et al.: Deep convolutional neural networks for multimodality isointense infant brain image segmentation. Neuroimage **108**, 214–224 (2015)
8. Moeskops, P., Viergever, M.A., Mendrik, A., De Vries, L.S., Benders, M.J.N.L., Isgum, I.: Automatic segmentation of MR brain images with a convolutional neural network. IEEE Trans. Med. Imaging **35**, 1252–1261 (2016)
9. Nie, D, Li, W, Gao, Y, Sken, D.: Fully convolutional networks for multi-modality isointense infant brain image segmentation. In: 13th IEEE International Symposium on Biomedical Imaging (ISBI), pp. 1342–1345 (2016)
10. Shin, H.-C., et al.: Deep convolutional neural networks for computer-aided detection: CNN architectures, dataset characteristics and transfer learning. IEEE Trans. Med. Imaging **35**(5), 1285–1298 (2016)
11. He, K., Zhang, X., Ren, S., Sun J.: Deep residual learning for image recognition. In: 2016 IEEE Conference on Computer Vision and Pattern Recognition (CVPR), pp. 770–778. IEEE (2016)
12. He, K., Zhang, X., Ren, S., Sun, J.: Identity mappings in deep residual networks. In: Leibe, B., Matas, J., Sebe, N., Welling, M. (eds.) ECCV 2016. LNCS, vol. 9908, pp. 630–645. Springer, Cham (2016). https://doi.org/10.1007/978-3-319-46493-0_38
13. Ronneberger, O., Fischer, P., Brox, T.: U-Net: convolutional networks for biomedical image segmentation. In: Navab, N., Hornegger, J., Wells, W.M., Frangi, A.F. (eds.) MICCAI 2015. LNCS, vol. 9351, pp. 234–241. Springer, Cham (2015). https://doi.org/10.1007/978-3-319-24574-4_28
14. Chen, L.-C., Papandreou, G., Kokkinos, I., Murphy, K., Yuille, A.L.: DeepLab: semantic image segmentation with deep convolutional nets, atrous convolution, and fully connected CRFs. IEEE Trans. Pattern Anal. Mach. Intell. **40**(4), 834–848 (2017)
15. Zhao, H., Shi, J., Qi, X., Wang, X., Jia, J.: Pyramid scene parsing network. In: 2017 IEEE Conference on Computer Vision and Pattern Recognition (CVPR). IEEE (2017)
16. Tustison, N.J., et al.: N4ITK: improved N3 bias correction. IEEE Trans. Med. Imaging **29**(6), 1310–1320 (2010)
17. Simard, P.Y., Steinkraus, D., Platt, J.C.: Best practices for convolutional neural networks applied to visual document analysis. In: Proceedings of the Seventh International Conference on Document Analysis and Recognition, p. 958. IEEE (2003)
18. MRBrainS2018 Homepage. http://mrbrains18.isi.uu.nl/. Accessed 11 Oct 2018

A Skip-Connected 3D DenseNet Networks with Adversarial Training for Volumetric Segmentation

Toan Duc Bui, Sang-il Ahn, Yongwoo Lee, and Jitae Shin[(✉)]

Department of Electrical and Computer Engineering, Sungkyunkwan University, Suwon 16419, Republic of Korea
jtshin@skku.edu

Abstract. In this paper, we propose a novel end-to-end adversarial training on volumetric brain segmentation architecture that allows to enforce long-range spatial label contiguity and label consistency. The proposed network consists of two networks: generator and discriminator. The generator network allows to take volumetric image as input and provides a volumetric probability map for each tissue. Then, the discriminator network learns to differentiate ground-truth maps from the probability maps of generator network. We design a discriminator in a fully convolutional manner to differentiate the predicted probability maps from the ground-truth segmentation distribution with the consideration of the spatial information on voxel level, which makes it difficult to learn the discriminator. In order to overcome it, the proposed discriminator provides a 3D confidence map which indicates corresponding regions of the probability maps close to the ground-truth. Based on the 3D confidence map information, the generator network will refine prediction output close to the ground-truth maps in a high-order structure.

Keywords: Adversarial training · Brain segmentation · Deep convolutional networks

1 Introduction

Automatic brain segmentation plays an importance rule for disease diagnosis, progression, and treatment monitoring [11]. With a success of deep learning [9], many algorithms have been proposed to obtain an accurate segmentation result [2,12,14]. For example, Moeskops et al. [12] introduced an automatic segmentation method using multiple patch 2D convolutional networks (CNN). Çiçek, Özgün et al. proposed 3D UNet using skip connected path to concatenate low and high level features to produce a full-resolution segmentation. Bui et al. [1] proposed a fully convolutional 3D DenseNet by concatenate multiple contextual information from different level features. Although these methods demonstrated good performance, the contour obtained from thresholding based on the score of

© Springer Nature Switzerland AG 2019
A. Crimi et al. (Eds.): BrainLes 2018, LNCS 11383, pp. 378–384, 2019.
https://doi.org/10.1007/978-3-030-11723-8_38

volumetric infant brain sometimes can be imprecise with small isolated regions or holes in the predictions.

Recently, Goodfellow et al. [4] proposed generative adversarial network (GAN) for generative image modeling. It consists of two networks, discriminator and generator. While a discriminator tries to distinguish ground-truth images from the outputs generated by the generator, the generator tries to generate as realistic outputs as the discriminator cannot differentiate from the ground-truth image. Inspired of GAN, Luc et al. [10] introduce an adversarial training approach for training semantic segmentation models. It shows that the adversarial training leads to improve segmentation accuracy.

In this paper, we extend the 3D DenseNet networks [1] by adding an end-to-end adversarial training for volumetric segmentation. While existing methods such as [6,10,13] which have both generator and discriminator designed for 2D segmentation problem, we focus on 3D segmentation problem. In order to achieve it, first we introduce a generator network that uses 3D features to explore the information of adjacent slices to enhance volumetric segmentation. The proposed generator network allows to take volumetric image as input and provides a volumetric probability map for each tissue. Then, the discriminator network learns to differentiate ground-truth maps from probability maps of generator network. Inspired of fully discrimator network [6], we extend it from 2D to 3D discriminator network by using $1 \times 1 \times 1$ convolution at final layer. We use skip connection that allows to capture multiple contexture from discriminator network. The generator and discriminator networks are trained in end-to-end manner in order to jointly optimize all the weights in the network using an efficient weight update technique.

2 Method

Figure 1 illustrates the overview of our proposal for volumetric infant brain segmentation. It consists of two networks: generator and discriminator network. The generator network uses 3D features to explore the information of adjacent slices to enhance volumetric segmentation. The proposed generator network allows to take volumetric images as inputs and provides a volumetric probability maps for each tissue. The probability maps through the discriminator network that uses a fully convolutional scheme to obtain spatial 3D confidence. The 3D confidence map indicates which regions of the probability maps are close to the ground truth. Based on the 3D confidence map information, the generator network will refine prediction output close to the ground truth maps in a high-order structure. The generator and discriminator networks are trained in end-to-end manner in order to jointly optimize all the weights in the network using an efficient weight update technique.

For generator network, we modify the network in [1] by adding a squeeze-and-excitation (SE) block [5] after each dense block. In this way, it allows to explore the interdependencies between the channels. Figure 1 illustrates our generator networks for brain MRI segmentation. It consists of contains 47 layers with

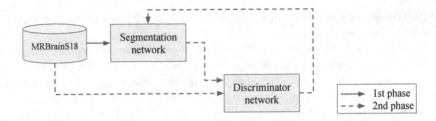

Fig. 1. Our proposed flow chart for brain MRI segmentation

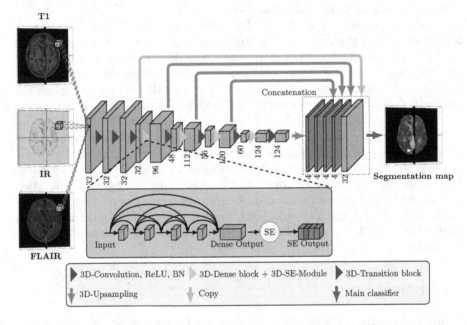

Fig. 2. Modified 3D-SkipDenseSeg network architecture for brain MRI segmentation.

downsampling and upsampling paths. The downsampling path aims to reduce feature resolution and to increase the receptive field. It is performed by four dense blocks with growth rate of $k = 16$. We used four $3 \times 3 \times 3$ convolutions in each dense block. After each dense block, we add an SE block to generate feature inter-dependencies between the channels. Then the output features from SE block are fed to transistion layers that use $1 \times 1 \times 1$ convolution to reduce features size. Meanwhiles, the upsampling paths tries to recover original resolution from the low feature resolutions. We upsample low-level features map directed to orignal resolution and concatenate them together through skip connection before feeding to a classifier. By this way, we can reduce the number of learning parameters. The network can take multi-modalities images, such as T1, IR and FLAIR, as input to generates segmented image as shown in Fig. 2.

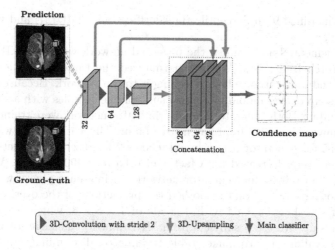

Fig. 3. Discriminator network for brain MRI segmentation. The brighter regions indicate that they are close to the ground truth distribution

Figure 3 shows the structure of discriminator network. The outputs of generator network are fed to the discriminator network that has three $4 \times 4 \times 4$ convolution with stride of 2 to reduce features map. We upsample the result of each convolution into original size and concatenate these feature together to capture multiple contexture information from discrimiatnor network. $1 \times 1 \times 1$ convolution is used to classify the concatenated feature into two classes: real or fake. The discriminator in a fully convolutional manner to differentiate the predicted probability maps from the ground-truth segmentation distribution with the consideration of the spatial information on voxel level, which increases the difficulty to learn the discriminator. The proposed discriminator provides a 3D confidence map which indicates which regions of the probability maps are close to the ground truth. Based on the 3D confidence map information, the generator network will refine prediction output close to the ground truth maps in a high-order structure.

3   Experimental Results

We used the public dataset from the MRBrainS18 Challenge to evaluate the performance of the proposed method. The dataset consists of 7 subjects for training and 23 subjects for testing. For each subject, three modalities are available that includes T1-weighted, T1-weighted inversion recovery and T2-FLAIR with voxel size of $0.958\,\mathrm{mm} \times 0.958\,\mathrm{mm} \times 3.0\,\mathrm{mm}$. Each subject was manually segmented into 11 classes by the challenge organizers. For each participant team, the 7 subjects for training with ground-truth was provided for fine-tuning the network, while 23 testing subjects were retained by the organizers for a fair comparison. The aim of this challenge is to automatically segment the images from each subject into 9 classes and compared with the manual labels using Dice coefficient

(DSC) [3], modified 95th percentile Hausdorff distance (HD) [7] and volumetric similarity (VS).

We implemented and trained the proposed network with an NVIDIA Titan X and Pytorch framework. We first normalized the three modalities input to zero mean and unit variance before feeding it to the network. Because of limited memory resource, we randomly cropped sub-volume samples with a size of 48 × 64 × 64 for input. Both generator and discriminator network were trained with an Adam optimizer [8] with a mini-batch size of four. The learning rate was initially set to 0.0005 for generator network and 0.000002 for discrimiator network. The weights then were decreased by a factor of 0.1 every 4000 epochs. We used a weight decay of 0.0001 for generator network. In inference phase, we used the majority voting strategy that smoothed the predictions of the overlapping sub-volumes of size 48 × 64 × 64 with a stride of 1 × 8 × 8, which resulted in an improved result. We use 6 subjects from MRBrainS18 challenge for training, 1 subjects for validation. To make stable training on discriminator network, we first only train the generator network in 500 epochs to provide a reasonable result from the generator network. After that, The generator and discriminator networks are trained in end-to-end manner in order to jointly optimize all the weights in the network. It spends about 12 h for training and 6 min to segment each subject on Titan X Pascal.

Table 1 shows performance results on validation set using with and without adversarial training. The adversarial training not only achives a better result on DSC but also on other metrics such as HD and VS. From the result, we can conclude that the adversarial training leads improve the segmentation accuracy.

Table 1. Performance comparison of proposed method with or without adversarial training on validation data (DSC: %, HD: mm, VS: mm).

Methods	DSC	HD	VS
Without adversarial training	86.87	3.1232	0.9555
With adversarial training	**87.16**	**2.9845**	**0.9645**

Figure 4 shows the comparison results between with or without adversarial training of a 2D slice on the validation set. From the figure, we can observe that the adversarial training yields a better performance than the case of without adversarial training.

Figure 5 shows our performance of 23 subject on the testing set. We achive rank of 14 over 22 participant teams. The reason is that the adversarial training is difficult for training small label such as white matter lesions.

3.1   Conclusion

In this paper, we proposed an adversarial training on 3D segmentation task. We extend the discriminator network in a fully convolutional manner to differentiate

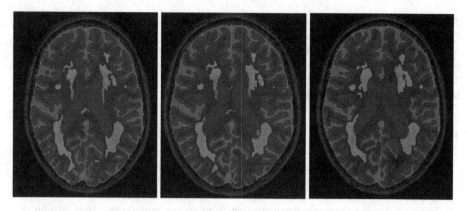

Fig. 4. An comparison of segmentation result on validation set: (left) without adversarial training, (middle) with adversarial training and (right) ground-truth image

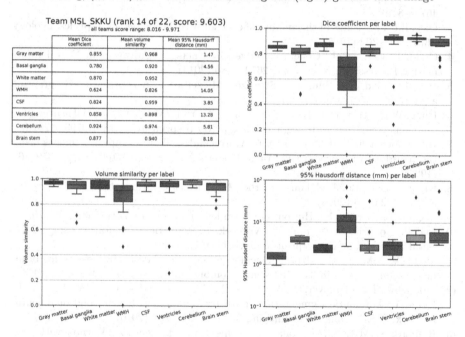

Team MSL_SKKU (rank 14 of 22, score: 9.603)
all teams score range: 8.016 - 9.971

	Mean Dice coefficient	Mean volume similarity	Mean 95% Hausdorff distance (mm)
Gray matter	0.855	0.968	1.47
Basal ganglia	0.780	0.920	4.56
White matter	0.870	0.952	2.39
WMH	0.624	0.826	14.05
CSF	0.824	0.959	3.85
Ventricles	0.858	0.898	13.28
Cerebellum	0.924	0.974	5.81
Brain stem	0.877	0.940	8.18

Fig. 5. Segmentation result of 23 subject on testing set

the predicted probability maps from the ground truth segmentation distribution with the consideration of the spatial information on voxel level, which makes it difficult to learn the discriminator. The experiment results show that the adversarial training leads to improve the segmentation accuracy not only on 2D segmentation task, but also on 3D segmentation task such as brain MRI segmentation.

Acknowledgment. This research was supported partly by the Basic Science Research Program through the National Research Foundation of Korea (NRF) funded by the Ministry of Education (No. 2018R1C1B6007472). This research was supported partly by the MSIT (Ministry of Science and ICT), Korea, under the ITRC (Information Technology Research Center) support program (IITP-2018-2018-0-01798) supervised by the IITP (Institute for Information & communications Technology Promotion).

References

1. Bui, T.D., Shin, J., Moon, T.: 3D densely convolution networks for volumetric segmentation. arXiv preprint arXiv:1709.03199 (2017)
2. Çiçek, Ö., Abdulkadir, A., Lienkamp, S.S., Brox, T., Ronneberger, O.: 3D U-Net: learning dense volumetric segmentation from sparse annotation. In: Ourselin, S., Joskowicz, L., Sabuncu, M.R., Unal, G., Wells, W. (eds.) MICCAI 2016. LNCS, vol. 9901, pp. 424–432. Springer, Cham (2016). https://doi.org/10.1007/978-3-319-46723-8_49
3. Dice, L.R.: Measures of the amount of ecologic association between species. Ecology **26**(3), 297–302 (1945)
4. Goodfellow, I., et al.: Generative adversarial nets. In: Advances in Neural Information Processing Systems, pp. 2672–2680 (2014)
5. Hu, J., Shen, L., Sun, G.: Squeeze-and-excitation networks. arXiv preprint arXiv:1709.01507, vol. 7 (2017)
6. Hung, W.C., Tsai, Y.H., Liou, Y.T., Lin, Y.Y., Yang, M.H.: Adversarial learning for semi-supervised semantic segmentation. arXiv preprint arXiv:1802.07934 (2018)
7. Huttenlocher, D.P., Klanderman, G.A., Rucklidge, W.J.: Comparing images using the hausdorff distance. IEEE Trans. Pattern Anal. Mach. Intell. **15**(9), 850–863 (1993)
8. Kingma, D.P., Ba, J.: Adam: a method for stochastic optimization. arXiv preprint arXiv:1412.6980 (2014)
9. LeCun, Y., Bengio, Y., Hinton, G.: Deep learning. Nature **521**(7553), 436 (2015)
10. Luc, P., Couprie, C., Chintala, S., Verbeek, J.: Semantic segmentation using adversarial networks. arXiv preprint arXiv:1611.08408 (2016)
11. Mendrik, A.M., et al.: Mrbrains challenge: online evaluation framework for brain image segmentation in 3T MRI scans. Comput. Intell. Neurosci. **2015**, 1 (2015)
12. Moeskops, P., Viergever, M.A., Mendrik, A.M., de Vries, L.S., Benders, M.J., Išgum, I.: Automatic segmentation of MR brain images with a convolutional neural network. IEEE Trans. Med. Imaging **35**(5), 1252–1261 (2016)
13. Radford, A., Metz, L., Chintala, S.: Unsupervised representation learning with deep convolutional generative adversarial networks. arXiv preprint arXiv:1511.06434 (2015)
14. Ronneberger, O., Fischer, P., Brox, T.: U-Net: convolutional networks for biomedical image segmentation. In: Navab, N., Hornegger, J., Wells, W.M., Frangi, A.F. (eds.) MICCAI 2015. LNCS, vol. 9351, pp. 234–241. Springer, Cham (2015). https://doi.org/10.1007/978-3-319-24574-4_28

Automatic Brain Structures Segmentation Using Deep Residual Dilated U-Net

Hongwei Li$^{(\boxtimes)}$, Andrii Zhygallo, and Bjoern Menze

Technical University of Munich, Munich, Germany
{hongwei.li,andrii.zhygallo,bjoern.menze}@tum.de

Abstract. Brain image segmentation is used for visualizing and quantifying anatomical structures of the brain. We present an automated approach using 2D deep residual dilated networks which captures rich context information of different tissues for the segmentation of eight brain structures. The proposed system was evaluated in the MICCAI Brain Segmentation Challenge (http://mrbrains18.isi.uu.nl/) and ranked 9th out of 22 teams. We further compared the method with traditional U-Net using leave-one-subject-out cross-validation setting on the public dataset. Experimental results shows that the proposed method outperforms traditional U-Net (i.e. 80.9% *vs* 78.3% in averaged Dice score, 4.35 mm *vs* 11.59 mm in averaged robust Hausdorff distance) and is computationally efficient.

Keywords: Brain structure segmentation · Deep learning

1 Introduction

Brain MRI segmentation is an important task in many clinical applications. Various approaches for brain analysis rely on accurate segmentation of anatomical regions. For example, it is commonly used for measuring and visualizing different brain structures, for delineating lesions, for analysing brain development, and for characterization of brain disorders such as Alzheimers disease, epilepsy, schizophrenia, multiple sclerosis (MS), cancer, and infectious and degenerative diseases. Manual segmentation is the gold standard for in-vivo images. However, it requires outlining structures slice-by-slice by neuroradiologist, which is highly time-consuming and prone to rater-bias. Therefore, there is a need for automated segmentation approaches to provide accuracy close to that of expert raters with a high reproducibility.

Early works on segmentation of normal brain structures focus on white matter (WM), gray matter (GM), and cerebrospinal fluid (CSF), which is important for studying early brain developments in infants and quantitative assessment of the brain tissue and intracranial volume in large scale studies. Atlas-based approaches [7,12], which match intensity information between an atlas and target images and pattern recognition approaches [10], which classify tissues based on a set of local intensity features, are the classical approaches that have been

© Springer Nature Switzerland AG 2019
A. Crimi et al. (Eds.): BrainLes 2018, LNCS 11383, pp. 385–393, 2019.
https://doi.org/10.1007/978-3-030-11723-8_39

used for brain tissue segmentation. The MRBrainS Challenge 2013 [8] was held to compare state-of-the-art segmentation algorithms on three brain structures in conjunction with the 16th International Conference on Medical Image Computing and Computer Assisted Intervention. Deep-learning based approaches have shown superior performances to the traditional state-of-art methods on the segmentation of brain stroke lesions, brain white matter lesions and brain tumors [5,6,9].

In this paper, we presented a deep-learning based method for segmenting eight brain tissues including cortical gray matter (GM), basal ganglia, WM, white matter lesions/hyperintensities (WMH), CSF, ventricles, cerebellum and brain stem. Deep dilated residual U-Net was adopted to learn context and texture information of different brain tissues. Multi-sequence data including T1, T1-IR and FLAIR which captures complementary information of different brain structures. The proposed 2-D network was more computationally efficient than 3D network and traditional U-Net. Experimental results showed that the proposed method outperforms traditional U-Net.

2   Materials

2.1   Dataset and Protocols

Dataset. Thirty MRI scans were acquired on a 3.0T Philips Achieva MR scanner at the University Medical Center Utrecht (Netherlands). The following sequences were acquired and used for the evaluation framework: 3D T1 (TR: 7.9 ms, TE: 4.5 ms), T1-IR (TR: 4416 ms, TE: 15 ms, and TI: 400 ms), and T2-FLAIR (TR: 11000 ms, TE: 125 ms, and TI: 2800 ms). The sequences were aligned by rigid registration using Elastix [3] and bias correction was performed using SPM8. After registration, the voxel size within all provided sequences (T1, T1-IR, and T2-FLAIR) was $0.96 \times 0.96 \times 3.00 \, mm^3$. Seven scans with annotations were released as a public training set, and the remaining twenty-three scans were used as hidden testing set. For more details on the method of ranking performance, please find the relevant information on the challenge website.

Evaluation Metric. Three types of measures were employed to evaluate the segmentation results. The Dice coefficient is used to determine the spatial overlap and is defined as:

$$Dice = \frac{2|G \cap P|}{|G| + |P|} \tag{1}$$

where G is the reference standard, P is the segmentation result.

The 95th-percentile of the Hausdorff distance is used to determine the distance between the segmentation boundaries. Hausdorff distance is defined as:

$$H(G, P) = max\{\sup_{x \in G} \inf_{y \in P} d(x, y), \sup_{y \in P} \inf_{x \in G} d(x, y)\} \tag{2}$$

where $d(x, y)$ denotes the distance of x and y, sup denotes the supremum and inf for the infimum.

The third measure is the volumetric similarity. Let V_G and V_P be the volume of lesion regions in G and P respectively. Then the volumetric similarity (VS) in percentage is defined as:

$$VS = \frac{|V_G - V_P|}{V_G} \tag{3}$$

3   Methodology

3.1   Image Preprocessing

A patient-wise normalization of the image intensities was performed both during training and testing. For the scan of each patient, the mean value and standard deviation were calculated based on intensities of all voxels. Then each image volume was normalized to zero mean and unit standard deviation. Rotation, shearing, scaling along horizontal direction (x-scaling), and scaling along vertical direction (y-scaling) were employed for data augmentation. After data augmentation, a four times larger training dataset was obtained.

3.2   2D Dilated Residual U-Net

We used Dilated Residual U-Net (DRUNet), which was originally proposed in [1] for nerve head tissues segmentation in optical coherence tomography images. DRUNet exploits the inherent advantages of the U-Net skip connections [11], residual learning [2] and dilated convolutions [13] to capture rich context information and offer a robust brain structure segmentation with a minimal number of trainable parameters.

DRUNet architecture is presented in Fig. 1. The model consists of downsampling and upsampling parts. In turn, each part includes one standard block and two residual blocks. Corresponding blocks in downsampling and upsampling parts are connected through skip connections. Convolution layers in both block types have 32 filters of size 3×3. In total the entire network consists of 156,105 trainable parameters.

3.3   Combination of Modalities

Multi-sequence data including T1-weighted (T1), T1-weighted inversion recovery (T1-IR) and FLAIR which captures complementary information of different brain structures were used for training the network. In clinical practice, the combination of FLAIR and T1 is beneficial for segmenting white matter lesions while the combination of T1 and T1-IR is helpful for annotating cerebrospinal fluid. We feed different combinations of modalities for multiple networks.

3.4   Ensemble Model

To improve the robustness of our model, an ensemble method was used in the testing stage. Then when given a new testing subject, each subject will be segmented based on the averaged probability maps by the ensemble model.

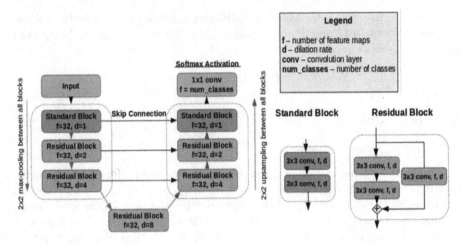

Fig. 1. Details of DRUnet architecture which contains residual blocks with dilated convolutions.

3.5 Our Submissions

Submission 1. We used only DRUNet for simultaneously segmenting the ten labels including infarction and pathologies were set to background label during the training of the network. We generated five DRUNet models with the same architecture but trained with shuffled batches. Then in testing stage, each subject was segmented based on the averaged probability maps by the ensemble models.

Submission 2. We used two Dilated Residual U-Nets (DRUNet) and one traditional U-Net for segmenting different labels. Since not all the labels were annotated in the same modalities, i.e., white matter lesions were annotated on the FLAIR scan and the outer border of the CSF was segmented using both the T1-weighted scan and the T1-weighted inversion recovery scan, we employed a multi-stage approach to segment different tissues from coarse to fine using different combinations of input modalities. Firstly, coarse segmentation including eight brain tissues (other labels including infarction and pathologies were set to background label) was performed using FLAIR and T1-weighted modalities by DRUNet (model 1). Secondly, CSF was independently segmented using T1 and T1-IR modalities by DRUNet (model 2). Thirdly, since segmentation of white matter lesions is a very challenging task, we used the pre-trained model of the winning method in MICCAI WMH challenge [4] (model 3) to perform segmentation independently. Finally we fused the multi-stage segmentation results. Five DRUNet models for model 1 and model 2, respectively, with the same architecture were trained with shuffled batches.

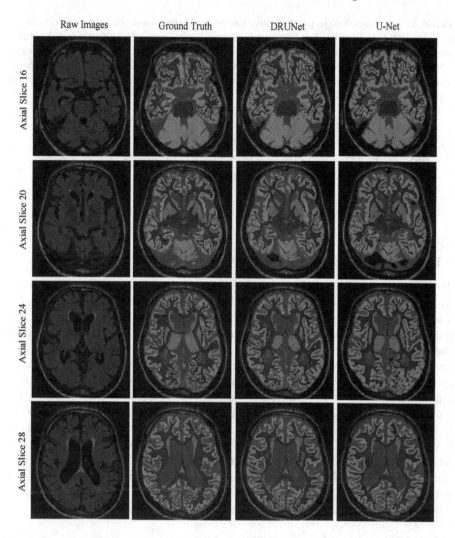

Fig. 2. Sample segmentation result on *Case 70*. From top to bottom: four axial slices of the same scan. From left to right: FLAIR MR images, the associated ground truth, segmentation result using DRUNet and segmentation result using U-Net. (Best viewed in colour). We can observed from the segmentation result of axial slice 16 that DRUNet achieved better performance on large continuous regions while U-Net generated some isolated false positives. It indicates that the dilated convolution in DRUNet helps to capture context information. On the other hand, for the segmentation of small tissues such as WMHs, DRUNet seems to generate more false positives than U-Net as observed from axial slice 28.

Table 1. Leave-one-subject-out evaluation of our submissions on the public training set containing seven subjects. The averaged Dice score, averaged H95, averaged volume similarity of eight tissues for each subject were shown in the table. The left and right values in each cell were the results of submission 1 and submission 2 respectively. The values in bold indicates the subject on which the two submissions has significant segmentation difference.

Metrics	Subject 1	Subject 2	Subject 3	Subject 4	Subject 5	Subject 6	Subject 7
Dice	0.86/0.85	0.82/0.82	0.77/0.77	**0.73/0.77**	0.85/0.84	0.80/0.80	0.83/0.81
H95	2.98/2.43	3.15/2.33	6.07/6.56	**8.25/5.87**	3.42/2.39	4.17/6.58	2.49/8.07
VS	0.97/0.98	0.92/0.91	0.86/0.87	**0.82/0.88**	0.94/0.92	0.88/0.89	0.92/0.90

4   Results

4.1   Leave-One-Subject-Out Evaluation

To test the generalization performance of our systems across different subjects, we conducted an experiment on the public training datasets (seven subjects) in a leave-one-subject-out setting. Specifically, we used the subject IDs to split the public training dataset into training and validation sets. In each split, we used slices from six subjects for training, and the slices from the remaining subject for testing. This procedure was repeated until all of the subjects are used as testing. The results were shown in Table 1. There exists significant segmentation difference on subject 4. We further observed the brain structures of subject 4 and found it was a heathy brain scan without WMHs, infarctions and other lesions. The reason for the performance difference could be that the models in first submission were trained on 10 labels including infarctions and other lesions while the models in the second submission were trained on 8 main structures excluding two other labels. When testing on healthy scans, the models trained with 8 main healthy tissues could be more effective since the data distributions among training and testing were similar.

4.2   Comparison with U-Net

We further compared the performance of the proposed method (submission 1) with traditional U-Net using the state-of-the-art architecture proposed in [4]. As shown in Table 2, generally our approach outperformed traditional U-Net, especially in segmentation of WM and CSF, with an improvement of 8% and 11% in Dice score. WM and CSF are both large structures in brains. We concluded that the use of dilated convolutions is beneficial for capturing the context information of large target. Furthermore, our model is with much fewer trainable parameters (156,105 *vs* 8,748,609). Thus the training of the network is computationally efficient. The segmentation results of both DRUNet and U-Net on test *case 70* was shown in Fig. 2.

Table 2. Comparison on each class with traditional U-Net under leave-one-subject-out setting. The performance on each class was averaged over seven subjects. The values in bold indicated significant improvement over traditional U-Net.

Metrics	GM	BG	WM	WMH	CSF	Ventricles	Cerebellum	Brain stem	*Averaged*
$Dice_{U-Net}$	0.83	0.84	0.70	0.79	0.43	0.89	0.9	0.88	*0.783*
$Dice_{DRUNet}$	0.84	0.85	**0.78**	**0.81**	**0.54**	0.88	**0.92**	0.85	*0.809*
$H95_{U-Net}$	1.26	1.8	43.5	1.78	23.09	3	15.63	2.67	*11.59*
$H95_{DRUNet}$	1.29	1.67	**5.82**	1.61	**14.8**	3.15	**2.97**	3.45	*4.35*
VS_{U-Net}	0.95	0.95	0.84	0.93	**0.71**	0.94	0.96	0.94	*90.25*
VS_{DRUNet}	0.96	0.94	**0.89**	0.94	0.66	0.93	0.97	0.92	*90.13*

4.3   Results on Hidden Testing Cases

Our submissions were independently evaluated by the challenge organizer. Figures 3 and 4 show the box plots of performance on eight labels on 23 testing scans. Submission 1 and submission 2 ranked 9th and 12th respectively out of 22 teams.

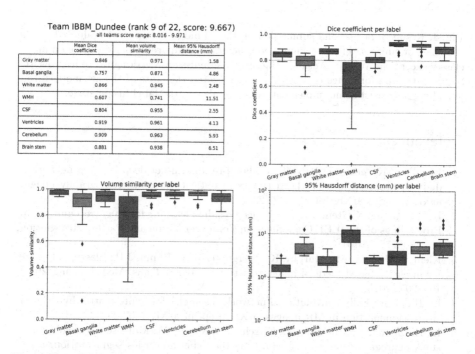

Fig. 3. Result of our first submission on the 23 hidden testing set evaluated by the challenge organizers. Our method achieved Dice scores of more than 80% and volume similarity of more 90% on the major classes while the segmentation performance on WMHs is relatively poor. This is because the WMHs are in small volumes and thus the most difficult structure to be segmented.

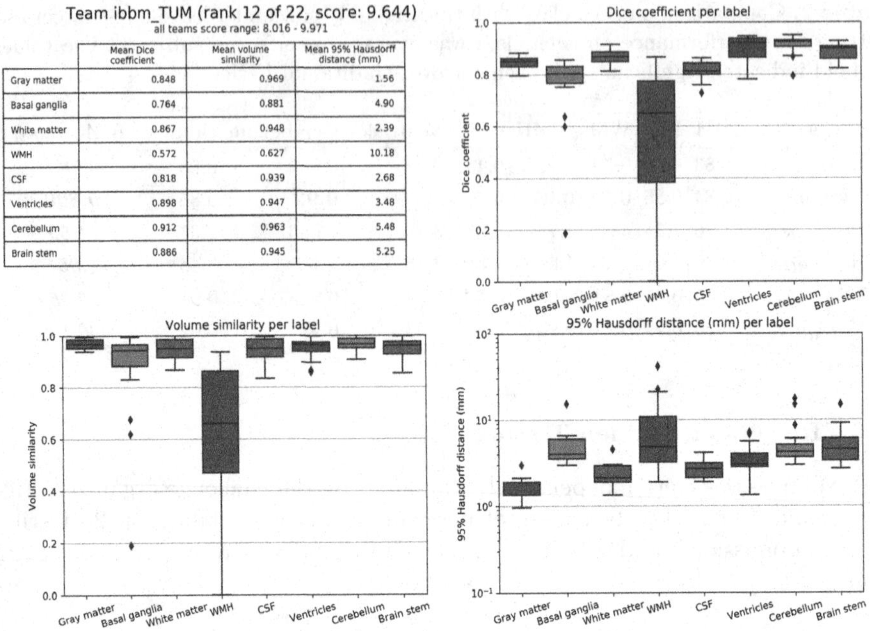

Team ibbm_TUM (rank 12 of 22, score: 9.644)
all teams score range: 8.016 - 9.971

	Mean Dice coefficient	Mean volume similarity	Mean 95% Hausdorff distance (mm)
Gray matter	0.848	0.969	1.56
Basal ganglia	0.764	0.881	4.90
White matter	0.867	0.948	2.39
WMH	0.572	0.627	10.18
CSF	0.818	0.939	2.68
Ventricles	0.898	0.947	3.48
Cerebellum	0.912	0.963	5.48
Brain stem	0.886	0.945	5.25

Fig. 4. Result of our second submission on the 23 hidden testing set evaluated by the challenge organizers. The two submissions achieved comparable performance on major classes except the WMHs. Actually the second submission was designed to improve the segmentation performance of WMHs and integrated the state-of-the-art models from [4]. There may exist some implementation mistakes in the label fusion stage.

References

1. Devalla, S.K., et al.: DRUNET: a dilated-residual u-net deep learning network to digitally stain optic nerve head tissues in optical coherence tomography images. arXiv preprint arXiv:1803.00232 (2018)
2. He, K., Zhang, X., Ren, S., Sun, J.: Deep residual learning for image recognition. In: Proceedings of the IEEE Conference on Computer Vision and Pattern Recognition, pp. 770–778 (2016)
3. Klein, S., Staring, M., Murphy, K., Viergever, M.A., Pluim, J.P.: Elastix: a toolbox for intensity-based medical image registration. IEEE Trans. Med. Imaging **29**(1), 196–205 (2010)
4. Li, H., et al.: Fully convolutional network ensembles for white matter hyperintensities segmentation in MR images. arXiv preprint arXiv:1802.05203 (2018)
5. Li, H., Zhang, J., Muehlau, M., Kirschke, J., Menze, B.: Multi-scale convolutional-stack aggregation for robust white matter hyperintensities segmentation. arXiv preprint arXiv:1807.05153 (2018)
6. Maier, O., et al.: ISLES 2015-a public evaluation benchmark for ischemic stroke lesion segmentation from multispectral MRI. Med. Image Anal. **35**, 250–269 (2017)
7. Makropoulos, A., et al.: Automatic whole brain MRI segmentation of the developing neonatal brain. IEEE Trans. Med. Imaging **33**(9), 1818–1831 (2014)

8. Mendrik, A.M., et al.: MRBrainS challenge: online evaluation framework for brain image segmentation in 3T MRI scans. Comput. Intell. Neurosci. **2015**, 1 (2015)

9. Menze, B.H., et al.: The multimodal brain tumor image segmentation benchmark (brats). IEEE Trans. Med. Imaging **34**(10), 1993 (2015)

10. Moeskops, P., et al.: Automatic segmentation of MR brain images of preterm infants using supervised classification. NeuroImage **118**, 628–641 (2015)

11. Ronneberger, O., Fischer, P., Brox, T.: U-Net: convolutional networks for biomedical image segmentation. In: Navab, N., Hornegger, J., Wells, W.M., Frangi, A.F. (eds.) MICCAI 2015. LNCS, vol. 9351, pp. 234–241. Springer, Cham (2015). https://doi.org/10.1007/978-3-319-24574-4_28. http://lmb.informatik.uni-freiburg.de/Publications/2015/RFB15a. arXiv:1505.04597

12. Vrooman, H.A., et al.: Multi-spectral brain tissue segmentation using automatically trained k-nearest-neighbor classification. Neuroimage **37**(1), 71–81 (2007)

13. Yu, F., Koltun, V.: Multi-scale context aggregation by dilated convolutions. arXiv preprint arXiv:1511.07122 (2015)

3D Patchwise U-Net with Transition Layers for MR Brain Segmentation

Miguel Luna and Sang Hyun Park[✉]

Department of Robotics Engineering, DGIST, Daegu, South Korea
{miguel,shpark13135}@dgist.ac.kr

Abstract. We propose a new patch based 3D convolutional neural network to automatically segment multiple brain structures on Magnetic Resonance (MR) images. The proposed network consists of encoding layers to extract informative features and decoding layers to reconstruct the segmentation labels. Unlike the conventional U-net model, we use transition layers between the encoding layers and the decoding layers to emphasize the impact of feature maps in the decoding layers. Moreover, we use batch normalization on every convolution layer to make a well generalized model. Finally, we utilize a new loss function which can normalize the categorical cross entropy to accurately segment the relatively small interest regions which are opt to be misclassified. The proposed method ranked 1st over 22 participants at the MRBrainS18 segmentation challenge at MICCAI 2018.

Keywords: Convolutional neural network · Brain MR image ·
Semantic segmentation · Transition layer · Normalized cross entropy

1 Introduction

Segmentation of brain magnetic resonance (MR) image is an important task for early diagnosis of brain diseases and for identifying the cause of diseases. Accordingly, many classical machine learning algorithms have been proposed [5], but the variability between subjects and the presence of abnormalities make these methods unreliable in large scale studies.

Recently, deep learning approaches significantly outperform the previous learning methods [1], and thus are used for the segmentation of overall brain structures, tumors, or other abnormalities. Shreyas et al. and Cui et al. extract 2D axial view slices [4,12] and then apply modified versions of U-net [11] or FCN [10] to segment brain tumors. Chen et al. [2] propose a 3D convolutional neural network (CNN) to perform volumetric brain segmentation. To address the memory limitation, they sample 3D patches that include relatively large area allowed by memory, and then input them to the CNN consisting of multiple convolution and pooling layers to take advantage of a large receptive field on their filters. On the other hand, Dolz et al. [6] extract smaller 3D patches to reduce memory requirements, maintain the volume resolution by avoiding

© Springer Nature Switzerland AG 2019
A. Crimi et al. (Eds.): BrainLes 2018, LNCS 11383, pp. 394–403, 2019.
https://doi.org/10.1007/978-3-030-11723-8_40

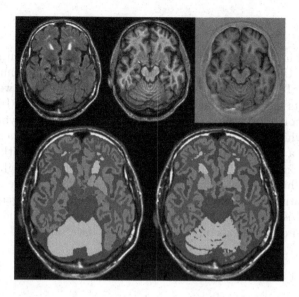

Fig. 1. MRBrainS18 challenge dataset and our result. Multi-modal MR images are shown on the top row (FLAIR image, T1 weighted image and Inverse Recovery T1 weighted image). The ground truth (left) and the prediction (right) generated by our method are shown on the bottom row.

pooling layers and to substantially increase the number of training samples. To use contextual information in both small and large scales, Kamnitsas et al. [8] sample two volume patches of the same region at two different resolutions and then train a multi scale CNN model.

In this paper, we propose a patch wise 3D convolutional neural network to perform segmentation of brain tissues. The network consist of encoding layers and decoding layers similar to conventional U-net [11], but use relatively small 3D volumes as inputs to learn various local patterns as in Dolz et al.'s method [6]. Unlike the U-net, we add transition layers between the encoding layers and the decoding layers to emphasize the impact of feature maps in the decoding layers. We also include batch normalization on every convolution layer to prevent over-fitting and make a well generalized model. Finally, we train our network with a normalized categorical cross entropy to balance the loss from small abnormalities such as white matter hyperintensity and that of large brain structures like white matter and gray matter.

The proposed model using the transition layer and the batch normalization can distinguish multiple brain tissues by effectively generalizing various local patterns. Moreover, the normalized loss function helps to accurately segment small abnormalities which are opt to be misclassified (Fig. 1).

2   Dataset

The data used in MRBrainS18 challenge was acquired from 30 subjects on a 3T scanner at the UMC Utrecht. For each subject, fully annotated multi-sequences (T1-weighted, T1-weighted inversion recovery and T2-FLAIR) were provided. The 30 subjects included patients with diabetes, dementia and Alzheimers, and matched controls (with increased cardiovascular risk) with varying degrees of atrophy and white matter lesions (age > 50). For training, 7 data samples were provided with a file containing the manual reference standard with following 11 labels: (0) Background, (1) Cortical gray matter, (2) Basal ganglia, (3) White matter, (4) White matter lesions, (5) Cerebrospinal fluid in the extracerebral space, (6) Ventricles, (7) Cerebellum, (8) Brain stem, (9) Infarction and (10) Other. The objective was to segment the 8 labels excluding the background, infarctions and other lesions on the remaining 23 data samples.

3   Method

We propose a patch wise segmentation method using 3D fully convolutional neural networks. The network takes as input a 3D volume and produces a 3D output prediction map of the same size as the input. The volume could be the entire MRI scan, but the size is too large to fit it in a modern GPU and the model is not generalized efficiently. Therefore, we use the CNN to make predictions on small 3D sections of the scan and finally use them to reconstruct a full size prediction map.

We use an encoder-decoder CNN structure since the pooling layers increase the receptive field of the filters and they allow the model to learn higher level features and detect larger volumes more accurately. We put skip connections from encoder to decoder to allow the model to use geometrical information from lower level features, but unlike U-Net [11], we use a convolutional transition layer from encoder to decoder to reduce the size of the low level feature maps in order to give more weight to the higher level features learned through deeper layers in the network. We also normalized our loss function to ensure every class, regardless of its size, has a similar impact during the training process.

3.1   Proposed Network

The proposed network consists of 17 convolutional layers divided into 3 stages; encoder, transition, and decoder (see Fig. 2). All convolutional layers have $3 \times 3 \times 3$ kernels, ReLU activation and a batch normalization [7] layer after each convolutional layer. The CNN works with 3D patches heuristically defined as $8 \times 24 \times 24$ voxels on the z, y and x coordinates by considering the spacing information.

The encoder takes 3D input patches from the set of input images, one for each MRI modality. The patches from the input images are concatenated (i.e., the block size is $8 \times 24 \times 24 \times 3$, where 3 is the number of input modalities)

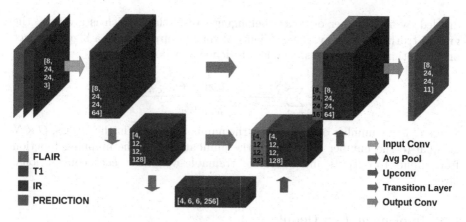

Fig. 2. Description of the proposed CNN model. The blue blocks represent feature maps at the encoding stage. They downsize through the pooling layers, but also connect with the decoding stage through the transition layers. Green feature maps are the results of up-convolution operation. They are concatenated with reduced channel maps from the transition layers. Green and yellow feature maps go through convolution operations twice before the next up-convolution operation. (Color figure online)

and then pass through a series of three convolutional layers with 64 output filters in cascade (Level 1). The output feature maps are reduced to the half of original size (i.e., $4 \times 12 \times 12$) by an average pooling and then pass through four convolutional layers with 128 filters in cascade (Level 2). In the same manner, the feature maps are reduced once more to the size of $4 \times 6 \times 6$ and pass through five convolutional layers with 256 filters (Level 3).

The decoder stage starts from the feature maps generated in Level 3 and applies a deconvolution layer with a kernel $1 \times 2 \times 2$ and stride $1 \times 2 \times 2$ to match the size of the features maps in Level 2. The 2 features maps are concatenated followed by two convolutional layers and the process is repeated to get features maps in Level 1 as it is suggested in the U-Net [11] architecture. With a final output of size $8 \times 24 \times 24 \times 64$, the last convolution operation is applied to reduce the number of channels to the number of labels defined by the dataset, followed by a softmax activation function for more than one output label or sigmoid if there is a single label.

Moreover, we add transition layers [3] between the encoder and the decoder. Specifically, the feature maps generated in the last layer at Level 1 on the encoder, pass through a convolutional layer with 16 filters and then are connected to the decoder. Similarly, the feature maps at Level 2 pass through a convolutional layer with 32 filters and then are connected to the decoder.

3.2 Loss Function

The loss function was defined as a normalized categorical cross entropy loss (1) [13], in other words, for each class the logarithmic errors are added and then divided by the number of voxels belonging to the class which is given by the ground truth. In this way, the gradients of smaller objects have a bigger impact on the optimization process. The loss is define as follows

$$\Psi = -\sum_{c=1}^{C} \frac{\sum_{n=1}^{N} y_{c,n} log(f(\Theta, x_{c,n}))}{\sum_{n=1}^{N} y_{c,n}}, \tag{1}$$

where C is the number of clases, N is the number of voxels in the c class, $C \times N$ is equal to the number of voxels in the minibatch, f it the mapping function from inputs to outputs and Θ are the trainable weights in each convolutional layer.

3.3 Implementation Details

In training stage, we randomly sampled 3D patches from the input MRI images in the training set to get a new minibatch each step and then learn the 3D CNN which can predict the labels in the patch. We trained the model with Adam optimizer, learning rate 1e-4, learning rate decay of 10% after 40 thousand steps for 500 thousand steps. Our model was implemented in TensorFlow and trained on Nvidia Titan XP GPUs.

In the inference stage, we extracted 3D image patches from each MRI modality, applied the intensity normalization and concatenated these 3D volumes to run CNN model and get a prediction map (see Fig. 3). 3D patches are extracted from the testing images with a stride of $4 \times 12 \times 12$. This configuration ensures that every voxel is predicted 8 times with a score assigned to every class. The class that scores the highest is selected as a final prediction for that specific voxel.

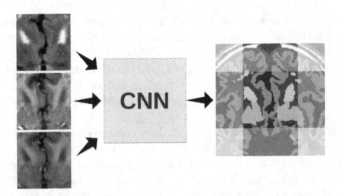

Fig. 3. 2D visualization of the patch extraction and prediction image reconstruction. FLAIR, T1, and IR patches are shown on the left and a partially reconstructed prediction image is shown on the right.

Table 1. Results of our method on the challenge dataset.

	Mean dice coefficient	Mean volume similarity	Mean 95% Hausdorff distance (mm)
Gray matter	0.860	0.976	1.27
Basal ganglia	0.834	0.959	3.64
White matter	0.882	0.955	2.12
WMH	0.652	0.849	10.25
CSF	0.837	0.956	2.22
Ventricles	0.931	0.967	2.81
Cerebellum	0.939	0.982	3.74
Brain stem	0.905	0.950	3.92

4   Experimental Results

The proposed method was evaluated on the MRBrainS18 challenge dataset [9]. Due to the limited training data, we first searched reasonable hyper-parameters by performing a leave-one-out cross validation on 7 training data samples. Thereafter, we trained a network with all 7 training data samples using the hyper-parameters. Including our method, the remaining challenge methods were evaluated on the hidden 23 test data samples. Most methods were based on the deep neural networks such as variations of 2D and 3D U-Net [11], DeepMedic network [8], multiple networks ensemble models, and only one method was based on a traditional generative model using Bayes theory on two priors, a probabilistic tissue atlas and their intensity distribution.

The segmentation accuracy was evaluated by multiple metrics such as mean Dice coefficient, mean volume similarity and mean 95% Hausdorff distance. Every metric was calculated for each label and then averaged across the test dataset. Table 1 shows the average scores for eight labels and Fig. 4 shows box plots representing the distributions of accuracy scores on 23 testing data.

The cerebellum was well segmented across the all testing subjects; the highest Dice coefficient and volume similarity were achieved without outliers in this structure. Although the basal ganglia, ventricles and brain stem were the structures with most number of outliers in all metrics, the average scores were relatively high. The gray matter, white matter and CSF achieved good performance in terms of Hausdorff distance and volume similarity metrics, but relatively lower Dice coefficient than other structures. Due to the large size of this structures, the volume similarity does not give much information, but the distance metric indicates that the misclassified voxels are close to the boundary. According to the Hausdorff distance, there are several structures with a score close to 3 mm, which happens to be the spacing at the z axis on the dataset. Those structures are more likely to have misclassify voxels on the z axis rather than on the x or y axis.

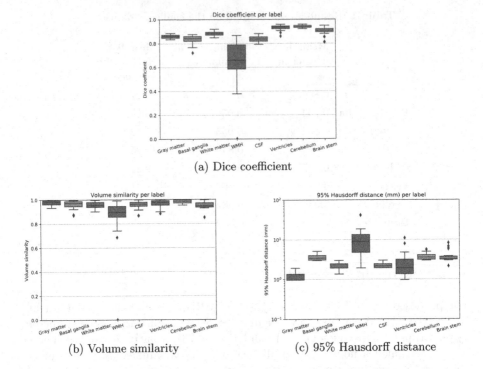

(a) Dice coefficient

(b) Volume similarity

(c) 95% Hausdorff distance

Fig. 4. Challenge metrics results on our method

The WMH class was the most challenging among all 8 structures. These are abnormalities in the white matter which means they do not have any regular shape, size or location. In general, our proposed model predictions were similar to the ground truth, volume similarity show that they were around 85% accurate, but a mean Hausdorff distance around 10 mm shows the presence of small False positive or False negative on entire WMH regions which also explains the lower Dice coefficient result. There are 2 outliers for the WMH category (see Fig. 4), the first one is a healty subject with a False positive region, or a subject with False negative regions, both examples will produce a 0 score in Dice coefficient and volume similarity. The second outlier seems to have predicted WMH regions outside the brain given its high Hausdorff distance and the outlier presence on the volume similarity.

To compare overall scores of submitted methods, a weighted version of each metric was calculated by a method described on the challenge website [9]. Specifically, all metrics were weighted, per segmentation label, to the 95% range of the original 22 participants in the challenge. Then, the weighted results were summed up to get a final score in the range of 8.016 to 9.971, where a higher score means a better performance.

Table 2 shows the performances of comparison methods. As shown in Table 2, our proposed method, i.e. MISPL, obtained the highest overall score on the

Table 2. Competition results.

Rank	Team	Weighted DC	Weighted VS	Weighted H95	Total
1	**MISPL**	4.44	5.85	−0.32	9.971
2	K2	4.42	5.84	−0.34	9.915
3	xuhuaren	4.43	5.8	−0.35	9.872
4	nanand2	4.33	5.82	−0.36	9.783
5	TencentXLab	4.38	5.77	−0.38	9.776
6	BIGR brain	4.32	5.8	−0.44	9.683
7	nic vicorob	4.34	5.75	−0.41	9.678
8	Coroflo	4.32	5.79	−0.44	9.672
9	IBBM Dundee	4.31	5.76	−0.41	9.667
10	NEUROPHET	4.34	5.74	−0.42	9.65
11	HUST-LRDE	4.32	5.73	−0.4	9.65
12	ibbm TUM	4.32	5.72	−0.39	9.644
13	KCL-PET	4.33	5.77	−0.48	9.624
14	MSL SKKU	4.33	5.76	−0.49	9.603
15	icip	4.3	5.77	−0.48	9.596
16	clong	4.32	5.75	−0.48	9.593
17	Biomediq	4.13	5.73	−0.41	9.448
18	bigr-igi	4.22	5.69	−0.6	9.305
19	CUMed	4.2	5.53	−0.5	9.225
20	sunrise14	4.09	5.5	−0.94	8.646
21	SPM	3.94	5.38	−0.75	8.578
22	Jazz1	3.78	5.32	−1.09	8.016

challenge. Most methods achieved good results on relatively large regions such
as cerebellum and ventricles, but achieved low scores on the WMH abnormalities because the WMH regions are usually much smaller than other tissues in
the brain. Thus, it decreased the overall score significantly. The conventional
categorical cross entropy gives more weight to bigger objects, resulting in an
under-segmentation or completely miss of those small objects. By normalizing
the loss function, our proposed method benefited from having fewer outliers than
other competitors and a better segmentation of the WMH regions, even though
it was also the lowest scored class among all classes in our result.

In summary, the proposed model ranked first in all three metrics, which
means it learned the most general features and also, it regularized well from the
very limited training data available. This points out the advantage offered by
the use of small 3D patches, a normalized loss function and pooling layers, even
if the 3D volume seemed to be already small.

5   Conclusion

We proposed a 3D patchwise convolutional neural network method to address the brain MRI segmentation task. Our network consists of encoder-decoder with transition layers, batch normalization and is trained with a normalized cross entropy. The transition layer and batch normalization regularized the model during training, which was vital given the very small amount of training data. The transition layer gives more weights to higher level features, and thus the model can learn the features required to segment large and small regions from a small 3D patch. We have also demonstrated the impact of the new loss function through the experiments. By analyzing three metrics on the MRbrainS18 challenge, we conclude that our method is effective to segment the whole brain region of interests.

Acknowledgement. This research was supported by Basic Science Research Program through the National Research Foundation (NRF) of Korea funded by the Ministry of Education (2018R1D1A1B07044473).

References

1. Akkus, Z., Galimzianova, A., Hoogi, A.: Deep learning for brain MRI segmentation: state of the art and future directions. J. Digit. Imaging **30**, 449 (2017)
2. Chen, H., Dou, Q., Yu, L., Heng, P.A.: VoxResNet: deep voxelwise residual networks for volumetric brain segmentation. arXiv:1608.05895 (2016)
3. Chen, L.C., Zhu, Y., Papandreou, G., Schroff, F., Adam, H.: Encoder-decoder with atrous separable convolution for semantic image segmentation. arXiv:1802.02611 (2018)
4. Cui, S., Mao, L., Jiang, J., Liu, C., Xiong, S.: Automatic semantic segmentation of brain gliomas from MRI images using a deep cascaded neural network. J. Healthcare Eng. **2018**, 14 (2018)
5. Despotovic, I., Goossens, B., Philips, W.: MRI segmentation of the human brain: challenges, methods, and applications. Comput. Math. Methods Med. **2015**, 23 (2015). Article ID 450341
6. Dolz, J., Gopinath, K., Yuan, J., Lombaert, H., Desrosiers, C., Ayed, I.B.: HyperDense-Net: a hyper-densely connected CNN for multi-modal image segmentation. arXiv:1804.02967 (2018)
7. Ioffe, S., Szegedy, C.: Batch normalization: accelerating deep network training by reducing internal covariate shift, pp. 448–456 (2015)
8. Kamnitsas, K., et al.: Efficient multi-scale 3D CNN with fully connected CRF for accurate brain lesion segmentation. Med. Image Anal. **36**, 61–78 (2016)
9. Kuijf, H.J., Bennink, E.: Grand challenge on MR brain segmentation at MICCAI 2018. http://mrbrains18.isi.uu.nl
10. Long, J., Shelhamer, E., Darrell, T.: Fully convolutional networks for semantic segmentation. In: Proceedings of IEEE Conference on Computer Vision and Pattern Recognition (CVPR), pp. 3431–3440 (2015)
11. Ronneberger, O., Fischer, P., Brox, T.: U-Net: convolutional networks for biomedical image segmentation. In: Navab, N., Hornegger, J., Wells, W.M., Frangi, A.F. (eds.) MICCAI 2015. LNCS, vol. 9351, pp. 234–241. Springer, Cham (2015). https://doi.org/10.1007/978-3-319-24574-4_28

12. Shreyas, V., Pankajakshan, V.: A deep learning architecture for brain tumor segmentation in MRI images. In: 2017 IEEE 19th International Workshop on Multimedia Signal Processing (MMSP) (2017)
13. Wu, Z., Shen, C., Hengel, A.: High-performance semantic segmentation using very deep fully convolutional networks. arXiv:1604.04339 (2016)

Computational Precision Medicine

Dropout-Enabled Ensemble Learning for Multi-scale Biomedical Data

Alexandre Momeni$^{(\boxtimes)}$, Marc Thibault$^{(\boxtimes)}$, and Olivier Gevaert$^{(\boxtimes)}$

Departments of Medicine and Biomedical Data Science,
Stanford University, Stanford, CA 94305, USA
{aamomeni,marcthib,ogevaert}@stanford.edu

Abstract. Leveraging information from multiple scales is crucial to understanding complex diseases such as cancer where this could have a significant impact in improving diagnoses, patient management and treatment decisions. Recent advances in Convolutional Neural Networks (CNNs) have enabled major breakthroughs in biomedical image analysis, in particular for histopathology and radiology images. Our main contribution is a methodology to combine independent CNN models built for these two types of images in order to improve diagnostic accuracy. We train separate CNN models and combine them using a Dropout-Enabled meta-classifier. Our framework achieved second place in the MICCAI 2018 Computational Precision Medicine Challenge.

Keywords: Biomedical imaging · Cancer ·
Computational Precision Medicine · Deep learning

1 Introduction

Recent breakthroughs in deep learning and Convolutional Neural Networks (CNN) have enabled the development of state-of-the art models for many image classification tasks [1,2], with wide applications in areas such as medical image analysis [3,4]. These models have the advantage that they automatically learn the appropriate image features, as opposed to traditional machine learning approaches which require hand-crafted features [5,6]. The features learned by each separate CNN model can also conveniently be stacked together to form the input of a joint model. We follow this approach, building two separate CNNs for radiology and pathology images for the desired task, and subsequently combining them.

Furthermore, the deep learning community has developed many regularization techniques in recent years to deal with the issue of overfitting. This is useful given the scarcity of medical images data. We incorporate these techniques throughout our training process. In particular, we use dropout [7] to generate

---

A. Momeni and M. Thibault—Both authors contributed equally to this manuscript.

A. Crimi et al. (Eds.): BrainLes 2018, LNCS 11383, pp. 407–415, 2019.
https://doi.org/10.1007/978-3-030-11723-8_41

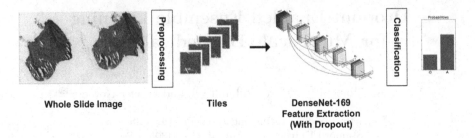

Fig. 1. Description of the histopathology prediction pipeline. The slides are tiles, features are extracted with a finetuned DenseNet, classification is made by a final fully connected network.

many variations of the base patients, augmenting the data available for our models. This method lets us leverage the complexity of the medical images, whilst still being able to generalize to unseen data.

Our method achieved second place in the MICCAI 2018 Computational Precision Medicine Challenge, in which participants were asked to classify a cohort of lower grade glioma tumor cases into two sub-types using radiology and histopathology images [8]. Here, we report this model and its results on this classification problem.

2 Methodology

2.1 Data Set

We evaluate our framework on the Computational Precision Medicine Combined Radiology and Pathology classification data set [8]. It contains histopathology and radiology images for 32 glioma patients with annotations for training and validation (16 oligendroglioma and 16 astrocytoma) and 20 glioma patients for testing. These images were collected from several medical centers.

2.2 Histopathology Image Data Preprocessing and Modeling

Analysis of histopathology slides is a critical step in oncology where it defines the gold standard for diagnosis, prognosis and treatment design. It largely consists of careful microscopic examination of hematoxylin and eosin (H&E) stained tissue sections by a highly skilled pathologist. This can be a tedious, time-consuming and sometimes subjective task. Advances in slide scanning technology and reductions in cost of digital storage capacity have enabled the widespread adoption of digital pathology over the past decade [9]. At the same time, the dramatic increase in computational power and the breakthroughs in deep learning have fueled the rich expansion of visual recognition research [1]. These developments together have led to the rapid emergence of computational histopathology. Most recent works have successfully leveraged state-of-the-art Convolutional Neural

Networks (CNNs) for tasks such as disease detection and diagnosis, highlighting the effectiveness and relevance of learned features in complex images such as histopathology slides [10–12].

Digital pathology images are massive data sets, which at highest zoom level can have a digital resolution upwards of 100k pixels in both dimensions. However, since localized annotations are very difficult to obtain, data sets may only contain whole slide level diagnosis labels, falling into the category of weakly-supervised learning. To deal with this, we modify existing CNNs architecture to incorporate a Multiple Instance Learning (MIL) framework [13]. This consists in dividing the histopathology slides into small high resolution patches, sampling randomly from these patches and applying patch level CNNs. The MIL framework is then used to combine patch level predictions intelligently and make an overall slide prediction. In the following sub-sections we present our preprocessing pipeline and provide a description of our model for histopathology image analysis.

Preprocessing. We perform the following preprocessing steps on the highest slide resolution available:

1. Region of Interest: tissue segmentation is necessary given that there are large areas of white background space in histopathology images which are irrelevant for analysis. We follow a threshold based segmentation method to automatically detect the foreground region. In particular, we first transform the image from RGB to HSV color space and apply Otsu's method [14] to find the optimal threshold in each channel. The masks are then combined to compute the final tissue segmentation.
2. Tiling: we tile the tissue region extracted from the original slides into 256×256 patches.
3. Color Normalization: stain normalization is essential given that the results from the staining procedure can vary greatly. Indeed, differences in slide scanners or staining protocol can materially impact stain color, which in turn can affect algorithm performance. Many methods have been introduced to overcome this well defined problem, including sophisticated end-to-end deep learning solutions [15]. For simplicity, we resort to a histogram equalization algorithm as proposed in [16].

CNN Model. Convolutional Neural Networks (CNNs) are very computationally expensive to train in practice and they require a large data set to avoid overfitting. We used the DenseNet-169 [17] architecture starting with initial pre-trained parameters from ImageNet [18] and we used fine tuning of these parameters to speed up convergence. We also replaced the last fully connected layers to be compatible with our classification task, using dropout for regularization. Our CNN is trained at the patch level and then produces an average slide level score by averaging all sampled patches for a given patient. The steps of the model are described schematically in Fig. 1.

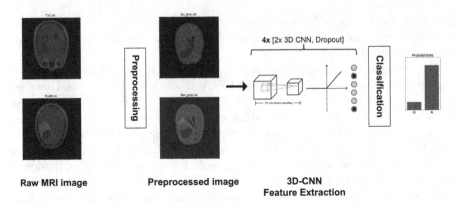

Raw MRI image   **Preprocessed image**   **3D-CNN Feature Extraction**

Fig. 2. Description of the radiology prediction pipeline. Preprocessing removes the skull and balances the channels, features are extracted with a 3D-CNN, classification is made with a fully connected network.

2.3   Radiology Image Data Preprocessing and Modeling

Our computational task consists in evaluating also the radiology images of a patient's lower grade glioma in the form of multi-modal Magnetic Resonance (MR) images. An automated algorithm for classification should analyze these 3D MR images, aggregate local information to understand where the tumor is located, and then compute metrics on these tumors (e.g. capturing size, intensity or texture). These features would then have to be evaluated so that a final decision rule can be devised.

A classical approach to this problem would be to compute hand-made features from the segmentation mask of the tumor [19,20] - this segmentation mask being obtainable in a completely computational fashion - and then to use these features to classify the patient's status using a linear classification algorithm in the feature space [4,20]. However, this method would have to rely on complex feature extraction methods [21], mimicking a radiologist's analysis, which can be time-consuming and difficult to hard-code.

Convolutional Neural Networks (CNNs), on the other hand, have proven effective in reaching state-of-the art results in computational analysis of biomedical images for disease detection and diagnosis [2,22]. Consequently, a newer, more data-driven way to tackle this classification task would be to rely on CNNs for feature extraction and image downsampling, letting us train a neural network through backpropagation to perform the classification task. This CNN can take into input either the raw image channels (i.e. MR image modalities), or the tumor segmentation mask, or both. However, in this contribution, we use the raw modalities as inputs, in order not to rely on external data, and to evaluate the stand-alone performance of our approach.

Preprocessing. We applied preprocessing steps which have been used for the tumor segmentation task, from brain MR images. It consists in the tasks of bias correction, skull-stripping, and registration:

1. Bias Correction: we used the FSL library to remove the bias fields from all modality images, as provided by [23];
2. Skull Removal: we used the Brain Extraction Tool (BET, [24]) from the same library to remove the skull structure from these images;
3. Co-registration: we used the reg_aladin command from the niftyreg library to co-register the modalities on the same standard grid [25].

We also enriched our data by randomly flipping scans along the x-axis at train time.

Furthermore, we chose to only rely on the contrast-enhanced T1 and FLAIR MR modalities for this radiology task. Indeed, T1+contrast is sufficient to magnify the grey matter; while the FLAIR modality lets the unusual brain structures appear explicitly. This resulted in 27 patients out of 32 available in the train set. The images were then re-sized to $320 \times 320 \times 24$.

CNN Model. The input of this model is a 3-D voxel image with three spatial dimensions times two modalities per voxel, the FLAIR and T1+contrast MR images. We used an 8-layers 3-D CNN to extract deep features from the images, with three 3-D maxpooling operations to reduce the sample size. The convolution layers have a receptive field of five cells from the previous layers, and extract from eight features on the first layer, to 64 channels on the last layer. The maxpooling layers downsample the images by a factor of two in each dimension.

After the last convolutional layer is applied, we averaged the extracted features over all the 3D space to have a unique 100-dimensional feature vector for the patient. This feature vector is then connected to a 1-dimensional output for a classification task with cross-entropy loss, and the whole network is trained with the back-propagation algorithm.

We inserted a total of 8 dropout layers throughout the network, both to avoid overfitting, and, most importantly for the ensemble learning step, to let us evaluate the network at test time. The overall pipeline is sketched in Fig. 2.

2.4 Meta-classifier Methodology Image Data Preprocessing and Modeling

Next, to combine the two CNNs modeling the histopathology and MR images, we introduce a meta-algorithm to combine the individual models. Our ensemble learning methodology allows each model to be trained separately, and combines their predictions into a single, more robust output.

Meta-classifier for Models Combination. Since these models rely on very different data sources, with different scales, batching methods and actual biological meaning, it is not always possible to train an end-to-end backpropagation algorithm combining these. A crucial limitation is the computational power required for these models to run simultaneously, especially in terms of storage space. Another inconvenience to the co-training the models is reduced modularity, where a big advantage is if more data from an extra level of biomedical data can be added without having to reconstruct and train the other models.

Our individually trained networks provide us with classification scores for each patient in our data set, quantifying the learned probability that the patient belongs to one category or the other. We then concatenate these individual scores into a two-dimensional vector and train a meta-classifier on these vectors of scores to make a combined prediction for the patient. The meta-classifier using information from both models is a classical classification algorithm, in this case we used a random forest. In order to quantify our ability to fit a meta-classifier, as well as its ability to generalize to unseen data, we used four-fold cross-validation combined with a random forest classifier with ten trees.

Dropout-Enabled Scores Consolidate the Models' Outputs. Next, we evaluated the use of regularization through dropout for the ensemble learning phase. The idea is to activate the models' random dropout layers at score extraction time, so that individual models produce multiple classification scores for each patient. This array of scores is then averaged into a global patient score. This approach is comparable to the reasoning of [26], where it is applied to RNNs for final classification, while we use it as the penultimate step for our classifier combination.

The rationale behind this choice is that individual classification CNNs were trained with dropout, such that they have learned several robust ways to classify a patient's status. Consequently, scores extracted with dropout capture the variability of the models' prediction. Averaging these scores into a global dropout-enabled patient score removes the noise which emerges when running a single score prediction. It yields more robust aggregated classification scores which leverages the networks' structure, without the need of an end-to-end joint training.

We quantified the extent to which this technique improves the class separation, by running the same analysis as before. We evaluated the performance of a random forest classifier using four-fold cross-validation on the dropout-enabled scores. We used accuracy and Area Under the ROC curve (AUC) to evaluate the models.

3   Results

Figure 3(a) shows the output classification scores of each patient, respectively from the pathology and the radiology models, depending on their status, on the training set. This shows that these scores let us define with some confidence some of the patients' status. However, there does not seem to be a simple separation between the two classes leading to a reliable meta-classifier.

Next, Table 1 presents the classification results of this method on the raw scores, as estimated via vross-validation. It appears that the pathology model is a good predictor of the patients' statuses (accuracy 78%, AUC 0.83); while radiology does not make a good contribution (accuracy 53%, AUC 0.54). The combined model performed well, but not significantly better than the pathology model, suggesting that the radiology model does not contribute to the performance (accuracy 81%, AUC 0.84).

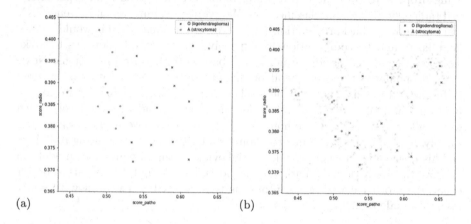

(a)   (b)

Fig. 3. Scatter plot of the patients classification scores, according to the pathology model on the x-axis, and the radiology model on the y-axis, colored by their status. (a): Raw classification scores. (b): Dropout-enabled classification scores; the patient-averaged scores are highlighted.

Next, Fig. 3(b) presents the classification scores of the patients of both classes, after the dropout-enabled aggregation phase has been applied. This shows that the patients which were in grey zones are now easier to classify, resulting in more separable classes. The performances of the dropout-enabled scores are summarized in Table 1. The classification scores of the histopathology model alone are

Table 1. Accuracy and AUC for classifying subtype of brain tumors using 4-fold cross-validation and a random forest classifier, both for raw scores and average dropout-enabled scores.

Model	Data	Accuracy	AUC
Raw scores	Radiology	53%	0.54
	Histopathology	78%	0.83
	Combined	81%	0.84
Dropout-enabled scores	Radiology	70%	0.70
	Histopathology	79%	0.83
	Combined	**85%**	**0.92**

similar to the previous ones (accuracy 79%, AUC 0.83). Most importantly, we see that this technique improves the performance of the radiology-based model (accuracy 70%, AUC 0.70), and contributes to enhancing the combined classifier (accuracy 85%, AUC 0.92).

4 Conclusion

We have proposed a new approach to combine individual CNN classifiers leveraging dropout to combine two types of biomedical image data: histopathology and radiology images, for classification. We avoided training an end-to-end CNN training through backpropagation on heterogeneous data, as this would reduce modularity and be computationally expensive. Our method allows us to make more relevant decisions for outlier cases, based on separate predictions. Next, using dropout to generate several predictions of a patient's data can be used in further directions, that will be studied independently in further works: we plan on using the same analysis at feature-level instead of score-level, which would allow us to analyze the non-linear interactions between multi-scale data. Finally, applying this technique to glioma subtype classification using radiology and histopathology images provides initial evidence that a multi-scale model can improve the prediction of a patient's diagnosis. Extending this analysis to bigger cohorts and other classification problems is warranted and will allow to study the multi-scale model in more detail.

Acknowledgements. Research reported in this publication was supported by the National Institute of Biomedical Imaging and Bioengineering of the National Institutes of Health under Award Number R01EB020527. The content is solely the responsibility of the authors and does not necessarily represent the official views of the National Institutes of Health.

References

1. LeCun, Y., Bengio, Y., Hinton, G.: Deep learning. Nature **521**(7553), 436 (2015)
2. Esteva, A., et al.: Dermatologist-level classification of skin cancer with deep neural networks. Nature **542**(7639), 115 (2017)
3. Ker, J., Wang, L., Rao, J., Lim, T.: Deep learning applications in medical image analysis. IEEE Access **6**, 9375–9389 (2018)
4. Yi, D., Zhou, M., Chen, Z., Gevaert, O.: 3-D convolutional neural networks for glioblastoma segmentation. arXiv preprint arXiv:1611.04534 (2016)
5. Gevaert, O., et al.: Predictive radiogenomics modeling of EGFR mutation status in lung cancer. Sci. Rep. **7**, 41674 (2017)
6. Bakr, S.H., et al.: Noninvasive radiomics signature based on quantitative analysis of computed tomography images as a surrogate for microvascular invasion in hepatocellular carcinoma: a pilot study. J. Med. Imaging **4**(4), 041303 (2017)
7. Srivastava, N., Hinton, G., Krizhevsky, A., Sutskever, I., Salakhutdinov, R.: Dropout: a simple way to prevent neural networks from overfitting. J. Mach. Learn. Res. **15**(1), 1929–1958 (2014)

8. MICCAI CPM Competition. http://miccai.cloudapp.net/competitions/82
9. Madabhushi, A., Lee, G.: Image analysis and machine learning in digital pathology: challenges and opportunities (2016)
10. Cireşan, D.C., Giusti, A., Gambardella, L.M., Schmidhuber, J.: Mitosis detection in breast cancer histology images with deep neural networks. In: Mori, K., Sakuma, I., Sato, Y., Barillot, C., Navab, N. (eds.) MICCAI 2013. LNCS, vol. 8150, pp. 411–418. Springer, Heidelberg (2013). https://doi.org/10.1007/978-3-642-40763-5_51
11. Liu, Y., et al.: Detecting cancer metastases on gigapixel pathology images. arXiv preprint arXiv:1703.02442 (2017)
12. Hou, L., Samaras, D., Kurc, T.M., Gao, Y., Davis, J.E., Saltz, J.H.: Patch-based convolutional neural network for whole slide tissue image classification. In: Proceedings of the IEEE Conference on Computer Vision and Pattern Recognition, pp. 2424–2433 (2016)
13. Dietterich, T.G., Lathrop, R.H., Lozano-Pérez, T.: Solving the multiple instance problem with axis-parallel rectangles. Artif. Intell. **89**(1–2), 31–71 (1997)
14. Otsu, N.: A threshold selection method from gray-level histograms. IEEE Trans. Syst. Man Cybern. **9**(1), 62–66 (1979)
15. Shaban, M.T., Baur, C., Navab, N., Albarqouni, S.: StainGAN: stain style transfer for digital histological images. arXiv preprint arXiv:1804.01601 (2018)
16. Nikitenko, D., Wirth, M., Trudel, K.: Applicability of white-balancing algorithms to restoring faded colour slides: an empirical evaluation. J. Multimedia **3**(5), 9–18 (2008)
17. Huang, G., Liu, Z., Van Der Maaten, L., Weinberger, K.Q.: Densely connected convolutional networks. In: CVPR, vol. 1, p. 3 (2017)
18. Deng, J., Dong, W., Socher, R., Li, L.-J., Li, K., Fei-Fei, L.: Imagenet: a large-scale hierarchical image database. In: IEEE Conference on Computer Vision and Pattern Recognition, CVPR 2009, pp. 248–255. IEEE (2009)
19. Wang, S., et al.: Central focused convolutional neural networks: developing a data-driven model for lung nodule segmentation. Med. Image Anal. **40**, 172–183 (2017)
20. Echegaray, S., et al.: Core samples for radiomics features that are insensitive to tumor segmentation: method and pilot study using CT images of hepatocellular carcinoma. J. Med. Imaging **2**(4), 041011 (2015)
21. Itakura, H., et al.: Magnetic resonance image features identify glioblastoma phenotypic subtypes with distinct molecular pathway activities. Sci. Transl. Med. **7**(303), 303ra138 (2015)
22. Akkus, Z., Galimzianova, A., Hoogi, A., Rubin, D.L., Erickson, B.J.: Deep learning for brain MRI segmentation: state of the art and future directions. J. Digit. Imaging **30**(4), 449–459 (2017)
23. Zhang, Y., Brady, M., Smith, S.: Segmentation of brain MR images through a hidden Markov random field model and the expectation-maximization algorithm. IEEE Trans. Med. Imaging **20**(1), 45–57 (2001)
24. Smith, S.M.: Fast robust automated brain extraction. Hum. Brain Mapp. **17**(3), 143–155 (2002)
25. Ourselin, S., Roche, A., Subsol, G., Pennec, X., Ayache, N.: Reconstructing a 3D structure from serial histological sections. Image Vis. Comput. **19**(1–2), 25–31 (2001)
26. Gal, Y., Ghahramani, Z.: A theoretically grounded application of dropout in recurrent neural networks. In: Advances in Neural Information Processing Systems, pp. 1019–1027 (2016)

A Combined Radio-Histological Approach for Classification of Low Grade Gliomas

Aditya Bagari[iD], Ashish Kumar[iD], Avinash Kori[iD], Mahendra Khened[iD], and Ganapathy Krishnamurthi[(✉)][iD]

Indian Institute of Technology Madras, Chennai 600036, India
gankrish@iitm.ac.in

Abstract. Deep learning based techniques have shown to be beneficial for automating various medical image tasks like segmentation of lesions and automation of disease diagnosis. In this work, we demonstrate the utility of deep learning and radiomics features for classification of low grade gliomas (LGG) into astrocytoma and oligodendroglioma. In this study the objective is to use whole-slide H&E stained images and Magnetic Resonance (MR) images of the brain to make a prediction about the class of the glioma. We treat both the pathology and radiology datasets separately for in-depth analysis and then combine the predictions made by the individual models to get the final class label for a patient. The pre-processing of the whole slide images involved region of interest detection, stain normalization and patch extraction. An autoencoder was trained to extract features from each patch and these features are then used to find anomaly patches among the entire set of patches for a single Whole Slide Image. These anomaly patches from all the training slides form the dataset for training the classification model. A deep neural network based classification model was used to classify individual patches among the two classes. For the radiology dataset based analysis, each MRI scan was fed into a pre-processing pipeline which involved skull-stripping, co-registration of MR sequences to T1c, re-sampling of MR volumes to isotropic voxels and segmentation of brain lesion. The lesions in the MR volumes were automatically segmented using a fully convolutional Neural Network (CNN) trained on BraTS-2018 segmentation challenge dataset. From the segmentation maps $64 \times 64 \times 64$ cube patches centered around the tumor were extracted from the T1 MR images for extraction of high level radiomic features. These features were then used to train a logistic regression classifier. After developing the two models, we used a confidence based prediction methodology to get the final class labels for each patient. This combined approach achieved a classification accuracy of 90% on the challenge test set (n = 20). These results showcase the emerging role of deep learning and radiomics in analyzing whole-slide images and MR scans for lesion characterization.

All authors contributed equally.

A. Crimi et al. (Eds.): BrainLes 2018, LNCS 11383, pp. 416–427, 2019.
https://doi.org/10.1007/978-3-030-11723-8_42

1 Introduction

1.1 Medical Image Analysis for LGG Classification

Medical imaging techniques like Magnetic Resonance Imaging (MRI) is used for detection and assessment of the abnormalities inside the body. The non-invasive and non-ionizing property of the MRI make them suitable for oncology imaging studies such as brain tumors. In addition to MR images, the gold standard for tumor assessment and grading usually employ whole-slide imaging of tissue biopsy under a microscope for assessing at cellular level. The assessment of these medical images are mostly done by visual inspection of trained radiologist or a pathologist. However, manual inspection of vast amounts of data is usually error prone, time consuming and introduces inter-rater variability. Hence, several research communities in medical image analysis are continuously working on developing methods to automate tasks such as segmentation and quantification of medical images.

Gliomas are one of the leading cause of brain cancer and usually associated with poor prognosis and lesser survival rates. The gold standard for grading of gliomas is mostly based on the pathology reports got from tissue biopsies. In, this work we propose a methodology of classification of Gliomas into Astrocytomas and Oligodendrogliomas based on MR images of the brain using radiomic features localized to segmented tumor from the T1 MR image. We also propose a methodology of refining the lower-confidence predictions from model based on MR images by combining the model's predictions with whole-slide image analysis.

2 Datasets Used

The dataset comprised of Radiology and Histopathology slides from 30 different patients. The dataset was equally distributed for the two classes of tumor-namely Astrocytoma and Oligodendroglioma. The Radiology data for each patient consisted of FLAIR, T1, T1C, T2 MR sequences. FLAIR and T2 sequences were missing in four cases provided, hence they were not included in the training set as our segmentation model required all the 4 MR sequences. The pathology data consisted of single whole-slide images for each of the 30 patients.

3 Histopathology Approach

3.1 Preprocessing

The pathology dataset contained 30 wholeslide images each for a single patient. Each pathology slide for a given patient was a large scale image typically spanning across 10–50k pixels across each of the 2 dimensions. A single pathology slide was acquired in multiple scales of resolution. The whole slide image (WSI) contained large areas of white space irrelevant for the training process. The first step in our pathology analysis was finding the Regions of Interest (RoI) in a WSI. The WSI was converted from RGB colorspace to HSV colorspace for better contrast enhancement. The lower and upper thresholds were applied on the pixel values to get binary masks from the WSI (Fig. 1).

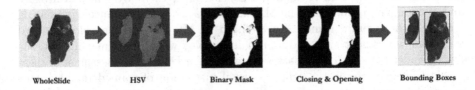

WholeSlide   HSV   Binary Mask   Closing & Opening   Bounding Boxes

Fig. 1. Finding the region of interest (RoI) from the Whole Slide Image

On the generated binary mask, morphological closing and opening operations were applied to fill small holes and remove scattered foreground pixels (of size less than 3×3) from the foreground pixels. From the processed binary masks, bounding boxes around all the discrete contours (each contour encompasses a connected region of tissue in the RoI) were obtained. The generated bounding boxes served as a blueprint for the patch extraction process.

From each of the WSIs, patches of size 224×224 were extracted from the entire RoI with a stride of $(224, 224)$ in both dimensions. To limit the number of patches from a slide, the maximum number of patches from a bounding box were limited to 2k.

We observed that the color intensity variation of WSI across different cases in the dataset was huge and hence stain normalization technique proposed by Reinhard et al. (2001) was employed to obtain uniform patches across multiple whole-slides. This method works by transferring an image from RGB space to a $l\alpha\beta$ color space where the correlation between the different color axes is minimal, hence transformation in color channels can be applied independently, without having any undesirable cross channel artifacts. The normalization technique ensured that the extracted patches from different WSIs had minimal variation in intensity (Figs. 2 and 3).

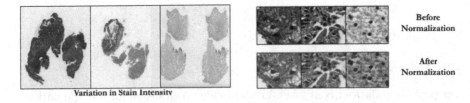

Variation in Stain Intensity

Before Normalization

After Normalization

Fig. 2. Stain normalization for getting uniform colour intensity patterns across different slides

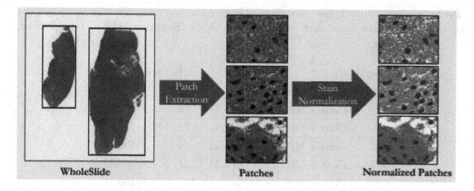

Fig. 3. Patch extraction from the Whole Slide Image followed by stain normalization

3.2 Feature Extraction and Anomaly Detection

Due to the presence of similar normal regions in the WSIs of the two classes, it was important to remove such patches which potentially came from regions that were non-tumour. To find out potential tumor patches, it was essential to obtain a good feature representation for each patch. Autoencoders are employed for extracting features from unlabelled data. We use an autoencoder based approach to extract features from each patch. We train an autoencoder using all the patches extracted from each of the slides. The autoencoder has five convolutional layers to downsample a patch followed by a single fully connected layer in the middle and again five convolutional layers to upsample the patch back to its original size. The central fully connected layer has 128 nodes giving a feature vector of size 128×1 for each patch. We use pixel wise reconstruction loss to train this autoencoder (Table 1).

After obtaining the feature vectors for each patch in a given Whole Slide Image, it was required to find a subset of these patches which can contain potential tumor regions. We treated the task of finding this subset as the problem of anomaly detection, where the tumor (anomaly) patches were required to be filtered out from the entire set of patches. We used the Isolation Forest Liu et al. (2008) technique to perform the task of anomaly detection. Isolation forest uses the feature vector representation of all patches from a WSI image and isolates anomaly patches based on the these features in a two stage process. The first stage builds isolation trees (random decision trees) and in the second stage, test instances are passed through the trees to obtain an anomaly score for each instance based on the path length required to isolate an observation.

After extracting patches exhaustively from a WSI, we selected anomaly patches and used only the selected patches for further training. All these filtered patches from a WSI were assigned the same label as of the WSI. A total of 60k patches combined for both the classes were obtained after selecting these anomaly patches. We further used these patches for training a classification model for the two classes, Astrocytoma and Oligodendroglioma (Fig. 4).

Table 1. The table below describes the architecture used for the auto-encoder to extract features from each patch. Note: Op - Operation, MP - Max-pooling, US - Up-sampling.

Layer	Op	Kernel	Stride	# Kernels	NonLin
1	Conv + MP	3×3	$(1,1)$	8	ReLU
2	Conv + MP	3×3	$(1,1)$	16	ReLU
3	Conv + MP	3×3	$(1,1)$	32	ReLU
4	Conv + MP	3×3	$(1,1)$	64	ReLU
5	Conv + MP	3×3	$(1,1)$	128	ReLU
6	FC	-	-	128	ReLU
7	US + Conv	3×3	$(1,1)$	64	ReLU
8	US + Conv	3×3	$(1,1)$	32	ReLU
9	US + Conv	3×3	$(1,1)$	16	ReLU
10	US + Conv	3×3	$(1,1)$	8	ReLU
11	US + Conv	3×3	$(1,1)$	3	ReLU

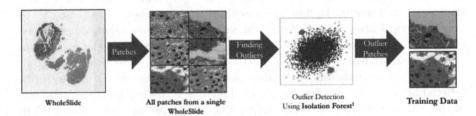

Fig. 4. Features were extracted using an autoencoder based approach and Isolation Forest was used to find anomaly patches from the entire set of patches for a single Whole Slide Image

3.3 Two-Class Classification

DenseNet-161 (Huang et al. (2017)) network was chosen to distinguish patches from Astrocytoma and Oligodendroglioma. The network was trained using binary cross entropy loss function (Fig. 5).

Testing: During testing, a stride based patch extraction was used to obtain all the patches in the region of interest, using a stride of $(224, 224)$. From these patches, anomaly patches were found using Isolation Forest, in a way similar to the training phase. All the filtered patches from a particular whole-slide were fed to the densenet. After obtaining the prediction scores for all the patches from a WSI, a majority class voting was performed among all the predictions to obtain the class label for the slide. The class which had a higher frequency in the predictions was assigned as the class label for the slide.

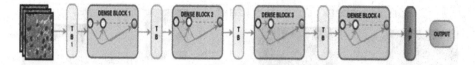

Fig. 5. A Densenet based network was used for the two-class classification task

4   Radiology

4.1   Understanding MR Images

Different pulse sequences of MR imaging system are used to enhance different parts of the tumor. For assessment of brain tumor the following pulse sequences are generally used (Fig. 6):

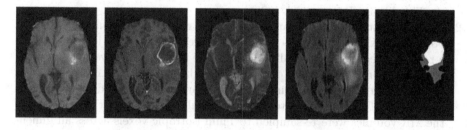

Fig. 6. From left to right: T1, T1c, T2, FLAIR, segmented tumor

1. **T1 weighted:** T1 image of tissues affected by brain tumors are of low signal intensity.
2. **T2 weighted:** T2 image of tissues affected by brain tumors are typically of high signal intensity. Calcifications due to tumor are mostly dark on T2.
3. **Fluid Attenuated Inversion Recovery (Flair):** Uses attenuation of intensity in Cerebro-spinal fluids (CSF) to differentiate between CSF and abnormalities. Generally gives the whole tumor region.
4. **T1-weighted post contrast imaging (T1c):** Gadolinium is used to enhance images and is useful in identifying vascular structures and break-down in the blood-brain barrier typically found in the Necrotic region of the brain.

4.2   Pre Processing of MR Images

Magnetic resonance images were pre-processed to remove structures that could interfere with image segmentation (Fig. 7).

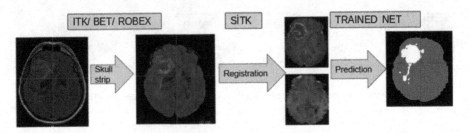

Fig. 7. MR image sequence pre-processing pipeline

1. **Skull Stripping**

   It is necessary to remove the skull from the MRI as the enhancement arising due to its presence can be wrongly interpreted as tumor. Most segmentation networks are trained using skull stripped images as input and hence it is important to maintain this while providing new data. This was done using the ITK library Ibanez et al. (2005).

2. **Co-registration and re-sampling to isotropic voxel spacing**

   Followed by skull stripping is the step of co-registering the MRI sequences to a reference sequence. Generally, there can be movement between scans if the patient does not remain still or if the scan is being done on a different day or while using a different machine. As a standard, we registered sequences T1, FLAIR and T2 with respect to T1c scan for all patients. After co-registration the MR volumes were re-sampled to 1 mm isotropic voxels across all dimensions.

4.3 Segmentation of MR Images

Segmentation of the tumor in Magnetic resonance imaging (MRI) is the first step towards diagnosis. Features like the size, location of the tumor can dictate the stage of the tumor and the appropriate treatment. We trained a segmentation network with the following properties:

1. **Network architecture:** 3-D CNN as shown in the Fig. 8 was used for the task of semantic segmentation.

2. **Data:** Our network architecture was inspired from Kamnitsas et al. (2017) and was trained using the data provided by BraTS Menze et al. (2015) 2018 challenge. 25^3 and 51^3 (re-sized to 19^3) sized patches were extracted from all the four sequences (T1, T2, Flair and T1c) and the network was trained to predict center 9^3.

3. **Training:** The weights of the network were initialized using Xavier initializer Glorot and Bengio (2010), and training was done using the weighted combination of weighted cross entropy and dice loss. Adam optimizer was used with initial learning rate of 0.001 and with the decay factor of 0.1.

4. **CNN based brain tumor segmentation:** The segmentation of the whole tumor region was done using a in-house fully convolution neural network trained on BraTS-2018 dataset (Menze et al. (2015)).

4.4   Radiomic Feature Extraction

The segmentations generated via the above-mentioned model were post processed to ensure to remove noise or any false positive region segmented. By applying connected component analysis, only the biggest tumor predication was kept. As low grade gliomas are small in size, a patch of size 64 * 64 * 64 of the segmentation and the corresponding T1 sequence image per case, centered around the predicted tumor region were extracted (Fig. 9).

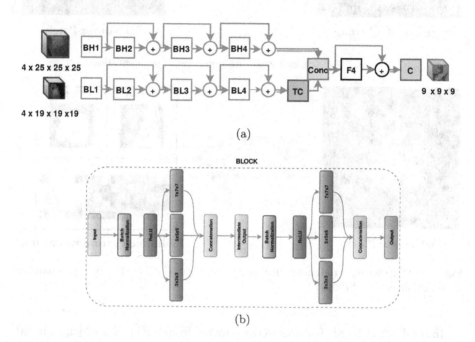

Fig. 8. Semantic segmentation network for segmentation of gliomas from MR volumes. (a) The top portion of the network accepts high-resolution patches (25^3) while the bottom pathway accepts low-resolution input (51^3 patches resized to 19^3) as input. Both the high and low-resolution pathway is composed of inception modules so as to learn multi-resolution features. TC in the network stands for transposed convolution and is used to match the features of the spatial dimension of the low-resolution pathway with those learned in the high-resolution path. The BL and BH refer to building blocks for low and high resolution pathway. (b) The building block of the network. In the block, the dimension of the feature map in an inception module is maintained by setting the padding to 0, 1, 2 for 3×3, 5×5 and 7×7 respectively.

As 3D images contain a lot of spatial and physical information, the 3D T1 image patches and the segmentation patches were used to extract features such as shape, texture, first order, second order and other higher order features using radiomics. The approach to extract high-through quantitative features relies on the radiomics platform provided by pyradiomics Griethuysen et al. (2017).

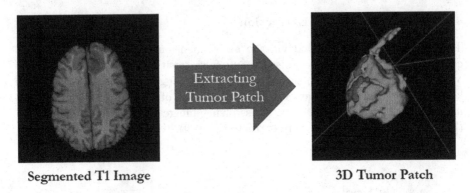

Segmented T1 Image   **3D Tumor Patch**

Fig. 9. 3D patch extraction from generated segmentation

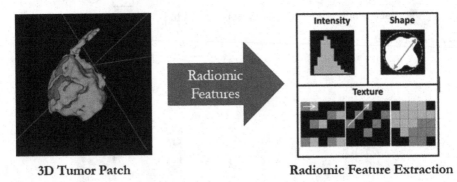

3D Tumor Patch   **Radiomic Feature Extraction**

Fig. 10. Extracting radiomic features from the 3D T1 image patches
www.radiomics.world (2018)

In total 105 radiomic features were extracted from the images which included 3 major kinds of radiomic features namely, Shape Features (13), First Order Features (18) and Texture Features (74). The complete list of extracted radiomic features are detailed below (Fig. 10).

1. **Shape Features (13):**
   Elongation, Flatness, Least Axis, Major Axis, Maximum 2D Diameter Column, Maximum 2D Diameter Row, Maximum 2D Diameter Slice, Maximum 3D Diameter, Minor Axis, Sphericity, Surface Area, Surface Volume Ratio, Volume

2. **First order statistics (18):**
   10 Percentile, 90 Percentile, Energy, Entropy, Interquartile Range, Kurtosis, Maximum,Mean, Mean Absolute Deviation, Median, Minimum, Range, Robust Mean Absolute Deviation,Root Mean Squared, Skewness, Standard Deviation, Total Energy, Uniformity, Variance

3. **GLCM (Gray Level Co-occurrence Matrix) (23):**
   Auto-correlation, Cluster Prominence, Cluster Shade, Cluster Tendency,

Contrast, Correlation, Difference Average, Difference Entropy, Difference Variance, ID, IDM, IDMN, IDN, IMC1, IMC2, Inverse Variance, Joint Average, Joint Energy, Joint Entropy, Maximum Probability, Sum Average, Sum Entropy, Sum Squares

4. **GLRLM (Gray Level Run Length Matrix) (30):**
   Dependence Entropy, Dependence Non-Uniformity, Dependence Non-Uniformity Normalized, Dependence Variance, Gray Level Non-Uniformity, Gray Level Variance, High Gray Level Emphasis, Large Dependence Emphasis, Large Dependence High Gray Level Emphasis, Large Dependence Low Gray Level Emphasis, Low Gray Level Emphasis, Small Dependence Emphasis, Small Dependence High Gray Level Emphasis, Small Dependence Low Gray Level Emphasis, Gray Level Non-Uniformity, Gray Level Non-Uniformity Normalized, Gray Level Variance, High Gray Level Run Emphasis, Long Run Emphasis, Long Run High Gray Level Emphasis, Long Run Low Gray Level Emphasis, Low Gray Level Run Emphasis, Run Entropy, Run Length Non-Uniformity, Run Length Non-Uniformity Normalized, Run Percentage, Run Variance, Short Run Emphasis, Short Run High Gray Level Emphasis, Short Run Low Gray Level Emphasis

5. **GLSZM (Gray Level Size Zero Matrix) (16):**
   Gray Level Non-Uniformity, Gray Level Non-Uniformity Normalized, Gray Level Variance, High Gray Level Zone Emphasis, Large Area Emphasis, Large Area High Gray Level Emphasis, Large Area Low Gray Level Emphasis, Low Gray Level Zone Emphasis, Size Zone Non-Uniformity, Size Zone Non-Uniformity Normalized, Small Area Emphasis, Small Area High Gray Level Emphasis, Small Area Low Gray Level Emphasis, Zone Entropy, Zone Percentage, Zone Variance

6. **NGTDM (Neighbouring Gray Tone Difference Matrix) (5):**
   Busyness, Coarseness, Complexity, Contrast, Strength

4.5   Training Methodology

MR Image based training methodology:

1. For a given MRI sequence, the 105-length deep feature vectors extracted from radiomics were reduced using Principal Component Analysis to a 16-length deep feature vector. This 16-length feature vector is then trained against the classification status present in the training data.
2. We took 27 training samples and generated (27, 16) shaped feature vectors and trained a logistic regression classifier with LIBLINEAR as the optimization algorithm on a 5-fold cross validation basis.
3. Prior to testing on the 20 test samples, the logistic regression model was fitted on the entire training data of 27 train samples. The features for the test set were extracted from the radiomic feature extraction system, normalized using the mean and standard deviation obtained during training and probabilistic classification predictions were obtained from the fitted logistic regression model.

5   Combining the Pathology and Radiology Predictions

For combining the predictions obtained by the pathology and radiology model, we use a simple higher confidence based voting criteria. Given the predictions from both the pathology based model and the radiology based model, we compare the probability values given by each of the model for a particular class and assign the final prediction based on the model which outputs the class label with a higher probability (Fig. 11).

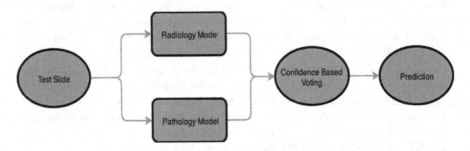

Fig. 11. Pipeline for combining the predictions from both the pathology and radiology model.

6   Results

Performance on Challenge Test Dataset
On testing the algorithms on a dataset containing 20 radiology and pathology images, an accuracy of 80% individually was observed both by the radiology and pathology models. The combined radiology and pathology model boosted the accuracy by 10% resulting in an accuracy of 90% on the entire dataset.

7   Conclusions

The results of our study show the feasibility of deep-learning based models for analyzing MR and whole-slide images. Our algorithm treats the histopathology and MR dataset separately and combines the individual predictions from both the models to output a single classification label. We demonstrated that anomaly detection based patch extraction can improve the classification results in Whole Slide Image based analysis. We also showed that radiomics based features aid in accurate classification of low-grade gliomas. One limitation of our work is that the MR based model requires all the 4 MR sequences of the patient for detecting the tumor region and even if a single sequence is missing the MR model fails to correctly narrow down on the tumor region. Another limitation could be the high processing time required to generate predictions during inference stage due to the heavy preprocessing and postprocessing required for both the models.

Additional work needs to be done to explore potential ways to combine the two models on a feature level earlier in the classification pipeline. Another area of future work involves further investigation of anomaly detection based approaches for extracting relevant patches where the pixel wise annotations are not available. The model also needs to be tested on a large cohort for establishing generalization.

References

Glorot, X., Bengio, Y.: Understanding the difficulty of training deep feedforward neural networks. In: Proceedings of the Thirteenth International Conference on Artificial Intelligence and Statistics, pp. 249–256 (2010)

Huang, G., Liu, Z., Van Der Maaten, L., Weinberger, K.Q.: Densely connected convolutional networks. In: CVPR, vol. 1, p. 3 (2017)

Ibanez, L., Schroeder, W., Ng, L., Cates, J.: The ITK software guide (2005)

Kamnitsas, K., et al.: Efficient multi-scale 3D CNN with fully connected CRF for accurate brain lesion segmentation. Med. Image Anal. **36**, 61–78 (2017)

Liu, F.T., Ting, K.M., Zhou, Z.-H.: Isolation forest. In: 2008 Eighth IEEE International Conference on Data Mining, pp. 413–422. IEEE (2008)

Menze, B.H., et al.: The multimodal brain tumor image segmentation benchmark (BRATS). IEEE Trans. Med. Imaging **34**(10), 1993–2024 (2015)

Reinhard, E., Adhikhmin, M., Gooch, B., Shirley, P.: Color transfer between images. IEEE Comput. Graph. Appl. **21**(5), 34–41 (2001)

van Griethuysen, J.J., et al.: Computational radiomics system to decode the radiographic phenotype. Cancer Res. **77**(21), e104–e107 (2017)

www.radiomics.world (2018)

Robust Segmentation of Nucleus in Histopathology Images via Mask R-CNN

Xinpeng Xie[1], Yuexiang Li[2], Menglu Zhang[1], and Linlin Shen[1(✉)]

[1] Computer Vision Institute, Shenzhen University,
Shenzhen, Guangdong, China
{xiexinpeng2017,1810272064}@email.szu.edu.cn,
llshen@szu.edu.cn
[2] Youtu Lab, Tencent, Shenzhen, China
vicyxli@tencent.com

Abstract. Nuclei segmentation plays an import role in histopathology images analysis. Deep learning approaches have shown its strength for histopathology images processing in various studies. In this paper, we proposed a novel deep learning framework for automatic nuclei segmentation. The framework adopts the Mask R-CNN as backbone and employs structure-preserving color normalization (SPCN) and watershed for pre- and post-processing. The proposed framework achieved a Dice score of 90.46% on the validation set, which demonstrates its competing segmentation performance.

Keywords: Nuclei segmentation · SPCN · Deep learning ·
Instance segmentation

1 Introduction

Nuclei morphology plays an import role in identifying aberrant phenotypes, which needs to detect and segment all the nuclear in hematoxylin and eosin (H&E) stained histology images. The features and distribution of nucleus can provide direct and reliable information for the diagnosis of cancer. The main challenges of nuclei segmentation are: (i) wide variation of cell appearance, such as color, shape, and texture; (ii) weak/missing cell boundaries; (iii) crowding and overlapping cells. With the development of deep learning technique, it is worthwhile to design and implement more accurate and efficient algorithms for nuclei segmentation. In [1], Bengtsson et al. proposed a seeded watershed transform to incorporate intensity, gradient, connectivity, and shape information for nucleus segmentation. Classical techniques are most procedural and require a large number of free parameters, Ronneberger et al. pretented the U-net, a novel FCN based network architecture, for biomedical image segmentation [2] and won the Grand Challenge for Computer-Automated Detection of Caries in Bitewing Radiography at ISBI 2015. In [3], Cui et al. designed a deep learning algorithm for one-step contour aware nuclei segmentation. They introduced a nucleus-boundary model to predict nuclei and boundaries simultaneously. Khoshdeli et al. [4] integrated boundary- and region-based information to distinguish touching or overlapping nuclei. They labeled foreground (nuclei), background and boundary pixels, to

© Springer Nature Switzerland AG 2019
A. Crimi et al. (Eds.): BrainLes 2018, LNCS 11383, pp. 428–436, 2019.
https://doi.org/10.1007/978-3-030-11723-8_43

train region- and boundary-based models, respectively. Then they trained another convolutional neural network to fuse the information from the two models to get the final result.

In this work, we propose a deep learning based framework for nuclei segmentation of Glioma whole slide tissue images. The main contributions of this work are listed in the following:

(1) We adopt structure-pre-serving color normalization (SPCN) [5] for sparse stain separation and color normalization, which can significantly boost the performance of deep learning framwork.
(2) We develop an effective framework based on Mask R-CNN to address the problem of overlapping nuclus segmentating, which achieves competitive results on MICCAI 2018 competition: Segmentation of Nuclei in Images.

2   Method

2.1   Dataset

The Digital Pathology dataset in [6] is adopted in this study. The dataset contains 15 pieces of about 668×583 annotated H&E stain images extracted from a set of Glioblastoma and Lower Grade Glioma whole slide tissue images. We separate the dataset to training and validation sets according to the ratio of 80:20. As the dataset is too small to train a deep learning network directly, two extra datasets, i.e. MICCAI 2017 segmentation challenge (MSC) [7] and 2018 Data Science Bowl (DSB) [8], are used for data augmentation. The MSC dataset includes 32 annotated H&E stain images from four categories, i.e. hnsc, lung, lgg and gbm. Each category has 8 images. The DSB dataset contains 664 fluorescent stain and H&E stain images. To alleviate color variation, we remove the fluorescent stain images in DSB dataset, which results in a dataset with 108 images.

2.2   Pre-processing

To overcome the color variation, structure-pre-serving color normalization (SPCN) [5] is adopted in our framework. For a source image S and a target image T, their color appearances and stain density maps is first estimated by factorizing V_s into $W_s H_s$, and V_t into $W_t H_t$ using sparse non-negative matrix factorization. Then, combined with color appearance of the target W_t, a scaled version of the density map of source H_s is calculated to generate the normalized source image, which can be described as follows:

$$H_s^{norm}(j,:) = \frac{H_s(j,:)}{H_s^{RM}(j,:)} H_t^{RM}(j,:), j = 1,\ldots,r.$$

where V_s, W_s, H_s repensent the observation matrix, stain color appearance matrix and stain density map matrix of source image s, respectively. $H_i^{RM} = RM(H_i) \in R^{r \times 1}, i = (s,t)$ and $RM(\cdot)$ compute robust pseudo maximum of each row vector at 99%.

In our experiment, we noticed that the E stain image contains tissue information, while the cell nucleus information is mainly contained in the H stain images. Therefore, we use the H stain images for network training. Some separated H&E stain images are presented in Fig. 1.

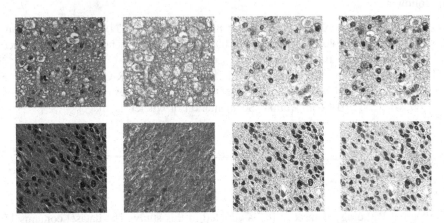

Fig. 1. Some examples of sparse stain separation for histological images. The first column is the original histological images, the second and third column are the E stain, H stain generated by SNMF, respectively. The last column is the H stain image after color normalization.

2.3   The Proposed Segmentation Framework

The flowchat of our segmentation framework is shown in Fig. 2. The Mask R-CNN [9] is employed as the backbone of our model. First, SPCN [5] is adopted to process original images. Each nuclei is separated from ground truth for RoI extraction. We slide a 512×512 window to crop patches from the whole slices and train the Mask R-CNN. We employ resnet101 as the feature extraction network. The watershed algorithm is employed as the post-processing to separate the joint cells. ImageNet [10] and COCO [11] datasets are adopted for network pre-training.

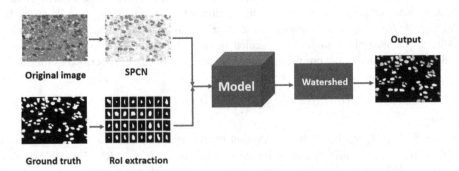

Fig. 2. The framework of our segmentation model.

Mask R-CNN. Figure 3 shows the Mask R-CNN [9] framework for nuclei segmentation, which has multiple architectures, i.e. (i) the convolutional backbone architecture for feature extraction, (ii) the network head for bounding-box recognition and mask prediction. The RoIAlign layer is developed to address misalignments between the RoI and the extracted features caused by RoIPool [12], which is simple but useful to predict pixel-level masks.

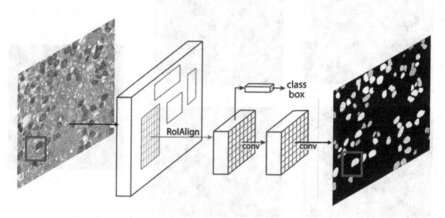

Fig. 3. The Mask R-CNN framework for nuclei segmentation.

2.4 Post-processing

Combining the information of detection, classification and bounding-box regression, MaskR-CNN [9] can efficiently segment each nuclei in images. However, not all of the overlapping nuclei are accurately separated. An example is shown in Fig. 4. Therefore, to address the problem, the Marker-based watershed [13] is involved in our framework. Given two nuclei S and G, their overlapping ratio $P_s = \frac{O}{S}$ and $P_g = \frac{O}{G}$, where O

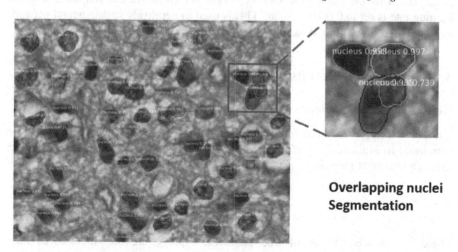

Overlapping nuclei Segmentation

Fig. 4. An example of overlapping nuclei segmentation.

represents the overlapping area of S and G, are calculated to decide the overlapping area of each nuclei. The watershed post-processing result is shown in Fig. 5 (Better seen in color).

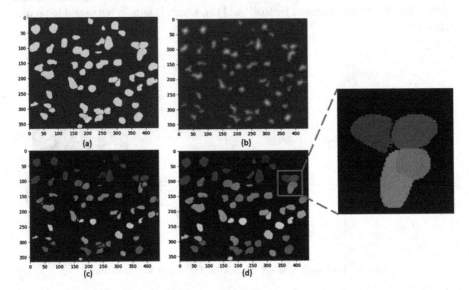

Fig. 5. Marker-based watershed post-processing. (a) is the original input mask; (b) is the marker from the minima computed from the distance transform of the binarized mask; (c) is the marker after binary_openning; (d) is the final output.

2.5 Implementation

The proposed framework is implemented using Keras toolbox, and trained with a mini-batch size of 64 on four GPUs (GeForce GTX TITAN X, 12 GB RAM). The initial learning rate is set to 0.0001. 'Adam' [14] is used to iteratively update neural network weights based on training data.

3 Experimental Results

3.1 Evaluation Criterion

The Dice score, a measure for comparison of binary segmentations S and G, is introduced to evaluate the performances on the validation set. It can be expressed in terms of statistical measures as:

$$D = \frac{2|S \cap G|}{|S| + |G|} = \frac{2\theta_{TP}}{2\theta_{TP} + \theta_{FP} + \theta_{FN}}$$

where θ_{TP} is the number of true positives, θ_{FP}/θ_{FN} are the numbers of false positives/false negatives.

3.2   Results

We first test the performance of our network trained using different training data. The dice scores on validation set trained with different datasets are summarized in Table 1. MSC and DSB represent the dataset from the MICCAI 2017 segmentation challenge [7] and the 2018 Data Science Bowl [8], respectively. Fixed DSB is the refined data after deleting fluorescent stain images from DSB. Compared with the network trained using the original Digital Pathology dataset [6], the network trained with MSC improves the dice score from 79.65% to 81.32%. While similar improvement was achieved for fixed DSB, the DSB actually decrease the dice score, which proves the necessarity to clean fluorescent stain images in DSB. Trained with the combination of original dataset, MSC and fixed DSB, the Mask R-CNN achieves the highest dice score, i.e. 85.01%.

Table 1. Dice score for different training data (%).

Training data	Dice score
Digital Pathology	80.16
MSC	81.32
DSB	80.09
fixed DSB	82.15
Digital Pathology + MSC + fixed DSB	**85.01**

Once the training dataset is fixed, we now test the performance of the Mask R-CNN, pretrained with different dataset and the effect of preprocessing and postprocessing. As listed in Table 2, the model pre-trained on COCO dataset performances better than that pre-trained on ImageNet dataset, i.e. a 1.44% dice score improvement was recorded. Furthermore, SPCN [5] and Watershed [13] improve the Dice score on validation set by 1.04% and 0.62%, respectively. Combined with SPCN [5] and Watershed [13], Mask R-CNN achieves the highest dice score with 90.46%.

Table 2. Dice score for different processing (%).

Model	Pre-training on	Dice score (%)
Mask R-CNN	ImageNet	86.99
Mask R-CNN	COCO	88.43
Mask R-CNN + SPCN	COCO	89.47
Mask R-CNN + Watershed	COCO	89.05
Mask R-CNN + SPCN + Watershed	COCO	**90.46**

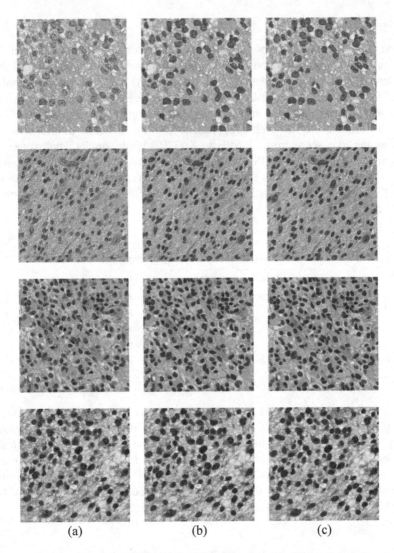

(a)   (b)   (c)

Fig. 6. Qualitative segmentation result on the data set. (a) The original input image (b) Visualizations of the ground truth (c) Visualizations of our model.

To visually assess our segmentation results, we overlapped the segmentation results and ground truths back to original images, as shown in Fig. 6.

CPM Competition Result. We achieved a competitive dice score of 86.10% on the MICCAI 2018 CPM competition testing set, while the highest score is 87.00%. Table 3 listed the Top-5 results on the leader-board.

Table 3. Dice scores for Top 5 (%).

Model	Dice score (%)
Top 1	87.00
Top 2	86.80
Top 3	86.20
Top 4 (ours)	**86.10**
Top 5	84.80

4   Conclusions

Diversity of phenotypes and overlapping nuclei are the two intrinsic challenges for nuclear segmentation in H&E stained images. In this paper, we proposed a Mask R-CNN based deep learning framework for nuclei segmentation. The overall framework consists of color normalization using SPCN [5], features extraction, and Watershed [13] postprocessing. The experimental results show that SPCN [5] and Watershed [13] can significantly improve segmentation performance, and our framework has strong strength in nuclei segmentation.

Acknowledgement. The work was supported by Natural Science Foundation of China under grands no. 61672357, 61702339 and U1713214, and the Science and Technology Project of Guangdong Province (Grant No. 2018A050501014).

References

1. Bengtsson, E., Wahlby, C., Lindblad, J.: Robust cell image segmentation methods. Pattern Recogn. Image Anal. **14**(2), 157–167 (2004). C/c of Raspoznavaniye Obrazov i Analiz Izobrazhenii
2. Ronneberger, O., Fischer, P., Brox, T.: U-Net: convolutional networks for biomedical image segmentation. In: Navab, N., Hornegger, J., Wells, W.M., Frangi, A.F. (eds.) MICCAI 2015. LNCS, vol. 9351, pp. 234–241. Springer, Cham (2015). https://doi.org/10.1007/978-3-319-24574-4_28
3. Cui, Y., Zhang, G., Liu, Z., et al.: A deep learning algorithm for one-step contour aware nuclei segmentation of histopathological images. arXiv preprint arXiv:1803.02786 (2018)
4. Khoshdeli, M., Parvin, B.: Deep leaning models delineates multiple nuclear phenotypes in H&E stained histology sections. arXiv preprint arXiv:1802.04427 (2018)
5. Vahadane, A., et al.: Structure-preserving color normalization and sparse stain separation for histological images. IEEE Trans. Med. Imaging **35**(8), 1962–1971 (2016)
6. MICCAI CPM: Digital pathology: segmentation of nuclei in images. https://miccai.cloudapp.net/competitions/83
7. MICCAI 2017 segmentation challenge. https://miccai.cloudapp.net/competitions
8. Kaggle Data Science Bowl. https://www.kaggle.com/c/data-science-bowl-2018
9. He, K., Gkioxari, G., Dollar, P., Girshick, R.: Mask R-CNN. In: International Conference on Computer Vision (ICCV) (2017)

10. Deng, J., Dong, W., Socher, R., Li, L.J., Li, K., Fei-Fei, L.: ImageNet: a large scale hierarchical image database. In: IEEE Conference on Computer Vision and Pattern Recognition, pp. 248–255 (2009)
11. Lin, T.-Y., et al.: Microsoft COCO: common objects in context. In: Fleet, D., Pajdla, T., Schiele, B., Tuytelaars, T. (eds.) ECCV 2014. LNCS, vol. 8693, pp. 740–755. Springer, Cham (2014). https://doi.org/10.1007/978-3-319-10602-1_48
12. Girshick, R.: Fast R-CNN. In: International Conference on Computer Vision (ICCV) (2015)
13. Dundar, M.M., et al.: Computerized classification of intraductal breast lesions using histopathological images. IEEE Trans. Biomed. Eng. **58**(7), 1977–1984 (2011)
14. Kingma, D.P., Ba, J.: Adam: a method for stochastic optimization. arXiv e-print arXiv:1412.6980 (2014)

Stroke Workshop on Imaging and Treatment Challenges

Perfusion Parameter Estimation Using Neural Networks and Data Augmentation

David Robben$^{(\boxtimes)}$ and Paul Suetens

Medical Image Computing (ESAT/PSI), KU Leuven, Leuven, Belgium
david.robben@kuleuven.be

Abstract. Perfusion imaging plays a crucial role in acute stroke diagnosis and treatment decision making. Current perfusion analysis relies on deconvolution of the measured signals, an operation that is mathematically ill-conditioned and requires strong regularization. We propose a neural network and a data augmentation approach to predict perfusion parameters directly from the native measurements. A comparison on simulated CT Perfusion data shows that the neural network provides better estimations for both CBF and Tmax than a state of the art deconvolution method, and this over a wide range of noise levels. The proposed data augmentation enables to achieve these results with less than 100 datasets.

1 Introduction

In stroke perfusion imaging [4] a series of 3D MR or CT images of the brain are acquired after injection of a contrast bolus. These images show the contrast agent – and hence the blood – flow in and out of the brain. As such, we have in each voxel a time series that shows the change in image intensity due the contrast agent. This intensity change can be converted to the concentration of contrast agent. This imaging modality plays a crucial role in stroke diagnosis, allowing to measure perfusion parameters such as cerebral blood flow (CBF), blood volume (CBV) and arrival time in each voxel of the brain. These perfusion parameters allow to assess to what extent the brain is affected by the stroke, distinguishing the core of the infarct (dead tissue), the penumbra (tissue at risk) and the healthy tissue. The volumes of the core and penumbra are essential to decide on the treatment plan of an acute stroke patient [2].

The native images acquired in perfusion imaging, be it CT perfusion or MR perfusion, are not directly interpretable. Typically, a deconvolution analysis is performed, where the concentration time series that are measured in each cerebral voxel (the time concentration curve or TCC), are deconvolved with the so-called arterial input function (AIF), which is the concentration time series measured in one of the large feeding arteries of the brain. The deconvolved time series are no longer influenced by the contrast injection protocol or the cardiac status of the subject. They are impulse response functions (IRF), the signal that

A. Crimi et al. (Eds.): BrainLes 2018, LNCS 11383, pp. 439–446, 2019.
https://doi.org/10.1007/978-3-030-11723-8_44

theoretically would be measured if the injection of contrast agent where a Dirac impulse directly into the feeding artery of the brain. It has been shown that, under reasonable assumptions, the maximum of this deconvolved time series is proportional to the cerebral blood flow (CBF). This parameter is strongly predictive for the health of the tissue and is currently used in clinical practice to identify the infarct core: voxels with a CBF less than 30% compared to the other side of the brain are considered to be core [1]. The time when the deconvolved time series reaches its maximum, is called Tmax, and increased values are indicative for tissue is at risk. A threshold of 6 s is currently clinically used to determine the perfusion lesion (i.e. the combined core and penumbra) of the infarct [1].

Indeed, deconvolution plays a central role in perfusion analysis. However, deconvolution is a mathematically ill-posed problem, and given the relatively low signal to noise ratio of perfusion images, a successful implementation of perfusion analysis requires duly attention to this problem. First, there is need for proper preprocessing: motion correction, temporal and spatial smoothing, and possibly spatial downsampling. Second, the deconvolution is regularized, suppressing the high frequency signal in the reconstructed impulse response function. This is typically done in singular value decomposition (SVD) based deconvolution by regularizing the singular values, e.g. using Tikhonov regularization. Nevertheless, the deconvolution-based perfusion parameters remain noise sensitive and research for improved algorithms [3] or even deconvolution-free summary parameters [8] remains ongoing.

Recently, several works have proposed to use machine learning techniques to estimate, based on the perfusion and treatment parameters, how the final infarct will look [6,7,9,12]. However, all these approaches first perform a deconvolution analysis – which suffers from the earlier mentioned problems – and then use the perfusion parameters as input features for the machine learning algorithms. One notable exception is [10] who use both the perfusion parameters and the native measurements as input to their method.

In this work, we show on simulated data that a neural network can learn to perform this deconvolution and achieve more accurate estimations of CBF and Tmax than a state of the art deconvolution technique can. Additionally, we show how a perfusion specific data augmentation can be used to learn this deconvolution from a relatively small number of training samples. Knowing that a neural network is able to learn how to perform the deconvolution, opens possibilities for new research, where the final infarct is predicted directly from the native images, bypassing the standard deconvolution as a preprocessing step.

2   Methods

We will compare a state of the art deconvolution approach with our proposed neural network-based approach.

2.1   Baseline: Tikhonov Regularized SVD-Based Deconvolution

As a baseline, we use Tikhonov regularized SVD-based deconvolution [4] with the Volterra discretization scheme [11] – to which we will simply refer as deconvolution. This method takes the AIF and TCC and will produce the IRF. From the IRF, we can infer both the CBF and the Tmax:

$$\text{CBF} = \frac{1}{\rho} \max_t \text{IRF}(t) \tag{1}$$

$$\text{Tmax} = \underset{t}{\text{argmax}} \, \text{IRF}(t). \tag{2}$$

The IRF has as unit $1/s$ and hence the CBF $ml/g/s$ (being ml blood per g of brain tissue per s). When reporting CBF values, we will follow the common practice of using $ml/100\,g/min$. The IRF is discretized, so to produce continuous estimates of Tmax, the IRF is fitted with a quadratic spline.

The deconvolution has one hyperparameter, the relative regularization parameter λ_{rel} which sets the amount of filtering. For each experiment, the optimal regularization strength is found by evaluating the performance on the training set for a range of possible values ($0.01 * 2^{0,1,2,...,9}$).

2.2   Proposed Neural Network

Through experimentation we found that even simple networks succeed at this task and the following network is used for all experiments. Our network has two input layers, one for the AIF, and one for the TCC. The two input layers are concatenated, followed by two fully connected layers with each 30 neurons and end in a single output neuron. The fully connected layers have a PreLU activation and the output layer has no activation function. The output is a single value, depending on the experiment, either the CBF or the Tmax.

Note that while it is possible to provide spatial context to the network – e.g. the TCC of the neighboring voxels, which would most certainly improve the performance – this is out of scope for this work and would make the comparison between the two approaches unfair.

The network is trained in a supervised fashion by feeding examples of an AIF and a TCC with known CBF or Tmax and minimizing the absolute difference between prediction and ground truth. The optimization is performed by stochastic gradient descent with learning rate 0.01, Nesterov momentum of 0.9 and mini-batches of 2048 samples.

2.3   Proposed Data Augmentation

To limit the amount of training data that is required to train this network, we propose a perfusion specific data augmentation. The key insight is that the perfusion measurements are a linear time invariant system. This means that, if the contrast injection was a bit later, both the AIF and the TCC would show the same delay. Similarly, if the injection was earlier, all curves should shift to the left. However, the IRF (and hence the Tmax and CBF) would remain the same in both cases. If the concentration of the iodine or gadolinium in the contrast agent were a fraction higher or lower, the concentration curves should be changed with the same fraction. But again the IRF would remain untouched.

Hence, we will create additional samples from a single training sample by applying a random time shift (earlier or later) and a random scaling to both AIF and TCC. In our experiments, the shift is randomly chosen between -1 and 2 time points (since our measurements are discrete) and the scaling has a uniform distribution between $[0.7, 1.3]$. This augmentation enhances the number of different training samples the network sees, reducing the required size of our actual training set.

3   Experiments

To compare the deconvolution with the proposed methods, we need to compare the predictions with the ground truth. However, ground truth measures of the CBF or Tmax are not possible in vivo. Hence we perform a series of experiments on simulated data.

3.1   Data

In the simulations, we model the arterial input function (AIF) and the impuls response function (IRF) as gamma variates:

$$\gamma(t; t_0, \alpha, \beta, A) = A * (t - t_0)^\alpha * e^{-\frac{t-t_0}{\beta}}. \tag{3}$$

The different parameters are chosen randomly such that the resulting curves resemble the curves measured with CTP. The values t_0, α, β are chosen from uniform distributions. For the AIF: $t_0 \in [0, 15]$, $\alpha \in [1.5, 3.5]$, $\beta \in [1.5, 3.5]$. For the IRF: $t_0 \in [0, 10]$, $\alpha \in [0, 0.5]$, $\beta \in [2.5, 4.5]$.

For the AIF, the value of A is chosen such that the AIF's maximum is uniformly distributed between 100 and 500 HU.[1] For the IRF, the value of A is chosen such that the integral of the IRF (i.e. the CBV) is uniformly distributed between 0.1% and 6%. For the experiments about Tmax estimation, a different range is chosen (between 2% and 6%) since the deconvolution-based Tmax estimation performs poorly on weak signals. Keeping the full range would move the comparison in favor for the proposed method while not being relevant in clinical practice.

The TCC is obtained by convolving the simulated AIF and IRF. Finally, the AIF and TCC are sampled (19 samples over a time span of 40 s) and Gaussian noise is added (σ varies from 0.1 to 3.2, and is given later for each specific

---

[1] In CTP, the change in intensity is proportional to the contrast agent concentration and hence deconvolution techniques yield the same results whether applied on the intensity changes or on the concentration. The former have the advantage of being more easily interpretable, and hence we will express the AIF and TCC in Hounsfield Units (HU).

experiment). Figure 1 shows a randomly simulated sample and the resulting distribution of CBF, CBV, MTT and Tmax.

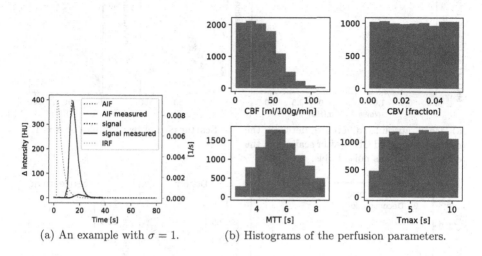

(a) An example with $\sigma = 1$.   (b) Histograms of the perfusion parameters.

Fig. 1. The simulated data.

3.2   SVD-Based Deconvolution Versus Proposed Neural Network

We aim to compare how well the deconvolution approach and the neural network are able to estimate the CBF and Tmax from the AIF and TCC. The performance of the Tmax estimation is measured using the mean absolute difference (MAD) between the true and estimated value. For the CBF, we also use the MAD, but only after scaling all the estimates with an optimal scaling factor (i.e. the one that minimizes the MAD). This is warranted since in clinical practice the relative CBF is used to predict the infarct core.

For a range of noise values, we generate a training set (1M samples) and a testing set (10k samples), where a sample consists the AIF, the TCC and the ground truth perfusion parameter (CBF or Tmax). For each noise level, the optimal amount of Tikhonov regularization is determined on the training set and that value is used on the test set. The neural network is trained for one epoch on all training samples and subsequently predicts the test set. Figure 2a and c summarize the results and show an improvement for both measures on all noises levels. Figure 2b and d show the distribution of the predictions at a single noise level.

3.3   Data Augmentation

In the previous experiment, we trained on 1 million samples, each with a different AIF and hence each corresponding to a different acquisition. Such large training

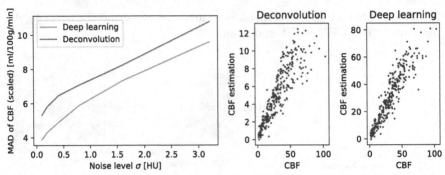

(a) Mean absolute difference between the true and estimated CBF (after scaling) in the test set for various noise levels.

(b) Scatter plot of test samples at $\sigma = 1$.

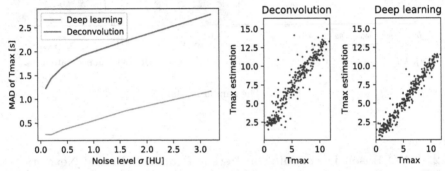

(c) Mean absolute difference between the true and estimated Tmax in the test set for various noise levels.

(d) Scatter plot of test samples at $\sigma = 1$.

Fig. 2. Comparison between estimations produced by the neural network and the SVD-based deconvolution.

sets are not realistic, and in this experiment we explore how many samples are necessary and whether our proposed data augmentation can lower that number. Using a fixed noise level of $\sigma = 1$, we create training sets of various sizes, by varying the number of AIFs (which corresponds the number of acquisitions) and the number of TCCs per AIF. The proposed data augmentation is used to increase the number of samples with a factor 10. We use the same performance metrics and training method as in the previous section, with the number of training epochs adapted such that each network is trained for the same number of iterations (corresponding to one epoch with 1M samples). Figure 3 summarizes the results, showing that less than 100 acquisitions are sufficient and that the proposed data augmentation can lead to large improvements, especially in situations with limited training data.

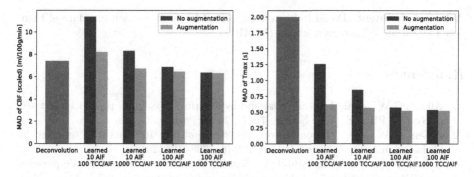

Fig. 3. Influence of the number of training samples and data augmentation on the performance of the neural network.

4   Discussion and Conclusion

We proposed a simple neural network and a data augmentation approach to predict perfusion parameters from native perfusion measurements (i.e. the AIF and TCC). A comparison on simulated data shows that the neural network provides better estimations for both CBF and Tmax than a state of the art deconvolution method, and this over a wide range of noise levels. Using the proposed data augmentation, it is feasible to achieve these results with less than 100 datasets.

Earlier, Ho et al. [5] showed that a neural network can learn how to deconvolve. They trained and tested a CNN on MR perfusion datasets while using as a ground truth the perfusion parameters obtained by an SVD-based deconvolution method. They showed that a neural network can produce a reasonable approximation for the various perfusion parameters. In this work, we go one step further, and show that a neural network can outperform deconvolution methods on simulated CT Perfusion data.

We see two main directions for future research. First, one could investigate how neural network based perfusion parameter estimation works on real data. As mentioned earlier, validation of such an approach is difficult since ground truth measurements are not available. However, it might be possible to produce convincing evidence by comparing different modalities from the same subject or by comparing results on high- and low-resolution versions of the same dataset. Second, one could revisit the earlier mentioned methods that attempt to estimate the final infarct from perfusion parameter maps. Especially for neural network based approaches, such as [9], it might be beneficial to provide the original time series to the network instead of the perfusion parameters. Having the network perform the deconvolution operation implicitly might lead to better estimates of the CBF and Tmax, or – even more promising – might lead the network to learn new perfusion parameters that are even more predictive.

Acknowledgement. David Robben is supported by an innovation mandate of Flanders Innovation & Entrepreneurship (VLAIO).

References

1. Albers, G.W., et al.: Ischemic core and hypoperfusion volumes predict infarct size in SWIFT PRIME. Ann. Neurol. **79**(1), 76–89 (2016)
2. Albers, G.W., et al.: Thrombectomy for stroke at 6 to 16 hours with selection by perfusion imaging. N. Engl. J. Med. **378**(8), 708–718 (2018)
3. Boutelier, T., Kudo, K., Pautot, F., Sasaki, M.: Bayesian hemodynamic parameter estimation by bolus tracking perfusion weighted imaging. IEEE Trans. Med. Imaging **31**(7), 1381–1395 (2012)
4. Fieselmann, A., Kowarschik, M., Ganguly, A., Hornegger, J., Fahrig, R.: Deconvolution-based CT and MR brain perfusion measurement: theoretical model revisited and practical implementation details. Int. J. Biomed. Imaging **2011**, 1–20 (2011)
5. Ho, K.C., Scalzo, F., Sarma, K.V., El-Saden, S., Arnold, C.W.: A temporal deep learning approach for MR perfusion parameter estimation in stroke. In: 2016 23rd International Conference on Pattern Recognition (ICPR), pp. 1315–1320. IEEE, December 2016
6. Kemmling, A., et al.: Multivariate dynamic prediction of ischemic infarction and tissue salvage as a function of time and degree of recanalization. J. Cereb. Blood Flow Metab. **35**(9), 1397–1405 (2015)
7. Maier, O., et al.: ISLES 2015 - a public evaluation benchmark for ischemic stroke lesion segmentation from multispectral MRI. Med. Image Anal. **35**, 250–269 (2017)
8. Meijs, M., Christensen, S., Lansberg, M.G., Albers, G.W., Calamante, F.: Analysis of perfusion MRI in stroke: to deconvolve, or not to deconvolve. Magn. Reson. Med. **76**(4), 1282–1290 (2016)
9. Nielsen, A., Hansen, M.B., Tietze, A., Mouridsen, K.: Prediction of tissue outcome and assessment of treatment effect in acute ischemic stroke using deep learning. Stroke **49**(6), 1394–1401 (2018)
10. Pinto, A., et al.: Enhancing clinical MRI perfusion maps with data-driven maps of complementary nature for lesion outcome prediction. In: Frangi, A.F., Schnabel, J.A., Davatzikos, C., Alberola-López, C., Fichtinger, G. (eds.) MICCAI 2018. LNCS, vol. 11072, pp. 107–115. Springer, Cham (2018). https://doi.org/10.1007/978-3-030-00931-1_13
11. Sourbron, S., Luypaert, R., Morhard, D., Seelos, K., Reiser, M., Peller, M.: Deconvolution of bolus-tracking data: a comparison of discretization methods. Phys. Med. Biol. **52**(22), 6761–6778 (2007)
12. Wu, O., et al.: Predicting tissue outcome in acute human cerebral ischemia using combined diffusion- and perfusion-weighted MR imaging. Stroke **32**(4), 933–942 (2001)

Synthetic Perfusion Maps: Imaging Perfusion Deficits in DSC-MRI with Deep Learning

Andreas Hess[1], Raphael Meier[1], Johannes Kaesmacher[1,2], Simon Jung[2], Fabien Scalzo[3], David Liebeskind[3], Roland Wiest[1], and Richard McKinley[1(✉)]

[1] Support Centre for Advanced Neuroimaging,
University Institute of Diagnostic and Interventional Neuroradiology,
Inselspital, Bern University Hospital, Bern, Switzerland
richard.mckinley@gmail.com
[2] Department of Neurology, Inselspital, University of Bern, Bern, Switzerland
[3] Department of Neurology, University of California Los Angeles (UCLA),
Los Angeles, USA

Abstract. In this work, we present a novel convolutional neural network based method for perfusion map generation in dynamic susceptibility contrast-enhanced perfusion imaging. The proposed architecture is trained end-to-end and solely relies on raw perfusion data for inference. We used a dataset of 151 acute ischemic stroke cases for evaluation. Our method generates perfusion maps that are comparable to the target maps used for clinical routine, while being model-free, fast, and less noisy.

1 Introduction

Dynamic susceptibility contrast-enhanced (DSC) magnetic resonance imaging (MRI) is an essential tool to assess perfusion deficits in acute ischemic stroke [6]. Given a perfusion sequence, traditional methods rely on an estimation of the arterial input function (AIF) and on a parametric model to generate perfusion maps. Automatic estimation of the AIF tends to be non-robust, whereas a manual selection of the AIF is impractical. Besides, the computation of perfusion maps costs time in the range of minutes, in situations where time is often critical.

In 2017, Song et al. [7] introduced temporal similarity perfusion (TSP) maps, together with a model-free, iterative process for their generation. The generated maps are compared to traditional time-to-peak (TTP) and mean-transit-time (MTT) maps to assess their clinical value for the detection of perfusion

Supported by SNSF grant no. 320030L_170060 and the Swiss Heart Foundation.

Electronic supplementary material The online version of this chapter (https:// doi.org/10.1007/978-3-030-11723-8_45) contains supplementary material, which is available to authorized users.

A. Crimi et al. (Eds.): BrainLes 2018, LNCS 11383, pp. 447–455, 2019.
https://doi.org/10.1007/978-3-030-11723-8_45

deficits. The proposed method operates without the need for AIFs, but it is fixed to the generation of TSP maps. Machine learning has been used extensively to post-process perfusion maps, e.g., to estimate tissue at risk (penumbra) [4]. Meanwhile, in McKinley et al. [5] regression of perfusion maps from raw DSC-MRI perfusion data in the presence of an externally provided AIF was demonstrated. Fully automatic end-to-end regression of DSC perfusion maps has not been approached so far.

In this paper, we present a model-free, convolutional neural network (CNN) based architecture to predict arbitrary perfusion maps in an end-to-end manner. There have already been applications of CNNs on dynamic contrast-enhanced (DCE) perfusion sequences [8]. However, our work is the first application on DSC-MRI perfusion sequences to the best of our knowledge.

2 Methods

In this section, we describe the proposed neural network architecture for predicting synthetic perfusion maps. Since the architectures we developed for predicting different types of perfusion maps are similar (see supplementary material, Fig. S1), we solely focus on the one used for predicting T_{max} perfusion maps.

2.1 Pre-processing and Augmentation

We pre-process both the raw perfusion data and the target perfusion maps. Each volume is padded with zeros to match the maximum size of any volume in the training dataset. Additionally, the perfusion sequence is complemented with volumes at the end to match the maximum sequence length of any perfusion sequence in the training dataset. For this, we reflect the data to generate frames for padding instead of using zero-filled volumes. The resulting perfusion maps are of size $24 \times 256 \times 256$, the perfusion sequences of size $80 \times 24 \times 256 \times 256$, where 80 is the number of frames and 24 the number of image slices.

After padding, we perform data standardization, i.e., we transform the data, so it has zero mean and unit variance. Finally, we apply voxel-wise temporal Gaussian smoothing for the perfusion sequence with $\sigma_t = 1.0$.

From inspecting the training portion of our dataset, we conclude that the time of perfusion sequence start is not in a global relation to the time of bolus arrival in the brain, i.e., there are cases where the bolus arrives much earlier or later than in most other cases. Even though such cases occur rarely, we want our model to be able to handle arbitrary delays of the bolus. To prevent the model from learning a global bolus arrival time, we augment the training dataset by randomly offsetting the perfusion sequence by -5 to 30 frames. A negative number means we remove frames at the beginning of the sequence and add padding frames at the end, a positive number indicates the opposite. The padding is generated via reflecting.

2.2 Deep Architecture for Perfusion Map Regression

The goal is to predict voxel-values in a target perfusion map P based on the raw perfusion sequence S. Our crucial assumption is that a voxel-value $p_{x,y,z}$ in

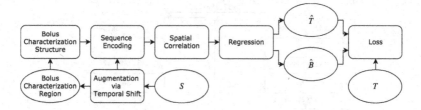

Fig. 1. An overview of the perfusion map prediction model.

perfusion map P mostly depends on the sequence of voxels at the same location in the raw perfusion sequence, i.e., on $s_{1:T,x,y,z}$, where T indicates the total number of frames in the perfusion sequence. There are obvious limitations to this assumption, which will be discussed in Sect. 4. Based on this assumption, we use a CNN to capture the temporal evolution of the raw perfusion sequence voxel-wise. Figure 1 shows an overview of our architecture. The individual substructures will be briefly explained in the following sections.

Bolus Characterization. Our model has neither knowledge of the arterial input function nor the time of bolus arrival in the brain. Ignoring these aspects would significantly impair the performance of our model. Therefore, we add a bolus characterization structure (BCS) which should help capture the time of bolus arrival in the brain as well as the AIF. Guided by the fact that those characteristics are captured best by large blood vessels entering the brain, we select the input to the BCS to be a patch sequence from the perfusion sequence, located at the transition between the basilar artery and the posterior cerebral artery. The location of this patch is globally fixed, i.e., it is not fine-tuned to the individual volume. Therefore, it may happen that this patch does not contain the desired blood vessels for specific instances in our data. The BCS processes the supplied patch sequence via two 3D convolutional layers, encoding each patch into a vector of size 16. The sequence of encoded patches is forwarded to the sequence encoder.

Sequence Encoding. The sequence encoder handles every voxel sequence $s_{1:T,x,y,z}$ independently, together with the additional information supplied by the bolus characterization structure. Note that this information is the same for all $s_{1:T,x,y,z}$ of a perfusion sequence S. For simplicity, we will describe how the sequence encoder handles one individual voxel sequence. In reality, the sequence encoder processes multiple voxel sequences concurrently.

Effectively, the sequence encoder works on three inputs: a sequence of voxel-values $s_{1:T,x,y,z}$, the sequence of frame times $\in \mathbb{R}^{80}$, and the sequence of encoded patches from the bolus characterization structure. These sequences are concatenated along their non-temporal dimension, resulting again in a sequence of length 80. This result is passed through three 1D convolutional layers, of which the first

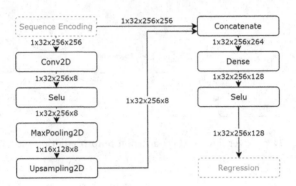

Fig. 2. The architecture we used for local spatial correlation. The dense layer only operates on the last dimension of the data. Note that the shape of the data passed from the sequence encoder to the 2D convolution stems from the fact that we process the volumes in patches of $1 \times 32 \times 256$ and that each voxel sequence $s_{1:T,x,y,z}$ is encoded into a vector of size 256. The first dimension of the data is redundant and is only depicted for clarity. *Selu* denotes the activation function used [3].

two are followed by a max-pooling layer. The output is a vector of size 256, capturing the evolution over time of one voxel in the perfusion sequence.

Spatial Correlation. So far, there is no flow of information between neighboring voxels, i.e., each voxel sequence $s_{1:T,x,y,z}$ is handled independently. Due to the low spatial resolution of the volumes along the axial axis, we omit spatial correlation between slices. To allow for some learned local filtering within slices, we apply 2D convolution to the voxel-wise output of the sequence encoder. Figure 2 illustrates this process.

Regression. Given the spatially correlated encoding of voxel sequence $s_{1:T,x,y,z}$, the prediction of the perfusion map voxel-value $\hat{p}_{x,y,z}$ and the estimated uncertainty $\hat{b}_{x,y,z}$ is performed via a fully connected layer with two output neurons and identity activation.

Loss. The training objective is based on heteroscedastic aleatoric uncertainty modeling [1], i.e., instead of a single prediction, the model outputs a mean and a measure of uncertainty per voxel. This effectively corresponds to a probability distribution per voxel. The loss is given by the negative log-likelihood of the observed target map data. We choose a Laplace distribution to assign probabilities to observed values since a Gaussian distribution is too light tailed to cope with the amount of noise in target perfusion maps. Equation 1 formally defines the negative log-likelihood l', given $p_{x,y,z}$, $\hat{p}_{x,y,z}$ and estimated uncertainty parameter $\hat{b}_{x,y,z}$.

$$l'\left(p_{x,y,z}, \hat{p}_{x,y,z}, \hat{b}_{x,y,z}\right) = \log \hat{b}_{x,y,z} + \frac{|p_{x,y,z} - \hat{p}_{x,y,z}|}{\hat{b}_{x,y,z}} \tag{1}$$

To be able to further focus the networks efforts onto voxel-values of high clinical importance and to simultaneously reduce the influence of noise, we apply a weighting scheme to the negative log-likelihood l' of each voxel. The weighting scheme is based on a predefined function I that assigns an importance to each value in P as shown in Eq. 2. The weight for a pair $(p_{x,y,z}, \hat{p}_{x,y,z})$ is given by the function W as shown in Eq. 3. The final voxel-wise loss l is given by Eq. 4. We use an average to compute the loss over multiple voxels.

$$I(z) = \begin{cases} 1.0, & \text{if } 0.0 \le z \le 40.0 \\ 0.1, & \text{else} \end{cases} \tag{2}$$

$$W\left(p_{x,y,z}, \hat{p}_{x,y,z}\right) = \max_{z=\min(p_{x,y,z}, \hat{p}_{x,y,z})}^{\max(p_{x,y,z}, \hat{p}_{x,y,z})} I(z) \tag{3}$$

$$l\left(p_{x,y,z}, \hat{p}_{x,y,z}, \hat{b}_{x,y,z}\right) = l'\left(p_{x,y,z}, \hat{p}_{x,y,z}, \hat{b}_{x,y,z}\right) \cdot W\left(p_{x,y,z}, \hat{p}_{x,y,z}\right) \tag{4}$$

Training. The perfusion maps are processed in patches of size $1 \times 32 \times 256$, the perfusion sequences in patches of size $80 \times 1 \times 32 \times 256$. These patches are gathered in batches of size four before being passed to the network. For optimization, we use Adam [2] with an initial learning rate of $5e^{-4}$. The learning rate is divided by two every four epochs. We rely on dropout regularization with a dropout rate of 0.5 for fully connected layers and do not use l_2-norm weight decay.

3   Experimental Setup and Results

The model was trained and evaluated on DSC-MRI perfusion data of patients with acute ischemic stroke. For a given perfusion sequence, the target perfusion maps were generated using *oscillation index singular value decomposition* in Olea Sphere® 2.3 with default settings as used in clinical routine. The complete dataset contains 189 cases, of which we excluded 38 because the detected AIF was inaccurate, leaving us with a dataset of 151 cases. Approval for this retrospective study was obtained from the local ethics committee (KEK Bern, Switzerland, approval number: 231/14). Written informed consent was waived according to the retrospective nature of this analysis.

We randomly partitioned the dataset into training set, validation set and test set with ratios of 0.5, 0.2 and 0.3 respectively. The training set was used for optimizing the network's weights, the validation set for model selection, and the test set for evaluating the final model. The selected hyperparameters are listed in Table S1 (see supplementary material).

The neural network was trained end-to-end, using the pre-processed perfusion sequences as input and the pre-processed perfusion maps from Olea as target

Table 1. The MAEC of the predicted T_{max} maps for different models. The lines below model A show the performance of models where the listed component was removed.

Model	MAEC@Validation	MAEC@Test
Model A	**0.513**	0.530
– Augmentation (B)	0.531	**0.524**
– Spatial correlation (C)	0.562	0.629
– Bolus characterization (D)	0.632	0.680
– Loss weighting (E)	0.683	0.738

maps. Training was performed on a Windows machine with an Intel Xeon E5-1630 v3 @ 3.7 GHz and a Nvidia GeForce GTX 1080, and on an Ubuntu machine with an Intel i7-4790K CPU @ 4.00 GHz and two Nvidia GeForce GTX 1070. All models were trained for 30 epochs, the selection of the final model was based on its performance on the validation set. The forward propagation took time in the range of seconds to compute a complete perfusion map.

3.1 Quantitative Results

To quantitatively evaluate and compare different models, we used a mean absolute error with clipping (MAEC) as performance measure. It is identical to a mean absolute error, except that voxel-values are clipped to the interval $[0, 20]$ before computing differences. The clipping was done since values below 0 mostly correspond to air and the ones above 20 definitely indicate a perfusion deficit or are part of the noise. The exact values 0 and 20 were chosen because $[0, 20]$ is a reasonable window for inspecting T_{max} maps.

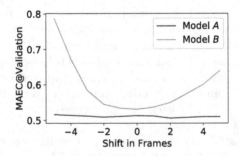

Fig. 3. The influence on the MAEC for different models when shifting the raw perfusion sequence by a given number of frames.

The model described in Sect. 2.2 is referred to as model A in the remainder of this section. Table 1 shows the performance of model A and compares it to variants of model A where critical components introduced in Sect. 2.2 were removed. While model A performed best on the validation set, model B performed best on the test set. When manually inspecting the perfusion sequences of the validation and test set, we observed that the time of contrast bolus arrival varies more in the validation set than it does in the test set, with a standard deviation of 2.76 frames in contrast to 2.09 frames. Hence, the cases in the validation set did potentially benefit more from data augmentation via temporal shift. However, this temporal shift made training harder, which made the model perform slightly worse on cases where the bolus arrival delay was close to the mean delay. To further investigate the effectiveness of our augmentation, we measured the MAEC on the validation data for model A and model B after shifting the perfusion sequences by a number of frames. The results are shown in Fig. 3 and clearly indicate that the augmentation successfully helped the model to compensate for different times of bolus arrival.

Further, we observed that removing the spatial correlation, the bolus characterization structure or the loss weighting significantly decreased the model's prediction accuracy on both the validation and the test set. It is evident that removing loss weighting increases the MAEC, since the loss weights assign high importance to the values that influence the MAEC.

3.2   Qualitative Results

Figure 4 shows the target map T_{max}, the model's prediction $\hat{T}_{max}$ and the estimated variance of the prediction $\hat{\sigma}^2$ for three samples from the test set.

For the first two cases, the model's prediction is very close to the target map. Compared to the target maps, the predictions tend to contain less high-valued noise and generally have a smoother appearance. We would have expected the model to predict high uncertainty where its prediction error is likely to be high, i.e., we would have hoped for a positive correlation between prediction error and $\hat{\sigma}^2$. The $\hat{\sigma}^2$ maps do not match those expectations. Instead, we observed a correlation between the predicted value and $\hat{\sigma}^2$, meaning the prediction of correct high values seems to be hard for the model. This observation makes sense since there is a considerable amount of high-valued noise in the target maps.

The third row of Fig. 4 shows an example where our model failed. From observing the corresponding raw perfusion sequence and comparing it to perfusion sequences of examples where the model performed better, we noticed that the signal attenuation caused by the contrast agent is comparably weak for this case. Also, the signal is very noisy, partially due to slight head movements, which are amplified by the low axial resolution of the volumes. Given this additional information, we assumed that the bolus characterization structure was unable to correctly capture the bolus arrival in the brain, which led to a poor prediction.

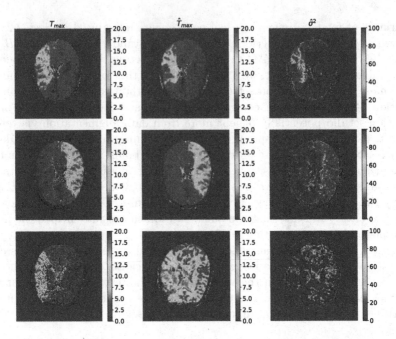

Fig. 4. The target map T_{max}, the predicted map $\hat{T}_{max}$ and the estimated variance $\hat{\sigma}^2$ of the prediction for three examples from the test set. Note that $\hat{\sigma}^2$ is given by $2\hat{b}^2$.

4   Discussion and Conclusion

We made some simplifying assumptions in the presented approach, the most crucial one being that voxels in a perfusion map mainly depend on the perfusion sequence of voxels at the same location. This simplification does not always hold, especially when there was head movement during the sequence acquisition. An obvious solution to this is to register the individual volumes of the perfusion sequence before processing them any further. However, this is hard due to the low resolution of the volumes along the axial axis, which can lead to significant interpolation artifacts. Furthermore, it does not fit the concept of an end-to-end learning model. Another possible solution would be to make the sequence registration part of the model.

In conclusion, we presented a model-free, CNN-based method for inferring perfusion maps in an end-to-end manner. We demonstrated our method's performance on an ischemic stroke dataset of 151 patients and have shown that the predictions are comparable to the target perfusion maps. We are currently working on a clinical evaluation of the synthetic perfusion maps in order to confirm the applicability of CNNs in real-world DSC-MRI perfusion imaging.

References

1. Kendall, A., Gal, Y.: What Uncertainties Do We Need in Bayesian Deep Learning for Computer Vision? (2017)
2. Kingma, D.P., Ba, J.: Adam: A Method for Stochastic Optimization (2014)
3. Klambauer, G., Unterthiner, T., et al.: Self-Normalizing Neural Networks (2017)
4. McKinley, R., Häni, L., et al.: Fully automated stroke tissue estimation using random forest classifiers (FASTER). JCBFM 37(8), 2728–2741 (2017)
5. McKinley, R., Hung, F., et al.: A machine learning approach to perfusion imaging with dynamic susceptibility contrast MR. Front. Neurol. 9, 717 (2018)
6. Olivot, J.M., Mlynash, M., et al.: Geography, structure, and evolution of diffusion and perfusion lesions in Diffusion and perfusion imaging Evaluation For Understanding Stroke Evolution (DEFUSE). Stroke 40(10), 3245–3251 (2009)
7. Song, S., et al.: Temporal similarity perfusion mapping: a standardized and model-free method for detecting perfusion deficits in stroke. PLoS ONE 12(10), e0185552 (2017)
8. Ulas, C., Tetteh, G., et al.: Direct Estimation of Pharmacokinetic Parameters from DCE-MRI using Deep CNN with Forward Physical Model Loss (2018)

ICHNet: Intracerebral Hemorrhage (ICH) Segmentation Using Deep Learning

Mobarakol Islam[1,2], Parita Sanghani[2], Angela An Qi See[3],
Michael Lucas James[4], Nicolas Kon Kam King[3], and Hongliang Ren[2(✉)]

[1] NUS Graduate School for Integrative Sciences and Engineering (NGS),
National University of Singapore, Singapore, Singapore
[2] Department of Biomedical Engineering,
National University of Singapore, Singapore, Singapore
`ren@nus.edu.sg`
[3] Department of Neurosurgery, National Neuroscience Institute, Singapore, Singapore
[4] Department of Anesthesiology and Neurology, Duke University, Durham, NC, USA

Abstract. We develop a deep learning approach for automated intracerebral hemorrhage (ICH) segmentation from 3D computed tomography (CT) scans. Our model, ICHNet, evolves by integrating dilated convolution neural network (CNN) with hypercolumn features where a modest number of pixels are sampled and corresponding features from multiple layers are concatenated. Due to freedom of sampling pixels rather than image patch, this model trains within the brain region and ignores the CT background padding. This boosts the convergence time and accuracy by learning only healthy and defected brain tissues. To overcome the class imbalance problem, we sample an equal number of pixels from each class. We also incorporate 3D conditional random field (3D CRF) to smoothen the predicted segmentation as a post-processing step. ICHNet demonstrates 87.6% Dice accuracy in hemorrhage segmentation, that is comparable to radiologists.

Keywords: Intracerebral hemorrhage · Stroke · Deep learning · Convolutional neural network · PixelNet · Conditional Random Field · Hypercolumn

1 Introduction

Intracerebral hemorrhage (ICH) is a form of brain stroke which is associated with high mortality and morbidity [1,16]. Most of the patients who survive a hemorrhagic stroke develop long-term disabilities as a result of the compression of the brain tissues around the affected region, caused by the edema [22]. Radiological imaging like Computed Tomography (CT) is typically used for diagnosis, treatment planning, and prognosis monitoring of ICH patients. Traditionally, radiologists visualize the hematoma by manual delineating on the CT scan and estimate its initial volume, which is used for predicting mortality and functional

A. Crimi et al. (Eds.): BrainLes 2018, LNCS 11383, pp. 456–463, 2019.
https://doi.org/10.1007/978-3-030-11723-8_46

outcome of the patient. The lengthy process of manually delineating associated with inter-rater variability and the need for highly trained radiologists at all times forms the limitations of this traditional process. In order to carry out a precise quantitative analysis of the hematoma, it is important to have accurate automated segmentation.

Recently, deep learning based automated segmentation approaches have gained momentum, as they possess the ability to perform complex tasks at a very fast rate and with high accuracy similar to the human specialist. Some recent examples include brain tumor segmentation [11], ischemic lesion segmentation [5], lung tumor segmentation [12], cardiac segmentation [25] and pancreas segmentation [4]. In particular, automated hemorrhage (stroke lesion) segmentation has received increasing attention in stroke management by dealing with a vast amount of data and supporting clinician to take numerous complex decisions. Choi et al. [8] propose an ensemble of deep neural networks for automated prognosis of post-treatment ischemic stroke. To overcome the computational burden of 3D Ischemic MRI scan, Kamnitsas et al. [17] devise 3D CNN with dense training scheme of adjacent image patches into one pass while automatically adapting to the inherent class imbalance. Chen et al. [4] exploit two CNN consists of DeconvNets [21] and a multi-scale convolutional label evaluation net to segment acute ischemic lesion from diffusion-weighted MR imaging (DWI). All these models try to achieve state of art performance by utilizing different architectures of 2D, 3D and dual path CNN with handcrafted features and CRF as post-processing. However, none of these methods have been applied on CT scans for intracerebral hemorrhage (ICH) segmentation. On the other hand, RADnet [9] uses recurrent attention DenseNet [14] with LSTM to segment and classify brain hemorrhage from CT scans but in the case of traumatic brain injury (TBI).

In this paper, we propose a novel deep learning model (ICHNet) with a brain mask training scheme to segment intracerebral hemorrhage (ICH). In Pixel-level segmentation with the convolutional predictor (for example CNN), stochastic gradient descent (SGD) considers the training data independently and predicts each pixel separately [6]. Besides, max-pooling and striding create spatial insensitivity in the higher layer which limits spatial accuracy in pixel-wise segmentation. To minimize this problem, Hariharan et al. [10] extract features of the same pixels from multiple layers and form a vector called "Hypercolumns". Bansal et al. [3] randomly sample a moderate number of pixels in the training phase to ensure memory bound and reduce overfitting due to feature correlation of spatially-neighboring pixels. PixelNet [2] exploits hypercolumns techniques [10] and random sampling [3] to form hypercolumn descriptor for a sampled pixel from multiple convolutional layers. Subsequently, Islam et al. [13] utilize multimodal PixelNet to segment brain tumor from MRI scan and achieve state of the art performance. DeepLab [6], PSPNet [27], and ICNet [26] adopt 'atrous convolution' [7] to explicitly control the resolution and incorporate larger context without increasing the number of parameters or the amount of computation. Our current work is inspired by multi-modal PixelNet [2,13] and atrous convolution [6,7,26,27] to design a computationally efficient and state of art learning model

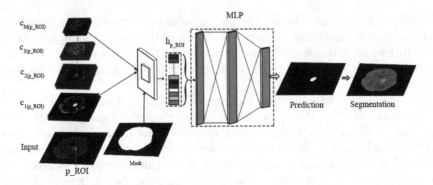

Fig. 1. ICHNet architecture. Hypercolumn features are extracted from multiple convolutional layers and feed to an MLP for prediction of hematoma segmentation.

for intracerebral hemorrhage (ICH) segmentation. The most significant contributions of our work are mainly in four aspects: (1) To our knowledge, this is the first work for automated intracerebral hemorrhage (ICH) segmentation from CT scans using deep learning; (2) Proposed model can train only by sampling a modest number of pixels from within the brain region, whereas conventional deep learning approaches use whole image or image patch including background. As it can ignore background and padding of the images from learning, so the model converges faster with better prediction rate; (3) Class imbalance in training dataset leads to a bias towards certain classes in the convolutional prediction. We deal with this problem by sampling an equal number of pixels for each class; (4) Comparing to multi-modal PixelNet [13], we adopt atrous convolution layer and dice loss layer for prediction and also 3D CRF and largest component analysis as post-processing.

2   Proposed Method

Our proposed model (Fig. 1) samples diverse pixels from a ROI (brain region) and constructs hypercolumn (hp) from multiscale convolutional and atrous convolutional layer features as in past work [2,10,13]. It contains total 15 convolutional layers where first 13 layers ($c_{i,j}$) similar to convolutional part of VGG-16 [23] (Convolution, ReLU, Pooling) and last 2 convolutional filters (c_i) followed by [19]. We integrate atrous convolution (ac) according to PSPNet [27]. To predict pixel-wise segmentation from hypercolumn features, we utilize a multi-layer perceptron (MLP) with 3 fully connected (fc) layers of size 4096 followed by ReLU activation functions. The convolutional and fully-connected layers of our architecture can be denoted as $\{c_{11}, c_{12}, c_{21}, c_{22}, c_{31}, c_{32}, c_{33}, c_{41}, ac_{42}, c_{43}, c_{51}, ac_{52}, c_{53}, c_6, c_7, h_p, fc_1, fc_2, fc_3\}$. Hypercolumn features are extracted from 6 convolutional layers of $\{c_{12}, c_{22}, c_{33}, c_{43}, c_{53}, c_7\}$. As our model can learn inside predefined ROI, so we can denote hypercolumn as h_{p_ROI}, where p_ROI is a random pixel inside ROI. Therefore, we can formulate the hypercolumn as:

$$h_{p\_ROI} = [c_{1(p\_ROI)}, c_{2(p\_ROI)}, ..., c_{M(p\_ROI)}], \tag{1}$$

where $c_{i(p\_ROI)}$ denotes the feature vectors of the p_ROI pixel from i^{th} convolutional layer. The main focus in our model is an extra layer called 'pixels' which carry the coordinates of the pixels we want to train. Due to this layer, it has the freedom to choose random pixels inside ROI. It can also select an equal number of pixels from each class which helps to overcome data skewness problem. If N is the number of sample pixels and there are K classes in our dataset then we choose N/K pixels from each class to from hypercolumn. We also adopt Dice loss function similar to [20] to overcome class imbalance problem.

3 Experiments

3.1 Dataset

The study cohort consists of CT scans of 89 patients with ICH from the Singapore General Hospital, aged 62.0 years (SD = 14.0), of whom 54 were men. Ethics approval was obtained from the SingHealth Centralized Institutional Review Board. The dataset also consists of annotations for the hematoma region marked by two blinded assessors from the neurosurgery department. Disagreements regarding the annotations were resolved with discussion with a third assessor for final consensus. The segmentation contours delineating hemorrhagic region with pixel label 1 and healthy tissue and background considered as pixel label 0. Finally, all scans are then resampled to isotropic 1 mm^3 resolution, skull-stripped as [24] and normalized to intensity range [0–255].

3.2 Training

As our model is capable of training with predefined pixels, the dataset is not resized in order to prevent shape and contextual information loss. However, we apply depth slicing along the axial plane and add padding to upsample all the slices to a common size of 250×250 for convolutional filters. The slices are augmented by randomly flipping the image horizontally, in order to make the model more generalized, We observe that slices with less than 2000 pixels of the brain region have no significance in training. Hence, we remove these slices along with all the blank slices. We sample 2000 pixels (N) per slice to extract multiscale convolutional features and form hypercolumn for MLP prediction during the training phase. To minimize class skewness, we randomly choose an equal number of pixels (1000 pixels per class) from each class. However, some slices do not exist hemorrhagic region, in which case, we sample 2000 (N) pixels randomly. In the testing phase, all the pixels from within the brain region were selected to form hypercolumn for MLP prediction.

The model is trained using stochastic gradient descent with a mini-batch size of 5. Its parameters are initialized by a pre-trained VGG-16 [23] model with learning rate 0.001 and momentum 0.9. Our model is implemented using a modified version of deep learning platform [6] based on CAFFE framework [15]. The time taken to train the model is around 30 h for 40 epochs on a single Nvidia GPU 1080Ti GPU.

3.3 Post-processing

To remove small spurious false positives and to smoothen the predicted segmentation, we utilize 3D fully connected Conditional Random Field (CRF) with Gaussian edge potentials as proposed in [18,19]. As the unary part of the CRF, we provide probability map generated from the softmax layer in prediction time. CRF regularizes the overall volume of the hemorrhage lesion leaving the internal structure of the lesion mostly intact. Further, remaining 3D-connected regions smaller than 1000 voxels are removed by using connected component analysis.

3.4 Results

To assess the performance of the model comprehensively, Dice coefficient, Hausdorff distance, Sensitivity, and Specificity are computed and presented in Table 1. The model evaluation is done using 5-fold cross-validation. Hence, the average value was considered. For example, the maximum dice accuracy obtained is 89.05%. However, the average value considering all folds is 87.60%. Our model also achieves average Hausdorff distance and sensitivity values of 11.76 and 91.51 respectively. The specificity obtained is almost 100% in all the cases. Table 2 shows the comparison of performance and computational efficiency of our model with other similar approaches. For a fair comparison, we consider the segmentation accuracy from the best trained model of the corresponding architecture. The same pre-processing and post-processing techniques are applied to all the different architectures compared. One of the most important observations is that our model requires almost half the time and the number of epochs to converge, as compared to the multi-modal PixelNet [13] and PSPNet [27]. It also achieves the best Dice coefficient compared to all other methods. However, PSPNet [27], one of the best performing techniques in computer vision application, obtains

Table 1. 5-fold cross validation result of Dice coefficients and Hausdorff distances for ICHNet.

Models	Dice	Hausdorff	Sensitivity	Specificity
Fold-1	88.67	11.01	91.83	**99.97**
Fold-2	**89.04**	**10.37**	**92.40**	**99.97**
Fold-3	86.48	12.87	91.02	99.77
Fold-4	87.51	11.61	91.34	99.85
Fold-5	86.32	12.98	90.98	99.65
Average	87.60	11.76	91.51	99.84

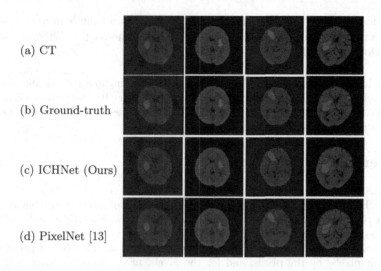

(a) CT

(b) Ground-truth

(c) ICHNet (Ours)

(d) PixelNet [13]

Fig. 2. Prediction of ICHNet (ours) and comparison with similar models. In segmentation, hemorrhage region denotes by red color. (Color figure online)

Table 2. Performance and computational efficiency comparison of our model and similar model multi-modal PixelNet [13] and PSPNet [27].

Models	Epochs	Time (Hour)	Dice	Hausdorff
ICHNet (Ours)	**40**	**30**	**87.60**	11.76
Multi-modal PixelNet [13]	70	55	85.92	15.91
PSPNet [27]	90	47	86.19	**11.50**

highest Hausdorff distance in this experiment. Figure 2 represents some predicted segmentations for our model with multi-modal PixelNet [13].

4 Discussion and Conclusion

We present a deep learning-based model: ICHNet, which predicts intracerebral hemorrhage (ICH) segmentation comparable to radiologists. In medical imaging, the anatomy of interest occupies only a very small region of the scan, which makes the prediction of the model strongly biased towards the background. ICHNet has the ability to train only using pixels obtained from within the brain region, which improves the optimization time and segmentation performance. Another advantage of using ICHNet is that it minimizes further skewness of the data by utilizing Dice coefficient as the objective function. Deep learning models with medical imaging applications using MRI consist of multi-channel (modalities) training data, unlike in the case of CT which comprises only one channel (therefore, lesser contextual information). This makes it challenging for the model to distinguish between the healthy region and the hemorrhage lesion. Future work incorporates building a 3D ICHNet model for different medical applications using MRI or CT data.

Acknowledgement. This work is supported by the Singapore Academic Research Fund under Grant R-397-000-227-112, NUSRI China Jiangsu Provincial Grant BK20150386 and BE2016077 and NMRC Bedside & Bench under grant R-397-000-245-511 awarded to Dr. Hongliang Ren.

References

1. van Asch, C.J., Luitse, M.J., Rinkel, G.J., van der Tweel, I., Algra, A., Klijn, C.J.: Incidence, case fatality, and functional outcome of intracerebral haemorrhage over time, according to age, sex, and ethnic origin: a systematic review and meta-analysis. Lancet Neurol. **9**(2), 167–176 (2010)
2. Bansal, A., Chen, X., Russell, B., Ramanan, A.G., et al.: PixelNet: representation of the pixels, by the pixels, and for the pixels. arXiv preprint arXiv:1702.06506 (2017)
3. Bansal, A., Russell, B., Gupta, A.: Marr revisited: 2D-3D alignment via surface normal prediction. In: Proceedings of the IEEE Conference on Computer Vision and Pattern Recognition, pp. 5965–5974 (2016)
4. Cai, J., Lu, L., Xie, Y., Xing, F., Yang, L.: Improving deep pancreas segmentation in CT and MRI images via recurrent neural contextual learning and direct loss function. arXiv preprint arXiv:1707.04912 (2017)
5. Chen, L., Bentley, P., Rueckert, D.: Fully automatic acute ischemic lesion segmentation in DWI using convolutional neural networks. NeuroImage: Clin. **15**, 633–643 (2017)
6. Chen, L.C., Papandreou, G., Kokkinos, I., Murphy, K., Yuille, A.L.: DeepLab: semantic image segmentation with deep convolutional nets, atrous convolution, and fully connected CRFs. IEEE Trans. Pattern Anal. Mach. Intell. **40**(4), 834–848 (2018)
7. Chen, L.C., Papandreou, G., Schroff, F., Adam, H.: Rethinking atrous convolution for semantic image segmentation. arXiv preprint arXiv:1706.05587 (2017)
8. Choi, Y., Kwon, Y., Lee, H., Kim, B.J., Paik, M.C., Won, J.H.: Ensemble of deep convolutional neural networks for prognosis of ischemic stroke. In: Crimi, A., Menze, B., Maier, O., Reyes, M., Winzeck, S., Handels, H. (eds.) BrainLes 2016. LNCS, vol. 10154, pp. 231–243. Springer, Cham (2016). https://doi.org/10.1007/978-3-319-55524-9_22
9. Grewal, M., Srivastava, M.M., Kumar, P., Varadarajan, S.: RADNET: radiologist level accuracy using deep learning for hemorrhage detection in CT scans. arXiv preprint arXiv:1710.04934 (2017)
10. Hariharan, B., Arbeláez, P., Girshick, R., Malik, J.: Hypercolumns for object segmentation and fine-grained localization. In: Proceedings of the IEEE Conference on Computer Vision and Pattern Recognition, pp. 447–456 (2015)
11. Havaei, M., et al.: Brain tumor segmentation with deep neural networks. Med. Image Analysis **35**, 18–31 (2017)
12. Hwang, S., Park, S.: Accurate lung segmentation via network-wise training of convolutional networks. In: Cardoso, M.J., et al. (eds.) DLMIA/ML-CDS -2017. LNCS, vol. 10553, pp. 92–99. Springer, Cham (2017). https://doi.org/10.1007/978-3-319-67558-9_11

13. Islam, M., Ren, H.: Multi-modal PixelNet for brain tumor segmentation. In: Crimi, A., Bakas, S., Kuijf, H., Menze, B., Reyes, M. (eds.) BrainLes 2017. LNCS, vol. 10670, pp. 298–308. Springer, Cham (2018). https://doi.org/10.1007/978-3-319-75238-9_26

14. Jégou, S., Drozdzal, M., Vazquez, D., Romero, A., Bengio, Y.: The one hundred layers tiramisu: fully convolutional densenets for semantic segmentation. In: 2017 IEEE Conference on Computer Vision and Pattern Recognition Workshops (CVPRW), pp. 1175–1183. IEEE (2017)

15. Jia, Y., et al.: Caffe: convolutional architecture for fast feature embedding. In: Proceedings of the 22nd ACM International Conference on Multimedia, pp. 675–678. ACM (2014)

16. Kalita, J., Misra, U., Vajpeyee, A., Phadke, R., Handique, A., Salwani, V.: Brain herniations in patients with intracerebral hemorrhage. Acta Neurol. Scand. **119**(4), 254–260 (2009)

17. Kamnitsas, K., et al.: Efficient multi-scale 3D CNN with fully connected CRF for accurate brain lesion segmentation. Med. Image Anal. **36**, 61–78 (2017)

18. Krähenbühl, P., Koltun, V.: Efficient inference in fully connected CRFs with Gaussian edge potentials. In: Advances in Neural Information Processing Systems, pp. 109–117 (2011)

19. Long, J., Shelhamer, E., Darrell, T.: Fully convolutional models for semantic segmentation. In: CVPR, vol. 3, p. 4 (2015)

20. Milletari, F., Navab, N., Ahmadi, S.A.: V-net: fully convolutional neural networks for volumetric medical image segmentation. In: 2016 Fourth International Conference on 3D Vision (3DV), pp. 565–571. IEEE (2016)

21. Noh, H., Hong, S., Han, B.: Learning deconvolution network for semantic segmentation. In: Proceedings of the IEEE International Conference on Computer Vision, pp. 1520–1528 (2015)

22. Saulle, M.F., Schambra, H.M.: Recovery and rehabilitation after intracerebral hemorrhage. In: Seminars in Neurology, vol. 36, p. 306. NIH Public Access (2016)

23. Simonyan, K., Zisserman, A.: Very deep convolutional networks for large-scale image recognition. arXiv preprint arXiv:1409.1556 (2014)

24. Smith, S.M.: Fast robust automated brain extraction. Hum. Brain Mapp. **17**(3), 143–155 (2002)

25. Tran, P.V.: A fully convolutional neural network for cardiac segmentation in short-axis MRI. arXiv preprint arXiv:1604.00494 (2016)

26. Zhao, H., Qi, X., Shen, X., Shi, J., Jia, J.: ICNet for real-time semantic segmentation on high-resolution images. arXiv preprint arXiv:1704.08545 (2017)

27. Zhao, H., Shi, J., Qi, X., Wang, X., Jia, J.: Pyramid scene parsing network. In: IEEE Conference on Computer Vision and Pattern Recognition (CVPR), pp. 2881–2890 (2017)

Can Diffusion MRI Reveal Stroke-Induced Microstructural Changes in GM?

Lorenza Brusini[1(✉)], Ilaria Boscolo Galazzo[1], Mauro Zucchelli[2], Cristina Granziera[3], and Gloria Menegaz[1]

[1] Department of Computer Science, University of Verona, Verona, Italy
lorenza.brusini@univr.it
[2] Inria, Sophia Antipolis Méditerranée, Biot, France
[3] Department of Neurology, Basel University Hospital, Basel, Switzerland

Abstract. The development of noninvasive techniques to image the human brain has enabled the demonstration of structural plasticity in response to motor learning. In the last years evidence has emerged on the potential of some measures derived from diffusion Magnetic Resonance Imaging (DMRI) as numerical biomarkers of tissue changes in regions involved in the motor network. In these works, the descriptors were extensively analysed in contralateral white matter (WM) along both single connections and networks relying on tract-based analyses and statistical evaluation. Though, their ability to detect changes in gray matter (GM) has been scarcely investigated. This work aims at the assessment of propagator-based microstructural indices in capturing GM changes and the relation of such changes to functional recovery at six months from the injury focusing on the Diffusion Tensor Imaging (DTI) and the three dimensional Simple Harmonics Oscillator based Reconstruction and Estimation (3D-SHORE) models.

1 Introduction

It is generally agreed that motor learning consists of several stages where both functional and structural changes occur (for review, [6]), although such neural changes do not necessarily translate into behavioral change. Understanding behaviorally relevant plasticity mechanisms, how they can be optimized in the intact and damaged brain and, most importantly, translated into behavioral improvements or higher performance, can inform and potentially improve interventions in clinical and expert populations. In [13], an extensive overview of the state-of-the-art in the emerging evidence for motor-learning-related structural plasticity and the implications for stroke rehabilitation is provided, highlighting that structural imaging will likely have a role in the future in providing measures that inform patient stratification for optimal outcomes. The most commonly used structural measures that have been considered so far target white matter (WM) microstructure and are derived from diffusion Magnetic Resonance Imaging (DMRI), such as fractional anisotropy (FA), complemented by macroscopic

© Springer Nature Switzerland AG 2019
A. Crimi et al. (Eds.): BrainLes 2018, LNCS 11383, pp. 464–471, 2019.
https://doi.org/10.1007/978-3-030-11723-8_47

measures of gray matter (GM) features such as the volume of GM regions [14,15]. Noteworthy, the characterization of GM microstructural changes is still largely underinvestigated [2]. Numerical biomarkers for microstructural changes in WM were proposed relying on tract-based analysis providing evidence of good sensitivity in capturing microstructural variations on the contralateral hemisphere to the infarcted region adding further evidence that changes in WM connectivity occur secondary to a stroke even in the contralesional motor cortical and subcortical areas [2–4,9]. Less well studied are structural changes in brain GM and how they relate to clinical outcomes. Studies of healthy individuals have demonstrated GM plasticity with learning (see above) or with loss of input (for example, due to limb loss) [7]. Stroke leads to reduced GM volume in distant but connected brain regions, and the extent of GM loss in these areas, for example, the contralesional motor cortex, is correlated with worse motor functioning as well as reduced response to rehabilitation. This finding suggests that some of the motor impairment seen after stroke may be due to losses in GM remote from the site of injury. Whether successful rehabilitation reverses this loss in GM remains an open question; however, a few small studies have started to explore this question further. Fan et al. [8] performed a longitudinal study and observed a GM volume increase in the contralesional hippocampus and precuneus correlating with better motor functioning. Cai et al. [5] also executed a longitudinal study, focusing on patients with subcortical stroke; they found that poor performance on motor assessments was predicted by severe atrophy of the ipsilesional precentral gyrus, whereas improvements were predicted by increases in GM volume in the contralesional frontal cortex. However, the analysis of microstructural changes in GM is still largely unexplored. The assessment of DMRI-derived indices sensitivity and specificity to GM modifications is inherently more complicate due to the high complexity of the microarchitecture of the so-called extracellular space. In fact, GM is comprised of complex shaped structures like brain cells (e.g. neurons and glia) densely packed together. In order to model in a neuroanatomically plausible way the GM, first of all it is necessary to have a realistic model for the different cells in the brain. This is extremely challenging, because brain cells are complex branched structures, comprised of different connected parts, like cell body (namely soma) besides cellular projections (namely neurites). Noteworthy, to date, only a few attempts to simulate more realistic brain cell structures for Diffusion Weighted (DW)-MRI applications have been published [12]. However, they still rely on a simplistic description of the cell structure, as for instance one dimensional branched structure and disconnected or connected cylinders. The present work aims at assessing the sensitivity of some microstructural descriptors that are available at the state-of-the-art in capturing the variations of the stroke-induced GM modulations in the cortical and subcortical contralateral regions to the injury involved in the motor networks, as well as their relation to the motor recovery score. Though being aware of the limitations mentioned above as well as of the inherent lack of specificity of DMRI, if such descriptors would convey information about the underlying tissue properties holding sensitivity to changes due to different potential causes, including pathology, they could be considered as fingerprints of microstructure modulations and thus be exploited for measuring structural plasticity. Here, the microstructural indices derived from the three dimensional Simple Harmonics Oscillator based Reconstruction and Estimation (3D-SHORE) model [11] and the classical Diffusion Tensor Imaging (DTI) model are derived and compared in their ability to predict the motor outcome at six months from the injury.

2   Methods

2.1   Dataset

Diffusion Spectrum Imaging (DSI) acquisitions were performed on ten subjects with ischemia (6 males, mean age: 60.3 ± 12.3 years) at three time points [within one week ($tp1$), one month ($\pm$ one week, $tp2$), and six months ($\pm$ fifteen days, $tp3$) after the infarct] and ten gender- and age-matched healthy subjects at two time points (one month apart, $tp1c$ and $tp2c$). The main acquisition parameters for the DSI were as follows: TR/TE = $6600/138$ ms, FOV = 212×212 mm^2, 34 slices, $2.2 \times 2.2 \times 3$ mm^3 resolution, GRAPPA = 2, scan time = 25.8 min, 515 points on a q-space 3D grid with radial grid size of 5, thirty-four different b-values (from 300 up to 8000 s/mm^2) and one image at $b0$. A high resolution 3D T1-weighted image was also acquired at each tp. The motor part of the National Institutes of Health Stroke Scale (NIHSS) score was measured at each tp for all patients and considered in the statistical analyses. All the subjects enrolled in the study signed a written informed consent in accordance to the Declaration of Helsinki.

2.2   Gray Matter Analysis

DTI-based FA and Mean Diffusivity (MD) were obtained from DSI data and the Generalized Fractional Anisotropy (GFA), Propagator Anisotropy (PA), Return To Axis Probability (RTAP), Return To Plane Probability (RTPP) and Mean Squared Displacement (MSD) were derived after fitting the 3D-SHORE model as in [2–4]. The Desikan-Killiany anatomical atlas was used to parcellate the T1-weighted images via Freesurfer resulting in 80 regions of interest (ROIs). GM volume maps were then derived using the SPM toolbox[1] and were binarised by applying a 0.95 probability threshold. These masks were then applied to the original Freesurfer ROIs, for a total of 72 regions surviving the thresholding. The mean GM value across each masked ROI in the contralateral hemisphere was calculated, leading to 36 ensemble average GM values (EAGM) for each index, subject and tp. For the list of the considered regions and relative abbreviations please refer to [2].

---

[1] http://www.fil.ion.ucl.ac.uk/spm/.

2.3   Statistical Analyses

For each patient, the z-score of the EAGM values with respect to the same measurement on the control group was calculated for all the indices, regions and tps as $z_i = (\mu - x_i)/\sigma$ where x_i is the EAGM value, μ and σ are the estimated mean and the standard deviation of the corresponding index on the control group. A one-way analysis of variance (ANOVA) for repeated measures was performed on the z-scores to test for significant differences across tps, considering each index and region separately. A *post-hoc* paired sample t-test was applied (p-value < 0.05) to the indices and regions with statistically significant ANOVA results and corrected for multiple comparison using false discovery rate (FDR). The overall Spearman's rank correlation coefficient between the EAGM values at all the tps and the respective NIHSS motor scores was calculated. Finally, three linear regression models were considered, as was done in [2] for tract-based analysis, that are (i) the Tensor-based model (TBM); (ii) the 3D-SHORE-based model (SBM); and (iii) the Global microstructural model (GBM). The respective set of predictors were the global EAGM values for FA and MD at *tp1*, plus age, stroke size and NIHSS at *tp1* for TBM; the whole EAGM values for GFA, PA, RTAP, RTPP and MSD at *tp1*, along with age, stroke size and NIHSS at *tp1* for SBM; and only the whole EAGM values of all the seven indices at *tp1* for the GBM.

3   Results

Figure 1 shows the DTI- and 3D-SHORE-derived indices maps at all tps in a sample subject. For simplicity of notations, the EAGM suffix is omitted in what follows. GFA, PA and FA were normalized to the respective maximum value and the square-root of the RTAP values was reported as in [1]. The ischemic injury was located in the right hemisphere and appeared hypointense in all the indices at *tp1* except for RTAP and RTPP, revealing higher values over this area. Considering the other two time points (*tp2* and *tp3*), contrast was reversed compared to *tp1* for RTAP, RTPP, MD, MSD, while the values in the lesion appeared to be constant for the three anistropy measures. For all the subsequent quantitative analyses we considered only the contralateral hemisphere where changes cannot be appreciated on the images. The distribution across patients of the z-score values is reported in Fig. 2 for four representative indices (GFA, MSD, FA and MD) and for a subset of the 36 ROIs [7 motor areas as in [2] plus frontal pole (FP), temporal pole (TP) and inferior parietal lobe (IPL)]. *Tp1* consistently revealed to be the time point with the largest deviation from the normal values, while z-scores tended towards zero over time. When statistically compared, the four indices reported in Fig. 2 showed significant differences across the three time points in a series of ROIs, e.g., MSD: $F(2,27) = 4.38$, p-value $= 0.02$ for thalamus (Thal) and $F(2,27) = 3.55$, p-value $= 0.04$ for sensory cortex (SC); FA: $F(2,27) = 4.59$, p-value $= 0.02$ for medial orbito frontal cortex (mOFC) and $F(2,27) = 3.57$, p-value $= 0.04$ for IPL. For the ROIs with significant ANOVA results, the t-test between pairs of time points demonstrated significantly higher

z-score values at *tp1* compared to *tp2* as well as *tp2* compared to *tp3* for the diffusivity measures (MD and MSD), confirming the longitudinal decrease of values ($p < 0.05$, uncorrected). For GFA and FA, the z-scores were initially negative at *tp1*, suggesting higher anistropy values in patients compared to controls just after the event, and gradually returned to zero over time. Of note, when the t-test results were FDR-corrected for multiple comparisons, only the statistical differences highlighted by MD over SC survived. The overall Spearman's correlation coefficients between EAGM values and NIHSS motor score are reported in Fig. 3 for each index and region. Significant coefficients (p-value < 0.05) were color coded. FA and MD resulted in more regions being significantly correlated than RTAP, RTPP, GFA, PA and MSD which showed a more selective pattern. Interestingly, considering the main regions involved in the motor networks, a significant correlation between FA and the clinical score was detected in the supplementary motor area (SMA). RTAP and RTPP resulted in a significant correlation with the motor score in Thal. Generally, all indices were positively correlated to NIHSS, except for MSD and MD for which a negative correlation was recorded in all regions, as well as for RTPP in FP and TP where significant negative correlations were found ($r_S = -0.39$ in both cases). Regarding the predictive models for GM, the TBM, which enclosed MD-FA at *tp1* plus the clinical variables, held only stroke size as relevant predictor and discarded the two microstructural indices ($R^2 = 0.52$; adjusted $R^2 = 0.46$; $p < 0.05$). Conversely, the SBM retained all the variables except stroke size as significant predictors, reaching a high correlation ($R^2 = 0.987$; adjusted $R^2 = 0.941$; $p < 0.05$). Finally, the GBM including only the DMRI-based indices revealed a good capability to timely predict the motor outcome at six months, keeping PA and FA as significant predictors ($R^2 = 0.882$; adjusted $R^2 = 0.848$; $p < 0.001$).

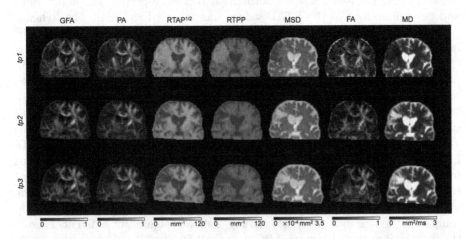

Fig. 1. DMRI-based indices calculated on a sample patient. Coronal slices are reported for each index (columns) and each time point (rows). Images are displayed in radiological convention.

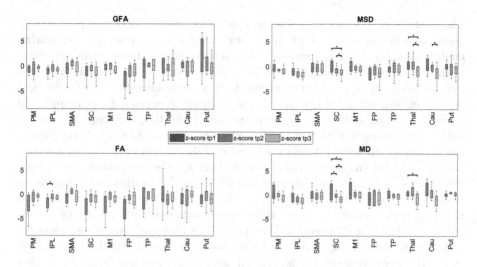

Fig. 2. Distribution of z-score values for a subset of the thirty-six contralateral regions, for each time point separately. Only the four indices revealing statistically significant results at the ANOVA test are reported, and the regions with statistical differences are highlighted ($p < 0.05$, uncorrected). (Color figure online)

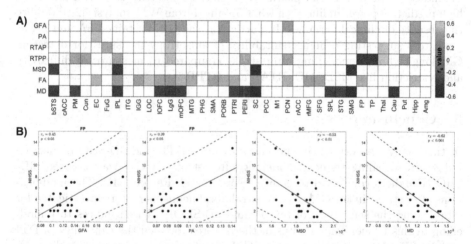

Fig. 3. (A) Spearman's rank correlation coefficient between each DMRI index and NIHSS motor score for the thirty-six contralateral regions. Only significant coefficients were color represented. (B) Scatter diagrams for two representative regions and four indices with significant correlation values [GFA and PA in the Frontal Pole (FP), MSD and MD in the sensory cortex (SC)]. (Color figure online)

4 Discussion

This work provides evidence in favour of the potential of the DTI- and 3D-SHORE-based indices as numerical biomarkers for GM remodelling after stroke.

The analysis of z-scores suggests the sensitivity of both DTI and 3D-SHORE descriptors to longitudinal GM modifications and thus their ability of capturing modulations across time. However, the interpretation of these results in terms of microstructural properties must be cautious because of both the inherent lack of specificity of DMRI measurements and the hidden degeneracies in MRI parameter estimation for neuronal tissue [10] that impedes a direct mapping of the measured parameters to tissue properties. Overall, the z-scores revealed a pattern in agreement with that illustrated in [2] based on ANOVA. In particular, the anisotropy indices (FA and GFA) were higher in patients than in controls with a trend leading to a small difference between the two groups across time while an opposite trend was observed for MSD and MD with mean values lower than control group. In addition, the analysis of *post-hoc* paired t-test resulted in a larger number of significant differences compared to [2] regarding regions involved in the motor networks. The Spearman's rank correlation coefficient for each descriptor and region revealed that both the 3D-SHORE indices and FA and MD were correlated to the clinical score. This provides hints for further investigating their clinical predictive power in GM as suggested also by the complementary analysis in WM in [2–4]. The analysis of the regression models revealed that the highest predictive power is given by SBM that retained all the indices as predictors except the stroke size. Noteworthy, the GBM reached a high predictive power in line with what obtained from WM analysis [2–4] despite the absence of clinical variables. In particular, the GBM retained FA and PA as predictors suggesting a higher suitability of anisotropy measures for detecting GM longitudinal remodelling after stroke compared to measures of diffusivity and restriction like MSD, MD, RTAP and RTPP.

5   Conclusion

In this study, the descriptive potential of both DTI- and 3D-SHORE-derived indices in revealing tissue remodelling after stroke in GM was assessed. 3D-SHORE indices correlated to clinical score as well FA and MD. The modifications after stroke in the GM of the contralateral hemisphere were investigated for establishing the sensitivity of tensor- and 3D-SHORE-based indices to remodelling in this tissue. The analysis of the z-scores revealed the capability of DTI and 3D-SHORE models in detecting GM modifications across time, finding significances in some regions devoted to motor tasks such as the SC. Linear regression emphasized the predictive power of all the indices for assessing the outcome after six months from the injury. This study provides evidence in support of the use of DMRI indices as potential fingerprints for GM modulations and calls for further investigation.

References

1. Avram, A.V., et al.: Clinical feasibility of using mean apparent propagator (map) MRI to characterize brain tissue microstructure. NeuroImage **127**, 422–434 (2016)
2. Boscolo Galazzo, I., Brusini, L., Obertino, S., Zucchelli, M., Granziera, C., Menegaz, G.: On the viability of diffusion MRI-based microstructural biomarkers in ischemic stroke. Front. Neurosci. **12**, 92 (2018)
3. Brusini, L., et al.: Ensemble average propagator-based detection of microstructural alterations after stroke. Int. J. Comput. Assist. Radiol. Surg. **11**(9), 1585–1597 (2016)
4. Brusini, L., et al.: Assessment of mean apparent propagator-based indices as biomarkers of axonal remodeling after stroke. In: Navab, N., Hornegger, J., Wells, W.M., Frangi, A.F. (eds.) MICCAI 2015. LNCS, vol. 9349, pp. 199–206. Springer, Cham (2015). https://doi.org/10.1007/978-3-319-24553-9_25
5. Cai, J., et al.: Contralesional cortical structural reorganization contributes to motor recovery after sub-cortical stroke: a longitudinal voxel-based morphometry study. Front. Hum. Neurosci. **10**, 393 (2016)
6. Dayan, E., Cohen, L.G.: Neuroplasticity subserving motor skill learning. Neuron **72**(3), 443–454 (2011)
7. Draganski, B., et al.: Decrease of thalamic gray matter following limb amputation. Neuroimage **31**(3), 951–957 (2006)
8. Fan, F., et al.: Dynamic brain structural changes after left hemisphere subcortical stroke. Hum. Brain Mapp. **34**(8), 1872–1881 (2013)
9. Lin, Y., et al.: The role of diffusion tensor imaging in the evaluation of ischemic brain injury-a review. Brain Connectomics **5**(7), 401–412 (2015)
10. Novikov, D., Veraart, J., Jelescu, I., Fieremans, E.: Rotationally-invariant mapping of scalar and orientational metrics of neuronal microstructure with diffusion MRI. NeuroImage **174**, 518–538 (2018)
11. Özarslan, E., et al.: Mean apparent propagator (map) MRI: a novel diffusion imaging method for mapping tissue microstructure. NeuroImage **78**, 16–32 (2013)
12. Palombo, M., Ligneul, C., Hernandez-Garzon, E., Valette, J.: Can we detect the effect of spines and leaflets on the diffusion of brain intracellular metabolites? NeuroImage (2017)
13. Sampaio-Baptista, C., Sanders, Z.B., Johansen-Berg, H.: Structural plasticity in adulthood with motor learning and stroke rehabilitation. Ann. Rev. Neurosci. **41**, 25–40 (2018)
14. Tardif, C.L., et al.: Advanced MRI techniques to improve our understanding of experience-induced neuroplasticity. NeuroImage **131**, 55–72 (2016)
15. Zatorre, R.J., Fields, R.D., Johansen-Berg, H.: Plasticity in gray and white: neuroimaging changes in brain structure during learning. Nat. Neurosci. **15**(4), 528 (2012)

Author Index

Printed in the United States
By Bookmasters